Lectins
Biology, Biochemistry, Clinical Biochemistry
Volume 4

Lectins

Biology, Biochemistry, Clinical Biochemistry
Volume 4

Proceedings of the
Sixth International Lectin Meeting
Poznań, Poland, September 2 – 6, 1984

Editors
T. C. Bøg-Hansen · J. Breborowicz

W DE G

Walter de Gruyter · Berlin · New York 1985

Editors
Thorkild Christian Bøg-Hansen, cand. scient., lic. techn.
The Protein Laboratory
University of Copenhagen
Sigurdsgade 34
DK-2200 Copenhagen N

Jan Breborowicz, M. D.
Department of Pathological Anatomy
University of Poznań
Przybysiewskiego 49
PL-60355 Poznań

Library of Congress Cataloging in Publication Data

International Lectin Meeting (6th: 1984: Poznań, Poland)
Proceedings of the Sixth International Lectin Meeting,
Poznań, Poland, September 2–6, 1984.
(Lectins-biology, biochemistry, clinical biochemistry; v. 4)
Includes bibliographies and indexes.
1. Lextins-Congresses. I. Bøg-Hansen, T. C. (Thorkild Christian), 1939
II. Breborowicz, J. (Jan) III. Title. IV. Title: Proceedings of the 6th
International Lectin Meeting. V. Series.
[DNLM: 1. Lectins-congresses. W3 LE32 v.4/QW 640 I61 1984p]
QP552.L42I57 1984 574.19'245 85-16088
ISBN 0-89925-081-5 (U.S.)

CIP-Kurztitelaufnahme der Deutschen Bibliothek

Lectins, biology, biochemistry, clinical biochemistry:
proceedings of the ... Lectin Meeting. – Berlin; New York: de Gruyter
NE: Lectin Meeting
Vol. 4 = 6. Poznań, Poland, September 2–6, 1984. –
1985. – XVI, 680 S.
ISBN 3-11-010298-6 (Berlin)
ISBN 0-89925-081-5 (New York)

ISBN 3 11 010298 6 Walter de Gruyter · Berlin · New York
ISBN 0-89925-081-5 Walter de Gruyter, Inc., New York

Printing: Gerike GmbH, Berlin. – Binding: Dieter Mikolai, Berlin. – Printed in Germany.

PREFACE

Since 1978 scientists involved in lectin research have been exchanging their ideas at annual lectin meetings. The first four meetings in this series were held at the Protein Laboratory of the University of Copenhagen. In recent years the lectin meetings have gained international recognition; it has become evident that organizing the meeting in collaboration with other research groups is very attractive. Thus two years ago in 1982 the fifth International Lectin Meeting, Interlec 5, was organized with G. Spengler in Bern. In 1984 Poznan became the location of Interlec 6.

Over 120 scientists from around the world participated in Interlec 6. Fifty of them also participated in two satellite workshops. These workshops concentrated on affinity electrophoresis (organized in Copenhagen) and on clinical applications of lectins (organized in Poznan).

Whilst it is very important that lectin researchers would meet and discuss, it is equally important that their discussion should be recorded and that this record be made available to a broader scientific community. This book serves this purpose and is a continuation of three pervious volumes of '<u>Lectins,</u>

Biology, Biochemistry, Clinical Biochemistry' published by W. de Gruyter, Berlin.

The first chapter of this book consists of papers devoted to the biological effects of lectins. Although in general the biological role of lectins is far from understood, some aspects do emerge. The application of lectins in biochemistry is the subject of the next two chapters. Growing interest in using lectins for demonstration and localization of glycoconjugates at cellular level can be observed. This subject was already well represented in volume 3 and became one of the main topics of this meeting. By now we can agree that lectins have proven to be precise and indispensable as histochemical tools. No doubt some of the results presented here will turn out to be of important practical value for clinical diagnosis.

The next chapter deals with other applications of lectins in biochemistry and in clinical biochemistry. Affinity electrophoresis is by tradition one of the topics of Interlec meetings; a number of papers are devoted to this topic. However, important model studies are presented on the glycosylation of proteins and the change and biological importance of the carbohydrate. Thus lectins serve as precise biochemical tools.

The field of lectin research is expanding quickly with better characterization of those lectins already known and with description of new lectins. This is well documented in the next three chapters dealing with animal and microbial lectins as well as with plant lectins. Some of these presentations stimulated a lively discussion on the criteria for classification of a substance as a lectin. Probably an acceptable working definition of a lectin is a protein of non-immune origin with the capacity to bind carbohydrate without entering into an enzymatic process.

This book ends with an appendix containing a list of some known lectins and their carbohydrate specificity and a report from the IUIS-WHO Subcommittee on Standardization of Lectins for Diagnosis. The report is a summary of the activities of an international group of workers trying to establish criteria for quality control of the lectin con A.

The editors of this volume wish to express their thanks to all friends who made it possible to organize the meeting in Poznan and to prepare this proceedings volume. We especially want to thank the contributing authors for their enthusiasm. It is our hope that this volume will serve not only as documentation of the Interlec 6 meeting in Poznan, but that it will also be useful to all those devoted to the progress of lectin research.

Further meetings are being planned in the hope that they will provide opportunities for personal contacts and for the exchange of new concepts. The next meeting, Interlec 7, to be held in Brussels, Belgium, in 1985 is being planned with E. van Driessche as host; J. L. Ochoa is to host Interlec 8 in La Paz, Mexico in 1986.

Jan Breborowicz
T.C.Bøg-Hansen

November 1984, Copenhagen

Table of Contents

I. BIOLOGICAL EFFECT OF LECTINS

The immunosuppresive nature of tomato lectin and its possible clinical relevance
D.C.Kilpatric, C.Graham, S.J.Urbaniak 3

The interaction of dietary lectin with porcine small intestine and the production of lectin-specific antibodies
R.Begbie, T.P.King 15

Lectin-induced arthritis of rabbit as a model of rheumatoid arthritis
R.Bräuer, K.Thoss, S.Henzgen, G.Waldmann 29

Mononuclear cell response to lectin: cAMP and chemiluminescense
M.Derwahl, J.Barth, K.G.Ravens, U.Schumacher 39

Peanut agglutinin and cell cooperation in interleukin-2 mediated mouse thymocyte stimulation
J.Favero, J.Marti, J.Dornand, J.C.Bonnafous, J.C.Mani 45

The effect of toxic lectins on histamine release from human basophil granulocytes
I.Sehrt, P.Luther, H.Franz, A.Kindt, R.Samtleben 53

II. HISTOCHEMISTRY AND CYTOCHEMISTRY WITH LECTINS

Lectin binding to non-malignant and malignant human uroepithelial cells in vitro
D.Dus, C.Radzikowski, H.Debray, J.Montreuil, B.Christensen J.Kieler 65

Lectin target cells in human brain tumors
K.Schwechheimer, I.A.Born, P.Kern, G.Weiss 75

Characterization of polycystic kidneys with lectins and differentiation markers of the nephron
H.Holthöfer, J.Rapola 83

PNA binding to human colorectal tissue as a marker of neoplastic transformation
T.F.Örntoft, S.E.Petersen 89

Intensive and selective staining of stored glycolipid in Sandhoff's disease by DBA, Dolichos biflorous agglutinin
J.Fischer, P.J.Klein 95

Differential diagnosis of anaplastic gastric cancers by lectins
J.Fischer, P.J.Klein 101

Cytochemical demonstration of cardiac glycosides in the heart muscle tissues using lectins and aldehyde-bisulfite-toluidine blue (abt) reaction
A.Neméth, E.Bayer, P.Sótonyi, J.Fischer. 109

Ulex europaeus I lectin binding glycoprotein in primary sensory terminals of human spinal cord
J.Fischer, P.J.Klein, B.Csillik 117

Lectin binding patterns in normal human salivary glands
I.A.Born, K.Schwechheimer, H.Maier, P.Moller 127

Lectin binding sites on bronchial carcinoma cells
A.Raedler, S.Böttcher, F.Stalling, E.Kaukel 137

Distribution of glycoconjugates in normal testis and in seminoma
R.Malmi, K.O.Söderström 145

Diagnosis of soft tissue tumors. Ulex europaeus lectin I is a marker for biphasic synovial sarcoma and mesothelioma
B.Borisch, U.Schumacher, D.Harms 155

The binding of Dolichos biflorus agglutinin to vascular endothelium and intestinal epithelium of mice, a genetic dimorphism
H.G.Uiterdijk, B.A.J.Ponder 161

Rat gastrointestinal glycocomjugates. Histological studies with lectins and monoclonal anti human blood group antibodies
P.Peschke, W.D.Kuhlmann 169

Lectin binding and uptake by human monomuclear cells
U.Schumacher, U.Welsch, J.Barth, K.G.Ravens, M.Derwahl 179

Ricin binding in rat brain
B.K.Groeger, P.L.Debbage 185

Lectin probes for visualization of cell differentiation markers on amphibian embryonic cells
A.M.Duprat, P.Rougé, S.Flottes, M.Bleys 193

Lectins as histochemical markers for helmintic antigens
M.Movsesijan, D.Borojević, M.Cuperlović 203

III. LECTIN TOOLS

Effect of pH and metal ions on the interactions between concanavalin A and carbohydrate.
An approach to standardization of affinity electrophoresis for con A - carbohydrate interaction. Part II
K.Takeo, M.Fujimoto, R.Suzuno, T.Tanaka, K.Nakamura, A.Kuwahara 213

Crossed affinoelectrophoresis with galactose specific lectins in the first dimension
J.A.Kint, J.G.Leroy 221

Neither ferritin nor its con A - binding fraction are tumor markers for renal carcinoma
M.M.Andersen, A.Lihme, S.Noack, T.C.Bøg-Hansen 229

Lectin blotting of normal and derivatized membrane proteins
S.Naaby-Hansen, A.Lihme, T.C.Bøg-Hansen, O.J.Bjerrum 241

Only the con A binding fraction of the β chain of haptoglobin is biologically active
I.Katnik, T.Guszczynski, W.Dobroszycka 253

Post synthetic modifications of proteins change during development. Acid phosphatases from different grass tissues
I.Lorenc-Kubis, B.Morawiecka 259

Glycoconjugate interactions in soybean root nodules
R.B.Mellor, D.Werner 267

A human broncial blood group A active glycoprotein. Structural characteristics studied by a quantitative lectin precipitation assay
A.M.Wu, C.E.Snyder, A.Herp, R.P.Hsieh 277

Lectin affinity of human placental opiate receptors
A.Valette, P.Rougé, J.Clos, E.J.Simon 289

Quantitative inhibition of lectin binding to immobilized carbohydrates
V.Hořejši, V.Matoušek, M.Tichá, T.Trnka 297

Quantitative study of lectin-mediated cell aggregation
M.Duszyk, M.Kawalec, J.Doroszewski 307

Proliferative response of human T lymphocytes to a mixture of two lectins
K.E.Wiktorowicz, W.Niedbala, A.Mackiewicz 313

Anti-H grouping with the Sambucus edulus lectin
M.Janicot, M.Lalaurie, J.Berlan, F.Miquel 321

Blood group substances for demonstration of lectins
H.Franz, J.Mohr, D.Neumann 327

IV. ANIMAL LECTINS

C-reactive protein as a model for how lectins may evolve
I.Kushner 337

Monocyte membrane lectin and in vitro phagocytosis of senescent human red blood cells
L.Gattegno, J.Vaysse, D.Bladier, D.Aminoff 345

Studies on lectins LXVI. Binding of serum glycoprotein asialooligosaccharides to the porcine liver and leucocyte membrane lectin
K.Bezouška, H.Karásková, O.Táborský, O.Kofronová, J.Voříšek, J.Kubrycht, J.Kocourek 353

Endogenous receptor for chick embryonic β-galactoside binding lectin
Y.Oda, K.I.Kasai 369

Studies on Lectins LXIV. Fish oocyte lectins. Further physiological characterization of lectins from oocytes of the brem, Abramis brama L., and the perch, Perca fluviatilis L. Interaction of the Perch lectin I with components of the nature egg jelly envelope
A.Krajhanzl, J.Kocourek, P.Pancoska, J.Nosek 377

Studies on Lectins LXV. Fish oocyte lectins. Physicochemical study of a lectin isolated from oocytes of the roach, Rutilus rutilus L. and partial characterization of its main polypeptide chains
A.Krajhanzl, A.Danisova, J.Kocourek, P.Pancoska 397

Marine invertebrate agglutinins: The lectin from Megapitaria squalida clam
F.Ascencio, J.L.Ochoa 409

V. MICROBIAL LECTINS

Sialic acid specific lectins of enterotexigenic Escherichia coli
T.Wadström, P.O.Sjöberg, M.Lindahl 417

K99 - a sialic acid specific haemagglutinin of enterotoxigenic E.coli ginds to glycophorin on human erythrocytes
M.Lindahl, R.Brossmer, T.Wadström 425

VI. PLANT LECTINS

Cacti lectins
E.Zenteno, J.L.Ochoa 437

Inhibition studies on the specificity of the Vicia graminea lectin
M.J.Prigent, V.Verez-Bencomo, P.Sinaÿ, J.P.Cartron 447

Ultrastructural localization of pea lectin in the embryo and cotyledons during development by the collodial-gold method
G.Smets, E.van Driessche, S.Beeckmans 453

Biological activities of misteltoe lectin I (ML 1) and its A and B chains. A review
H.Franz 463

Determination of lectin content in commercial misteltoe preparations for cancer therapy using the ELISA technique
P.Ziska, H.Franz 473

Carbohydrate-binding proteins from Datura stramonium
W.F.Broekaert, W.J.Peumans, A.K.Allen 481

Studies on lectins LXIII. Isolation and characterization of a lectin from Phallus impudicus, the stinkhorn mushroom
G.Entlicher, K.Jesenská, L.Jarosová-Dejlová, M.Jarnik, J.Kocourek 491

Studies on lectins LXII. A non-specific erythro-agglutinating lectin and a blood group a specific lectin in the mushroom Agrocybe
M.Tichá, V.Dudová, J.Kocourek, J.Volc 505

Lectin involved in fast mobilization of seed phosphorus reserves. Rye germ agglutinin (RGA) activate the endogenous acid phosphatase
M.Ferens, B.Morawiecka 515

Lectin as a storage protein in elder (Sambucus nigra) bark
W.Peumans, M.Nsimba-Lubaki, W.Broekaert 523

Removal of Amaranthua leucocarpus lectin affects the nutritional value of the seed meal
A.M.Calderón de la Barca, J.L.Ochoa, E.Zenteno, M.Valencia, R.Carbajal 531

Mitogenic, immunosuppressive and phagocytic activity of Machaerocereus eruca and Amaranthus leucocarpus lectins
E.Zenteno, J.L.Ochoa, C.Parra, L.F.Montaño, I.Rayon, G.Maldonado, B.Ruiz, R.Carvajal 537

Further studies on sperm-reactive lectins in Indian plants
R.S.Sandhu, S.K.Chopra, J.S.Arora, R.S.Reen 547

New lymphocyte-stimulating lectins from Indian plants
R.S.Sandhu, S.S.Kamboj, R.S.Reen, R.J.Tatake,
D.Subrahmanyan, S.Somasundaram, S.G.Gangal 559

Is potato lectin a precursor of cell wall extensin?
E.van Driessche, S.Beeckmans, R.Dejaegere, L.Kanarek 567

The occurrence of lectin during the life-cycle of Pisum sativum L.
M.Hosselet, E.van Driessche, M.van Poucke, L.Kanarek 583

Soybean seed lectin (SBA) binds to soil humic acids and clays
H.Causse, P.Rougé, L.Labroue, J.P.Spilthooren 591

Fetuin binding py Phaseolus coccineus var 'Alubia' seed agglutinin
H.Nolasco, J.L.Ochoa 599

Leaves of the wild grass Agropyrum repens contain an Gal-NAc specific lectin
B.Cammue, H.M.Stinissen, W.J.Peumans 607

Characterization of the different lectins from Viscum album (misteltoe) and their structural relationships with the agglutinins from Abrus precatorius and Ricinus communis
R.Samtleben, M.Kiefer, P.Luther 617

APPENDIX

Table of lectin binding specificities
A.Wu, A.Herp 629

Third report of the International Working Party-IUIS-WHO Subcommittee on Standardization of Lectins for Diagnosis. Progress in the standardization of concanavalin A
J.Breborowicz, H.Franz, M.Caron, A.Faure, K.Gryska,
J.A.Kint, L.M.J.Sabbe, J.Kohn, I.Lorenc-Kubis, H.Rudiger,
K.Takeo, K.Toftager-Larsen, M.Wu, T.C.Bøg-Hansen 637

Quality control of concanavalin A: Proposal for requirements
H.Franz, P.Ziska, C.Flemming, V.Horejsi, M.Tichá 649

Comparative studies of different tumor cell lines and som D-type retroviruses with lectins
M.Rudolph, R.Samtleben 653

Comparison of the structure of a galactosyl-binding lectin from the tunicate Didemnum candidum with other putative recognition molecules from invertebrates and vertebrates
G.R.Vasta, J.J.Marchalonis 663

AUTHOR INDEX 671

SUBJECT INDEX 675

Part I

Biological Effects of Lectins

THE IMMUNOSUPPRESSIVE NATURE OF TOMATO LECTIN AND ITS POSSIBLE CLINICAL RELEVANCE

David C Kilpatrick, Catherine Graham, and Stanislaw J Urbaniak.
Blood Transfusion Centre, Royal Infirmary, Edinburgh, Scotland, U.K.

The first reference in the literature to some form of lectin-like activity in the tomato (Lycopersicon esculentum) plant may have been that of Marcusson-Begun(1). Many years later, Hossaini (2) reported an A,B,O blood group non-specific haemagglutinin in tomato seeds, but did not purify or characterise the active principle. This finding conflicted with the survey of 262 plant sources by Boyd & Reguera(3): they found no haemagglutinating activity in saline extracts of tomato seeds tested against A,B or O erythrocytes. Later, Howard and co-workers(4) referred to "haemagglutinin activity in every part of the tomato plant except the fruit and seed" which they believed to be non-lectin in nature on account of properties which included high thermal stability and resistance to proteolytic enzymes. Yeoman et al (5) also found haemagglutinating activity towards human erythrocytes in extracts from several tissues of the tomato plant, including the seeds. Finally, Kilpatrick(6) showed that although extracts from various tissues, including the seed, would agglutinate human erythrocytes, the highest specific activity was found in the juice of ripe fruits, and this activity was isolated in the form of a glycoprotein lectin with specificity for oligomers of N-acetylglucosamine ($[GlcNAc]_n$). Nachbar et al(7) purified and characterised apparently the same glycoprotein. The tomato seed haemagglutinin also appears to be a lectin, but seems to be more closely related to the seed lectin from the related plant, Datura stramonium, than to the lectin from the fruit of its own species(8,9). It is the fruit-derived agglutinin that is normally understood by the

Lectins, Vol. IV

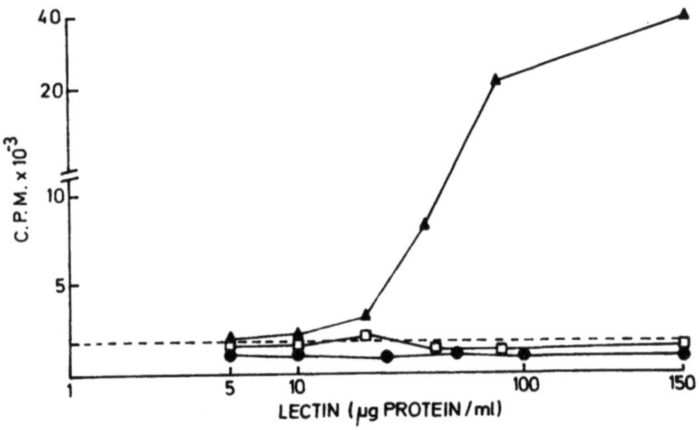

Fig. 1. Effect of solanaceous lectins on lymphocyte transformation Human lymphocytes were cultured for three days in the presence of tomato (●), potato (□) or Datura (▲) lectins. After pulsing for the final 16h with 1μC_i of ^{3}H-thymidine, the cultures were harvested using an automatic cell harvester and counted for radioactivity. Results shown are the means of triplicates. The dotted line indicates the value of a negative (lectin-free) control.

term "tomato lectin", and which shall be referred to as such in this article. The tomato lectin is one of several serologically related, $[GlcNAc]_n$-binding lectins from the Solanaceae family(10).

Knowledge and exploitation of lectins have been closely associated with the history of immunology(11). In recent years certain lectins have been widely used to stimulate lymphocyte transformation and to isolate or identify lymphocyte subsets. Mitogenic lectins appear to be in a minority (11,12), and the reason why some lectins but not others stimulate lymphocytes is far from clear despite promising hypotheses that have subsequently been found wanting(13). In this regard, the solanaceous lectins are of great interest and could prove useful tools for studying the mechanism of lymphocyte transformation. Although these lectins have a related structure and a similar saccharide specificity, the Datura lectin is mitogenic for human lymphocytes(14), while the tomato and potato lectins are not (Fig 1). Nachbar et al(7) found that the tomato lectin

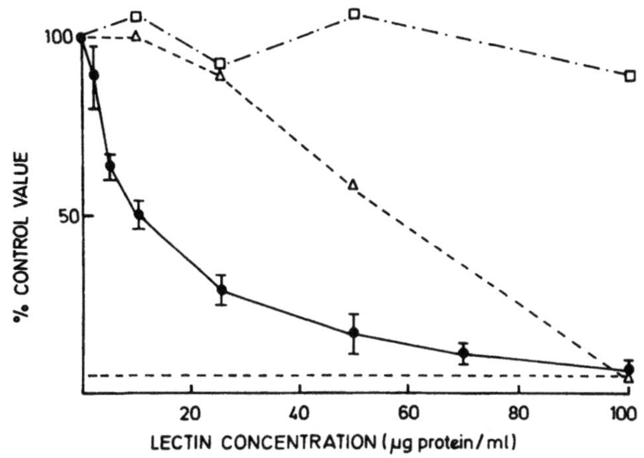

Fig. 2. Inhibition of PPD-mediated transformation by tomato lectin. Lymphocytes were cultured for five days with PPD and pulsed with 1 μC_i of ^{3}H-thymidine during the final 16h. The cultures also contained varying amounts of tomato lectin (●), potato lectin (□) or WGA (△). All cultures were performed in triplicate. After harvesting with an automatic cell harvester onto filters, the latter were counted for radioactivity. The results (mean ± 1 SD) are expressed as a proportion of a positive control containing PPD but no added lectin. The dotted line indicates the value obtained with a negative control (without PPD or lectin).

was actually anti-mitogenic (ie suppressed mitogen-mediated transformation) for chicken lymphocytes. We have found that tomato lectin is also antagonistic towards human lymphocyte transformation induced by a variety of stimuli in vitro, and are investigating the possible mechanisms of action of this effect.

Lymphocyte Transformation Experiments

When human lymphocytes were cultured in the presence of tomato lectin and pulsed with ^{3}H-thymidine, the incorporation of radioactive label into nucleic acid often appeared less than in controls without tomato lectin. This impression could be

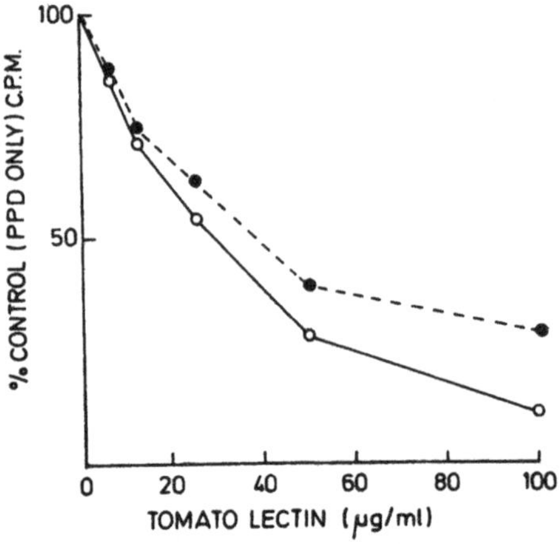

Fig. 3. Comparison of isoeletins X and Y with respect to inhibition of antigen mediated lymphocyte transformation. Experimental details were as described in the legend to Fig. 2
X:●- -●; Y:○—○

confirmed by increasing the lymphocyte number in the culture wells: as the absolute number of counts increased, so the depression of spontaneous lymphocyte transformation in the presence of tomato lectin became more obvious. Transformation mediated by the mitogenic lectins, concanavalin A, pokeweed mitogen and PHA from Phaseolus vulgaris, were also sometimes inhibited by tomato lectin, although very variable results were obtained. There is no reason to believe that tomato lectin binds directly to Con A, PHA or PWM. It is particularly interesting that tomato lectin can inhibit transformation induced by the lectin from D. stramonium: this inhibition is not observed with potato lectin, although the latter is structurally related, has a similar carbohydrate specificity, and also readily agglutinates lymphocytes.

Although inhibition of mitogen-induced transformation was variable, more consistent results were obtained by studying blastogenesis stimulated by recall antigens. We found the purified protein derivative (PPD) of tuberculin the most useful; typical results relating transformation inhibition to tomato lectin concentration are illustrated in Fig 2. Interestingly, potato

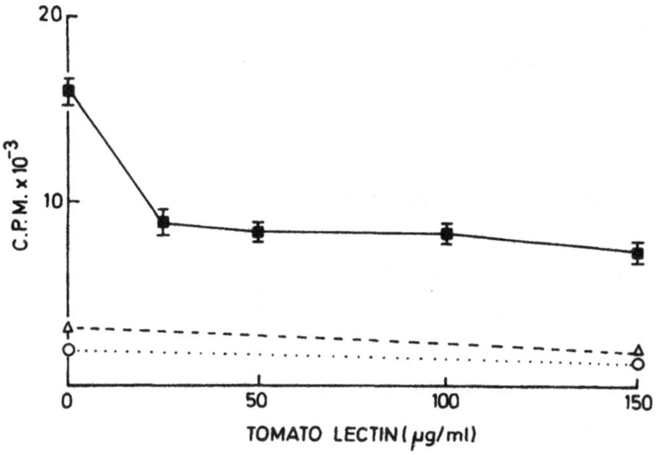

Fig. 4. Inhibition of a mixed lymphocyte reaction by tomato lectin mononuclear cells were prepared from two HLA-DR incompatible individuals (A and B), and were cultured either separately or together with a concentration range of tomato lectin for 6 days. The cultures were pulsed with 1 μC_i of ^3H-thymidine for 6h before the cells were harvested and counted for radioactivity. Results are expressed as mean ± 1SD of triplicates. ■—■ A + B; △- -△ A + A; ○···○ B + B.

lectin has little inhibitory effect, but wheat germ agglutinin (WGA), an unrelated (taxonomically) lectin with a broadly similar saccharide specificity, was also inhibitory, as has been reported by other workers(15,16). The tomato lectin can be partially separated into two major isolectins (X and Y) by chromatofocusing(17); both forms were found to be inhibitory, although form Y(pI 8.8) was slightly more potent (Fig 3). The tomato lectin can also inhibit the allogeneic mixed lymphocyte reaction(MLR). Inhibition of a two-way MLR over a range of tomato lectin concentrations is illustrated in Fig 4.

The basis of the inhibitory effect of tomato lectin on lymphocyte transformation is unlikely to be simple toxicity or inhibition of tritiated thymidine uptake for the following reasons: (1) lymphocytes cultured for several days with high concentrations of tomato lectin still exclude the vital dye, trypan blue; (2) tomato lectin-induced inhibition of transformation can be prevented by the simultaneous addition of

$[GlcNAc]_n$; (3) tomato lectin added to antigen cultures immediately before pulsing with ^{3}H-thymidine was entirely without effect.

Possible Mechanisms of Action

There are several possible ways by which the tomato lectin might effect its inhibitory action on lymphocyte transformation. For example, it might act simply by binding to a necessary growth factor present in the human serum used to supplement the lymphocyte culture medium. However, culture medium containing serum that has been passed through a column of tomato lectin-Sepharose can support lymphocyte culture almost as well as that containing unfractionated serum, so removal of serum growth factors cannot account for the tomato lectin effect.

Secondly, the tomato lectin might bind to growth factors produced by the stimulated mononuclear cells themselves, such as interleukins, or might prevent the production of interleukins or other lymphokines/monokines indirectly as a consequence of binding to lymphocyte receptors or receptors on antigen-presenting cells. These possibilities are under active investigation. Certainly both lymphocytes and monocytes can be readily agglutinated by tomato lectin (at a concentration of 5 µg lectin/ml or less) and since the lectin's inhibitory action is not observed in the presence of $[GlcNAc]_n$, it appears a strong possibility that the lectin might act via binding to a cell surface receptor.

Identification of Lectin Receptors on Human Lymphocytes

The tomato lectin might act by binding to any one of several cell surface glycoproteins (T cell receptor, T3 antigen, T4 or T8 antigens, HLA Class I or Class II molecules) involved in the activation of T lymphocyte clones. Alternatively, it might act by binding to a hypothetical suppressor receptor as Green et al have suggested for WGA and the mushroom (Agaricus bisporus) lectin (18).

To investigate the nature of the lymphocyte cell surface receptors for tomato lectin, human lymphocytes were labelled vectorially by lactoperoxidase-catalysed iodination(19) or by oxidation with sodium metaperiodate followed by reduction with

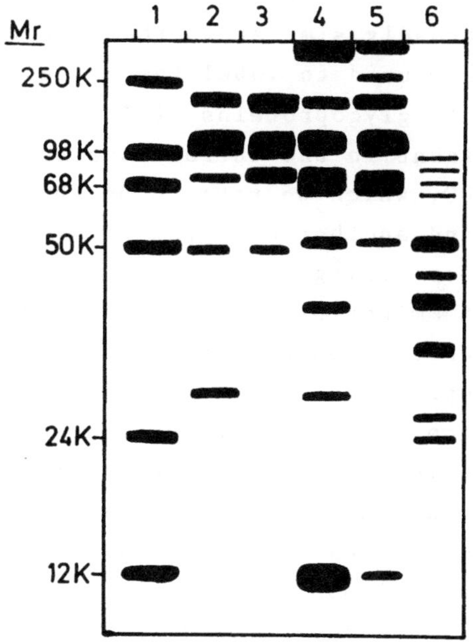

Fig. 5. Identification of lymphocyte receptors for tomato lectin. Purified (≥ 90%) peripheral blood buffy coat lymphocytes were labelled with surface reactive agents and extracted with non-ionic detergent. The extract was subjected to affinity chromatography on tomato lectin-Sepharose. This sketch summarises the typical results of several experiments. Lane: (1) Coomassie blue pattern of molecular weight markers; (2) autoradiographic pattern of the unfractionated extract from ^{3}H-labelled lymphocytes; (3) autoradiographic pattern of the ^{3}H-labelled material eluted from the affinity column with $(GlcNAc)_n$; (4) ^{125}I-labelled equivalent of the second lane; (5) ^{125}I-labelled equivalent of the third lane; (6) Coomassie blue pattern of proteins eluted from the affinity column with $(GlcNAc)_n$, the cell surface components of which are visualised in lanes 3 and 5.

tritiated borohydride(20). The cells were then extracted with non-ionic detergent, and the extract was applied to a column of tomato lectin-Sepharose. After washing, the column was eluted with $[GlcNAc]_n$, and the material thus obtained was analysed by discontinuous SDS-PAGE(21). Protein bands were visualised using Coomassie blue, while presumed cell surface glycoproteins were visualised by autoradiography.

A composite sketch of the results obtained is given in Fig 5. The pattern of tomato lectin-binding proteins visualised by autoradiography was broadly similar whether iodination or tritiation methods were used to label the surface components. The lectin bound mainly glycoproteins of large Mr, including two of the most prominent in the unfractionated extract, one of apparent Mr 200 000, which we take to be the leukocyte common antigen(22), and another of apparent Mr 100 000, which we take to be the major sialoglycoprotein of T lymphocytes(23). With the exception of a minor band of apparent Mr just above 50 000, the tomato lectin does not appear to bind appreciably to cell surface glycoproteins of subunit size below 70 000. This is important since it appears to exclude most of the surface molecules mentioned above (but not the T8 antigen) known to participate in lymphocyte activation. In particular, the iodination method appears to label HLA-Class 1 molecules (apparent Mr 44 000 and 12 000) but those bands are not apparent in the autoradiograph obtained after affinity chromatography on tomato lectin-Sepharose. It seems unlikely, therefore, that the lectin acts by binding to and blocking the function of any of the antigens known to be involved in antigen recognition. It is possible, however, that by binding to one or more of its receptors, the lectin triggers a suppressor signal, possibly thereby turning off interleukin production.

Clinical Relevance

The immunosuppressive nature of tomato lectin in vitro obviously raises the exciting possibility that the lectin might have a similar effect in vivo. If so, it might find therapeutic application in clinical transplantation or in the treatment of patients with autoimmune disease.
The delayed-type hypersensitivity response to intradermal injection of antigens to which an individual is already sensitised is thought to be an in vivo manifestation of the T lymphocyte response detected by lymphocyte transformation in culture. Several recall antigens were therefore injected into the arm of a human subject either with or without tomato lectin, and the extent of the response was measured two days later (Table 1). No difference was found, however, between the induration due to antigens administered with or without tomato lectin.

Table 1

Skin tests

Sample injected	Extent of Induration
Saline only	Nil
Tomato lectin only	Nil
PPD only	21 x 20 mm
PPD + tomato lectin	23 x 20 mm
Candida antigen only	16 x 15 mm
Candida antigen + tomato lectin	14 x 14 mm
Mumps antigen only	13 x 12 mm
Mumps antigen + tomato lectin	14 x 11 mm

The listed antigens (in 0.1 ml) were injected intradermally into the arm of a human subject either diluted into physiological saline or into tomato lectin solution in saline (final concn. 0.5 mg/ml). The extent of induration was measured 48 hours later. PPD = Purified protein derivative of tuberculin.

Despite these negative skin test results, it is still possible that the lectin might suppress some other facet of the immune response _in vivo_. It is particularly interesting to speculate about the possible effects of dietary consumption of tomato lectin. Tomato fruits are consumed raw or lightly cooked in considerable quantities by some individuals, and it is conceivable that ingested tomato lectin could influence the immune system, possibly by interaction with gut associated lymphoid tissue. This presupposes that the lectin would survive the normal digestive process. This possibility is compatible with the properties of the lectin _in vitro_: it is very resistant to heat treatment, to extremes of pH and to treatment with proteolytic enzymes (7,17).

Two lines of evidence support the suggestion that tomato lectin is not fully digested in the alimentary canal. (1) Rats fed on a diet rich in tomato juice produced faeces containing a product which formed a precipitation line after Ouchterlony double diffusion against rabbit anti-(tomato lectin) serum (but not pre-immune rabbit serum). (2) Experiments in which fasted rats were fed ^{125}I-labelled tomato lectin revealed that even after 3h in the stomach, most of the radioactivity was still present in a TCA-precipitable form. A substantial proportion of the radioactivity recovered from the small intestine and colon was also precipitated by TCA. Moreover, only a small proportion ($<5\%$) of the radioactivity administered could be recovered from the blood or internal organs after 3 hours when most of the ^{125}I-lectin had passed through the stomach. A more limited corresponding experiment on a human subject produced broadly similar results: the proportion of radioactivity detected in the bloodstream and in the thyroid after ^{125}I-tomato lectin consumption indicated that the bulk of the tomato lectin ingested was not fully broken down.

Conclusions

The tomato lectin has been found to be immunosuppressive for human lymphocytes *in vitro*, a property reported for very few other lectins. If the lectin were also able to depress lymphocyte transformation *in vivo*, it could have important applications in medicine as a non-toxic immunosuppressive drug. The lectin was found to have no effect on delayed-type hypersensitivity in a human subject, but its relative stability to conditions in the mammalian alimentary canal encourages the speculation that tomato lectin consumed in the diet might have interesting biological consequences which could be exploited.

Acknowledgements

Some of the work referred to was carried out in collaboration with Dr Arpad Pusztai and Mr George Grant at the Rowett Research Institute, Aberdeen, Scotland. We are grateful for helpful advice and assistance from Dr A E Dewar, Dr A H Maddy, Dr S Smith and Dr P L Yap, all of Edinburgh. We also thank the Chief Scientists Organisation of the Scottish Home and Health Department for providing financial support for the project, and Miss Audrey Wilson for careful typing of the manuscript.

References

1. Marcusson-Begun, H.:Z.Immunitatsforsch 45, 49-73 (1926)
2. Hossaini, A.A.: Vox Sang. 15, 410-417 (1968)
3. Boyd, W.C., Reguera, R.M.: J. Immunol. 62, 333-339 (1949)
4. Howard, I.K.,Sage, H.J., Horton, C.B.: Arch. Biochem. Biophys. 149, 323-326 (1972)
5. Yeoman, M.M., Kilpatrick, D.C., Miedzybrodzka, M.B., Gould, A.R.: Symp. Soc. Exp. Biol. 32 139-160 (1978)
6. Kilpatrick, D.C.: Biochem. J. 185, 269-272 (1980)
7. Nachbar, M.S., Oppenheim, J.D., Thomas, J.O.: J. Biol. Chem. 255, 2056-2063 (1980)

8. Kilpatrick, D.C., Jeffree, C.E.,Lockhart, C.M., Yeoman, M.M.: FEBS Lett 113, 129-133 (1980)

9. Kilpatrick, D.C., Graham, C., Urbaniak, S.J., Jeffree, C.E., Allen, A.K.: Biochem. J. 220, 843-847 (1984)

10. Kilpatrick, D.C.: in Chemical Taxonomy, Molecular Biology and Function of Plant Lectins (Goldstein, I.J., Etzler, M.E., eds.) pp 63-70, Alan R Liss, New York (1983)

11. Lis, H.,Sharon, N.: in The Antigens (Sela, M.,ed.), Vol. 4, pp 429-529, Academic Press, London/New York (1977)

12. Barker, B.E.: In Vitro 4, 64-79 (1969)

13. Pandolfino, E.R., Namen, A.E., Munske, G.R., Magnuson, J.A.: J. Biol. Chem. 258, 9203-9207 (1983)

14. Kilpatrick, D.C.:Plant. Sci. Lett. 15, 279-284 (1979)

15. Green, W.C., Waldman, T.A.: J. Immunol. 124, 2979-2987 (1980)

16. Barrett, D.J., Edwards, J.R., Pietrantuono, B.A., Ayoub, E.M.: Cell. Immunol. 81, 287-297 (1983)

17. Kilpatrick, D.C., Weston, J.,Urbaniak, S.J.: Anal. Biochem. 134, 205-209 (1983)

18. Green, W.C., Fleisher, T.A., Waldmann, T.A.: J. Immunol. 126, 580-586 (1981)

19. Hubbard, A.L., Cohn, Z.A.: J. Cell. Biol. 55, 390-405 (1972)

20. Gahmberg, C.G.,Andersson, L.C.:J.Biol.Chem. 252, 5888-5894 (1977)

21. Laemmli, U.K.: Nature 227, 680-685 (1970)

22. Dalchau, R., Kirkley, J., Fabre, J.W.: Eur. J. Immunol. 10, 737-744 (1980)

23. Dalchau, R., Kirkley, J., Fabre, J.W.: Eur. J. Immunol. 10, 745-749 (1980)

THE INTERACTION OF DIETARY LECTIN WITH PORCINE SMALL INTESTINE AND THE PRODUCTION OF LECTIN-SPECIFIC ANTIBODIES

R. Begbie and T. P. King
Rowett Research Institute, Bucksburn, Aberdeen AB2 9SB
Scotland

Cell-surface receptors for lectins have been identified in a variety of cell types (see for review (1)). In many instances their presence has been acknowledged merely in a diagnostic context where the exquisite carbohydrate specificities of particular lectins have been utilised as structural probes and relatively little attention has been paid to the possible occurrence and consequences of *in vivo* interactions of lectins and cell surfaces. One area where such encounter may occur with virtual daily frequency is to be found in the gut of animals consuming diets rich in plant-derived components. This is particularly so in non-ruminant production where substantial proportions of the protein components of the feed may be provided from lectiniferous seed meals. The implication of seed lectins in the role of naturally-occurring toxicants acting *via* the small intestine is now established (2) and recent work from these laboratories (3,4,5) concerned with the nutritional toxicity of phytohaemagglutinin(s) (PHA) from kidney beans (*Phaseolus vulgaris* L.) has done much to indicate probable modes of action of these lectins in the laboratory rat. Recent work has established that kidney beans are also toxic for pigs (13) and that toxicity may be mediated by the disruptive effects of dietary PHA on the architecture of the small intestine. We describe here some observations on the

Lectins, Vol. IV

relative effects of the lectins of two cultivars of dry beans (*Phaseolus vulgaris* L.) and of peas (*Pisum sativum*) included in the diets of growing pigs.

Materials and methods

White kidney beans (*P. vulgaris* L.), described as cv. 'Processor'-type, were obtained from Hurst, Gunson, Cooper, Taber, Ltd, Witham, Essex. Pinto beans (*P. vulgaris*) were purchased from a local supplier and marrowfat peas (*Pisum sativum*), cv. 'Maro', from Messrs. Nickerson, Ltd.

The fractional precipitation method of Hall (6) was employed to isolate three major fractions from kidney beans; G1, the reserve globulin, phaseolin; G2, globulin lectins and A, albumins. Sub-fractionation of the kidney bean lectins was carried out by affinity chromatography on a fetuin-Sepharose 4B column (7) and by salt gradient chromatography on SP-Sephadex A-50 (8). A sample of purified lectin from pinto beans, cv. 'Pinto 111', was kindly provided by Dr. A. J. Pusztai. Pea agglutinin was purified by dextran-affinity chromatography (9). Antibodies against purified L4 and E4 *P. vulgaris* lectins (for nomenclature see Felsted *et al*. (10)) and against isolated pea agglutinin were raised in rabbits and purified by the procedure of Harboe and Ingild (11). Polyspecific antibodies against the saline-extractable seed proteins of kidney beans were produced in pigs by parenteral hyperimmunisation with four doses of 15 mg protein injected intramuscularly in the thigh at 14 day intervals. Serum glycoprotein(s) which persisted in the Harboe-Ingild immunoglobulin fraction and which precipitated with PHA, were removed by chromatography on TSK 3000 SW (Anachem, Ltd). Specific antisera against porcine IgG, IgM and IgA were kind gifts from Dr. M. E. J. Barratt. Blood samples (approximately 10 ml) were withdrawn into Vacutainer collection tubes *via* the anterior vena cava of experimental animals at 0, 2.2, 6.2, 13 and 20 days exposure to diets. After 27 days a final sample of 200-300 ml was collected at slaughter. Serum was isolated by centrifuging after clot retraction and was stored at -15°C until examination. Otherwise purified porcine immunoglobulin fractions were isolated as above.

The specific anti-lectin immunoglobulins were isolated by immunoabsorption on a column of 'deactivated' PHA-Sepharose 4B. The absorbent was prepared by treating PHA-Sepharose 4B with 8 M urea: non-covalently bound sub-units were thereby dissociated from the gel and the absorbent lost the capacity to bind glycoproteins although retaining the ability to bind lectin-specific antibodies, presumably, by virtue of the residual sub-units containing the requisite epitopes. An aliquot (1.0 ml) of the crude immunoglobulin fraction (x 5 concentrated over original serum) was applied to the column (20 x 1.0 cm) and the non-binding components were washed through with 0.05 M Tris-HCl buffer, pH 8.0. Specific anti-PHA antibodies were desorbed with 3 M lithium thiocyanate and the thiocyanate was removed by exhaustive dialysis against 0.1 M NaCl.

Line immunoelectrophoresis was performed in agarose C (Pharmacia GB, Ltd) gels by the method of Kroll (12). Double diffusion analyses were carried out in agar gels containing 1.2% w/v Ionagar (Oxoid, Ltd) in 0.04 M Tris - 0.10 M glycine buffer, pH 8.7.

Immunocytochemistry and electron microscopy. The procedures used for examination of tissue from the small intestine of experimental animals (killed by captive bolt pistol after 27 days consumption of diets) have been described in detail elsewhere(13). Representative tissue samples, fixed for 3 h in 2.5% paraformaldehyde at pH 7.3, were embedded in paraffin wax and dewaxed sections (5-7 m) were stained according to an indirect immunofluorescence method in which the primary antibody was monospecific for E4-PHA (20). For transmission electron microscopy, tissue samples were fixed in 2.5% glutaraldehyde at pH 7.3, postfixed in 0.2% osmium tetroxide at pH 7.3 and stained with 5% uranyl acetate and lead citrate.

Animals and growth studies. Four castrate male pigs Large White x (Landrace x Large White) of 30 kg starting weight were offered *ad lib* a diet containing raw kidney beans (400 g kg^{-1} dry weight of diet). Ground barley (590 g kg^{-1}) and mineral-vitamin supplement (10 g kg^{-1}) made up the remainder of the diet. Food intakes were recorded daily and the pigs were weighed at weekly intervals. Each experimental animal was weight-matched with a control which received an equivalent amount of a standard stock diet with a 48 h delay interval. The stock diet contained barley (590 g kg^{-1}), white fish meal (75 g kg^{-1}), undecorticated heated soybean meal (75 g kg^{-1}), weatings (100 g kg^{-1}) and a mineral-vitamin supplement. Water was available on demand. Similar incorporation rates and experimental strategies were employed in feeding trials with pinto beans and with peas.

Results

Growth performance and clinical effects. 'Processor'-type beans. Although intake was sub-maintenance in both groups of animals as a consequence of marked appetite depression in the bean-fed animals, growth rate was much more severely impaired in the pigs receiving beans than in the controls (Table 1). Bean-fed pigs also suffered bouts of diarrhoea with apparent intermittent recovery.

Pinto beans. Acceptance of the diet containing pinto beans was rather better than for the Processor-type bean diet and animals were able to maintain weight but not to exhibit normal growth.

Table 1

Growth performance of growing pigs fed experimental and stock diets for 27 days.

Diet	Mean food intake (kg)	Mean weight change (kg)	Wt change/ unit food consumed	Weight change (kg/d)
Stock diet[a]	12.14±1.34 (4)	-2.68±1.07 (4)	-0.22	-0.10
'Processor'-type beans[a]	12.14±1.34 (4)	-7.83±0.52 (4)	-0.65	-0.29
Pinto beans[b]	24.13±2.07 (3)	-0.33±0.80 (3)	-0.014	-0.012
Stock diet[b]	26.00(1)	+6.98(1)	+0.27	+0.26

a. Number of observations is given in parentheses
b. Animals were restricted to a daily ration of 1 kg feed

Immunocytochemistry and cytopathology. The results of the studies with 'Processor'-type beans have been described in detail elsewhere (13) but the altered morphology of the small intestinal microvilli and the binding of PHA to the brush border and its internalisation in the enterocytes is illustrated in Figs 1A and 1B for reference. Extension of these studies to the small intestines of three pigs which had

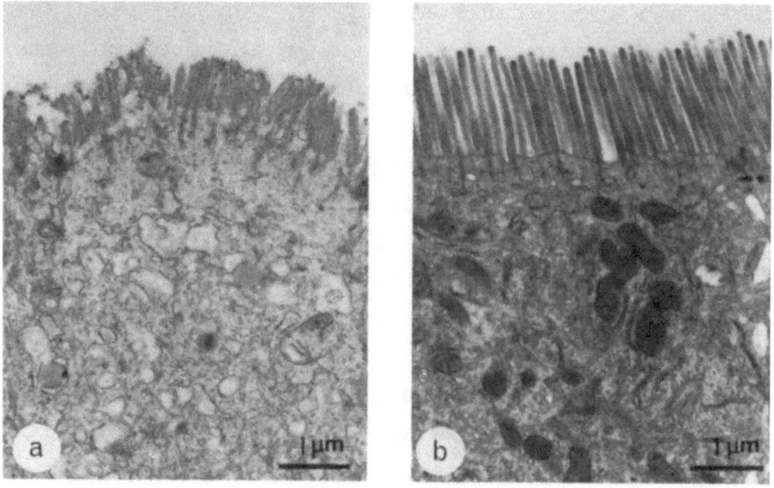

Fig. 1A. Transmission electron micrograph of section through the jejunum of a pig fed a) the "Processor"-type bean diet and b) the stock diet, showing disruption of microvilli.

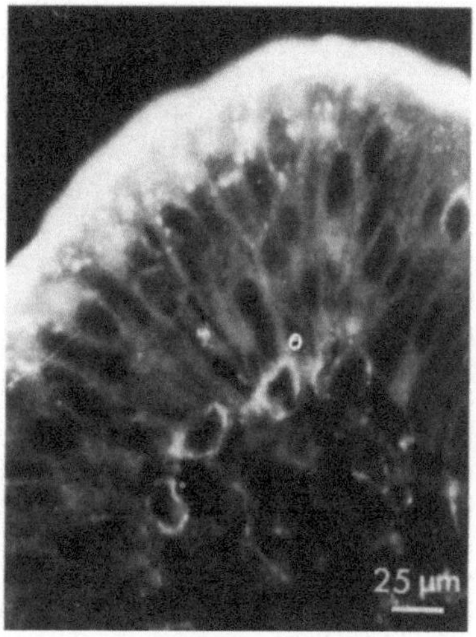

Fig. 1B. Immunofluorescence micrograph of section through jejunum of pig fed "Processor"-type bean diet. Fluorescence indicates presence of PHA bound to the brush border and within apical cytoplasm of mature enterocytes.

been maintained on the pinto bean diet revealed changes similar to those described above but at a more muted level. Some disruption was evident in the proximal part of the small intestine in two of the three animals where the microvilli were stunted and irregularly arranged. In the ileal tissue of all three pinto bean-fed pigs and in the proximal small intestine of the least-affected animal the microvilli were regularly arranged but were shorter and less abundant than on cells in similar locations in pigs fed on the stock diet. There were no detectable defects of cytomorphology in the gut of pigs which had consumed the diet containing marrowfat peas. The small intestinal tissue was histologically and ultrastructurally indistinguishable from that of animals which had been fed the control diet. Furthermore there was no evidence of binding of pea agglutinin (PSA) to the gut mucosal surfaces although PSA applied to pre-fixed sections of duodenal and jejunal samples of control animals showed evidence for binding sites in the apical tips of the microvilli.

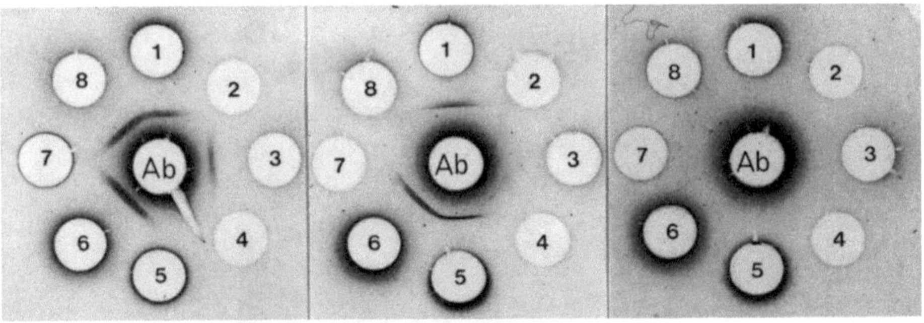

Fig. 2. Specificity of circulating antibodies induced in growing pigs after consuming a diet containing raw "Processor"-type beans for 27 days. For fig. 2A, B and C: 1, pH 8-soluble fraction from "Processor"-type beans (30 μg); 2, glycoprotein I + L4-PHA (20 μg); 5, E4-PHA (20 μg); 6, as well 1 (90 μg); 7, blank; 8, as well 3 (60 μg). For fig. 2A; Ab, Ig fraction produced by parenteral hyperimmunisation of pig with seed extract (40 μl, 2 x concentrated). For fig. 2B, Ab, Ig fraction from pigs consuming kidney beans (40 μg). Fig. 2C, Ab, Ig fraction from pigs consuming stock diet (40 μl).

Immunological studies. 'Processor'-type beans. Parenteral hyperimmunisation of pigs with a kidney bean extract induced polyspecific antibodies against the main seed antigens with the predominant response directed to the major reserve protein, phaseolin (Fig. 2A). Although the presence of humoral antibodies against seed antigen(s) was also demonstrated by double diffusion in the sera of pigs which had consumed 'Processor'-type kidney beans for 27 days, further examination using a panel of purified seed proteins indicated that these antibodies were exclusively directed against the lectin components of the seed (Fig. 2B). The major response was towards the E_4 lectin and other E sub-unit containing lectins, but activity was also observed towards the L sub-unit as evidenced by the precipitation line spurring with that against E_4 (Wells 4 and 5 in fig. 2B).

The specific anti-lectin immunoglobulins were isolated by affinity chromatography on PHA-Sepharose 4B and were shown to belong to the IgG class by double diffusion analysis against an antiserum specific for pig IgG. No IgA or IgM with anti-lectin activity could be detected.

The levels of PHA-specific serum antibody in the orally-immunised pigs which had consumed a total of 140-170 g lectin (calculated) per os over 27 days were shown by line immunoelectrophoresis (Fig. 3) to be significantly higher than those of animals hyperimmunised parenterally over 48 days with a total dose of about 10-12 mg lectin contained in a seed extract (Fig. 3B).

The time course of the lectin-specific antibody response was monitored for four individual animals given a diet containing raw beans. Examination of time sequential samples showed that, while the intensity of individual responses was variable, in general a significant level of antibodies to the bean lectin was detectable within 6 days after first exposure to the diet (Fig. 4). In one pig no PHA-specific antibodies could be detected by the methods employed.

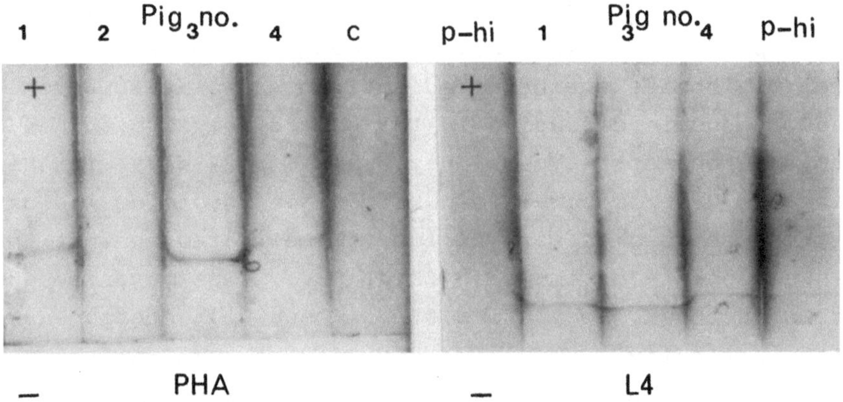

Fig. 3. Line immunoelectrophoresis of lectin against serum immunoglobulins from pigs consuming either the "Processor"-type bean diet or the stock diet (27 days). Antibody-containing strips contained equivalent volumes of Ig fractions (approximately 2-fold concentrated over parent serum, 200 µl per gel strip). Antigen (50 µg) was driven into the antibody containing gels under an applied voltage of 2 V cm^{-1}. For Fig. 3A, antigen was PHA (G2 fraction); C, Ig fraction from pig fed stock diet. For Fig. 3B, antigen was L4 lectin. End strips (p-hi) contained Ig fractions from pigs hyperimmunised with extracted seed proteins.

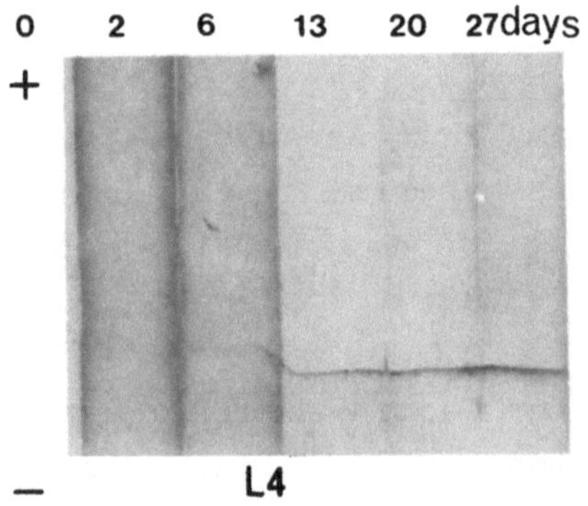

Fig. 4. Time course of appearance of antibodies against lectin. Antigen was L_4 lectin. Blood samples were withdrawn from a single bean-fed pig at intervals (days after first exposure) as indicated above each lane (the lower the line, the higher the antibody concentration).

Pinto beans. The development of significant levels of antibody to the pinto seed lectin (PSL) was also demonstrated in pigs which had consumed pinto beans. Thus when sera were with various seed fractions, antigenic activity was confined to the seed albumins and the specific antigen was identified as PSL by double diffusion against the purified protein (Fig. 5).

Peas. When sera from pigs which had consumed pea-containing diets were tested with the pea agglutinin diffuse precipitation lines were observed: these lines were common to both control sera from animals fed the stock diet and to the test sera and it was possible to prevent the precipitation by incorporation in the agar of 0.1% w/v methyl α-D-mannoside, a specific inhibitor of PSA-glycoprotein interaction. No evidence for the presence of true immunoglobulin precipitins specific for any pea protein was obtained.

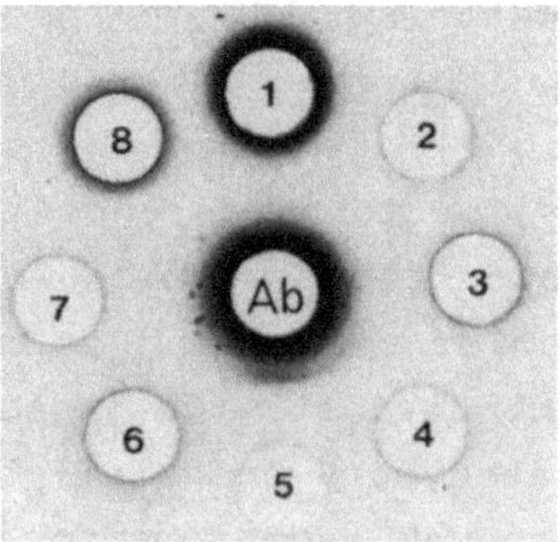

Fig. 5. Specificity of circulating immunoglobulins induced by feeding pinto beans to pigs. Ab, Ig from a pig fed the pinto bean diet (28 d), 40 µl; 1, pH 8-soluble beans (250 µg); 3, G2 fraction pinto beans (250 µg); 4, A fraction pinto beans (250 µg); 5, PSL (100 µg); 6, G1 fraction Processor beans (500 µg); 7, G2 fraction Processor beans (100 µg); 8, A fraction Processor beans (500 µg).

Discussion

The damage induced in the brush border of the small intestine of pigs receiving raw kidney beans in their diet (13) is consistent with that described by others for laboratory rats (15,16,17) and is recognised as a primary factor in the mechanisms underlying the growth-depressing effects of kidney beans. The gut lesion has been described in detail previously (13) but is characterised by disruption of the normal palisade arrangement of the apical microvilli of the enterocytes in the proximal small intestine with the most severe effects confined to the villus tips and correlating with the distribution of lectin-receptor activity (Fig. 1A and 1B). Small lectin-positive vacuoles were consistently observed in the apical cytoplasm of mature enterocytes in duodenum and jejunum. Thus the gross changes in morphology appear to be consequent on the binding of active PHA supplied in the diet, although the precise mechanism leading to the observed changes remains to be elucidated. Similar effects have been observed in animals ingesting pinto beans but the damage is much less severe and it is not clear whether this is attributable to the smaller amounts of lectin presented to the gut (pinto beans contain less than 1/100th of the amount of lectin protein contained in 'Processor'-type beans) or to differences in receptor affinity/frequency since it has been shown that PSL differs from PHA in both molecular weight and in carbohydrate specificity (18). In contrast, the ingestion of PSA in diets containing unheated, dry peas failed to produce any lesions of the gut although receptors for the lectin were demonstrable in fixed sections of pig small intestine. This may indicate either that PSA fails to survive transit through the stomach in active form or that competitive glycoconjugates in gut secretions prevent access of the PSA to the brush border receptors. It is significant that unheated, dry peas (Pisum sativum) provide a satisfactory protein source for growing pigs(14) when supplied in the diet at levels similar to those for kidney beans.

Another specific consequence of the interaction of dietary PHA lectin with the gut is the stimulation of humoral antibodies exclusively directed towards the lectin. No other seed protein in raw beans elicited a response although at least some of these survive as far as the terminal ileum in sufficiently large fragments (> 100 K daltons) to be immunogenic (Begbie, unpublished results). Such selective response to the lectin component of the diet suggests that, although the integrity of the mucosal barrier may be compromised, the uptake of antigenic lectin occurs by a specific route and may involve its capacity to bind to cellular receptors in the gut. While binding to the brush border of and internalization by the columnar absorptive cells has been demonstrated in these studies, the particularly rapid appearance of anti-lectin antibodies in the blood may result from stimulation of lymphoid follicles underlying M cells which bear lectin receptors in their apical membranes. Nicolson et al. have recently shown that ricin and wheat germ agglutinin bind to both M cells and absorptive cells but are transported across the epithelium by M cells only (19). Whatever the route of presentation of antigen to immunocompetent cells their polyclonal products recognise both L and E sub-units of the kidney bean isolectins indicating that the immunogen may be a substantially intact polypeptide.

A humoral immune response was also induced in pigs in response to ingestion of pinto beans. Again the antibodies were specific for the seed lectin which, in this P. vulgaris cultivar, is a very minor protein constituent. In contrast, no immune response could be detected to PSA when raw peas were incorporated in pig diets. Thus there emerges some degree of correlation between the intensity of antibody response and both the severity of disruption of the intestinal microvilli and the relative growth performance exhibited by the animals on the three different legume-containing diets although it must be recognised that the contributions of other antinutritional factors may influence growth response. While the nature and degree of toxicity experienced as a result of consumption of raw legume seeds appear to be multifactorial, present indications are that the ability of the constituent lectins to survive gastric and pancreatic secretions and to generate a humoral immune response after intestinal binding/absorption may be important contributory factors in determining the relative toxicity of various legumes.

References

1. Nicolson, G.L.: Int. Rev. Cytol. 39, 89-190 (1974).
2. Jaffe, W.G.: Toxic Constituents of Plant Foodstuffs, Second edition, (Liener, I.E., ed.) 1980, Academic Press, New York, London, Toronto, Sydney, San Francisco, pp. 73-102.
3. Pusztai, A.J., Clarke, E.M.W., Grant, G., King, T.P.: J. Sci. Food Agric. 32, 1037-1046 (1981).
4. Pusztai, A.J., Greer, F., da Guia Silva Lima, M., Prouvost-Danon, A., King, T.P.: Chemical Taxonomy, Molecular Biology and Function of Plant Lectins (Goldstein, I.J. and Etzler, M.E. eds), Alan R. Liss, New York, 1983, pp 271-272.
5. Pusztai, A.J., Greer, F.: Protides of the Biological Fluids (Peeters, H., ed.) 32, (1984) in press.
6. Sun, S.M., Hall, T.C.: J. Agric. Food Chem. 23, 184-189 (1975).
7. Pusztai, A.J., Stewart, J.C.: Biochim. Biophys. Acta 536, 38-49 1978).
8. Manen, J.-F.: Saussurea 9, 23-44 (1978).
9. Trowbridge, I.S.: J. Biol. Chem. 249, 6004-6012 (1974).
10. Leavitt, R.D., Felsted, R.L., Bachur, N.R.: J. Biol. Chem. 252, 2961-2966 (1977).
11. Harboe, N., Inglid, A.: Scand. J. Immunol. 2, Suppl. 1, 161-169 (1973).
12. Kroll, J.: Scand. J. Immunol. 2, Suppl. 1, 61-67 (1973).
13. King, T.P., Begbie, R., Cadenhead, A.: J. Sci. Food Agric. 34, 1404-1412 (1983).
14. Fowler, V.R.: Vicia faba : Feeding value, processing and viruses (Bond, D.A. ed.) Martinus Nijhoff, The Hague, 1980, pp. 31-41.
15. King, T.P., Pusztai, A., Clarke, E.M.W.: J. Comp. Path. 90, 585-595 (1980).
16. King, T.P., Pusztai, A., Clarke, E.M.W.: J. Comp. Path. 92, 357-373 (1982).
17. Lorenzsonn, V., Olsen, W.A.: Gastroenterol. 82, 838-858 (1982).
18. Pusztai, A.J., Grant, G., Stewart, J.C.: Biochim. Biophys. Acta 671, 146-154 (1981).
19. Neutra, M.R., Guerina, N.G., Hall, T.L., Nicolson, G.L.: Gastroenterol 82, 1137 (1982) Abstr.
20. King, T.P., Pusztai, A., Clarke, E.M.W.: Histochem. J. 12, 201-208 (1980).

LECTIN-INDUCED ARTHRITIS OF RABBIT AS A MODEL OF RHEUMATOID ARTHRITIS

Rolf Bräuer, Klaus Thoss, Steffen Henzgen, Günter Waldmann
Friedrich Schiller University, Institute of Pathology,
DDR - 6900 Jena, German Democratic Republic

The aetiology and pathogenesis of many chronic inflammatory diseases are still unknown and therefore it is pertinent to model such conditions in experimental animals. A good example of this is the human rheumatoid arthritis. During the last two decades several models of experimentally induced arthritis have been developed, mainly for the search of antiinflammatory and anti-arthritic drugs (1). The antigen-induced arthritis of rabbits (2,3) has been widely used as a model for studying the pathogenesis of rheumatoid arthritis because its striking similarities to the human disease including the histological feature, the chronicity and the response to drug treatment (4,5). Chronic inflammatory arthritis is induced by injecting a foreign protein (human fibrin, ovalbumin or bovine serum albumin) into the knee joint of an animal previously sensitized with the same antigen in Freund's complete adjuvant. In our model of lectin-induced arthritis we use different lectins for the induction of chronic joint inflammations in rabbits (6,7). In the present studies we have tested a series of immune reactions in the course of lectin-induced arthritis to get further insights into the pathogenic mechanisms of the experimental model, especially concerning the significance of the cell-mediated immunity for the chronicity of inflammation and for the destruction of joint tissues, and on the possible relation to the human rheumatoid arthritis.

Lectins, Vol. IV

Materials and Methods

Lens culinaris lectin (LCA), *Pisum sativum* lectin (PSA) and wheat germ agglutinin (WGA) were prepared as described earlier (7-9). Concanavalin A (Con A) was purchased from Pharmacia Fine Chemicals (Uppsala, Sweden), phytohaemagglutinin (PHA) from Wellcome Research Laboratories (Beckenham, England).
Non-sensitized rabbits of the Californian strain, weighing 2.5-3.5 kg were injected intraarticularly with 2.5 mg LCA, PSA, Con A or WGA in 0.5 ml saline into the right knee. Control animals received a saline injection in the same manner.
For preimmunization the animals were injected subcutaneously with 10 mg LCA or ovalbumin in 0.5 ml saline emulsified with 0.5 ml Freund's complete adjuvant. Two weeks later a booster injection was given, and after further two weeks the arthritis was induced by injection of 5 mg LCA or ovalbumin in 0.5 ml saline into the right knee joint cavity.Control animals were injected subcutaneously with saline in adjuvant and intraarticularly with saline.
Specific antibodies in the sera were quantitated by the reserve single radial immunodiffusion method of Mancini (10).
Peripheral blood lymphocytes were isolated from heparinized blood of the ear vene by density centrifugation with Ficoll-Paque (Pharmacia) (11) and resuspended to a concentration of 1×10^6 cells/ml in HEPES-buffered RPMI 1640 medium (GIBCO Labs, Grand Island, USA) containing 10 % pooled inactivated normal rabbit serum.Samples of 0.2 ml were incubated in quadruplicates for 72 h at 37 ^{o}C with different concentrations of LCA, Con A or PHA. For the last 18 h the cultures were radiolabeled with 0.5 µCi ^{3}H-thymidine per tube. The stimulation index represents the ratio of the radioactivity of stimulated to unstimulated samples.
For testing the influence of serum on the lymphocyte proliferation, peripheral blood lymphocytes of an untreated rabbit were cultured in similar way in medium supplemented with 20 % non-inactivated pooled serum from arthritis or control animals.
For the activation of suppressor cells by Con A, peripheral blood lymphocytes were incubated for 24 h with 25 µg/ml Con A, followed by treatment with Mitomycin C (50 µg/ml, 30 min) and methyl-α-D-mannopyranoside (0.1 M, 15 min). Its suppressive effect on the proliferative response of autologous peripheral blood lymphocytes to Con A stimulation was tested by culturing these cells (100 µl, 2×10^5 cells) with freshly prepared lymphocytes from the same animal (100 µl, 1×10^5 cells) at the conditions described above.

Results

A single injection of LCA, PSA or Con A into the knee joint of non-sensitized rabbit induces the development of a severe arthritis with a tendency to chronification. Similarly as in the human rheumatoid arthritis, the histomorphological feature of the persistant inflammation is characterized by lympho-plasmacellular infiltration,fibroblast and histiocyte proliferation, proliferative vasculitis and villous hyperplasia of the synovial membrane. Local aggregations of lymphocytes in form of secondary lymph follicles, marked sclerosis by neoformation of collagen fibres and degradations of articular cartilage could be demonstrated. The non-mitogenic WGA, however, induces only minimal inflammatory alterations without chronification. Because LCA has the strongest arthritic potential and we have at disposal a large scale preparation, this lectin was used for further model investigations. Evidenced by fluorescence with FITC-labeled LCA, the lectin persits in the joint of non-sensitized animals for many weeks presumably by its binding to glycoproteins of the connective tissue and induces the acute inflammation directly, probably by altering the permeability and microcirculation. Moreover, antibodies against LCA are present in the serum 14 days after injection, thereafter a slow reduction was observed. After six months antibodies to LCA could no longer be demonstrated, likewise they never were detected in sera of control animals.

Following the acute phase of inflammation, therefore, the LCA-induced arthritis in non-sensitized animals is comparable to the antigen-induced arthritis: both are characterized by locally retained antigen within the joint connective tissue in a specifically sensitized animal. However, in our model these conditions are further modified or amplified by the mitogenic activity of LCA itself, leading to unspecific stimulation of T-lymphocytes _in vivo_ (in preparation).

Preimmunization of the animals with LCA produces a high level of specific antibodies, already available at the moment of arthritis induction (Fig. 1). After the intraarticular injection of LCA a further rise of the antibody level takes place. In immunized controls LCA-specific antibodies never were detected.

The preimmunization also increases the severity and persistence of the arthritic reaction. This reaction, which is stronger than in the

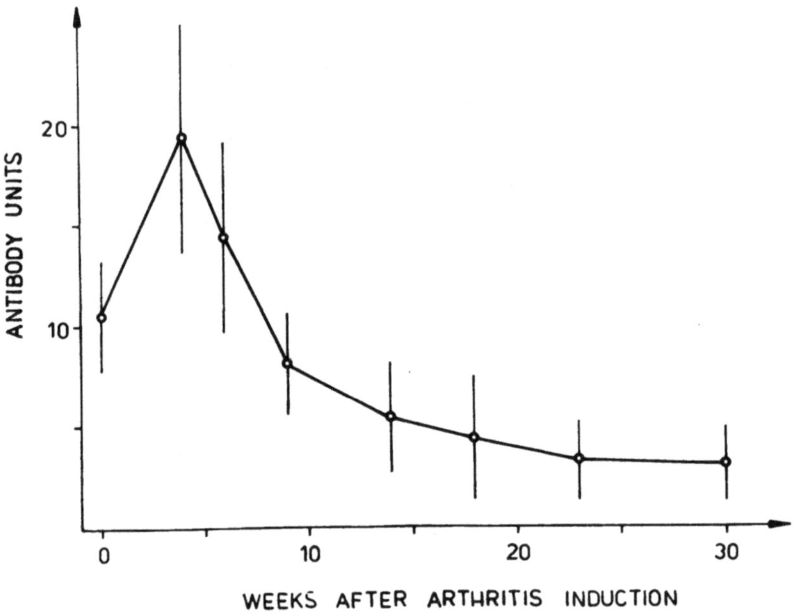

Fig. 1. Titer of Precipitating Antibodies against LCA in the Course of LCA-induced Arthritis in Presensitized Animals. Mean ± S. D. of 10 Animals.

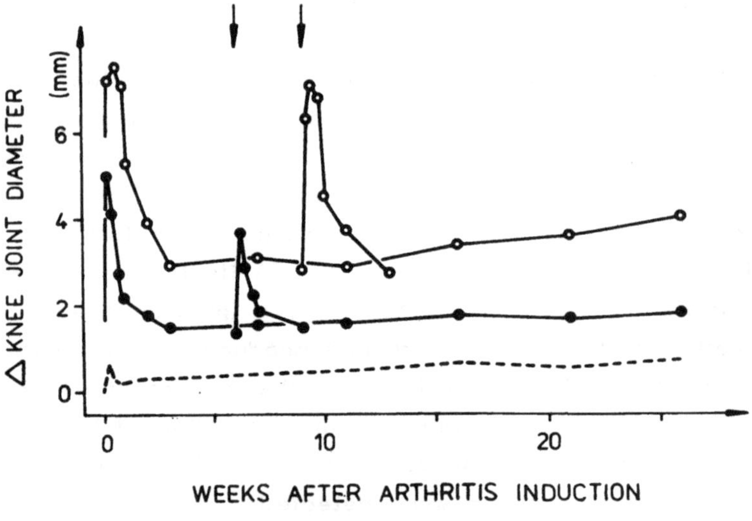

Fig. 2. Arthritic Reaction Measured as Change of the Knee Joint Diameter of Rabbits after a Single and a Repeated Intraarticular Injection of LCA (o—o) or Ovalbumin (●—●) in Specifically Sensitized Animals. Immunized Controls (----) were Intraarticularly Injected with Saline. Mean of 12 Animals. The arrows indicate the second intraarticular injection.

ovalbumin-induced arthritis, is also observed with repeated intra-articular injection (Fig. 2).
The cell-mediated immunity of the rabbits was tested continuously in the course of LCA-induced arthritis by the _in vitro_ reactivity of peripheral blood lymphocytes to mitogenic concentrations of LCA and Con A. In the control animals the mitogenic responsiveness decreased with time (Fig. 3).

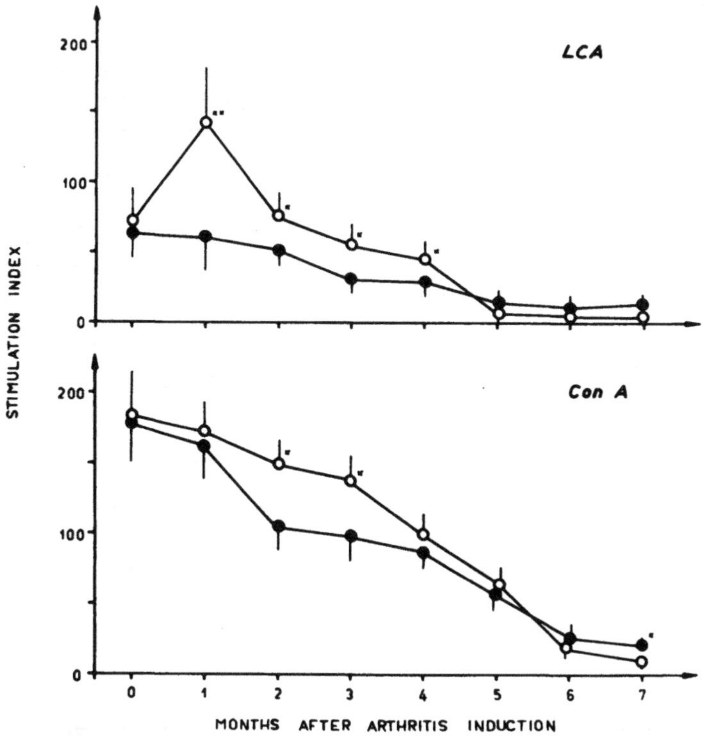

Fig. 3. Responsiveness of Peripheral Blood Lymphocytes of Non-sensitized Rabbits with LCA-induced Arthritis (o—o) and Controls (•—•) to _in vitro_ Stimulation with LCA (10 µg/ml) and Con A (5 µg/ml). Mean ± S. D. of 12 Arthritis and 10 Control Animals.
x: $p < 0.05$; xx: $p < 0.01$.

On the other hand, in the early phase of inflammation the arthritis animals showed significantly higher LCA stimulation. The hyper-reactivity of lymphocytes of the same animals could also be demonstrated to Con A two to three months after the arthritis induction with LCA, though not in that degree as to LCA.
In the late phase of arthritis (7 to 9 months after induction), however, a hyporesponsiveness of lymphocytes from arthritis animals to LCA and Con A was striking (Table 1).

Table 1. Hyporesponsiveness of Peripheral Blood Lymphocytes in the Late Phase of LCA-induced Arthritis to Mitogenic Stimulation.

	Stimulation Index				
	LCA (µg/ml)				Con A (µg/ml)
Group	1	10	20	40	10
Control	10.8	48.3	51.1	35.6	121.6
Animals	± 4.9	± 23.8	± 15.0	± 17.6	± 43.6
Arthritis	3.0	16.3	24.7	13.5	49.9
Animals	± 1.2	± 8.8	± 6.2	± 9.0	± 15.5
p (%)	<5	<5	<5	<5	<1

Lymphocytes of non-sensitized rabbits 8 months after arthritis induction with LCA were stimulated _in vitro_ with LCA and Con A. Control animals received only a intraarticular saline injection. The stimulation index is given as mean ± S. D. of 9 control and 11 arthritis animals.

As there is evidence for a suppressor cell defect in autoimmune diseases including rheumatoid arthritis (12,13), we tried to detect suppressor cell activity in the peripheral blood of experimental and control animals. A suppressor cell population can be stimulated in rabbits by incubation of peripheral blood lymphocytes for 24 h with Con A _in vitro_ (10). We demonstrated the effect of these Con A-induced suppressor cells on the mitogenic stimulation of autologous peripheral blood lymphocytes. The arthritis animals in the early phase of inflammation exhibited a significantly decreased suppressor cell potential (Table 2). In the late phase, on the other hand, the Con A-induced suppressor cell potential was identical in arthritis and control animals.

Table 2. Suppressor Cell Activity in LCA-induced Arthritis.

Time after Arthritis Induction	Suppressor Index (%) 4 Months		8 Months	
Con A-concentration (μg/ml)	1	5	1	5
Control Animals	35.4±10.7	45.2±18.4	53.0±16.0	48.2±21.5
Arthritis Animals	-14.4±43.6	5.2±20.3	56.0±14.2	58.0±17.6
p (%)	<5	<5	n.s.	n.s.

Suppressor cells in the peripheral blood of non-sensitized rabbits with LCA-induced arthritis were induced in vitro with Con A. The effect of these cells on the proliferative response of autologous peripheral blood lymphocytes to Con A stimulation was tested by culturing with freshly prepared lymphocytes from the same animal. Control animals were injected intraarticularly with saline.

Furthermore, we have tested the effect of pooled sera from arthritis or control animals on the mitogenic stimulation of peripheral blood lymphocytes from normal untreated rabbits. Sera obtained from animals in the early phase of arthritis significantly enhance the Con A and PHA stimulation, whereas the mitogenic responsiveness to low concentration of LCA was decreased (Table 3).

We believe, that LCA-induced mitogenic factors are responsible for this stimulatory effect. However, the decreased stimulation with LCA is related to specific antibodies against LCA, still existing in the serum at this time and binding parts of the added mitogen. By using sera from animals in the late phase of arthritis, the stimulation of lymphocytes was significantly lower than with control sera in all mitogen concentrations tested. Serum inhibitory factors, recently detected in some different chronic diseases including rheumatoid arthritis (14-16) could be responsible for this effect. Moreover, at the end of experiments we have tested the response of spleen lymphocytes from animals with LCA-induced arthritis to cartilage antigens _in vitro_, i. e. extracts of rabbit costal cartilage and proteoglycans from bovine nasal cartilage. In some arthritis animals with strong cartilage degradations, a proliferative response of the spleen lymphocytes to these cartilage antigens could be discovered. This suggests the development of autoimmunity to cartilage macromolecules in the course of chronic inflammatory joint lesions.

Table 3. Influence of Serum from Rabbits with LCA-induced Arthritis on the Mitogenic Stimulation of Peripheral Blood Lymphocytes from Untreated Animals.

	Change of Stimulation Index (%)								
	LCA (µg/ml)			Con A (µg/ml)			PHA (µg/ml)		
	1	5	10	1	5	10	1	5	10
4 Months	-67	-47	+16	+61	+46	+22	+17	+22	+165
p (%)	<1	<1	n.s.	<1	<1	n.s.	<5	<5	<1
8 Months	-26	-21	-38	-31	-39	-40	-49	-55	- 18
p (%)	<5	<5	<1	<1	<1	<1	<1	<1	n.s.

Peripheral blood lymphocytes from untreated animals were cultured in medium containing 20 % pooled serum from arthritis or control animals. The relation of the stimulation index by culturing with arthritis serum/control serum is given as % - 100. Mean of 20 estimations.

Conclusions

The experimental lectin-induced arthritis in rabbits is characterized by a biphasic course of the cell-mediated immunity:

1) the early phase: hyperresponsiveness of peripheral blood lymphocytes to mitogenic stimulation, decreased suppressor cell activity, and occurence of stimulatory serum factors.

2) the late phase: hyporesponsiveness of peripheral blood lymphocytes to mitogenic stimulation, normal suppressor cell activity, and inhibitory serum factors.

The hyperresponsiveness may be attributed to the decreased suppressor cell activity; the hyporesponsiveness in the late phase must be based on other mechanisms, possibly on the inhibitory serum factors which could be bound to lymphocytes.

Therefore, in spite the differences between lectin-induced arthritis and the human rheumatoid arthritis (mainly concerning the initiation and the systemic manifestation) there are surprising similarities not only in the histopathological features and the chronicity, but also in some cell-mediated immune reactions.

Regulatory disorders on the T-cell level, mainly the defective suppressor cell activity in combination with the acute inflammation may probably account for the autoimmune phenomena against cartilage and could be responsible for the chronicity of joint inflammation as well as for destructions of articular tissues. This may play a fundamental role in the pathogenesis of experimental and human arthritis. Therefore we consider the lectin-induced arthritis as a valuable experimental model for further investigations on the pathogenesis of human rheumatoid arthritis.

References

1. Billingham, M. E. J.: Pharmac. Ther. 21, 389-428 (1983).

2. Dumonde, D. C., Glynn, L. E.: Br. J. Exp. Pathol. 43, 373-383 (1962).

3. Consden, R., Doble, A., Glynn, L. E., Nind, A. P.: Ann. Rheum. Dis. 30, 307-315 (1971).

4. Hunneyball, I. M., Stanworth, D. R.: IRCS J. Med. Sci. 9, 1-4 (1981).

5. Hunneyball, I. M.: Adv. Inflamm. Res. 7, 249-261 (1984).

6. Thoss, K., Raabe, G., Müller, P.: All. Immunol. 22, 311-316 (1976).

7. Waldmann, G., Henzgen, S., Thoss, K.: Exp. Pathol. 16, 194-201 (1978).

8. Roth, J., Thoss, K.: Exp. Pathol. 10, 258-267 (1975).

9. Thoss, K., Roth, J.: Exp. Pathol. 14, 215-219 (1977).

10. Bräuer, R., Henzgen, S., Thoss, K., Waldmann, G.: Exp. Pathol. 24, 117-131 (1983).

11. Bøyum, A.: Scand. J. Immunol. 5, Suppl. 5, 9-15 (1976).

12. Abdou, N. I., Lindsley, H. B., Racela, L. S., Pascual, E., Hassanein, K. M.: J. Rheumatol. 8, 9-18 (1981).

13. Sakane, T., Takada, S., Murakawa, Y., Kotani, H., Honda, M., Ueda, Y.: J. Immunol. 129, 1972-1977 (1982).

14. Horwitz, D. A., Garrett, M. A., Craig, A. H.: Clin. Exp. Immunol. 27, 100-110 (1977).

15. Lohrmann, A., Berg, P. A., Goethe, S., König, G.: Klin. Wchschr. 57, 1225-1228 (1979).

16. Highton, J., Panayi, G. S., Shepherd, P., Wooley, P.: Ann. Rheum. Dis. 41, 563-568 (1982).

MONONUCLEAR CELL RESPONSE TO LECTINS: cAMP AND CHEMILUMINESCENCE

Michael Derwahl, Jürgen Barth, Kurt-Günther Ravens
I. Medizinische Klinik der Universität Kiel,D-2300 Kiel

Udo Schumacher
Pathologisches Institut der Universität Kiel, D-2300 Kiel

The stimulation of mononuclear cells by plant lectins is an important method of studying the early events of the initiation of cell proliferation in vitro (1, 2). Several biological effects follow the binding of lectins to the membranes of leucocytes, for example an increase of chemiluminescence and an increase in membrane adenylate cyclase activity (3, 4). Some authors have described an increase of cyclic 3'5'-adenosine monophosphate (cAMP) in human mononuclear cells in response to lectins, while others have been unable to confirm these results (5, 6, 7, 12). Increases in cAMP have been reported with both mitogenic and nonmitogenic lectins, suggesting that it may be unrelated to the activation signal (8).

Lectins also effect the oxidative metabolism of cells. After exposure to lectins, human mononuclear cells respond with rapid fluctuation in the electrical cell membrane potential. This is connected with a respiratory burst and increased oxygen consumption which is related to a release of oxidant radicals, those can be measured as chemiluminescence (10).
Aim of our study is to evaluate the metabolic effects of lec-

Lectins, Vol. IV

tins after exposure to mononuclear cells. Another question is whether there is a correlation between chemiluminescence response to lectins and cyclic AMP levels.

Materials and Methods

The plant lectins concanavalin A, wheat germ agglutinin and soy bean agglutinin were obtained from Sigma Chemical Company Munich, W-Germany. Luminol (5-amino-2,3-dihydro-1,4-phalazinedion) was acquired from Sigma Chemical Company Munich, W-Germany. Dulbecco's modified Eagle medium was obtained from Gibco Laboratories, Detroit, Michigan, USA.

Cell preparation: Venous blood samples (20 ml) from healthy male donors were collected in heparinized containers (10 IU heparin/ml blood). Mononuclear cells were isolated by Ficoll-Hypaque gradient centrifugation. These cell preparations were washed three times with Dulbecco's modified Eagle medium.

Chemiluminescence-assay: Chemiluminescence was measured in a Berthold Biolumat Counter LB 9505 (10). After adding Luminol (20µg/ml), suspensions of mononuclear cells (1000 cells/µl) were incubated in the dark for 10 min at 37°C. The base chemiluminescence was measured, and lectins of various concentrations added to the assay. Following the addition of the lectins the time course of photon count rate was recorded. For purposes of comparison, the maximum cpm rates were calculated, after correction by substraction of the baseline counts.

cAMP-assay: An aliquot of the cell suspension was prepared in the following way: Lectins were added to the suspension of mononuclear cells (10^5 cells/ml) and incubated for 5 min at 37°C. The cells were then homogenized at 4°C by ultrasound using a Biosonic III (Bronwill Scientific, New York). For deproteinization ice-cold 6% trichloroacetic acid was added. Suspensions were then centrifuged at 5000 g for 15 min at 4°C, and supernatants were washed five times with hydrated acidified ether to remove trichloroacetic acid. The solutions were then evaporated at 65°C under a stream of nitrogen. After reconstitution of samples in bidistilled water, cAMP concentrations were analysed with a commercially available radioimmunoassay (Becton-Dickenson, Heidelberg, W-Germany).

Results

Fig. 1 shows the chemiluminescence after incubation with different amounts of Con A, WGA and SBA. After adding 10µg Con A to the cell suspension of mononuclear cells we observed a significant stimulation which is 2.9-fold ($p < 0,01$). A concentration of 50µg Con A caused a 12-fold increase of chemiluminescence and 200µg Con A caused an 18-fold increase. Thus increases induced by Con A are dose dependent.
Biochemically analysed cAMP concentrations after the addition of various amounts of Con A are given in Fig. 2. Using 10µg Con A we found a rise of cAMP of about 1.7-fold. Further increases of 2.9-fold were found after stimulation with 50µg Con A when compared to non-activated cells. Further increases in the Con A concentration did not result in an accompanying further rise of the cAMP.
The addition of WGA was followed by a significant increase of chemiluminescence as demonstrated in Fig. 1. After adding 1µg WGA we found an 8.2-fold rise of chemiluminescence. Using higher amounts of WGA there was a continuous dose-dependent rise. After adding 100µg WGA we found an 11.5-fold increase.

Similar stimulation of the cells with WGA resulted in an increase of the cAMP concentration, which was also dose-dependent ranging between 0.83 and 3.67-fold rise. (Fig. 2)
Only a modest rise of both chemiluminescence and cyclic AMP occurred after incubation with a nonmitogenic lectin SBA:
The rise of chemiluminescence ranged between 0.26-fold and 1-fold. The rise of the cAMP concentration was found between 0.54 and 0.77-fold (Figs. 1 and 2).
The main peak of chemiluminescence was measured 90 to 120 sec after adding lectins. Different from that we found the highest cAMP concentrations 5 min after adding the lectins.

Discussion

Our study is the first one which connected two of several biologic effects which occur following the binding of plant lectins (Con A, WGA and SBA) to mononuclear cells: chemiluminescence and concentration of intracellular cAMP. We observed increases of chemiluminescence after stimulation with the mitogenic lectins Con A and WGA, as well as with the nonmitogenic SBA. Our results confirmed other reports (3, 11). A significant rise of intracellular cAMP was measured after incubation with the mitogenic lectins Con A and WGA. These results are in accord with published reports (4, 12, 13). A modest increase of cAMP was found after stimulation with nonmitogenic lectin SBA. Former reports have noted increases in cAMP with both mitogenic and nonmitogenic lectins, suggesting that these effects may be unrelated to the activation signal (8). Because of the dichotomy of the cAMP action, (i.e. mitogens increase intracellular cAMP, while other agents which increase intracellular cAMP inhibit lectin-induced activation), it was suggested that cAMP might be compartmentalised within mononuclear cells: cAMP localised in one part of a cell induces activation, while that in another part inhibits the activation process (14). These data suggest that activating and inhibiting processes in the same cell are cAMP dependent. This may explain our finding that the cAMP concentration increasedafter stimulation with mitogenic as well as nonmitogenic lectins. The different peak time of chemiluminescence (90 to 120 sec after adding lectins) and cAMP concentrations (5 min after adding lectins) suggests that cAMP is no activation signal for chemiluminescence. The highest peak of chemiluminescence response was obtained after stimulation with Con A, while the highest cAMP concentration measured was in response to WGA. These data suggest that chemiluminescence response is independent of adenylate cyclase activation in peripheral blood mononuclear cells. It is known that chemi-

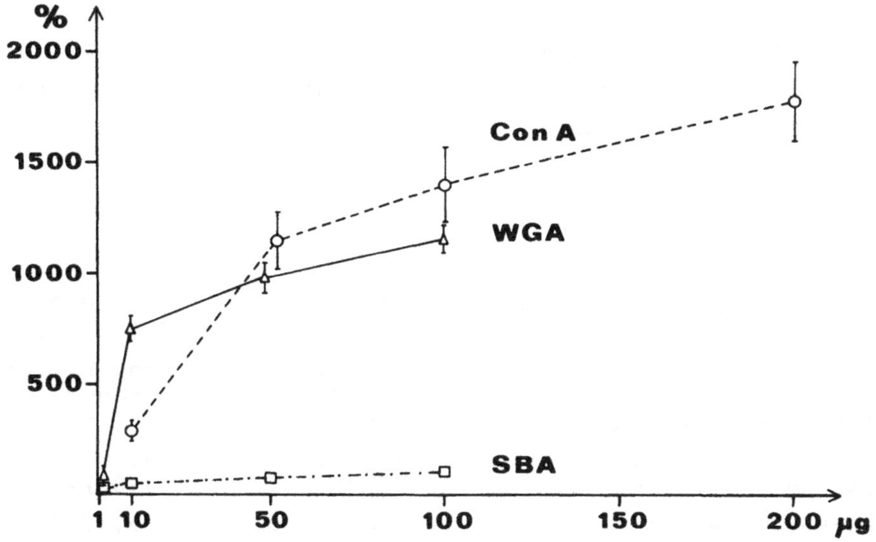

Fig. 1. Effect of different amounts of lectins (Con A, WGA, SBA) to chemiluminescence. Maximal increase above base chemiluminescence is demonstrated in % ± SD. Peak levels were seen 90 - 120 sec after adding lectins.

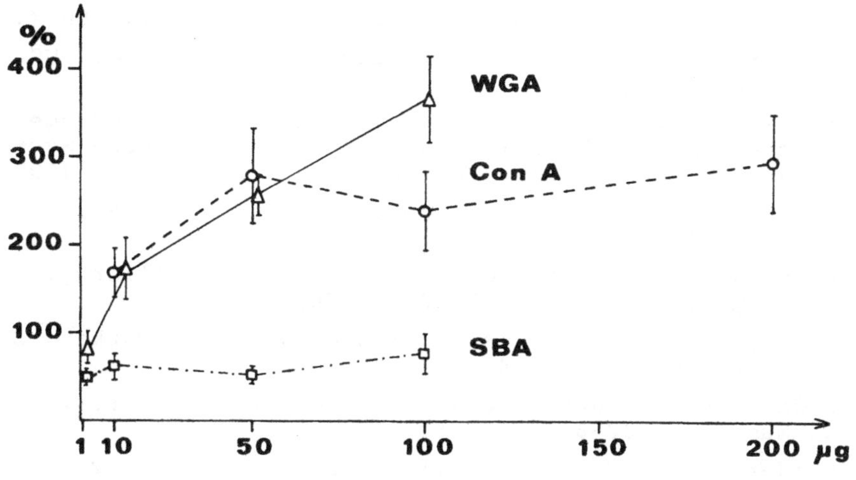

Fig. 2. Effect of different amounts of lectins (Con A, WGA, SBA) to cAMP concentration. Maximal increase above base cAMP is demonstrated in % ± SD. Peak levels were seen 5 min after adding lectins.

luminescence of mononuclear cells is mainly generated by the monocyte-fraction. Further investigations have to show which part of cAMP in mononuclear cells is due to lymphocytes respectively to monocytes.

References

1. Hadden, J.W., Hadden, E.M., Sadlik, J.R., Coffey, R.G.: Proc. Natl. Acad. Sci. Vol. 73, No. 5, 1717-1721 (1976).
2. Ferber, E., Reilly, L.E., de Pasquale, G., Resch, R. in: Proceedings of the 8th Leukocytes Culture Conference, Academic Press Inc., New York (1974).
3. Mookerjee, B.K., Ferberg, E., Ernst, M., Sharon, N., Fischer, H.: Immunol. Commun. 9 (7), 653-676 (1980).
4. Parker, C.W.: Annals New York Academy of Science, 332, 255-261 (1979).
5. Goffstein, B.J., Gordon, C.K., Wedner, H.J., Arkinson, J.P.: Clin. Med., 96 (6), 1002-1014 (1980).
6. De Rubertis, F.R., Zeuser, T.: Biochem. Biophys. Acta 428, 91-103 (1976).
7. Krishnaraj, R., Talwar, G.P.: J. Immunol. 111 (4), 1010-17 (1973).
8. Burleson, D.G. Sage, H.J.: J. Immunol. 116 (3), 696 (1976).
9. Korchak, H.M., Weissman, G.: Proc. Nat. Acad. Sci. USA 75, 3118-3822 (1978).
10. Wrogeman, K., Weidemann, M.J., Peskar, B., Staudinger, H.J. Fischer, H. in: Function and Structure of Immune System, Plenum Publishing Corporation, New York, 395-399 (1979).
11. Peterhaus, E.: Biochem. Biophys. Res. Commun. 91, 383-392 (1979).
12. Wedner, H.J., Parker, C.W. in: Progress in Allergy, S. Karger, Basel 20, 195-300 (1976).
13. Smith, J.W., Steiner, A.L., Newberry, W.M., Parker, C.W.: J. Clin. Invest. 50, 432-441 (1971).
14. Wedner, H.J., Parker, C.W.: Biochem. Biophys. Res. Commun. 62, 808-814 (1975).
15. Smider, D.R., Parker, C.W.: Biochem. J. 162, 472-482 (1972).

PEANUT AGGLUTININ AND CELL COOPERATION IN INTERLEUKIN-2 MEDIATED MOUSE THYMOCYTE STIMULATION

Jean Favero, Jacques Marti, Jacques Dornand, Jean-Claude Bonnafous and Jean-Claude Mani

Laboratoire de Biochimie des Membranes, ER CNRS 228, ENSCM, 8 rue de l'Ecole Normale, F-34075 Montpellier (France)

Different lectins, such as concanavalin A (Con A) or phytohemagglutinin, and different oxidizing agents, sodium periodate or galactose oxidase combined with neuraminidase which unmasks galactosyl moieties, have been largely used for the study of T cell activation. Many reports have shown that the final mitogenic response, evidenced by [^{3}H]-thymidine incorporation, is related to the direct action of the lectin or the oxidizing agent on specific cell surface molecules and to cellular cooperation between different cell types (lymphocytes and macrophages) and different T cell subsets (helper, suppressor and responder T cells). This cellular cooperation has been largely studied through the different growth factors synthesized by the cells and defined under the general term of cytokines: mainly interleukin-1, IL-1, a monokine released by monocytes, and interleukin-2, IL-2, a lymphokine released by T lymphocytes. Another cellular cooperation evidenced in mitogenic stimulation is a cooperation by cell-cell contact. This type of cooperation was assessed by the fact that only multivalent lectins, leading to cellular agglutination, were able to display mitogenic activity; in the case of oxidizing agents it was postulated that Schiff bases could be formed between free $-NH_2$ groups of surface membrane proteins on accessory cells and the aldehyde groups formed by galactose oxidation (galactose oxidase stimulation) or sialic acid oxidation (periodate stimula-

Lectins, Vol. IV

tion) on lymphocytes. Peters showed that agglutination of the cells actually facilitates cellular cooperation (1). However several lectins agglutinate lymphocytes and do not display any mitogenic activity. In order to evaluate the importance of cell-cell contact in T cell activation we examined the influence of peanut agglutinin (PNA), a galactosyl directed lectin which agglutinates but does not stimulate mouse thymocytes (2), on the responsiveness of these cells to concanavalin A or galactose oxidase.

Materials and Methods.

Cells. In all experiments 3 week-old male Swiss mice were used. Thymic cells were prepared as described in (3). Mature and immature thymocyte populations were isolated by agglutination with PNA as in (4). Peritoneal macrophages were prepared by injecting RPMI 1640 in the peritoneal cavity of 2 month-old male Swiss mice and gathering the peritoneal exudate with a syringe. In order to diminish the adherence of the cells to plastic vessels, all manipulations were carried out at 4°C.

Neuraminidase or neuraminidase + PNA treatment of the cells. The cells (2×10^7) were incubated with either neuraminidase alone (0.04 U) or neuraminidase + PNA, in RPMI 1640, for 30 min at room tempe-

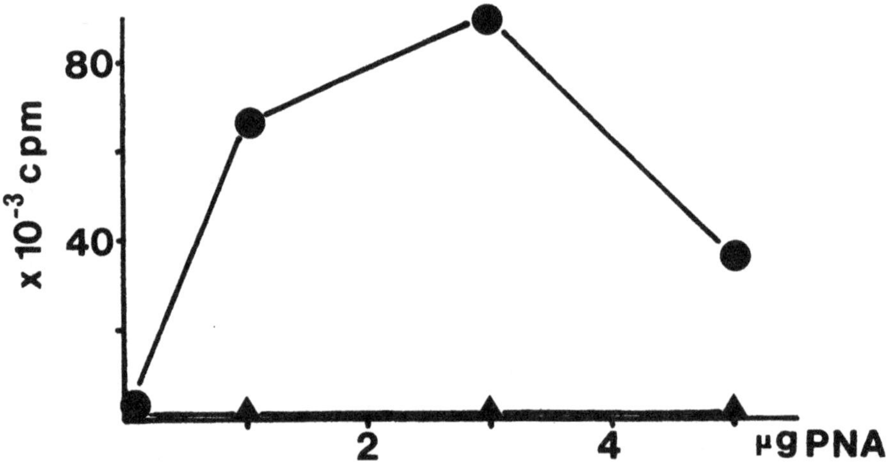

Fig. 1. Influence of the treatment with increasing amounts of PNA on the responce of neuraminidase-treated thymocytes cultured without mitogen (▲-▲) or with galactose oxidase (●-●). (^3H)-Thymidine incorporation as a function of PNA.

rature, then extensively washed. The lectin concentration was 3 µg per 2×10^7 cells, except for initial experiments in which the optimal PNA concentration was determined.

Culture conditions. Treated or untreated thymocytes (2.5×10^6 per ml) were cultured in 96 well microtest plates for 48 hr at 37° C under 5% CO_2 humidified atmosphere, in RPMI 1640 supplemented with 10% FCS. Stimulation was induced by either 5 µg Con A or 0.07 U galactose oxidase. [^{3}H]-Thymidine incorporation was determined after 4 hr pulse. Coculture of thymocytes + peritoneal macrophages was carried out by incubating 2.5×10^6 thymocytes with 125,000 macrophages per ml medium.

Conditioned medium preparation and IL-2 activity determination. Supernatants of thymocyte cultures were gathered after 24 hr incubation and used as conditioned media. Residual Con A was blocked by 20 mg/ml methyl α-D-mannopyranoside. The IL-2 activity of these media was assessed by their ability to promote the growth of an IL-2 dependent cytotoxic T cell line kindly provided by Dr Dupuy d'Angeac (Centre Paul Lamarque, Montpellier).

Results

Effect of PNA binding on the response of neuraminidase-treated thymocytes stimulated with Con A or galactose oxidase. Novogrodsky has shown that neuraminidase treatment of T lymphocytes, which removes sialic acid from the cell surface, enhances their response to mitogenic agents (5). This can be explained by the diminution of the cell surface net negative charge which allows better membrane contact between the cells. In order to increase cell contact we have treated the cells with PNA, a lectin which does not stimulate thymocytes pretreated or not with neuraminidase (2). When low amounts of PNA were bound on neuraminidase-treated thymocytes the mitogenic response to Con A or galactose oxidase was strongly enhanced as compared to that of untreated cells (Fig.1). At high lectin concentration the response was inhibited. This inhibition is probably related to steric hindrance which prevents the fixation of the mitogenic agents on their specific receptors. Indeed we have previously shown (6) that the cell surface molecules which are the best

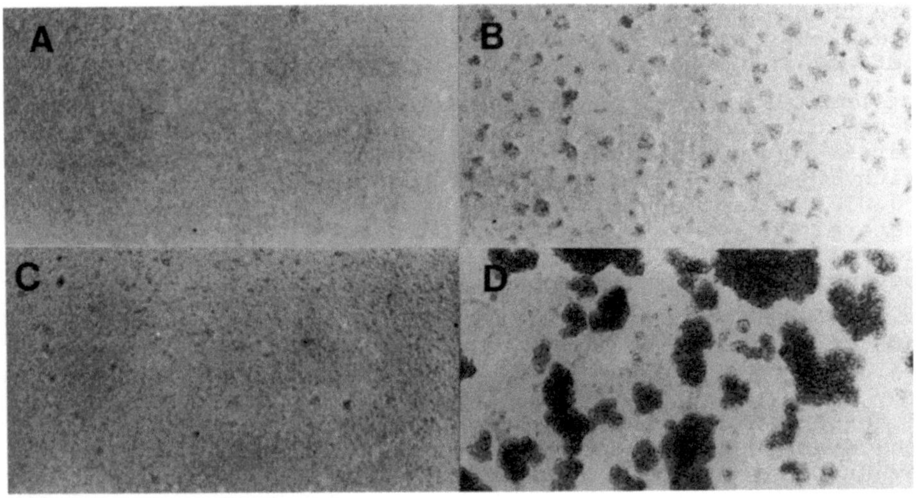

Fig. 2. Optical microscopy of thymocyte cultures: untreated (A), Con A stimulated (B), neuraminidase + PNA treated (C), neuraminidase + PNA treated and Con A stimulated (D).

candidates for Con A or galactose oxidase action in T cell stimulation are glycoproteins of high molecular weight (200-230 kd); we have also shown that these molecules are PNA receptors of neuraminidase-treated thymocytes (7). The increased response is neither related to a shift of the response maximum as a function of culture time nor to a modification of optimal mitogen concentrations (data not shown). The mitogenic stimulation is accompanied by the formation of cell clusters which can easily be visualized by optical microscopy in the culture wells (Fig. 2). PNA treatment of neuraminidase-treated cells did not induce visible cell clusters; however when Con A was added to the culture medium the clusters were far larger than those obtained with untreated cells stimulated with Con A. Neuraminidase treatment of thymocytes did not significantly increase the size of the clusters formed in the presence of Con A (data not shown).

Influence of neuraminidase + PNA treatment on IL-2 production by Con A stimulated thymocytes. In this type of experiments only Con A

Table 1

Response of cytotoxic T cell line to conditioned media prepared from untreated or treated thymocytes

Conditioned media prepared from	$[^3H]$-Thymidine incorporation cpm
Control	2000
Untreated thymocytes	2419
Con A stimulated thymocytes	10821
Neuraminidase + PNA treated thymocytes	2575
Neuraminidase + PNA treated and Con A stimulated thymocytes	53363

was used as mitogenic agent. Unseparated mouse thymocytes, previously treated or not with neuraminidase + PNA, were cultured for 24 hr in the presence of Con A. The supernatants of such cultures used as conditioned media were tested on the growth of an IL-2 dependent cytotoxic T cell line (Table 1). Conditioned media from neuraminidase + PNA pretreated thymocytes displayed high IL-2 activity whereas those from untreated cells showed very low activity: this low activity can be explained by the fact that mature cells which synthesize IL-2 (8) are the minor population in unseparated thymocytes. Supernatant from neuraminidase + PNA treated cells not stimulated with Con A does not show any IL-2 activity. We carried out experiments using PNA^+ and PNA^- separated thymocytes which showed that the increase of IL-2 activity occured only in conditioned media from mature PNA^- thymocytes which are actually the IL-2 producing cells. As observed with unseparated cells, neuraminidase + PNA treatment of PNA^- cells by itself - i.e. without subsequent Con A stimulation - did not lead to IL-2 production; no IL-2 could be detected in conditioned media from PNA^+ immature thymocytes whether the cells were pretreated with neuraminidase + PNA or not.

Influence of neuraminidase + PNA treatment on the cooperation between thymocytes and macrophages in Con A stimulation. Our hypothesis for the increase of IL-2 production by Con A stimulated thymocytes pretreated with neuraminidase + PNA is that the presence of low

Table 2

Con A response of adherent cell depleted thymocytes cocultured with peritoneal macrophages: each cell type has been pretreated (or not) with neuraminidase or neuraminidase + PNA. ^{3}H -Thymidine incorporation (cpm).

	No thymocytes	Untreated thymocytes	Neuraminidase treated thymocytes	Neuraminidase + PNA treated thymocytes
No macrophages		17644	22266	50855
Untreated macrophages	495	87465	152229	137593
Neuraminidase treated macrophages	1508	86031	225190	202158
Neuraminidase + PNA treated macrophages	1538	95510	343169	460622

amounts of PNA bound on the cells favors cellular cooperation by cell-cell contact between thymocytes and macrophages. We carried out experiments in which thymocytes were depleted from adherent cells by filtration through nylon wool and G-10 columns. Peritoneal macrophages were also prepared. Each population was treated (or not) with neuraminidase or neuraminidase + PNA, and reconstituted populations were cultured. The Con A response was enhanced when neuraminidase treatment had exposed galactosyl moieties on one cell population and PNA was bound on the other; it reached a maximum when both populations were treated with neuraminidase + PNA (Table 2). We think that the response increase is related to a better cross-linking established by PNA between thymocytes and macrophages. We have shown that neuraminidase + PNA treatment does not induce IL-1 production by Con A stimulated macrophages; for this reason we hypothesize that an enhancement in cell-cell contact could allow better diffusion of IL-1 leading to an increase of IL-2 production by immunocompetent PNA^{-} cells. The lack of effect of macrophage treatment on the response of untreated thymocytes might be explained by the low amount of PNA receptors on the IL-2 producing PNA^{-} cell population (7), which minimizes the cooperation with macrophages.

Discussion

Pretreatment of murine thymocytes with neuraminidase and low amounts of PNA (a non mitogenic lectin) strongly enhanced their response to galactose oxidase or Con A. This enhancement is correlated with the formation in the culture medium of clusters far larger than those obtained with Con A or galactose oxidase alone or with PNA alone. We have established that the increase of the response evaluated by [^{3}H]-thymidine incorporation is related to an increase of IL-2 production by the mature PNA$^-$ thymocyte population. Experiments using separated macrophage-depleted thymocytes cocultured with peritoneal macrophages as accessory cells have led to the hypothesis that neuraminidase + PNA treatment increases the cooperation between the two cell types, probably by enhancing cell-cell contact. We have hypothesized that increasing cell contact between accessory cells and thymocytes led to a better diffusion of IL-1 in the micro-environment of the large clusters formed in the culture medium, leading to the enhancement of IL-2 production by PNA$^-$ cells. This hypothesis is coroborated by the fact that neuraminidase + PNA treatment does not induce IL-1 production by macrophages in the presence of Con A, and by the fact that the enhancement of the response only occurs when the cells have been pretreates with neuraminidase (data not shown). Indeed we have shown in a previous report (7) that PNA$^-$ cells have PNA receptors on their cell membrane but in low amount: the best cross-linking will only be obtained after neuraminidase action which unmasks other PNA receptors on PNA$^-$ cells. Peters has already described that agglutination has a function in facilitating cellular cooperation (1); the results reported here show that cell contact between different cell types is a prerequisite event in lymphocyte stimulation; the enhancement of cell contact through PNA binding leads to an increase of IL-2 production by immunocompetent cells and accordingly to an enhancement of the mitogenic response. One could say that the same effect could be obtained simply by increasing Con A concentration which leads to an extensive agglutination of the cells. Ravid *et al* (9) have shown that increasing Con A concentration actually leads to an enhancement of IL-2 production but the too high lectin concentration inhibits the mitogenic response by a mechanism not yet well defined. In our system optimal concentrations of Con A for the mitogenic response are used; the combination of the two lectins (low amounts of PNA + Con A) bound on the cells leads to a synergistic effect on the formation of cell clusters allowing increased cell contact and avoiding the inhibitory effect of high Con A concentrations.

References

1. Peters, J.H.: Exp. Cell Res. 74, 179-186 (1972).
2. Novogrodsky, A., Lotan, R., Ravid, A., Sharon, N.: J. Immunol. 115, 1243-1248 (1975).
3. Favero, J., Bonnafous, J.C., Dornand, J., O'Brien, R.L., Mani, J.C.: Proceedings of the Fifth International Lectin Meeting: Lectin Biology, Biochemistry, Clinical Biochemistry (Bog Hansen, T.C., Spengler, G.A., eds.) Vol. 3, Walter de Gruyter, Berlin 1983, pp. 283-292.
4. Reisner, Y., Linker-Israeli, M., Sharon, N.: Cell. Immunol. 25, 129-134 (1975).
5. Novogrodsky, A.: Nature 250, 788-790 (1974).
6. Favero, J., Gordon, I.L., Parker, J.W., O'Brien, R.L.: Cell. Immunol. 68, 38-52 (1982).
7. Favero, J., Bonnafous, J.C., Dornand, J., Mani, J.C.: Cell. Immunol. 86, 439-447 (1984).
8. Bödeker, B.G.D., Van Eijk, R.V.W., Mühlradt, P.F.: Eur. J. Immunol. 10, 702-707 (1980).
9. Ravid, A., Rubin, A.L., Novogrodsky, A., Stenzel, K.H.: Exp. Cell Res. 145, 239-244 (1983).

THE EFFECT OF TOXIC LECTINS ON THE HISTAMINE RELEASE FROM HUMAN BASOPHIL GRANULOCYTES

I. Sehrt (1), P. Luther (1), H. Franz (2), A. Kindt (2), R. Samtleben (3)

(1) Research Institute of Lung Diseases and Tuberculosis

(2) State Institute for Immuno Preparations

(3) Institute for Molecular Biology of the Academy of Sciences of the GDR
Berlin, GDR

Since ancient times the seeds of *Abrus precatorius* L. (Semen jequiriti) and *Ricinus communis* L. (castor beans) have been used in folk medicine against a wide variety of diseases in various parts of the world.
In Europe castor beans were used in the classical Greek medicine. In the Eighteen eighties, extracts from Semen jequiriti were introduced into European medicine from South America, where they had been used traditionally against chronic eye diseases (1).

The high toxicity of the seeds, especially of castor beans, has been known for thousands of years, but it was only in 1887 that Dixson realized that the toxic properties were due to a protein (1).

Stillmark first described the agglutinating activity of ricin for red blood cells (2) and shortly afterwards Hellin observed the agglutinating ability of abrin (3). Previously, it was

Lectins, Vol. IV

thought that the toxic action of these proteins was due to their strong haemagglutinating properties, but later it was shown that the agglutinating activity is associated with non-toxic lectins (RCA I and APA I) present in the same seeds (4, 5). Highly purified ricin (RCA II) and abrin (APA II) possess little agglutinating ability.

On comparing abrin and ricin with the lectin of *Viscum album* L. (mistletoe) VAA I (ML I), a great deal of resemblance is observed in spite of their origin from unrelated plants. All three lectins are glycoproteins consisting of two peptide chains (A and B chain) of about the same mol. wt., connected by disulfide bonds. They share a similar carbohydrate specificity and inhibit protein synthesis in a cell-free system as well as in whole cells (4, 6-10). In addition they have in common some anti-neoplastic activity (11-15).

In our earlier investigations it was found that viscum lectin (VAA I) was a potent inhibitor of allergen-induced histamine release (HR) from basophil granulocytes of atopic patients (16). The present experiments were undertaken to study whether abrin and ricin have a similar effect on the immunologically induced release of histamine from basophils of patients with allergic bronchial asthma. It was also investigated whether there was a difference between the highly toxic lectins in comparison to the related non-toxic abrus and ricinus agglutinins. In order to gain more knowledge of the mechanisms involved, investigations on the effect of the lectins on non-immunologically mediated histamine release were included, using the ionophore A23187 as an inducer. Antigen and anti-IgE trigger the mediator release by opening receptor-operated calcium channels, whereas ionophore A23187 is thought to act as a calcium carrier, which bypasses these channels and transfers the ions directly across the cell membrane (17).

Materials and methods

The lectins from *Abrus precatorius* L. and *Ricinus communis* L. were isolated from extracts of their seeds by affinity chromatography on acid-treated Sepharose 4B and eluted with lactose (18). The separation of the agglutinins (APA I, RCA I) from the toxins (APA II, abrin; RCA II, ricin) was performed by gel filtration, using Sephadex G75 or G100 columns. The viscum lectin VAA I was isolated as described by Luther et al. (19); VAA II was purified by Samtleben et al. (20). The preparation of the A and B chain VAA I (ML I) was described by Franz et al. (21).

The following materials were purchased: Allergens from VEB Sächsisches Serumwerk, Dresden, GDR; rabbit anti-human IgE from Behringwerke AG, Marburg, FRG; protamine sulfate from VEB Berlin-Chemie, GDR ("Protamin forte"); ionophore A23187 from Boehringer, Mannheim, FRG.

The method used for the estimation of histamine release from washed leukocyte suspensions has been described earlier (16, 22). The lectins to be tested were pre-incubated with the cells for 5 min prior to the addition of the histamine releasing agents.

Ionophore A23187 was dissolved in dimethylsulfoxide (DMSO), 1 mg/ml, and further diluted with Tris buffer. The final concentration of DMSO in the assay mixture had no effect on the release of histamine.

The histamine release was expressed as % of total histamine (HR %).

Results

Cells obtained from patients with allergic bronchial asthma were challenged with specific allergens (pollen, flour) or with anti-IgE. The histamine release ranged from 20 to 70% with an average of 40%. All four lectins tested are shown to be potent inhibitors of immunologically induced histamine release (Table 1). The inhibitory effect ranged from 13 to 100%. The mean values of the inhibition were 83% for APA I, 78% for both RCA I and RCA II, and 55% for APA II. The smaller average inhibition of APA II in these experiments was not statistically significant (U-test).

Table 1a

Comparison of the effects of the lectins from Abrus precatorine (APA I and APA II) on the immunologically induced histamine release (HR) from human basophil granulocytes

	without lectin	With APA I		With APA II	
Inducer	HR %	HR %	Inhibiton %	HR %	Inhibition %
pollen	38	11	72	0	100
pollen	48	0	100	41	13
pollen	20	0	100	6	69
pollen	70	27	60	40	43
pollen	39	0	100	15	61
flour	25	13	50	21	17
anti-IgE	38	1	97	8	80
x + s	40 ± 16	7 ± 10	83 ± 22	19 ± 16	55 ± 32

Table 1b

Comparison of the effects of the lectins from Ricinus communis (RCA I and RCA II) on the immunologically induced histamine release (HR) from human basophil granulocytes

	without lectin	With RCA I		With RCA II	
Inducer	HR %	HR %	Inhibition %	HR %	Inhibition %
pollen	38	0	100	0	100
pollen	48	0	100	14	71
pollen	20	5	77	0	100
pollen	70	31	55	43	37
pollen	39	8	81	0	100
flour	25	16	38	13	50
anti-IgE	38	3	92	5	86
x ± s	40 ± 16	9 ± 11	78 ± 24	11 ± 15	78 ± 26

The lectins were present at a concentration of 42 µg/ml

Table 2

Effect of viscum lectin (VAA II) on histamine release (HR) from basophil granulocytes

Inducer	without lectin HR %	With VAA II VAA II µg/ml	HR %	Inhibition %
pollen	38	8.3	16	57
pollen	84	12.5	36	57
pollen	86	12.5	33	63
pollen	86	42.0	0	100
anti-IgE	77	12.5	12	84
anti-IgE	77	42.0	0	100
protamine	68	8.3	40	41

Viscum agglutinin II (VAA II; ML II) was active as an inhibitor of allergen- as well as anti-IGE-induced histamine release (Table 2). At a VAA II concentration of 42 µg/ml a complete inhibition was observed. The protamine-mediated histamine release was also inhibited (Table 2). We investigated also the effect of the A and B chain, respectively, of viscum lectin VAA I (ML I) on the mediator release of basophils of atopic patients (using pollen as the specific antigen) as well as of a non-allergic donor (protamine sulfate as releasing agent). No inhibitory effect of the A chain was registered independently of the type of inducing agent used (Table 3).

The B chain, however, acted as an inhibitor of both the immunologically and non-immunologically mediated histamine release. Pollen-induced histamine release was completely inhibited, whereas as an inhibitor of protamine-mediated histamine release the B chain was only moderately active (27% inhibition).

Table 3

Effect of the A and B chain from VAA I (ML I) on histamine release (HR) from human basophil granulocytes

Inducer	HR %	Inhibition of HR (%)	
		A chain	B chain
pollen	38	0	100
anti-IgE	38	0	56
protamine	68	0	27

The A chain concentration was 6 μg/ml and the B chain was 7 μg/ml

Table 4

Effect of viscum lectin (VAA I, ML I) on ionophore A23187-mediated histamine release (HR) from human basophil granulocytes (Four separate experiments are shown)

A23187 μg/ml	VAA I μg/ml	HR (%)				HR (%)
		1	2	3	4	mean value $x \pm s$
0.2	0	33	9	3	6	13 ± 14
0.2	4	51	12	17	11	23 ± 19
0.2	8	75	21	53	30	45 ± 24
0.2	21	82	34	73	51	60 ± 22
0.2	42	86	67	105	89	87 ± 16
0.4	0	50	43	67	-	53 ± 12
0.4	4	96	69	88	-	84 ± 14
0.4	8	104	83	97	-	95 ± 11

Table 5

Effect of the lectins from Abrus precatorius (APA I and APA II) and Ricinus communis (RCA I and RCA II) on the ionophore A23187-mediated histamine release (HR) of human basophil granulocytes (Four separate experiments are shown)

Lectin	concentration µg/ml	HR (%)				HR (%)
		1	2	3	4	mean value x ± s
APA I	o	13	6	3	30	13 ± 12
	8.4	87	48	29	-	55 ± 29
	16.8	97	71	56	56	70 ± 19
APA II (abrin)	o	13	6	3	30	13 ± 12
	8.4	56	34	14	-	35 ± 21
	16.8	81	49	68	47	61 ± 16
RCA II (ricin)	o	13	6	3	30	13 ± 12
	8.4	38	19	14	-	24 ± 13
	16.8	35	26	29	38	32 ± 5

The concentration of ionophore A23187 was o.2 µg/ml

In further experiments to study the effects of the lectins on mediator release, the ionophore A23187 was used as a releasing agent. VA I (ML I) did not show any inhibitory effect on A23187-mediated histamine release, it even produced an augmentation (Table 4).

The augmentation occurred at all levels of the induced release of histamine, whether it was less than 10% or as high as 67%. This effect is concentration dependent. The release process does not appear to be cytotoxic, because there was no loss of lactate dehydrogenase from the cells.

APA II and RCA II, the toxic lectins as well as APA I and RCA I, the non-toxic agglutinins also produced an augmentation of A23187-mediated histamine release, and this effect was concentration dependent too (Table 5).

The extent of enhancement for each lectin group was different (Table 5); with abrin and ricin respectively, the augmentation occurred to a lesser extent than with the agglutinins APA I and RCA I. This difference is statistically significant (RCA II; U-test).

Discusstion

Our results show that the highly toxic lectins abrin (APA II) and ricin (RCA II) are potent inhibitors of allergeninduced histamine release as was discovered recently for viscum lectin (VAA I) (16). We also show that the related non-toxic abrus and ricinus agglutinins APA I and RCA I have a strong inhibitory effect too, and VAA II revealed a similar inhibitory effect on immunologically (allergen, anti-IgE) as well as non-immunologically (protamine) mediated histamine release. As discussed earlier for the action of VAA I, we assume that the galactose-specific lectins bind to the carbohydrate-containing Fc-protion of the IgE; molecule, resulting in a block of the interaction of the allergen with the Fabportions of the IgE, the starting signal for the mediator release. The blocking is probably brought about by steric hindrance (16).

A further insight into the mechanism of lectin action was gained by using the ionophore A23187 as a releasing agent. Our results show that the toxic lectins as well as the related non-toxic agglutinins do not inhibit the A23187 mediated histamine release, but provoke an enhancement. This finding appears to exclude the hypothesis that lectin binding to galactose residues of the glycolipids and glycoproteins in the basophil membrane causes a stabilization of the cell membranes which prevents cell activation. A possible explanation is that capping of the receptors with lectins (23) enables a better acessibility of the relatively small A23187-molecule to the cell membrane, resulting in a higher level of histamine release.

The non-toxic agglutinins APA I and RCA I were more potent in enhancing A23187-mediated histamine release than the toxic APA II (abrin) and RCA II (ricin), as it was demonstrated by a greater agumentation at similar concentrations of the lectins. To date we cannot give an explanation for this finding.

The highly toxic A chain VAA I (ML I) did not have any effect on histamine release, whereas the B chain was inhibitory to both allergen- and anti-IgE as well as protamine-induced histamine release. This is in accordance with the finding, that the two chains possess different functions as was first shown for the A and B chain of abrin and ricin (7). The A chain is not able to bind carbohydrates and therefore cannot attach itself to the cell membrane. The B chain as the carbohydrate-binding moiety represents the actual lectin component effecting the inhibition of histamine release.

References

1. Olsnes, S., Pihl, A.: Receptors and Recognition, Series B, 1, Chapman & Hall, London, 1976, 129-173.
2. Stillmark, H.: Inaug.-Diss., Dorpat, 1888.
3. Hellin, H.: Inaug.-Diss., Dorpat, 1891.
4. Olsnes, S., Pihl, A.: Eur.J.Biochem 35, 179-185 (1973).
5. Olsnes, S., Pihl, A.: Biochemistry 12, 3121-3125 (1973).
6. Lin, J.-Y., Liu, K., Chen, C.-C., Tung, T.-C.: Cancer Res. 31, 921-924 (1971).
7. Olsnes, S.; Refsnes, K., Pihl, A.: Nature 249, 627-631 (1974).
8. Olsnes, S., Pihl, A.: FEBS Letters 28, 48-50 (1972).
9. Franc, H., Kindt, A., Ziska, P., Bielka, H., Benndorf, R., Venker, L.: Acta biol. et med. germ. 41, K9-K16 (1982).
10. Stirpe, F., Legg, R.F., Onyon, L.J., Ziska, P., Franz, H.: Biochem.J. 190, 843-845 (1980).
11. Lin, J.-Y., Tserng, K.-Y., Chen, C.-C., Lin, L.-T., Tung, T.-C.: Nature 227, 292-293 (1970).
12. Shionoya, H., Arai, H., Koyanagi, N., et al.: Cancer Res. 42, 2872-76 (1982).

13. Lutsik, M.D.: Dopov.Akad.Nauk Ukr. R.S.R. Ser. B 541-544 (quoted in Chem.Abstr. 37, 112216 (1975).

14. Luther, P., Franz, H., Haustein, B., Bergmann, K.-C.: Acta biol. et med. germ. 36, 119-125 (1977).

15. Ziska P., Franz, H.: Lectins - Biology, Biochemistry, Clinical Biochemistry, eds. Bøg-Hansen, T.C., Breborowicz, J., Vol IV, Walter de Gruyter, Berlin, New York.

16. Sehrt, I., Luther, P.: Lectins - Biology, Biochemistry, Clinical Biochemistry, ed. Bøg-Hansen, T.C., Vol. II, Walter de Gruyter, Berlin, New York, 1982, 45-56.

17. Foreman, J.C., Mongar, J.L., Gomperts, B.D.: Nature 245, 249-251 (1973).

18. Olsnes, S.: Methods Enzymol. 50 323-335 (1978).

19. Luther, P., Theise, H., Chatterjee, B., Karduck, D., Uhlenbruck, G.: Internat.J.Biochem. 11, 429-433 (1980).

20. Samtleben, R., Kiefter, M., Luther, P.: Lectins - Biology, Biochemistry, Clinical Biochemistry, eds. Bøg-Hansen, T.C., Breborowicz, J., Vol. IV, Walter de Gruyter, Berlin, New York.

21. Franz, H., Kindt, A., Ziska, P.: Lectins - Biology, Biochemistry, Clinical Biochemistry, ed. Bøg-Hansen, T.C., Vol. II, Walter de Gruyter, Berlin, New York, 1982, 771-776.

22. Sehrt, I., Bergmann, K.-C., Dehnert, I., Christ, R.: Dt. Gesundh.-Wesen 32, 303-307 (1977).

23. Singer, S.J., Nicolson, G.L.: Science 175, 720-731 (1972).

Part II

Histochemistry and Cytochemistry with Lectins

LECTIN BINDING TO NON-MALIGNANT AND MALIGNANT HUMAN URO-EPITHELIAL CELLS IN VITRO

Danuta Dus, Czesław Radzikowski

Tumor Immunology Department, Institute of Immunology and Experimental Therapy, Polish Academy of Sciences, 53114 Wrocław, Poland

Henri Debray, Jean Montreuil

Laboratoire de Chimie Biologique et Laboratoire Associé au CNRS n°217, Université des Sciences et Techniques de Lille I, 59655 Villeneuve d'Ascq Cédex, France

Britta Christensen, Jørgen Kieler

The Danish Cancer Society, Laboratory of Environmental Carcinogenesis, The Fibiger Institute, DK-2100 Copenhagen, Denmark

Progressive growth, invasion, and metastasis are the main criteria by which malignant tumors are distinguished from benign tumors in vivo. The malignancy of in vitro propagated cell lines can be demonstrated by their tumorigenic properties in susceptible hosts and by their invasive growth in various in vitro models.

Human epithelial cells explanted from non-malignant and malignant urothelium have been used for the study of a number of problems including cellular transformation and chemical carcinogenesis (see 1-3). In the Fibiger Institute in Copenhagen the following grades of transformation (TGr) have been described (3): TGr 0 - Non-malignant urothelial cell lines with a short life span; TGr I - Non-malignant urothelial cell lines with a prolonged, but finite life span; TGr II - Non-malignant urothelial cell lines with an infinite life span;

Lectins, Vol. IV

TGr III - Immortal urothelial cell lines which produce progressively growing tumors in nude mice and invade fragments of embryonic chick hearts in vitro.

Further studies of these four categories of human urothelial cell lines have revealed characteristic differences in morphology and chromosomal pattern (3,4), antigenic expression (4,5,6), angiogenic potential (7), and growth pattern (B. Christensen et al. - to be published).

The involvement of the specific cell surface glycoconjugates in the progressive growth of tumor cells has been studied by using various tumor models. The results indicate that these compounds can play a crucial role in cell-to-cell recognition and interaction and thus in the control of their social behaviour (8). Lectins have proven to be useful tools for investigations of the molecular differences between tumorigenic, metastatic and normal cells by detecting distinct changes in the cell surface glycoproteins (9,10).

In the present work we have studied the binding of four lectins to a number of human urothelial cell lines classified as TGr II and III. The affinity of TGr III cells to two of these lectins was found to be considerably increased as compared to TGr II cells.

Table 1.

Origin and Classification of Human Uroepithelial Cell Lines.

Cell line	Origin	Life-span	Invasiveness in vitro/ Tumorigenicity in nude mice	TGr Classification
HCV 29*	N.-M.	Infinite	-	II
Hu 609	N.-M.	Infinite	-	II
Hu 1734	TCC	Infinite	-	II
HCV 29T	N.-M.- HCV 29	Infinite	+	III
Hu 609T	N.M.- Hu 609	Infinite	+	III
Hu 456	TCC	Infinite	+	III
Hu 549	TCC	Infinite	+	III
Hu 1703He	TCC	Infinite	+	III

N.-M. = Non-malignant
TCC = Transitional Cell Carcinoma
**Obtained from dr. J. Fogh, Sloan Kettering Institute, New York.*

Material and Methods

A panel of eight cell lines were studied. As shown in Table 1 two cell lines of non-malignant origin (Hu 609 and HCV 29) and one (Hu 1734) derived from a transitional cell carcinoma (TCC) were classified as TGr II, i.e. immortal but non-invasive and non-tumorigenic. Out of five cell lines classified as TGr III (immortal, tumorigenic, invasive), two (Hu 609T and HCV 29T) were derived from two TGr II cell lines (Hu 609 and HCV 29) by apparently spontaneous transformation, and three (Hu 456, Hu 549, Hu 1703He) from TCC.

The cells were grown in monolayer cultures in FIB 41B culture medium supplemented with 10% heat-inactivated fetal bovine serum (FBS) and transferred after brief (2 min) exposure to trypsin-EDTA mixture. For immunofluorescence study cells were plated in 10-well multitest slides and incubated for 48 h in 5% CO_2/95% air atmosphere at 37°C. FBS, trypsin-EDTA and 10-well multitest slides were purchased from Flow (Irvine, Scotland).

Treatment of cells with neuraminidase. Vibrio cholerae neuraminidase (500 U/ml) was purchased from Serva (Heidelberg, FRG). Cells growing on multiwell slides were washed twice with calcium and magnesium-free PBS (CMF PBS) and incubated with neuraminidase (50 U/ml) for 30 min at 37°C. After incubation cells were washed twice with CMF PBS and used for lectin binding experiments.

Lectins. Source, abbreviation and nominal sugar specificity of the lectins used were as follows: Triticum vulgaris agglutinin, WGA (α-N-acetyl-D-glucosamine); Arachis hypogaea agglutinin, PNA (α-D-galactose); Canavalia ensiformis agglutinin, Con A (α-D-mannose or α-D-glucose); and Ricinus communis agglutinin. I, RCA I (α-D-galactose). Fluorescein-substituted WGA and Con A were purchased from Miles International (Austria). FITC-PNA was from Pharmindustrie (I.B.F. Reactifs, Clichy, France). RCA I has been purified by H. Debray and conjugated with FITC according to the method of Loor (11).

Binding of FITC-lectins to cells. Forty-eight hr cultures on multitest slides, intact or neuraminidase-treated, were washed twice in CMF PBS supplemented with 0.1% bovine serum albumin (Serva) - CMF PBS/BSA - and incubated with various concentrations of FITC-conjugated lectin diluted with CMF PBS/BSA, 5 µl per well, for 30 min at 22°C. After incubation the cells were washed twice with CMF PBS. The third washing was done with CMF PBS containing 0.05% sodium azide. The cells were examined under a fluorescence microscope (Leitz, Orthoplan). For checking the specificity of the reaction, lectin binding was also performed in the presence of specific sugar inhibitors at 0.2 M concentration.

Results and Discussion

In order to study the correlation between lectin binding and the grade of transformation of human epithelial cells four FITC-conjugated lectins were examined for their ability to bind selected uroepithelial cell lines classified as TGr II and TGr III. The results are shown in Table 2 to 5. All cell lines tested showed positive binding when high concentrations of lectins were used. Differential binding was observed when low and intermediate staining concentrations were used. In inhibition tests in which lectins were used with specific sugars at 0.2 M concentration, there was no staining.

Wheat germ agglutinin (WGA) is the most frequently used lectin in studies of tumor cell membrane characteristics. Reaction of this lectin with the cell surface has been reported to correlate with metastatic potential (12,13). Our results have shown that WGA binding to human uroepithelial cell lines correlates well with tumorigenicity and invasiveness. TGr III cells with these traits showed 4 to 8 times higher affinity to WGA than non-tumorigenic, non-invasive TGr II cells. The range of staining concentrations was 6.25 - 50 μg/ml (Table 2).

Peanut lectin (PNA) did not bind to any cell line prior to neuraminidase treatment. Neuraminidase-treated cells showed strong positive reaction (Fig. 1), indicating that the cell membrane in all cell lines examined contained only sialic acid-substituted PNA binding sites. PNA used in the concentration interval of 7.8 - 62.5 μg/ml showed 4 or even 8 times higher affinity with TGr III cells than with TGr II cells (Table 3).

Table 2.

Staining Reaction of Human Urotehlial Cell Lines to WGA-FITC.

Cell lines	Origin	Concentrations of WGA-μg/ml					
		100	50	25	12.5	6.25	3.12
TGr II:							
HCV 29	N.-M.	++	+	-	-	-	-
Hu 609	N.-M.	++	+	-	-	-	-
Hu 1734	TCC	++	++	+	-	-	-
TGr III:							
HCV 29T	N.-M.	++	++	+	+	-	-
Hu 609T	N.-M.	++	++	++	+	-	-
Hu 1703He	TCC	++	++	++	++	+	-
Hu 456	TCC	++	++	++	++	+	-
Hu 549	TCC	++	++	++	++	+	-

++: Strong; +: positive; -: negative
Abbreviations as for Table 1.

Ricinus communis I (RCA I) lectin and Con A showed binding to all cell lines (Tables 4 and 5), not changed by neuraminidase. The binding pattern did not allow a clear distinction between TGr II and TGr III cells, although a tendency towards an increased affinity of TCC derived TGr III cells to Con A and a decreased affinity of TGr III cells to RCA I was seen. Staining concentrations of these lectins were relatively high: 60.5 - 242 μG/ml for Con A and 25 - 100 μg/ml for RCA I. This result supports the observation that lectins can recognize different comples saccharide sequences belonging to the same N-glycan

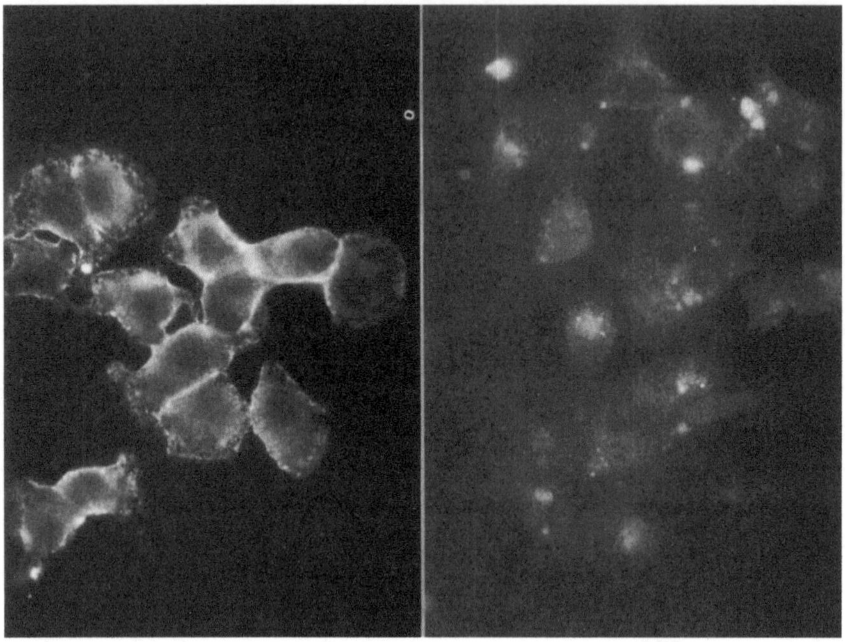

Fig. 1. Fluorescence staining of Hu 456 cells with FITC-PNA (15 μg/ml). 1A: Neuraminidase-treated cells; 1B: Intact cells. Magnification x 320.

Table 3.

Staining Reactions of Neuraminidase-Treated Human Urothelial Cell Lines to PNA-FITC

Cell lines	Origin	62.5	Concentrations of PNA-μg/ml 31.2	15.6	7.8	3.9
TGr II:						
HCV 29	N.-M.	++	+	-	-	-
Hu 609	N.-M.	++	+	-	-	-
Hu 1734	TCC	+	+	-	-	-
TGr III:						
HCV 29T	N.-M.	++	++	++	-	-
Hu 609T	N.-M.	++	++	+	-	-
Hu 1703He	TCC	++	++	+	-	-
Hu 456	TCC	++	++	++	+	-
Hu 549	TCC	++	++	++	-	-

Before the staining cells were treated with neuraminidase (50 units/ml, 30 min at 37°C).
Abbreviations as for Table 1.

Table 4.

Staining Reactions of Human Urothelium Cell Lines to Con A-FITC

Cell lines	Origin	Concentrations of Con A-μg/ml 242	121	60.5	30.2
TGr II:					
HCV 29	N.-M.	++	+	-	-
Hu 609	N.-M.	+	+	-	-
Hu 1734	TCC	++	+	-	-
TGr III:					
HCV 29T	N.-M.	++	+	-	-
Hu 609T	N.-M.	++	+	-	-
Hu 1703He	TCC	++	++	-	-
Hu 456	TCC	++	++	+	-
Hu 549	TCC	++	++	+	-

Abbreviations as for Table 1.

Table 5.

Staining Reactions of Human Urothelium Cell Lines to RCAI-FITC.

Cell Lines	Origin.	Concentrations of RCA I-μg/ml 100	50	25	12.5	6.25
TGr II:						
HCV 29	N.-M.	++	++	+	-	-
Hu 609	N.-M.	++	+	+	-	-
Hu 1734	TCC	++	++	+	-	-
TGr III:						
HCV 29T	N.-M.	++	+	-	-	-
Hu 609T	N.-M.	++	-	-	-	-
Hu 1703He	TCC	++	+	-	-	-
Hu 456	TCC	++	+	-	-	-
Hu 549	TCC	++	+	-	-	-

Abbreviations as for Table 1.

(14). Although PNA and RCA I lectins used in these studies have an identical monosaccharide specificity for D-galactose, they can recognize more complex saccharide sequences belonging either to 0-glycosidically linked glycans-Gal (β 1-3) GalNAc in the case of PNA (15,16) or to N-glycosidically linked

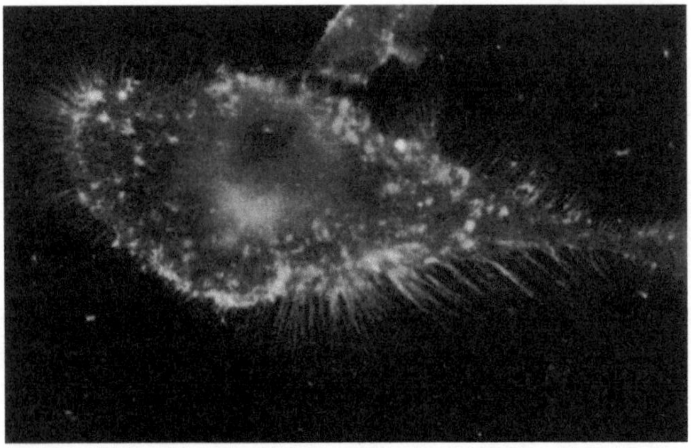

Fig. 2. Fluorescence staining of Hu 609 cell with FITC-WGA (50 μg/ml). Magnification x 660.

glycans - Gal(β 1-4)GlcNAc in the case of RCA I (14). This can explain the different binding patterns observed with these two lectins. The known specificity of WGA for N-acetylneuraminic acid-rich O-glycans (17) may cause the strong reactivity of this lectin observed with tumorigenic, invasive TGr III cells which also react with PNA, but only after neuraminidase treatment.

It has previously been shown (6) that TCC derived TGr III cells differ from urothelial TGr II cells by the presence of a surface antigen detectable by a monoclonal antibody. The present results are in line with observations made by other authors and indicate that selection of appropriate lectins allows the demonstration of differences in composition of cell surface glycoprotein accompanying malignant phenotype (18-22). However, there is no specific lectin binding limited to malignant cells, and it has been shown that alteration in cellular glycoconjugates occurs also during normal cell differentiation (23,24).

The agglutinating effect of Con A on cancer cells has been associated with the presence of microvilli on the surface of transformed cells (25). However, malignant transformation is not always accompanied by the development of microvilli (26), and microvilli can be demonstrated on non-malignant HCV 29 (3) and on Hu 609 TGr II cells as illustrated in Fig. 2 from which it is seen that fluorescent conjugated lectins may be used as a method for microvilli detection.

Acknowledgment: This research was supported by the Polish Academy of Sciences (Grant no. 10.5), the Universite des Sciences et Techniques de Lille I, The CNRS, the INSERM (Contract CRE no 823-029), and the MIR (Contract no 82-109).

References

1. Hepburn, P.J., R.W. Masters. 1983. In: The Pathology of bladder cancer, vol. 2 (G.T. Bryan, S.M. Cohen, eds.). CRC Press Inc., Boca Raton, Florida, p. 213-227.

2. Kieler, J. 1984. Cancer Metastasis Reviews.(in press)

3. Christensen, B., J. Kieler, M. Vilien et al. 1984. Anticancer Res. (in press)

4. Vilien, M., B. Christensen et al. 1983. Eur. J. Cancer Clin. Oncol. 19, 775-789.

5. Vilien M., H. Wolf, F. Rasmussen. 1981. Eur. J. Cancer 17, 321-327.

6. Bubenik, J., J. Kieler, P. Perlmann et al. Eur. J. Cancer Clin. Oncol. (in press)

7. Kaminski, M.J., J. Kieler, B. Christensen. 1983. Proc. Soc. Exp. Biol. Med. 174, 383-391.

8. Nicolson, G.L. 1983. Exper. Cell Res. 150, 3-22.

9. Sharon, N. 1980. Immunology 80, Progress In Immunology IV, (M- Fougereau and J. Dausset, eds.). Academic Press, New York, p. 254.

10. Montreuil, J., H. Debray, P. Debeire, P. Delannoy. 1983. Structural carbohydrates of the liver. (H. Popper et al. eds.). MTP Press Ltd., p. 239.

11. Loor, F. 1973. Exp. Cell Res. 82, 415-425.

12. Burger, M. 1989. Biology of the cancer cell. (K. Letnansky ed.). Kugler Publications, Amsterdam, p. 193.

13. Willmott, N. S. Simpson. 1983. Anticancer Res. 3, 401-406.

14. Debray, H., D. Decout, G. Strecker, G. Spik, J. Montreuil. 1981. Eur. J. Biochem. 117, 41-55.

15. Lotan, R., E. Skutelsky, D. Danon, N. Sharon. 1975. J.Biol. Chem. 250, 8518-8523.

16. Örntoft, T.F., S.E. Petersen: These proceesings.

17. Monsigny, M., A.C. Roche, C. Sene, R. Maget-Dana, F. Delmotte. 1980. Eur. J. Biochem. 104, 147-153.

18. Shiba, M., T. Ohiwa, A.J.P. Klein-Szanto. 1984. J.N.C.I. 72, 43-51.

19. Fisher, J. P.J. Klein, M. Vierbuchen, R. Fisher, G. Uhlenbruck. 1983. Cancer Detection and Prevention 6, 137-147.

20. Raedler, A., S. Böttcher, F. Stalling, E. Kaukel: These proceedings.

21. Schwechheimer, K., I.A. Born, P. Kern: These proceedings.

22. Fisher, J., P.J. Klein: These proceedings.

23. Malmi, R., K.O. Söderström: These proceedings.

24. Holthöfer, H., J. Rapola: These proceedings

25. Kolata, G.B. 1975. Science 188,819-820.

26. Kieler, J., P. Briand, M.C. Van Pateghem, M. Mareel. 1979. In Vitro 15, 758-771.

LECTIN TARGET CELLS IN HUMAN BRAIN TUMORS

Karl Schwechheimer[1], Ivan Antonio Born[2], Petra Kern[1],
Gabriele Weiss[1]
[1]Institute of Neuropathology and [2]Institute of Pathology,
University of Heidelberg, D - 6900 Heidelberg, GFR

Lectins are now extensively used in characterizing cell and tissue glycoconjugates. An increasing number of reports deal with lectins as markers for normal and neoplastic tissue (see refs. 2-15). Using lectins as probes in normal brain tissue and intracranial neoplasma have shown that lectin binding is specific and cytotypic (6, 16-18).

The present study was undertaken to map the lectin target cells in routinely processed biopsy specimens of about 200 human intracranial neoplasms including neuroectodermal, meningial, nerve sheath, neuronal and primitive neuroectodermal tumors.

Lectins, Vol. IV

Materials and Methods

Various lectins with different carbohydrate affinities were chosen to map their target cells in routinely processed biopsy specimens of about 200 human intracranial neoplasma including neuroectodermal tumors. Concanavalin A (Con A), Ulex europaeus agglutinin I (UEA I), Ricinus agglutinin I (RCA I), Soybean agglutinin (SBA), Helix promatia agglutinin (HPA), Wheat germ agglutinin (WGA) and Limulus polyphemus agglutinin (LPA) were examined. The well known affinity of UEA I to vascular endothelial cells (7, 10, 17) stimulated a study of vascular tumors of the central nervous system. Whereas Con A binding sites were revealed by a modified method (16) first described by Avameas (1), a four-step PaP technique (11) was adapted for the other lectins. Neuraminidase pretreatment sometimes enhanced the effect (HPA, RCA I, SBA, WGA).

Results

Lectin binding patterns are quoted in Table 1.

Con A, HPA, LPA, RCA I and WGA showed a variable binding to the different cellular elements of astrocytomas. Granular intracytoplasmic binding as well as decoration of fibrillary processes was observed (Fig. 1 a). Small cellular foci, which occured in anaplastic gliomas and glioblastomas, were always negative. The most ependymoma cells were negative or showed only weak intracytoplasmic binding. Papillae-lining and rosettes-forming cells exhibited intracytoplasmic lectin acceptors for Con A and LPA, brush border binding for SBA, HPA and RCA I (Fig. 1 b), and a combination of intracytoplasmic and luminal binding for WGA. Similar mapping was observed in plexus papillomas (Fig. 1 c).

In meningiomas,foci of weak to strong intracytoplasmic binding was observed with all lectins examined except of UEA I. Xanthomatous cells within meningiomas had a high lectin acceptor density (Con A, HPA, RCA I, SBA, WGA, LPA). This feature was pronounced in a xanthomatous meningioma. Epitheloid areas

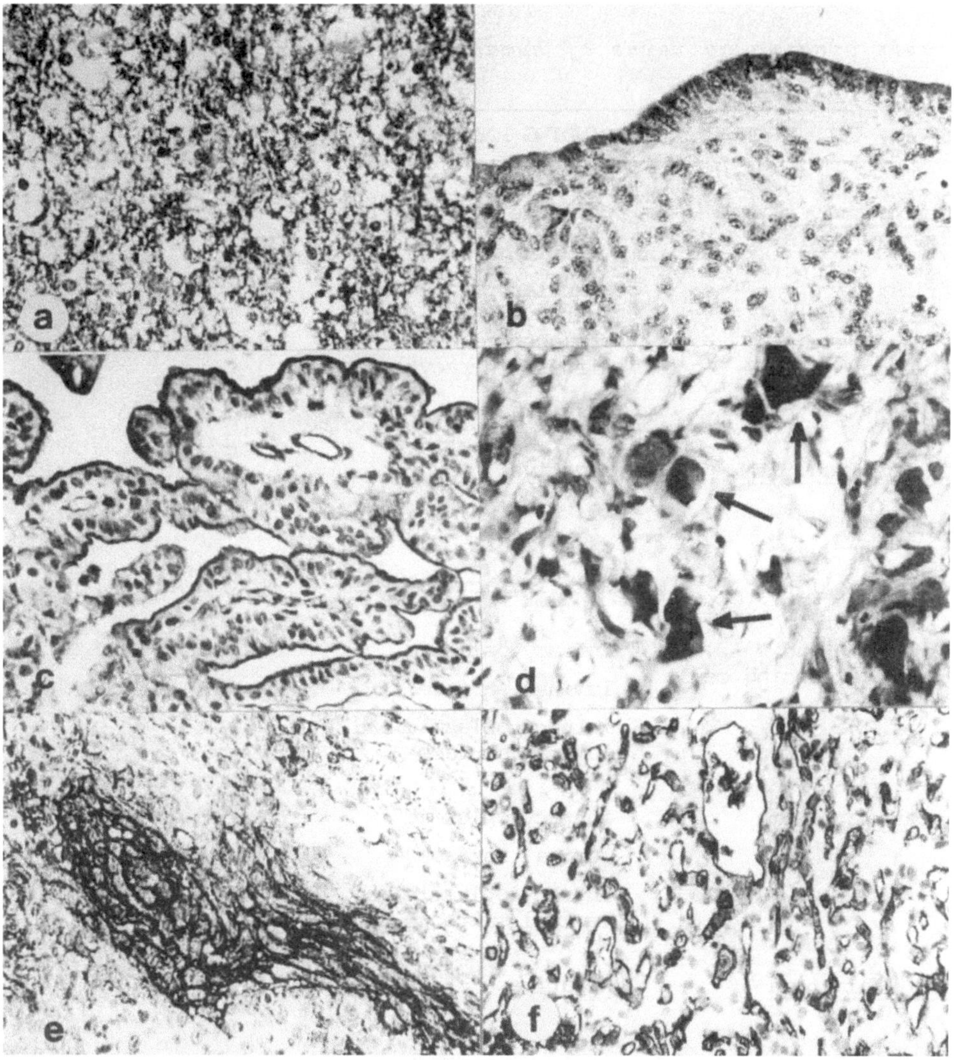

Fig. 1. a: Con A, astrocytoma, strong intracytoplasmic and fibrillary binding, b: Con A, papillae-lining cells in an ependymoma, strong intracytoplasmic granular binding, c: SBA, neuraminidase pretreatment, plexus papilloma, decoration of the apical surface membrane, d: Con A, ganglioglioma, high Con A-acceptor density in differentiated neurons, e: HPA, meningioma, focal intracytoplasmic and pronounced dotted membrane binding, f: UEA I, haemangioblastoma, strong reaction to vascular endothelium, negative stromal cells; hematoxylin counterstain, about 200 x.

Table 1

Lectin binding patterns of human intracranial tumors

TUMOR / CELL TYPE	BINDING PATTERN	LECTIN(S)
Astrocytoma, glioblastoma cells	variable	
- fibrillary, - protoplasmic, - gemistocytic, - piloid	granular perinuclear, intracytoplasmic, processes; diffuse intracytoplasmic	Con A, HPA, LPA, RCA I, WGA
- anaplastic	negative	-
Oligodendroglioma	negative	-
Ependymoma		
- tumor cell	weak intracytoplasmic	Con A, HPA, LPA, RCA I
- rosette- and papillae-lining cell	intracytoplasmic	Con A, LPA
	luminal surface	HPA, SBA, RCA I
	intracytoplasmic and luminal surface	WGA
Plexus papilloma	intracytoplasmic	Con A, LPA
	luminal surface	HPA, SBA_n
	intracytoplasmic and luminal surface	RCA I, WGA
Medulloblastoma	negative	-
Meningioma		
endotheliomatous type, fibroblastic type	weak intracytoplasmic	Con A, HPA, LPA, RCA I, SBA, WGA
"epitheloid" cell	granular, cyto-membrane, intracytoplasmic	HPA, RCA I, SBA, Con A, WGA
xanthomatous cell	intracytoplasmic, diffuse or granular	Con A, HPA, RCA I, SBA, WGA, LPA

Table 1: continued

Neurinoma		
type A-, type B-cell	weak, perinuclear, processes	Con A, HPA, LPA, RCA I, SBA, WGA
foamy cell	granular	Con A, HPA, RCA I, SBA
Neuronal tumor		
differentiated neuron	strong, granular	Con A, (LPA)
neuroblast	negative	-
Vascular malformation, vascular tumor		
endothelium	linear, apical, intracytoplasmic	UEA I > HPA, SBA, WGA, RCA I
stromal cell	negative	-

n = neuraminidase pretreatment

within endotheliomatous meningiomas expressed a high intracytoplasmic and very prominent membrane binding for HPA, RCA I, SBA and - to a lower extent - Con A and WGA (Fig. 1 d). Type A- and type B-cells of neurinomas demonstrated variable binding of all lectins tested except for UEA I.

In neuronal tumors, which included gangliocytomas, gangliogliomas and primitive neuroectodermal tumors strong granular Con A binding was found in differentiated neurons (Fig. 1 e), whereas undifferentiated neuroectodermal cells ("neuroblasts") were always negative. The same was true for medulloblastomas, which are regarded as primitive neuroectodermal tumors, too. HPA, SBA, WGA, RCA I are facultative markers for vascular endothelium; UEA I always shows a strong and constant reaction. Vascular malformations and vascular tumors of the central nervous system, which were additionally tested with UEA I, decorated nearly exclusively the vascular endothelium cells

(Fig. 1 f). Stromal cells, however, of haemangioblastomas were always negative.

Discussion

Lectin binding variability apparently indicates cellular heterogenity of human brain tumors, which seems to express a certain functional differentiation of the tumor cells. This was very conspicious in gliomas, meningiomas and neurinomas. On the other hand, it is very interesting that, despite different sugar affinities, diverse lectins can show similar binding patterns. This finding can be explained by the presence of diverse carbohydrate residues in the lectin acceptor molecules. A certain overlap of sugar affinities between lectins should be alternatively discussed. Thus, the demonstration of lectin target cells cannot be regarded as an absolutely reliable rule. In contrast, a intra- and interindividual heterogenity in the panel of lectin acceptor molecules must be born in mind.

The detection of circumscribed epitheloid areas in endotheliomatous meningiomas exemplifies the capacity of lectins to reveal intraindividual tumor differentiation.

Lectin mapping of rosette-lining cells in ependymomas and plexus papillomas gave very similar results (18). These corresponded well to the feature observed in non-neoplastic ependymal and plexus epithelium cells (17). The finding is suggestive to postulate a close cellular relationship between these two cell types.

Strong intracytoplasmic binding of Con A in differentiated neurons (16) was preserved in neoplastic neuronal cells, too. Primitive neuroectodermal cells ("neuroblasts") in neuroblastomas, sympathicoblastomas and medulloblastomas were always Con A-negative. This result shows lectin binding to be dependent on cellular differentiation. A similar relationship was observed in ependymomas, where the ependymoma cells were mostly lectin-negative, while the rosette-forming or papillae-lining cells exhibited a lectin binding pattern identical with the non-neoplastic ependymal cells (17, 18).

Although HPA, SBA, WGA and RCA I decorated vascular endothelium cells, UEA I was the most reliable marker. Thus, UEA I can be used to demonstrate capillary endothelium cells in haemangioblastomas. This result could be of practical importance in the differential diagnosis of haemangioblastoma and metastasis of renal cell carcinoma.

This work was supported by a grant of the Deutsche Forschungsgemeinschaft, Bonn - Bad Godesberg, GFR.

References

1. Avrameas, S., Karsenti, E., Bornens, M. In: Bittiger, H., Schnebli, H. P. (eds): Concanavalin A as a tool. John Wiley and Sons, London, New York, Sydney, Toronto, 85 -93 (1976).

2. Born, I. A., Schwechheimer, K.,Maier, H., Möller, P.. In: Bøg-Hansen, T. C., Breborowicz, J. (eds.): Lectins. Biology, Biochemistry, Clinical Biochemistry, Vol. 4. De Gruyter, Berlin, New York (1984) pp 127

3. Dus, D., Radzikowski, C., Debray, H., Christensen, B., Kieler, J.. In: Bøg-Hansen, T. C., Breborowicz, J., (eds): Lectins. Biology, Biochemistry, Clinical Biochemistry, Vol. 4. De Gruyter, Berlin, New York (1984) pp 65

4. Fischer, J., Klein, P.-J.. In: Bøg-Hansen, T. C., Breborowicz, J. (eds): Lectins. Biology, Biochemistry, Clinical Biochemistry, Vol. 4. De Gruyter, Berlin, New York (1984) pp 95

5. Fischer, J., Klein, P.-J.. In: Bøg-Hansen, T. C., Breborowicz, J. (eds): Lectins. Biology, Biochemistry, Clinical Biochemistry, Vol. 4 De Gruyter, Berlin, New York (1984) pp 101

6. Fischer, J., Klein, P.-J., Csillik, B.. In: Bøg-Hansen, T. C., Breborowicz, J. (eds): Lectins. Biology, Biochemistry, Clinical Biochemistry, Vol. 4. De Gruyter, Berlin, New York (1984) pp 117

7. Holthöfer, H., Virtanen, I., Kariniemi, A. L., Horima, M., Linder, E., Miettinen, A.: Lab. Invest. 47, 60-66 (1982).

8. Holthöfer, H., Rapola, J.. In: Bøg-Hansen, T. C., Breborowicz, J. (eds.): Lectins. Biology, Biochemistry, Clinical Biochemistry, Vol. 4. De Gruyter, Berlin, New York (1984) pp 83

9. Malmi, R., Söderström, K.-O.. In: Bøg-Hansen, T. C., Breborowicz, J. (eds): Lectins. Biology, Biochemistry, Clinical Biochemistry, Vol. 4. De Gruyter, Berlin, New York (1984) pp 145

10. Miettinen, M., Holthöfer, H., Lehto, V. P., Miettinen, A., Virtanen, I.: Am. J. Clin. Pathol. 79, 32-36 (1983).

11. Möller, P.: Virchows Arch. (Pathol. Anat.) 396, 313-317 (1980).

12. Nemeth, A., Sotonyi, P., Bayer, E., Fischer, J.. In: Bøg-Hansen, T. C., Breborowicz, J. (eds): Lectins. Biology, Biochemistry, Clinical Biochemistry, Vol. 4. De Gruyter, Berlin, New York (1984) pp 109

13. Örntoft, T. F., Petersen, St. E.. In: Bøg-Hansen, T. C., Breborowicz, J. (eds): Lectins. Biology, Biochemistry, Clinical Biochemistry, Vol. 4. De Gruyter, Berlin, New York (1984) pp 89

14. Raedler, A., Böttcher, S., Stalling, F., Kaukel, E.. In: Bøg-Hansen, T. C., Breborowicz, J. (eds): Lectins. Biology, Biochemistry, Clinical Biochemistry, Vol. 4. De Gruyter, Berlin, New York (1984) pp 137

15. Rudolph, M., Samtleben, R. As communicated at Interlec 6, the Sixth International Lectin Meeting, Poznan, Poland, 2-6 September, 1984

16. Schwechheimer K., Weiss, G., Möller, P.: Virchows Arch. (Pathol. Anat.) 402, 297-306 (1984 a).

17. Schwechheimer, K., Weiss, G., Schnabel, P., Möller, P.: Histochemistry 80, 165-169 (1984 b).

18. Schwechheimer, K., Weiss, G., Möller, P. J. Neurol. Sci. 63, 393-401 (1984 c).

CHARACTERIZATION OF POLYCYSTIC KIDNEYS WITH LECTINS AND DIFFERENTIATION MARKERS OF THE NEPHRON

Harry Holthöfer[1], Juhani Rapola[2]
[1]Department of Bacteriology and Immunology, Univ of Helsinki
[2]Department of Pediatrics, University Central Hospital, Helsinki, Finland

Cystic diseases of the kidney have a variable etiology and pathogenesis. Some cystic kidneys are a result of abnormal morphogenesis and differentiation, whereas in others the cysts develop in the already formed nephrons and collecting ducts by several different mechanisms. Clinical and radiological investigations are helpful in the classification of the cystic kidneys, but the final diagnosis rests upon the histological findings. From the histological picture it is, however, often difficult to decide what part of the nephron is dilated or whether the cyst is resulting from abnormal nephrogenesis. Microdissection studies, on the other hand, have provided information of the origin of the cysts and a widely used classification of polycystic kidneys is based on the use of this technique (11). However, microdissection is difficult and laborious to perform and it is not suitable for routine diagnostics of the cystic lesions.

Very little is known of the biochemical and histochemical signs of differentiation of renal cysts, a knowledge that could aid in better understanding of the origin and characteristics of these lesions.

We have previously shown that glycoconjugates are comparted

Lectins, Vol. IV

in various parts of the nephron in a characteristic way, as shown by differences in lectin binding reactions in different nephron segments (5,7). It also seems that the expression of cellular glycoconjugates undergo profound changes during different stages of nephrogenesis, during maturation of different segments of the nephrons and also during the differentiation of other tissues (6,10). On the other hand, we have found changes in the expression of lectin binding sites in diseased kidneys (8). These findings prompted us to investigate if lectins can be used to characterize the stage of differentiation of glycoconjugates in various types of cystic kidney diseases.

Materials and Methods

Kidney samples. Kidney samples of patients with Meckel syndrome (MS) and infantile polycystic kidney disease (IPCD) were obtained from autopsies at the Department of Pediatrics (University Central Hospital, Helsinki, Finland). Tissue samples were fixed in 4% buffered formaldehyde solution and routinely processed in parafin. Light microscopic examination of the kidney samples revealed the typical cystic kidney parenchyma in both disorders studied (Fig. 1a).

Antisera and lectins. Rabbit antibodies against rat brush border (BB) and against human Tamm-Horsfall (TH) glycoprotein were raised as described earlier in detail (2,3), in order to identify proximal and distal tubules, respectively. Fluorescein isothiocyanate (FITC) or tetramethyl rhodamine isothiocyanate (TRITC) conjugated antirabbit IgG (Cappel laboratories, Cochranville, Pennsylvania) were used as second antibodies in the indirect immunofluorescence experiments. FITC and TRITC labeled Helix pomatia (HPA) and Arachis hypogaea (PNA) lectins were purchased from Vector Lab. (Burlingame, California). The specificity of the lectin binding was tested as described earlier (7). Staining procedure. 5 μm paraffin sections of tissues were deparaffinized in xylene and graded alcohols and treated with trypsin 0.1 mg/ml for 30 min, washed and further incubated with the respective second antibodies. For lectin stainings the sections were incubated with the lectins (100 μg lectin/ml) for 30 min and mounted in polyvinylalcohol mounting medium. A zeiss Standard microscope equipped with high pressure mercury vapour light source, epi-illuminator and interference filters for FITC and TRITC fluorescence was used for microscopy.

Results

The cystic kidneys of Meckel syndrome showed subcapsular non-dilated tubules, some of which were positive for PNA (Fig. 1b) and HPA. Some of these tubules were positive for anti Tamm-Horsfall antibodies, and some showed brush border specific fluorescence. HPA and PNA stained strongly the epithelial cells of the cysts of various size. Part of the cysts were also positive for antibodies against proximal brush border (BB) and distal Tamm-Horsfall (TH) nephron. In IPCD, infanfile polystic kidneys, the cells of the cyst walls were readily stained

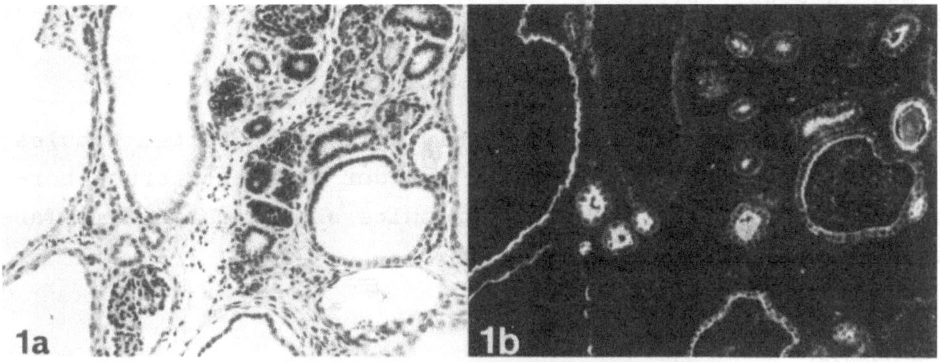

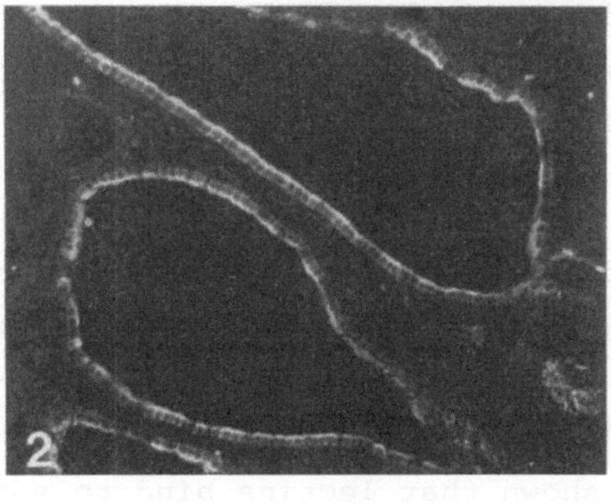

Fig. 1. Histological staining of Meckel syndrome (MS) kidney (a) and the corresponding area stained for PNA (b). PNA decorates the cyst walls and normal collecting ducts. x120.

Fig. 2. HPA specific staining of the epithelial cells of the cysts is seen in the Infantile polycistic kidneys (IPCD). x260.

Table 1

Staining reactions of labelled lectins and differentiation markers of the nephron on the cyst walls in cystic kidneys

Type of cystic lesion	Stained for			
	PNA	HPA	BB	TH
Meckel syndrone	++	++	++	++
Infantile polycystic kidney disease (IPCD)	++	++	-	+

BB and TH denote proximal tubular brush border antigens and distal tubular Tamm-Horsfall antigens, respectively.

with HPA (Fig. 2) and PNA, and only some less dialted tubules were positive for Tamm-Horsfall, but not with anti brush border antibodies (Fig. 3). These results are summarized in Table I.

Discussion

The classical methods to study the site of origin of the polycystic kidneys have been light microscopy and microdissection of the kidneys (1). The cystic kidneys of Meckel syndrome have been classified as dysplastic due to the basic failure of the metanephric parenchyma to differentiate (1). Similarly infantile polycystic kidneys are characterized by cysts in the cortical collecting ducts (1). The etiology of both of these lesions is unknown.

We have earlier shown that lectins bind to specific segments of the nephron in a typical way (5,7). Thus PNA and HPA bind exclusively to the cells of the distal nephron and to collecting ducts (7). In this study both PNA and HPA bound to the cells of the cyst walls and to the dilated tubules. This

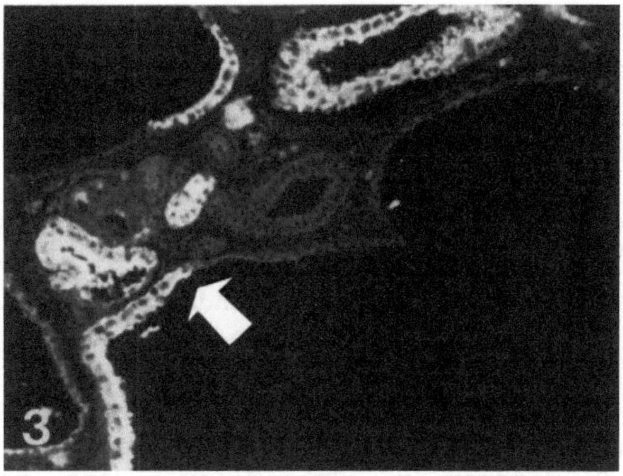

Fig. 3. Anti brush border (BB) antibodies react with part of the cells of the cyst walls (arrow) and with normal proximal tubules. x260.

indicates that HPA and PNA are useful probes to demonstrate glycoconjugates typical for the terminally differentiated epitehlial cells of the distal nephron. Moreover, the proximal and distal tubulus specific antibodies bound to the epithelial cells of the dilated nephron segment. This shows that the cyst walls have undergone differentiation at least to some extent. On the other hand, it is widely accepted that in the IPCD kidneys the site of the lesion is the cortical collecting duct (1). Our results are in line with these results by showing that lectins can not only be used to demonstrate features of differentiation but, in addition, to demonstrate the site of origin of these lesions. Changes in the structure and distribution of saccharide components have been recently shown to characterize certain skin diseases (9), kidney diseases (6) and cells and tissues undergone malignant transformation (4). The development and differentiation of cells seems to involve also changes in the structure of glycoproteins and glycolipids (10).

The results of this study show that lectins can provide a useful tool to show maturation of glycocompounds in association with differentiation of cells and tissues. Moreover, lectins specific for strict tissue components can be used to localize the exact site of the lesion.

Acknowledgements. The skillful technical assistance of Ms Riitta Väisänen is kindly acknowledged.
This study was supported by The Finnish Foundation for Cancer research, The Finnish Cultural Foundation and Sigrid Juselius Foundation.

References

1. Bernstein, J.: In: Cystic diseases of the kidney. Ed. K.D. Gardner Jr., John Wilay & Sons, New York 1976.

2. Ekblom, P., Miettinen, A., Saxen, L.: Induction of brush border antigens of the proximal tubule of the developing kidney. Dev. Biol. 74, 263-274 (1980).

3. Ekblom, P., Miettinen, A., Virtanen, I., Wahlström, T., Dawnay, A., Saxen, L.: In vitro segregation of the metanephric nephron. Dev. Biol 84, 880-895 (1981).

4. Hakomori, Si.: Glycosphingolipids in cellular interacton differentiation and oncogenesis. Annu. Rev. Biochem. 50, 733-764 (1981).

5. Holthöfer, H.: Lectin binding sites in the kidney. A comparative study of 14 animal species. J. Histochem. Cytochem. 31, 531-537 (1983).

6. Holthöfer et al., unpublished observations.

7. Holthöfer, H., Virtanen, I., Pettersson, E., Alfthan, O., Linder, E., Miettinen, A.: Lectins as fluorescence microscopic markers for saccharides in the human kidney. Lab. Invest. 45, 391-399 (1981).

8. Holthöfer, H., Virtanen, I., Törnroth, T., Miettinen, A.: Alterations in glomerular glycoconjugates of human kidneys as detected with labeled lectins. Submitted to Nephron.

9. Kariniemi, A.-L., Holthöfer, H., Miettinen, A., Virtanen, I.: Altered binding of Ulex europeaus I lectin to psoriatic epidermis. Brit. J. Dermatol. 109, 523-529 (1983).

10. Maylie-Pfenninger, M.-F., Jamieson, J.: Development of cell surface saccharides on embryonic pancreatic cells. J. Cell. Biol. 86, 96-103 (1980).

11. Potter, E.L.: Normal and abnormal development of the kidney. Year Book Medical Publishers, Chicago 1972.

PNA BINDING TO HUMAN COLORECTAL TISSUE AS A MARKER OF NEOPLASTIC TRANSFORMATION.

Torben F.Örntoft, Steffen E.Petersen.
Institute of Cancer Research, Dep. of Surgery I,
Aarhus Amtssygehus. DK - 8000 Aarhus, Denmark.

The transformation of colonic mucosa from normal to neoplastic is associated with several cellular alterations. A genetic defect leads to increased cellular proliferation and a desorganization of crypt structure with pseudostratification, and the appearance of cells with atypical nuclei (1). These alterations are identified by light-microscopy. However, neoplastic transformation also comprises biochemical alterations of cellular glycoprotein secretion and altered saccharide structures of glycoproteins and glycolipids in cellular membranes (2,3). These can be demonstrated by the use of lectins and antibodies.

Neoplastic conditions in the human colon are very common in industrialized countries. Approximately 4-10 per cent of individuals 40 years of age or older have colonic polyps. Knowledge of the biological potential of these alterations and hence the ability to identify people at high risk for colorectal cancer would be helpful. In this study we have focused on the interrelation between tissue dedifferentiation and altered saccharide structures as demonstrated by peanut agglutinin binding.

Lectins, Vol. IV

Materials and methods

We have examined 47 colorectal carcinomas and 22 colorectal polyps larger then lo mm. PNA binding was demonstrated by an indirect immunoperoxidase technique described in detail elsewhere (4). A crude PNA preparation or chromatographically purified PNA (E.Y.Lab., San Mateo, Cal.) was used as first layer, rabbit anti-PNA (E.Y.Lab) as second layer and swine anti-rabbit conjugated with peroxidase (Dakopatts, Copenhagen,Denmark) as third layer. Carbazole was used as chromatogen and the tissue sections were counterstained with Meyers hematoxylin. Some sections were treated with neuraminidase 1 U/ml (Behringwerke, Marburg, FRG) to demonstrate sialic acid substituted PNA binding sites. The DNA content was measured in a biopsy from 2o polyps as described earlier, trout erytrocytes were used as internal standard in a microscope based flow cytometer (5).

The grading (I-V) of the adenomas was based on the degree of epithelial pseudostratification and on nuclear morphology.

Results

PNA which has a high affinity for Galβ(1-3)-GalNAc (6) showed a cytostructural localization that altered with increasing grade of dedifferentiation.

In the stalk of adenomas we observed morphologically normal crypts which showed PNA binding in cells located in the basal third of the crypt. The PNA binding structures were in the supranuclear region and at the luminal plasma membrane (Fig. 1). This binding was absent in the upper 2/3 of the crypt. In grade I and II dedifferentiated crypts the binding pattern was as in the basal third of morphologically normal crypts (Fig. 1). This pattern was changed in grade III crypts where some cells showed diffuse cytoplasmic PNA binding. In grade IV crypts the supranuclear binding was absent and PNA only bound to the luminal plasma mambrane and here and there diffusely to the cytoplasm (Fig. 2). In addition there was an increasing number of unstained crypts with increasing dedifferentiation. Adenomas with a hyperdiploid DNA content more often expressed an abnormal staining pattern, than did those with a normal diploid DNA content.

In carcinomas the binding pattern of PNA was as in grade IV adenomas with PNA localized to the luminal plasma membrane and here and there diffusely bound to the cytoplasm (Fig. 2). Generally there was binding to secreted mucus, when the cells surrounding this showed PNA binding. Mucosa surrounding the carcinomas bound PNA to secreted mucus and in many cases to luminal plasma membranes. We examined mucosa at a distance from the colon cancer (mean 16 cm) and found PNA binding crypts with a slightly abnormal morphology, the normal crypts being PNA negative (7).

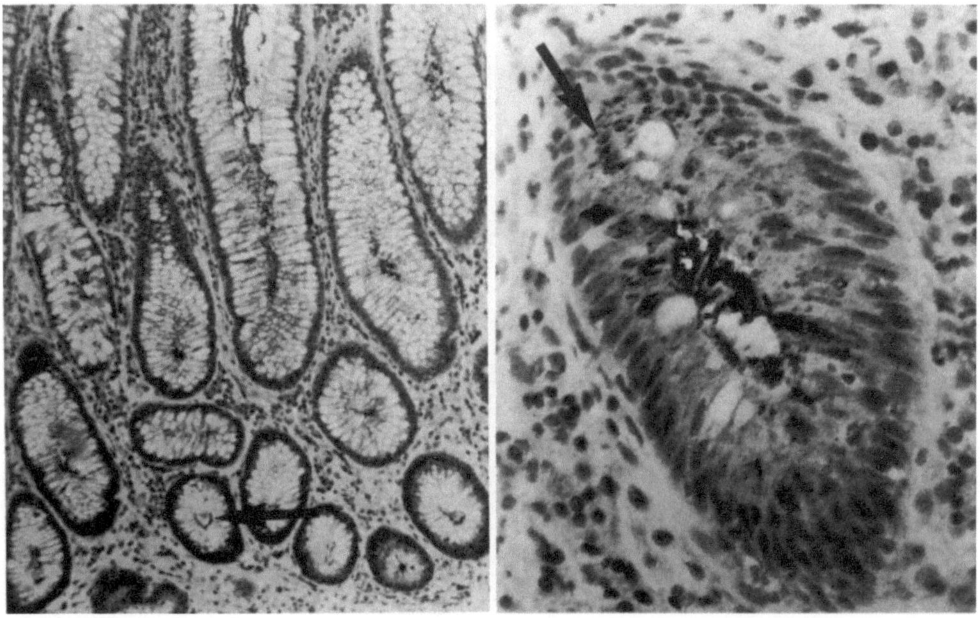

Fig. 1. Left: Morphologically normal crypts; PNA binds (black) to secreted mucus, the supranuclear region and luminal membrane (arrow) x 100.

Right: Grade II dedifferentiated adenoma crypt with PNA binding to secreted mucus, luminal membrane and supranuclear region (arrow) x 400.

Discussion

The lower third of normal crypts in the adenoma stalk which bound PNA is considered to be the major zone of cell proliferation (1). Cells in this region are synthesizing membrane glycoproteins/lipids and mucus for secretion. The PNA binding to the supranuclear region probably reflects identification of carbohydrates being elongated in the Golgi apparatus, see (1o). Transport of these to the plasma membrane and incorporation into this, or secretion into the lumen would then result in the PNA binding we demonstrate at the luminal plasma membrane and in the secreted mucus.

In dedifferentiated adenoma crypts the zone of proliferation has moved upwards (1) which correlates with the PNA binding

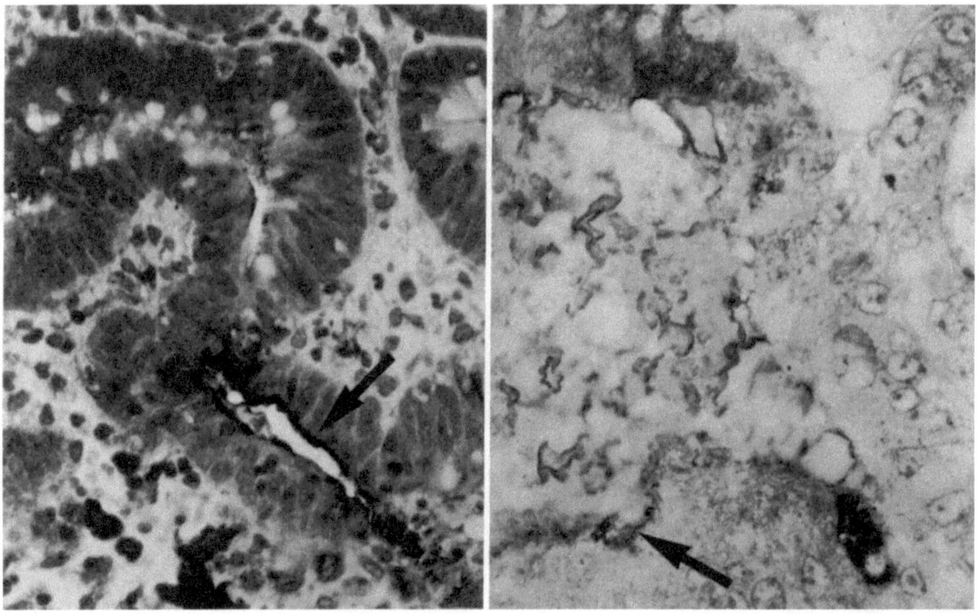

Fig. 2. Left: Grade IV adenoma crypt, PNA binds to luminal membranes (arrow) x450.

Right: Colorectal carcinoma, PNA binds diffusely to the cytoplasm of three cells and to secreted mucus and luminal membranes (arrow) x600.

being no longer restricted to the lower third of the crypts. Normal differentiation of crypt cells leads to a loss of PNA binding ability. Some adenoma cells do not lose this ability. This could reflect a defect in sialylation of Galβ(1-3)GalNAc or incomplete glycosylation of this (8).
Grade IV and V adenoma cells resembles carcinoma cells in the PNA binding pattern. This pattern could be explained by a reduced sialylation of blood group MN antigens leading to accumulation of precursorstructures as T and Tn antigens in the cytoplasm and in the plasma membrane (9). T antigens bind PNA with a high affinity (6). This precursor hypothesis is supported by the fact that neuraminidase treatment of the tissue sections increases this binding pattern. The localization of PNA binding is very similar to the localization of carcino embryonal antigen (CEA) (11) and we are currently investigating the ability of PNA to identify this antigen.

References

1. Deschner, E.E., Lipkin, M.: Cancer 35, 413-418 (1975).
2. Fischer, J., Klein, P.J., Vierbuchen, M., Fischer, R., Uhlenbruch, G.: Human Tumor markers: Biological basis and clinical relevance, Alan R.Liss, New York, 137-148 (1983).
3. Feizi, T.: Contr. Oncol, 19, 51-63 (1984).
4. Örntoft, T.F.: Virchows Archiv, A, in press.
5. Petersen, S.E.: Cytometry, 3, 3o5-3o7 (1983).
6. Lotan, R., Skutelsky, E., Danon, D., Sharon, N.: J.Biol. Chem., 25o, 8518-8523 (1975).
7. Örntoft, T.F.: Cancer Res., In press.
8. Boland, C.R., Montgomery, C.K., Kim, Y.S.: Proc.Natl.Acad Sci. USA, 79, 2o51-2o55 (1982).
9. Springer, G.F.: Science, 224, 1198-12o5 (1984).
lo. Fischer, J., Klein, P.J.: "Differential diagnosis of anaplastic gastric cancers by lectins", this issue.
11. Wagner, C., Hain, F., Breuer, H., Olude, S., Cremer, H., Müller-Walraf, R.: Oncodevelopmental Biology and medicine 2, 331-343 (1981).

INTENSIVE AND SELECTIVE STAINING OF STORED GLYCOLIPID IN SANDHOFF's DISEASE BY DBA (*Dolichos biflorus agglutinin*)

J. Fischer
Department of Pathology, University Medical School, Pécs
H-7643 Pécs, Hungary

P.J. Klein
Institute of Pathology, University of Freiburg
D-7800 Freiburg, F.R.G.

A close histochemical characterization of stored glycoconjugates in various storage diseases was impossible until recently due to the failure of an appropriate histochemical procedure. The periodic acid-Schiff (PAS) reaction or its modifications, i.e. the aldehyde-bisulfite-toluidine blue (ABT) reaction (1) were only capable of demonstrating free vicinal glycols in the carbohydrate side chains but could not give information on the monosaccharide composition of the reacting compound. Metachromatic stainings were also useful only when the stored component was highly sulfated, i.e. metachromatic leukodystrophy. In the very recent literature, labelled RCA I and Con A lectins were used for the demonstration of stored glycolipids in cells of Krabbe's (2) and Gaucher's (3) diseases. In the former, a galactocerebrosidosis, the terminal galactose molecules reacted strongly with RCA I, while in the latter, a glucocerebrosidosis, the terminal glucose molecules were marked heavily by Con A lectin. In the present study the lectin binding of the stored material within the ganglion cells was explored in a case of Sandhoff-Jatzkewitz-Pilz's disease, in which the hexosaminidase A and B deficiency had been verified biochemically. The observations clearly demonstrated that la-

Lectins, Vol. IV

belled lectins are useful tools in the histochemical differential diagnosis of some storage diseases.

Material and Methods

Frozen, as well as formaline fixed paraffin embedded, brain tissues from a case of Sandhoff's disease and 10 control brain tissue samples were used for the investigations. Horseradish peroxidase (HRP) labelled lectins: Conavalia ensiformis (Con A), Ricinus communis I (RCA I), Ulex europaeus I (UEA I), wheat germ (WGA), Dolichos biflorus (DBA) and peanut agglutinin (PNA) were purchased from Medac (Hamburg, F.R.G.).

Lectin histochemical staining was performed as follows: Frozen or deparaffinized tissue sections were incubated with the HRP-labelled lectins (10 μg/ml) for 30 min at room temperature. After washing (3X) with PBS containing 1 mM Ca^{++}, lectin binding sites were localized by the peroxidase enzyme reaction using 3-amino-9-ethylcarbazole as a substrate. Counterstaining with Majer's haematoxylin was followed by mounting in glycerol jelly. Parallel sections were stained with haematoxylin, periodic acid-Schiff (PAS) reaction for carbohydrates and Oil-red O for lipids.
Lipid extraction was performed using chloroform-methanol (3:1, v/v) or ether-ethanol (1:1, v/v) mixtures. In order to show the specificity of the lectin binding, lectin histochemical studies were also performed in the presence of the appropriate inhibitory sugars: 1-O-α-methyl-D-mannopyranoside and N-acethyl-D-galactosamine (Serva, Heidelberg, F.R.G.) at 0.2 M concentrations.
PNA and DBA binding were also studied after neuraminidase (Vibrio cholerae, Behringwerke, Marburg, F.R.G.) pretreatment (0.1 U/ml for 30 min at 37°C) of the tissue sections.

Results and Discussion

The lectin binding characteristics of neurones in the pons region of the brain in Sandhoff's disease are summarized in Table 1. The stored glycolipid component, which was strongly stained by Oil-red O and only faintly by the PAS reaction, showed a very intensive staining after labelling with DBA lectin. The greatly enlarged neurons were almost filled with heavily stained granular material (Fig. 1 a).

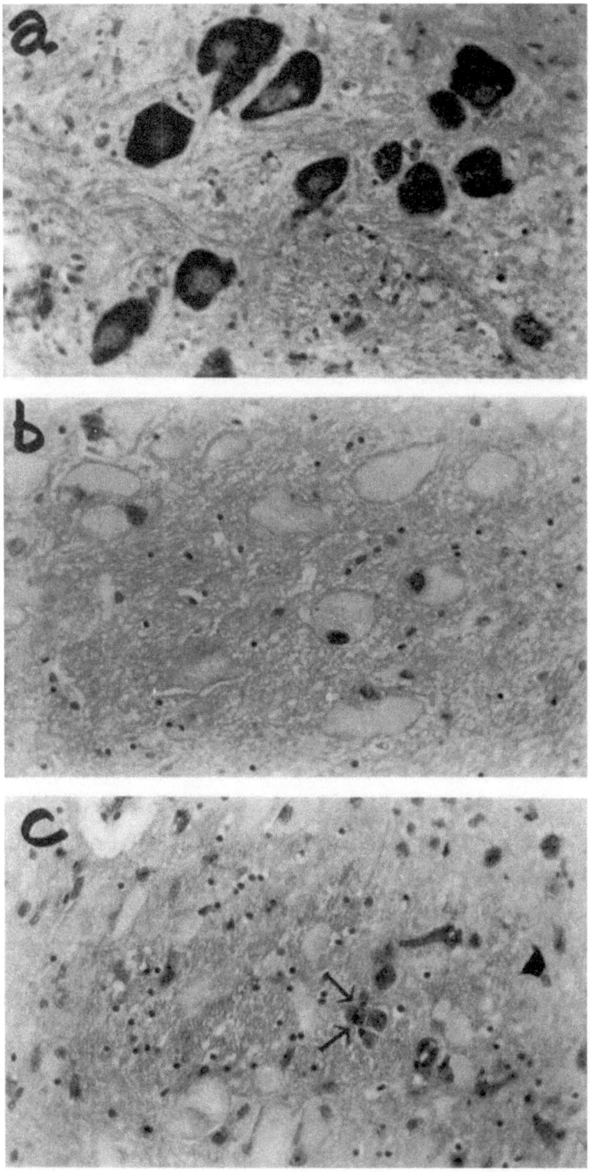

Fig. 1. Brain tissue in the pons region of Sandhoff's disease. (a) DBA-HRP staining. Stored glycolipid material show an intensive lectin binding. (b) UEA I-HRP staining. In the unstained neurons the typical signs of storage diseases, the balloned appearance of cell bodies and the displaced nuclei are seen. (c) Neuraminidase-PNA-HRP staining. Reactive glial elements show a positive reaction (arrows).

Table 1.

Lectin binding pattern of stored glycolipid in Sandhoff's disease

Lectins used	Sugar specificity	Reaction of neurons
Con A	mannose;glucose	+
RCA I	galactose	-
UEA I	fucose	-
WGA	N-acetyl-glucosamine	-
DBA	N-acetyl-galactosamine	+++
PNA	gal- 1,3-galNAc	-

Con A showed a more diffuse and moderate staining within the same cells. The high affinity of the neuronal cytoplasm in Sandhoff's disease to DBA as well as the moderate affinity to Con A was also exhibited by the paraffin embedded material, indicating that the main part of the stored component resisted dehydration procedure.

On the other hand, extensive lipid extraction of the tissue sections, either with chloroform-methanol or with ether-ethanol, almost completely abolished lectin binding.
The large neurons of the pons region in Sandhoff's disease remained completely unstained after labelling with WGA, RCA I and UEA I (Fig. b) and PNA lectins. PNA, on the other hand, showed a very intensive staining with the reactive glial elements (Fig. 1 c arrows). This finding corresponds to recent observation of Schwechheimer et al. (4-5) in normal and malignant nervous tissue. The lectin binding of neurons in Sandhoff's disease proved to be highly specific to the sugar residues of the stored glycolipid material, since the DBA binding was abolished by adding N-acethyl-D-galactosamine, and Con A binding was abolished by adding 1-O- methyl-D-mannopyranoside to the incubation solution.

In the 10 control brain tissue samples neurons remained unstained after DBA-HRP or Con A-HRP treatment and peroxidase reaction, indicating that the positive staining in Sandhoff's disease is a characteristic of the stored glycolipid molecules.

Table 2.

Molecular structure of carbohydrate chains in storage disease

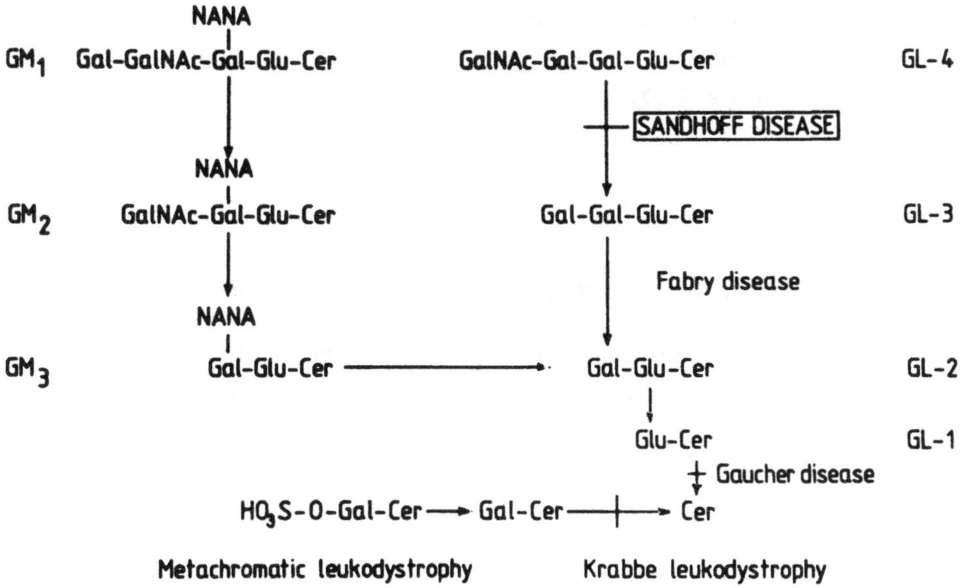

The lectin binding pattern found in Sandhoff's disease correlated well with the chemical structure of the stored glycolipid, since, as known, a tetrahexosylceramide with terminal N-acetyl-D-galactosamine accumulates in the neurons as a result of a combined hexosaminidase A and B deficiency (Table 2). Therefore the very intensive DBA binding of the neurons may be related to the terminal N-acetyl-D-galactosamine residues of the accumulated glycolipid, whereas, for the moderate Con A binding of the same structure, probably the interchain glucose molecules are responsible. Since neurons in the normal central nervous system did not possess affinity to DBA lectin, the intensive and selective DBA binding of them in Sandhoff's disease is of diagnostic significance and can help in the histological differentiation of hexosaminidase deficiency syndromes from other lipid storage diseases. As shown in Table 2, the majority of lipid storage diseases are characterized by other terminal sugar residues and therefore they may be selectively labelled by other lectins, as it has alredy been demonstrated for the terminal glucose molecule in Gaucher's (3) and for the terminal galactose molecule in Krabbe's diseases (2).

References

1. Romhányi,Gy., Deák, Gy., Fischer, J.: Histochemistry 43, 333. 1975.
2. Schroeder, R., Klein, P.J.: (In preparation)
3. Bianco,P., Bonucci, E.: Path. Res. Pract. 168, A133. 1983.
4. Schwechheimer, K., Weiss, P., Schnabel, P., Möller, P.: Histochemistry 80, 165. 1984.
5. Schwechheimer, K., Born, I.A., Kern, P., Weiss, G.: These proceedings

DIFFERENTIAL DIAGNOSIS OF ANAPLASTIC GASTRIC CANCERS BY LECTINS

J.Fischer

Department of Pathology, Univ.Med.School, Pécs
H-7643 Pécs, Hungary

P.-J. Klein

Institute of Pathology, Univ. of Freiburg
D-78oo Freiburg, F.R.G.

According to their histological appearance and biological behaviour, gastric neoplasms are subdivided into intestinal and diffuse type carcinomas (1). The former are characterized by papillary or tubular growth, and by cell types resembling that of the intestinal columnal epithelium. The latter are characterized by small round cells, by the typical signet ring cells, and sometimes by scirrhous infiltrations. Beside these morphological criteria, diffuse gastric carcinomas are best marked by the mucous glycoprotein secretion within their cytoplasms, which can be visualized by various carbohydrate histochemical methods (PAS-staining, Hale-reactions, alcian blue and high iron diamine staining). These methods are, therefore, used frequently for the functional characterization of gastric cancer cells (2,3). However, there are some highly undifferentiated anaplastic gastric carcinomas in which neither glandular differentiation nor any sign of glycoprotein synthesis can be seen by the above mentioned carbohydrate histochemical methods. Often these anaplastic gastric carcinomas, represent diagnostic problems in the pathohistological practice (4). It may be exceptionally difficult to dis-

Lectins, Vol. IV

tinguish them from malignant lymphomas of the stomach. Recently it has been demonstrated that more detailed information can be achieved about the mucus composition of normal and malignant and other epithelial cells by using labelled lectins (8-14). For example some undifferentiated gastric carcinomas, without any sign of mucus biosynthesis by the periodic acid-Schiff reaction showed an intensive lectin binding (15). An intensive PNA binding of medullary carcinomas was also demonstrated by Kondo et al. (16).
In the Pathological Department of University Medical School, Pécs, which is also the Reference Center of Malignant lymphomas in Hungary, three cases of highly anaplastic gastric carcinomas were observed recently, where the diagnosis could be set up only by investigating the more differentiated metastases of the tumors. In these carcinomas a very intensive PNA binding was observed, besides the completely negative PAS and negative or only faint Hale stainings. In two cases the binding of PNA was manifold increased after neuraminidase pretreatment of the tissue sections. Since cells of histiocytic origin possess similar PNA binding (17) we examined the PNA binding characteristics of the anaplastic gastric cancers and compared our findings to observations on histiocytes of normal lymph nodes, granulomas and malignant histiocytosis, as well as to other tumors of ectodermal origin. The investigations were completed by immunohistochemical studies using polyclonal antibodies directed against the mucus fraction of normal gastric mucosa isolated by PNA-affinity chromatography (18).

Material and Methods

Histochemical investigations were performed on formol fixed, paraffin embedded tissue specimens. Several tissue samples of the primary tumors and the lymph node metastases of the three cases of highly anaplastic gastric carcinomas were separately analysed. For comparative purposes, tissue spe-

cimens from normal lymph nodes, lymph nodes with activated histiocytes, giant cell granulomas, spleen with sarcoidosis, malignant histiocytosis and skin biopsies of melanoma and Merkel cell tumors were also studied.
Lectin-histochemical investigations were made by using horse-radish peroxidase labelled PNA /Sigma, St.Louis, U.S.A./, as described earlier (5,6,15). Parallel sections were always pretreated with neuraminidase (Vibrio cholerae, Behringwerke, Marburg, F.R.G.) 0.1 U/ml for 30 min at 37°C.
For elektronmicroscopical studies small tissue blocks from routinely fixed material were used. Tissue blocks were treated with HRP labelled PNA (diluted to 10 µg/ml in PBS) for 2 h, after neuraminidase pretreatment and washing in PBS. Enzyme was visualized using 3'3 diaminobenzidine as a substrate. After washing and post-osmification samples were embedded in Araldite. Ultrathin section were investigated in a Jeol 100 microscope without further staining.
Purification of PNA binding mucus component from normal human gastric mucosa was made using PNA-affinity column (18).
Production of antisera. 100 µg pure PNA-binding normal human gastric mucus glycoprotein was injected intraperitoneally to 4 weeks old Balb/c mice. After two boosters the animals were bled. Antisera were treated at 56°C for 30 min.
Immunohistochemical staining. Re-hydrated tissue sections were preincubated with bovine serum albumine (1%) after the endogenous peroxidase had been blocked by 3% H_2O_2 in methanol. Sections were then incubated with the antiserum (diluted 1:100 in PBS) for 1 h. Controls were treated with non-immune sera of normal mice. After washing in PBS, the bound antibodies were visualized by incubation with anti-mouse antibody (1:100) (Dako, Copenhagen, Denmark) for 30 min, followed by peroxidase reaction.
Controls were stained by haematoxylin-eosine, PAS and Hale reactions

Results and Discussion

Table 1 outlines the results of PNA binding prior to and after neuraminidase pretreatment as well as the binding of antibody directed against the PNA binding mucus component of normal human gastric mucosa. Experiments were performed with three anaplastic carcinomas, histiocytic cell types and some ectodermal tumors.

Table 1

Binding of PNA and antibody to anaplastic gastric carcinomas, histocytes and some ectodermal tumors.

		PAS	Hale	PNA	neuraminidase PNA	Antibody against PNA binding mucus
Cases of anaplastic gastric carcinomas	1	-	-	-	+++	++
	2	-	+	+	++	++
	3	-	-+	++	++	++
Histiocytes in normal lymph node		-+	-	+	++	+
Histiocytes in reactive lesion		+	-	++	+++	-
Sarcoid granuloma		+	-	-	++	-
Giant cell granuloma		+	-	+	++	-
Malignant histiocytosis		-+	-	-+	++	-
Melanoma		-	-	-	++	-
Merkel cell tumor		-	-	-	+	-

All three highly anaplastic gastric carcinomas had the same morphological appearance. They were characterized by solid growth, apparently small round cells and negative PAS reaction (Figure 1 a). After PNA-HRP labelling a few tumor cells showed a dot-like staining in the perinuclear cytoplasm. This type of PNA-binding corresponds to the findings of Kondo et al. (16) on 2o cases of medullary gastric carcinomas. However, on tissue sections pretreated by neuraminidase prior to PNA-HRP labelling, almost all tumor cells showed an intensive staining (Figure 1 b). Thus it became apparent that the majority of the cells possess large cytoplasms filled with glycoprotein secretions. Ultrastructurally, the lectin binding sites were detected in the Golgi-lamellae and in the secretion granules (Figure 1 c).

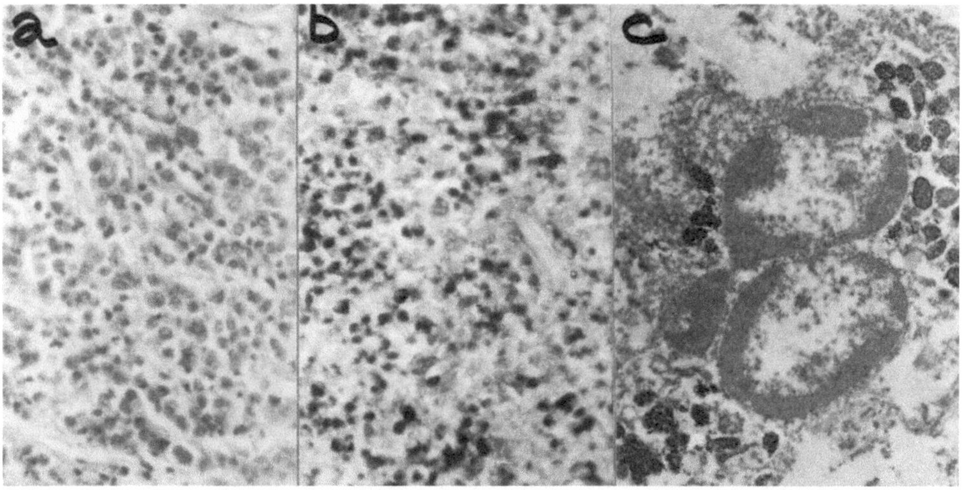

Fig. 1. Anaplastic gastric cancer. (a) PAS-reaction, (b and c) Neuraminidase pretreatment PNA-HRP labelling. Beside the negative PAS staining (a) and intensive perinuclear staining is seen in (b), which corresponds to the lectin binding of secretory granules (c).

Rarely, intracytoplasmic vacuolisation could be seen in the primary tumors.

In all three cases lymph node metastases showed a higher degree of differentiation characterized by intracytoplasmic vacuolisation and small microtubular structures. Neither the cytoplasmic vacuoles nor the apical membranes of microtubular structures were stained by the PAS or the Hale reaction (Figure 2 a) but were heavily marked by PNA lectin (especially after neuraminidase pretreatment) and by antibody directed against the PNA-binding component of normal human mucin (Figure 2 b).

These findings indicate that glycoprotein synthesis also occurs in this type of gastric carcinomas, but the chamical structure of the synthetized glycoprotein significantly differs from that of the normal and of intestinal and

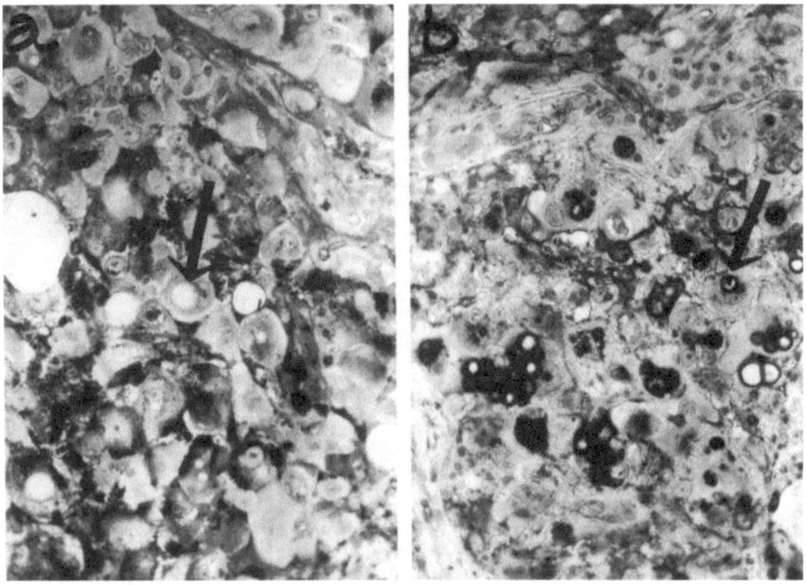

Fig. 2. Lymph node metastasis of anaplastic gastric cancer. (a) PAS-reaction, (b) indirect immunoperoxidase staining using antibody against the PNA-binding component of normal human gastric mucin. Vacuoles which are unstained in (a) show selective and intensive staining in (b), (arrows).

diffuse type cancer cells. This suggestion is supported by our earlier finding, demonstrating increased hexosamine content in mucus components isolated from adenocarcinoma (18). Intracytoplasmic PNA binding has never been observed in malignant lymphomas. But in all three cases of anaplastic gastric carcinomas we have found a very intensive intracytoplasmic lectin binding to small round cells. This indicates that lectin-histochemistry is an easy and valuable method for the differential diagnosis of these tumors and malignant lymphomas. On the other hand, it should be taken into consideration that perinuclear PNA binding is not a unique characteristic of anaplastic gastric cancer cells, since similar PNA binding was shown in the deeper part of the normal foveolar epithelium, in hyperplastic gastric

mucosa, and in activated intestinal columnar epithelial cells, as well (6,19).
Moreover, in some ectodermal tumors: Merkel cell tumor (Figure 3), melanoma, and in a large scale of cells of histiocytic origin a very intensive perinuclear PNA binding was observed, especially after neuraminidase pretreatment of tissue sections, indicating that carbohydrate chains with similar structure also occur in these cells at a defined stage of differentiation.

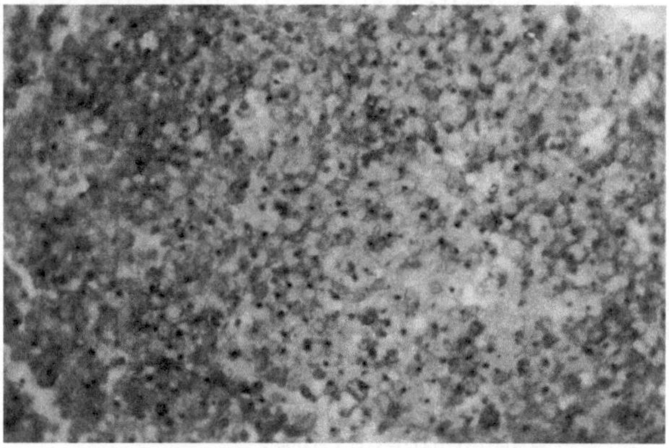

Fig. 3. Merkel cell tumor, neuraminidase, PNA-HRP. An intensive dot-rike perinuclear staining is seen in the majority of tumor cells.

Our findings on the PNA binding of histiocytic cell types correspond to the findings of Howard and Baktsakis (17). It should be emphasized, however, that in these latter cell types PNA binding was restricted to the perinuclear cytoplasm. Granular staining, intracytoplasmic vacuolization and microtubular structures, similar to those seen in anaplastic gastric cancer cells, could not be observed.
As shown in Table 1, the use of an antibody against the PNA-binding mucus component of normal gastric mucosa may be of great importance in doubt cases, since this antibody reacts strongly with anaplastic gastric carcinomas but shows an almost negative staining with ectodermal and histiocytic cell lines, reflecting antigenic differences

between the reacting structures.

References

1. Lauren, P. 1965. Acta Pathol.Microbiol. Scand. 64, 31.
2. Filipe, I.M. 1979. Invest. Cell Pathol. 2, 195.
3. Hounsell, E.F., Feizi, T. 1982. Med. Biol. 6o, 227.
4. Ming, S. 1977. Cancer 39, 2475.
5. Fischer, J., Klein, P.J., Vierbuchen, M., Fischer, R., Uhlenbruck, G. 1983. Lectins: Biology, Biochemistry, Clinical Biochemistry. Vol. III. Walter de Gruyter, Berlin - New York p. 167.
6. Fischer, J., Klein, P.J., Vierbuchen, M., Fischer, R., Uhlenbruck, G. 1983. Cancer Detection and Prevention 6, 137.
7. Kuhlmann, W.D., Peschke, P., Wurster, K. 1983. Virchows Arch. (Pathol Anat) 398, 319.
8. Peschke, P., Kuhlmann, W.D., Wurster, K. 1983. Experientia 39, 286.
9. Malmi,R., Söderström,K.O. These proceedings.
1o. Dus, D., Radzikowski, Cz., Debray, H., Christensen, B., Kider, J. These proceedings.
11. Born, A., Schwechheimer, K., Maier, H., Möller, P. These proceedings.
12. Örntoft, F.T., Petersen, S.E. These proceedings.
13. Holthöfer, H., Rapola, J. These proceedings.
14. Reader, A., Böttcher, S., Stalling, F., Kaukel, E. These proceedings.
15. Fischer, J., Klein, P.J., Vierbuchen, M., Skutta, B., Uhlenbruck, G., Fischer, R. 1984. Histochem.Cytochem. 32, 681.
16. Kondo, M., Watanabe, M., Takeda, Z., Urano, Y., Kobe, J. 1982. Med. Sci. 28, 241.
17. Howard, D.R., Batsakis, J.G. 1982. Am.J.Clin.Pathol. 77, 4o1.
18. Fischer, J., Uhlenbruck, G., Klein, P.J., Vierbuchen, M., Fischer, R. 1984. J.Histochem.Cytochem. 32, 69o.
19. Stoward, P.J., Spicer, S.S., Miller, R.L. 198o. J. Histochem. Cytochem. 28, 979.

CYTOCHEMICAL DEMONSTRATION OF CARDIAC GLYCOSIDES IN THE HEART MUSCLE TISSUES USING LECTINS AND ALDEHYDE - BISULFITE - TOLUIDINE BLUE (ABT) REACTION

Arpád Németh, Elizabeth Bayer
Department of Forensic Medicine of University Medical School, Pécs, Hungary H-7643

Péter Sótonyi
Department of Forensic Medicine of University Medical School, Budapest, Hungary

János Fischer
Department of Pathology of University Medical School, Pécs, Hungary H-7643

Digitalis glycosides and their related compounds are widely used in the treatment of various cardiac disorders. Experimental data suggest a number of possible explanations of their action (4). It is generally accepted that the sarcolemmal membrane (Na^{+}+ K^{+}) ATP-ase enzyme is the specific receptor for the mediation of their inotropic effect. In this way cardiac glycosides can inhibit the Na^{+}pump and activate Ca^{2+}passage into heart muscle cells (Fig. 1). The glycoside effect appears to be reversible: it can be washed out within 5 minutes (14). Recently published papers present new pharmacokinetic and biochemical information in this field (2,3,7,8,13) but without the morphological data which is necessary to explain correctly the experimental findings. Our recent paper has offered an advanced method, adapting Romhányi's (9) aldehyde-bisulfite-toluidine blue (ABT)

Lectins, Vol. IV

reaction (Fig. 2) to demonstrate membrane-bound glycoside oligosaccharides and to analyse their molecular orientation (12). Since glycoside pyranose (digitoxose) rings can be presumed to bind several oligosaccharide-specific plant lectins (1), the issue to be investigated is whether they are suitable for the morphological demonstration of local accumulation of cardiac glycosides in heart muscle cells.

Materials and Methods

Unifixed cryostat sections, formol-calcium fixed frozen sections and paraffin embedded human heart and rat heart tissue samples were studied, using the ABT reaction and lectin cytochemical methods. For model experiments Digoxin (Richter, Hungary) was chosen as a glycoside bocouse it contains sufficient sugar rings containding reactive OH groups in its digitoxose molecules for ABT reaction and lectin binding. Fixed and unfixed tissue sections were saturated with Digoxin (10^{-3} - 10^{-5} M, 30 min, 20°C), washed three times with physiologic saline solution and prepared for ABT reaction (10) or for binding of marked lectin compounds. A preliminary red blood cell agglutination test with Digoxin and different lectins showed that concanavalin A (Sigma) might be suitable for demonstration of membrane-bound glycosides. To obtain this finding we used fluorescein isothiocyanate labeled concanavalin A (con A-FITC) on control and Digoxin pretreated slides (20-40 µg/ml in PBS, 30 min, 4 °C). The significance of digitoxose sugar compounds in con A binding was controlled by periodate oxidation.
The ABT reaction, employed for topooptical staining, was analysed in a Leitz-Ortholux polarisation microscope: the optical sign of birefringence and retardation was measured by Köhler's rotating compensator. Submicroscopic localisation of ABT reaction product was analysed in frozen ultrathin sections by JEOL 100 B transmission electron microscope as described elsewhere (12). The con-A-FITC binding was examined under the fluorescent microscope (Zeiss Jena, Fluoval epifluorescence, 435-490 nm spectrum filter, HBO 200 W mercury lamp).

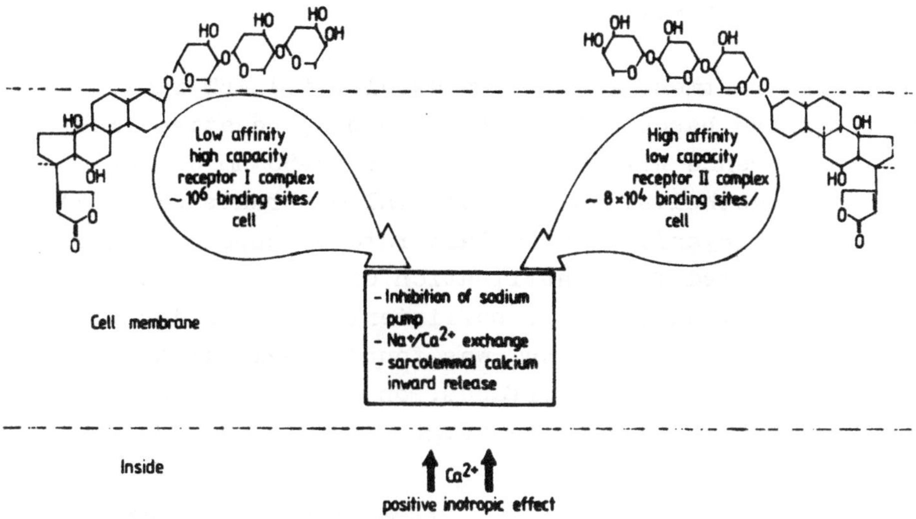

Fig. 1. The model of glycoside binding and action on cell membranes. The external sites of membrane $Na^+ + K^+$ -ATP-ase isoenzymes are supposed to be both the low affinity and high affinity receptors for glycosides in cardiac sarcolemmal membranes. In this way oligosaccharide chains of bound glycosides can be exposed, their OH groups can be available for histochemical reactions.

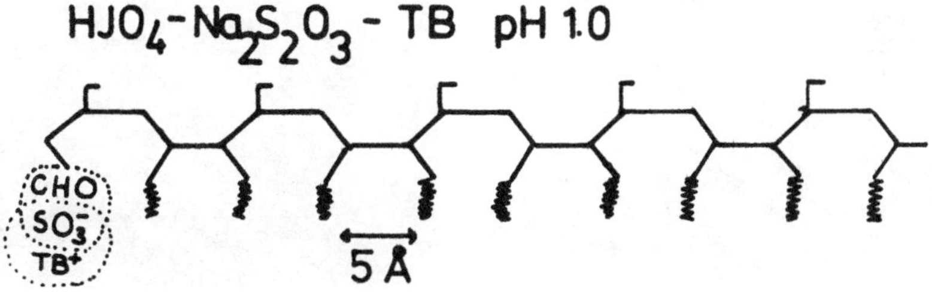

Fig. 2. Aldehyde-bisulfite-toluidine blue (ABT) reaction of oligosaccharide rings: Periodic acid (HJO_4) treatment can oxidize vicinal OH groups to dialdehyders, which, by the addition of bisulfite are rendered negative and bind toluidine blue (TB) in an oriented pattern. The reaction can produce negative birefringence with respect to the axis of the oligosaccharide chains (for light of wave lengths over λ 240 nm).

Results and Discussion

Digoxin pretreatment induced an accelerated ABT reaction with intensive basophilia, metachromasia, negative dichroism and birefringence with retardation of 30-50 nm on the capillary endothelial and sarcolemmal membranes (Fig. 3). When a K-ferricyanide - K/Na tartarate - cupper sulfide solution was used for stabilisation of the ABT reaction (12), the reaction product was of sufficient electron density for subcellular localisation and was found finely distributed on the endothelial cell surfaces, on sarcolemma membranes and in sarcolemmal cysternae (Fig. 4).

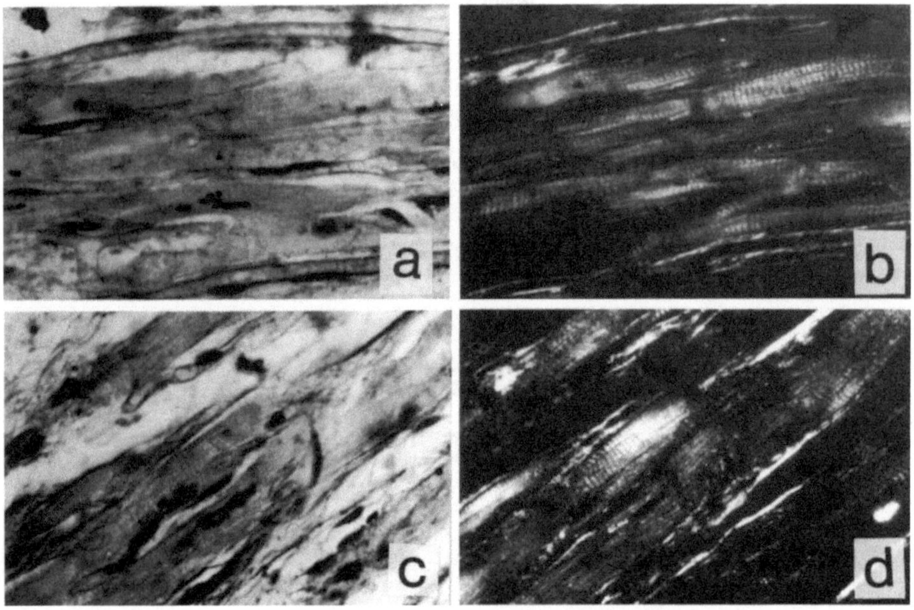

Fig. 3. Light- and polarization microscopic appearance of ABT reaction on formol-calcium fixed, Digoxin-pretreated human heart muscle tissue sections: intensive metachromatic toluidine blue staining (Fig. 3a) are birefringence of capillaries (Fig. 3b) and sarcolemma membranes (Fig. 3c, d) x 320.

Con A-FITC solution produced intensive fluorescence of capillary and sarcolemmal membranes on Digoxin pretreated tissue sections of the heart muscle. Formaldehyde fixation did not inhibit Digoxin binding and con A-FITC fluorescence. However periodate oxidation (10^{-3}M, 30 min, 20°C) of Digoxin containing samples completely prevented the selective binding of con A-FITC to the capillary and sarcolemmal membranes (Fig. 5). Mild periodate oxidation of glycosides has been employed as a useful method to modify receptor-glycoside binding by cleaving the pyranose rings and by producing reactive aldehydes on the glycoside oligosaccharides (5).

In our histochemical experiments periodate oxidation of glycosides was performed after their binding on specific membrane receptors.

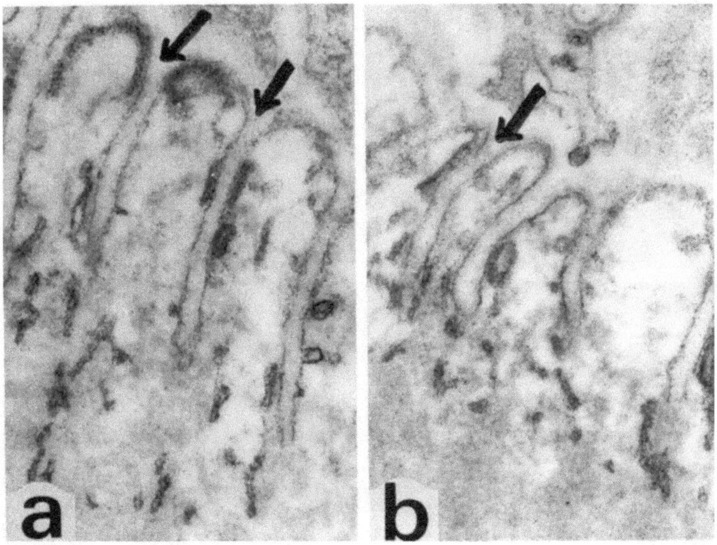

Fig. 4. Electronmicroscopic adaptation of ABT reaction: the arrows show the reaction product on the outer surface of sarcolemma membrane and inside the sarcolemmal cysternae. x 30.000.

In this way ABT reaction and labeled lectin (con A-FITC) binding proved to be suitable methods for the microscopical and submicroscopical localisation of cardiac glycosides, especially Digoxin, in heart muscle tissues. Furthermore, polarisation optical analysis of ABT reaction product could give information about the linear molecular arrangement of reactive carbohydrate compounds of glycosides on the specific receptor containing capillary and sarcolemmal membranes.

Monoclonal anti-Digoxin antibody currently employed in radioimmunoassay and potentially useful for treatment of

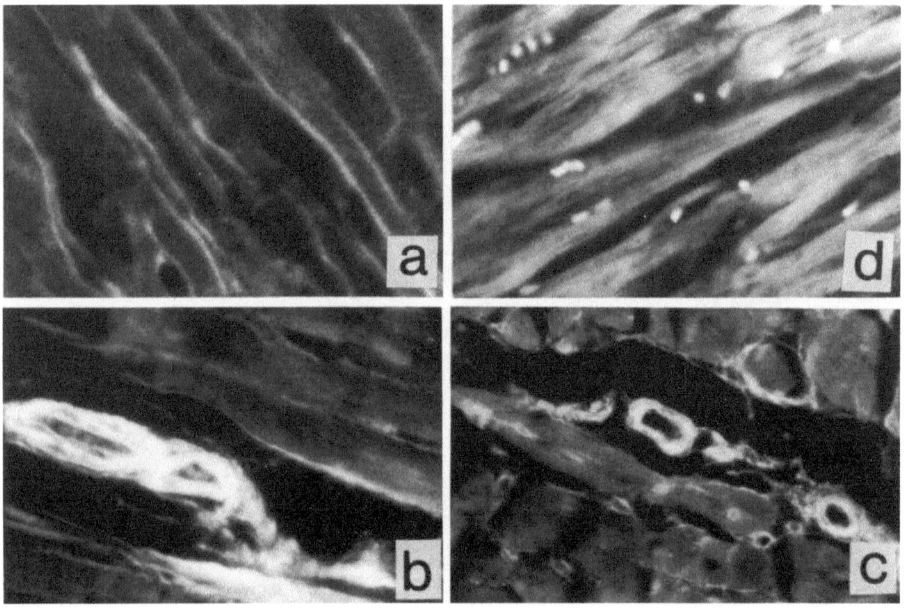

Fig. 5. Con A-FITC binding on Digoxin pretreated, formol-calcium fixed human heart muscle tissue slides: intensive fluorescence on capillaries and on sarcolemmal membranes (Fig. 5a,b, c), which was completely prevented by periodate oxidation. The red blood cells and heart muscle cells show autofluorescence following HJO_4 treatment (Fig. 5 d). x 120 and x 320.

Digoxin intoxication (6,11) may permit the development of further immunocytochemical methods which will permit further morphological demonstration of glycoside localisation in different tissues.

References

1. Bittiger, H., Schnebli, H.P.: "Concanavalin A as a Tool" Publ. John Wiley and Sons, London, 1976.

2. Brown, L., Erdmann, E., Thomas, R., 1983. Biochem Pharmacol 32. 2767-2774.

3. Erdmann, E., Brown, L. 1983. Eur Heart J 4 Suppl D 61-65.

4. Erdmann, E. 1984. "Cardiac Glycoside Receptors and Positive Inotropy" Suppl. to "Basic Research in Cardiology" Vol. 79 Steinkopf Publ. Darmstadt (FRG)

5. Heller, M. 1984. Basic Res in Cardiol 79. Suppl. 44-49

6. Lechat, Ph., Mudgett-Hunger, M., Margilies, MN., Haber, E., Smith, TW. 1984. J Pharmacol and Exp. Therap. 229 210-213.

7. Lüllmann, H., Peters, T., Preuner, J. 1982. Eur Heart J 3. Suppl. D 45-51.

8. Peters, U. 1982. Eur Heart J 3, Suppl D 65-78.

9. Romhányi, Gy., Molnár, L., Németh A. 1974. Histochemistry 39. 261-276.

10. Romhányi, Gy., Deák, Gy., Fischer, J. 1975. Histochemistry 43. 338-348.

11. Smith,TW., Butler, VP., Haber, E., Fozzard, H., Marcus, FI., Brenner, FW., Schulman, IC., Philips, A., N Engl J Med 307 1357-1362. 1982.

12. Sotonyi, P., Somogyi, E. 1983. Acta histochem 72. 117-122.

13. Sotonyi, P., Juhász-Nagy, A. 1984. Acta Physiol Hung 63. 153-160.

14. Werden, K., Zwissler, B., Wagenknecht, B., Krawietz, W., Erdmann, E. 1984. Basic Res in Cardiol 79, Suppl 35-43.

ULEX EUROPAEUS I LECTIN-BINDING GLYCOPROTEIN IN PRIMARY SENSORY TERMINALS OF HUMAN SPINAL CORD

J. Fischer

Department of Pathology, University Medical School, Pécs
H-7643 Pécs, Hungary

P.J. Klein

Institute of Pathology, University of Freiburg
D-7800 Freiburg, F.R.G.

B. Csillik

Department of Anatomy, University Medical School, Szeged
H-6701 Szeged, Hungary

Lectins, with binding specificity to various mono-, di- or oligosaccharides are extensively used for the histochemical characterization of cell and tissue glycoconjugates. In the last years several papers have been published on the lectin binding pattern of normal and neoplastic cells in the human central nervous system (CNS) (1-5). It has been shown also that Con A receptor-bearing high mannose-type oligosaccharides are present in high amounts in the synaptic membranes of the brain, especially within the cerebellum (6-8). Whilest studying lectin binding in Sandhoff's tetrahexosylceramide storage disease (9) we accidentaly observed that a well defined group of axon terminals in the dorsal horn of human spinal cord is selectively marked by horseradish peroxidase (HRP) labelled UEA I lectin. Since the staining proved to be highly specific and was restricted to a small part of primary sensory nerve terminals only (inner zone of layer II and layer III), our finding suggested that like some enzymes (10-

Lectins, Vol. IV

12), and some neuropeptides (13,14), a distinct part of sensory neurons can be selectively visualized by a unique glycoconjugate content. Moreover the newly discovered glycoconjugate, which is responsible for the UEA I lectin binding, may possess a specific function in the transmission or regulation of sensory impulses.

Material and Methods

Autopsy material derived from 10 cases of different ages were used for the investigations. Tissue samples from different areas of the brain, pons, medulla oblongata, spinal cord as well as spinal root ganglia, Gasserian and sympathetic ganglia were fixed either in 10% formaline or in 0.1 M cacodylate buffered (pH 6.3) glutaraldehyde (0.25%) for 1-2 hr and subsequently in cacodylate buffered formaline (4%) for 12 hr. Frozen and paraffin embedded tissue sections were examined for lectin binding. Paraffin sections were dewaxed as usual.

Lectin histochemical staining was performed using peroxidase labelled UEA I lectin, purchased from Medac (Hamburg, F.R.G.). In order to block endogenous peroxidase, tissue slices were pretreated with 1% hydrogen peroxide in methanol or in distilled water. After extensive washing in PBS, slices were incubated with UEA I-HRP (diluted to 10 µg/ml in PBS) for 30 to 60 minutes at room temperature. Lectin binding sites were visualized either by 3-amino-9-ethyl carbazole or by 3'3 diaminobenzidine as chromogen.

Control slices were treated with UEA I-HRP in the presence of the complementary sugar: 0.1 M L-fucose (Sigma, St. Louis, U.S.A.). Lectin labelling was also performed both on frozen and paraffin embedded tissue sections after extensive lipid extraction either in chloroform-methanol (2:1) or in ether-ethanol (1:1) mixtures.

All slices were counterstained with haematoxyline and mounted in glycerol jelly.

Results

Distribution of the UEA I binding substance in the CNS

Figure 1 a and b demonstrate the typical staining pattern of the human spinal cord following UEA I-HRP labelling at dif-

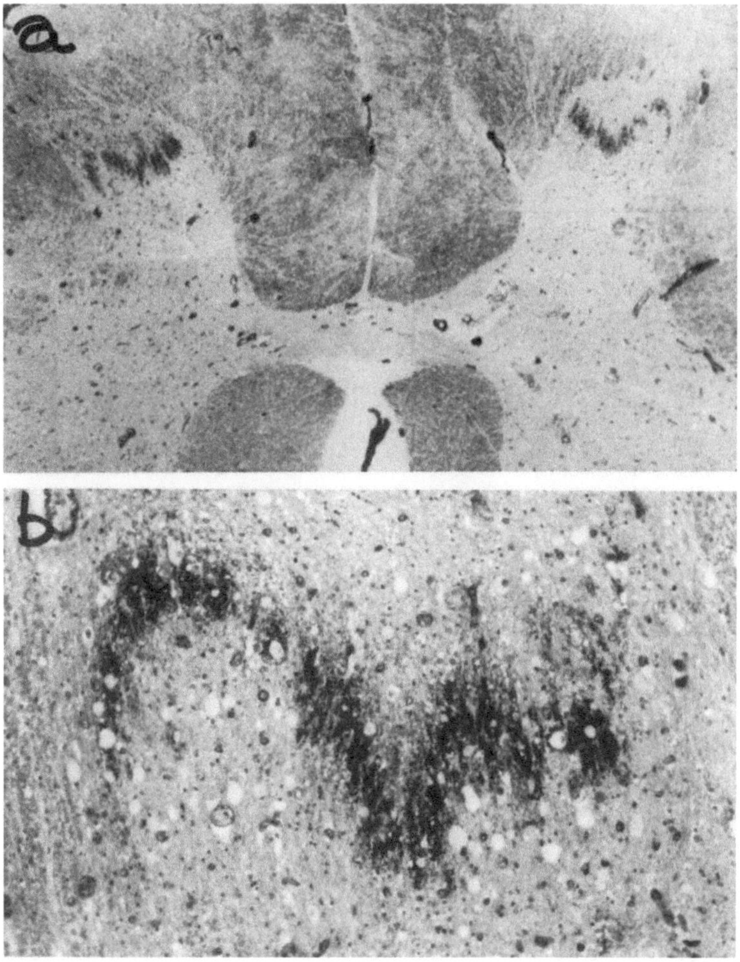

Fig. 1. Cross section of human spinal cord, UEA I-HRP. (a) Overall view on UEA I binding sites in the grey matter of dorsal horns (about 20 x).
(b) At higher magnification a granular to fine fibrillar staining is seen in the inner zone of layer II and in layer III (about 180 x).

ferent magnifications. The substance which was intensively stained was highly concentrated in the inner zone of layer II and in layer III of the grey matter of the dorsal horns. A more finely dispersed granular-to-fine fibrillar staining was also apparent in Lissauer's tract, which could be seen

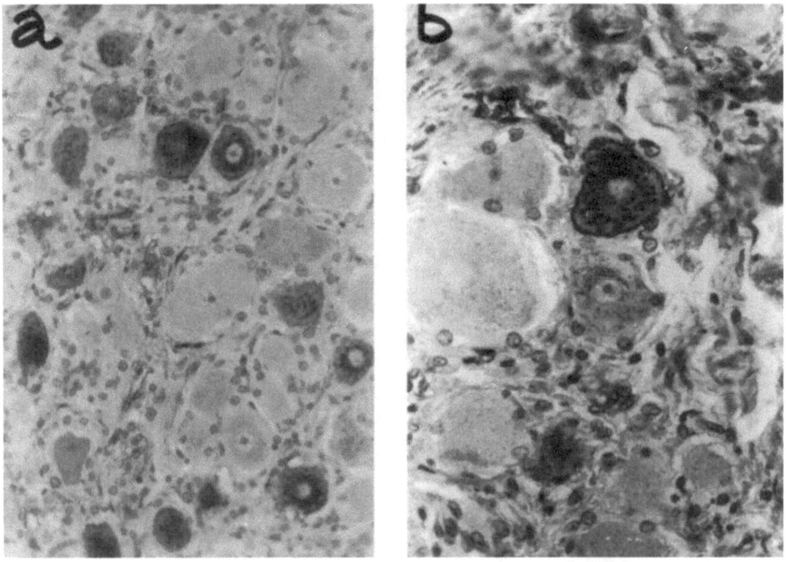

Fig. 2. Dorsal root ganglion, UEA I-HRP labelling. (a) UEA I binding glycoprotein is localized mainly in medium sized nerve cell bodies (about 150 x).
(b) The staining intensity of the individual neurons shows a high variation, also cell membranes are sometimes intensively stained (arrow). The reacting compound is localized in the form of large granules within the cytoplasm of the cells.

at higher magnifications only. In the medulla oblongata lectin binding was localized in the nucleus tractus spinalis nervi trigemini and in a part of nucleus fasciculi cuneati. Other nervous tissue elements of the CNS were negative after UEA I-HRP labelling. Besides nervous tissue structures the endothelial cell membranes and connective tissue elements within the vascular walls, pia mater and dorsal roots also showed very intensive UEA I binding. However the staining of connective tissue elements proved to be unspecific, as it has not been inhibited by adding 0.1 M L-fucose to the labelled lectin before incubation.

Distribution of UEA I binding substance in peripheral ganglia
About 25-35% of the sensory neurons in the dorsal root gang-

Table 1

Distribution of UEA I-HRP positive perikarya in human sensory ganglia

Size of neurons	10-30um	30-50um	50-100um	UEA I$^+$
	(UEA I$^+$/ UEA I$^-$ from 100 cell)			cells %
Dorsal root ggl.				
Th 2	4/18	25/35	6/12	35
Th 4	3/12	18/27	1/29	22
Th 10	5/20	19/38	4/14	28
Th 12	7/18	28/24	5/18	40
UEA I$^+$ cells %	19,5%	42%	18%	31%
Gasserian ggl.	4/27	12/35	0/22	16
	5/20	14/43	2/16	21
UEA I$^+$ cells %	16%	25%	5%	18%

lia showed a positive staining after UEA I-HRP labelling. The lectin binding component was localized in almost regularly dispersed large granules within the cell bodies of medium sized neurons (Figure 2 a). The staining intensity showed a wide variation between the individual perikarya, which might contain a few faint positive granules only or might be filled by numerous heavily stained ones (Figure 2 b). Sometimes the cell membranes were also stained (Figure 2 b arrow).

The majority of nerve cell bodies possessing high affinity for UEA I lectin also contained pigment granules within their cytoplasm. Although the pigment granules were smaller and situated around the nuclei or at one pole of the cell bodies in contrast to the large and equally distributed lectin stained granules, a clear distinction between them was only possible by using 3-amino-9-ethylcarbazole substrate for the peroxidase reaction. In this case the reddish reaction product can be distinguished excellently from the black pigment granules. Though the UEA I-HRP positive neurons were also present in the Gasserian ganglia, their relative proportion was lower (about 15-20% of total cell number). The size distribution of positively stained cell bodies in Gasserian as well as in dorsal root ganglia are compared in Table 1.

In contrast to our findings with the sensory ganglia the sympathetic ganglia were always completely negative after UEA I-HRP staining.

Control investigations

In control slices the UEA I binding of the sensory cell bodies in dorsal root ganglia and of the afferent terminals within layers II (inner zone) and III as well as of the endothelial cells was completely inhibited by adding 0.1 M L-fucose to the labelled lectin, indicating that lectin binding occurred on the specific receptor sites of the lectin. In contrast, the strong UEA I-HRP reaction of connective tissue elements could not be blocked by the inhibitory sugar, indicating an unspecific lectin binding.

Both fixation methods were equally effective in the preservation of the reactivity of UEA I binding sites in tissue sections. Lipid extraction did not reduce the binding of UEA I-HRP to the specific binding sites. Moreover, as seen on frozen section (generally with a low staining intensity) the UEA I-HRP binding was greatly enhanced after dissolution of the lipid components. This observation indicates that lipids may interfere with the lectin binding.

Discussion

Our investigations indicate that a glycoprotein with a unique localization to a subgroup of primary sensory neurons is discovered by our serendipitous finding. The glycoprotein nature of the lectin binding compound is emphasized, since it can be demonstrated in tissue sections also after extensive lipid extraction. UEA I lectin is known to be highly specific for terminal fucose residues of glycoconjugates, including blood group substances (15). Since UEA I-HRP binding to

sensory neurons and primary afferent terminals could be inhibited by L-fucose, the reacting glycoprotein has to possess large amounts of terminal fucose residues.
The observation that UEA I binding glycoprotein characterizes only a restricted population (25-35%) of the sensory nerve cell bodies, and they central terminals in the inner zone of layer II and in layer III of the grey matter of human spinal cord, indicates that only a distinct part of the sensory neurons and their terminals are selectively marked by UEA I.
The fact that primary sensory neurons can be divided into more or less definite subgroups has been shown by means of pure morphological (16,17) and of morpho-functional criteria, using marker enzymes, i.e. fluoride resistant acid phosphatase (18,19) antibodies against enzymes, anti-tyrosine hydroxylase, anti-dopamine-3-hydroxylase (12) or neuropeptides, somatostatin and substance P (13, 14). Unfortunately these data are almost exclusively restricted to sensory ganglia of animals, thus it is difficult to compare them with our findings. For example, sensory neurons of the rat are easily subdivided into large A and small B cells, but the size distribution of human dorsal root ganglion cells is more heterogenous, varying between large, medium sized and small ones (20). UEA I binding glycoprotein was found in all of these morphologically different cell types but it characterizes mainly the medium sized cell population, whereas both the marker enzymes and neuropeptides were demonstrated exclusively within the small B neurons of animals (14). Localizations of the reacting substances at the axon terminals in the grey matter seems to be more comparable. Whilest substance P was demonstrated mainly in layer I (21), somatostatin and fluoride resistant acid phosphatase in the central zone of layer II of animals (14), UEA I-binding glycoprotein was mainly localized to layer III and only to the inner zone of layer II in human spinal cord. Therefore the localization of UEA I-binding glycoprotein seems to be independent from that of the hitherto studied marker enzymes and neuropeptides.

Even the functional significance of the discovered lectin-binding glycoprotein remains to be solved. This substance is apparently transported from its site of production (i.e. the perikarya) to the synaptic terminals of one class of sensory nerves. There are at least three possible explanations for the functional significance of this glycoprotein, it may play a role: 1) in fast axonal transport of the molecule, which was shown to be highly dependent on the degree of glycosylation of some neuronal proteins (22); 2) in neuronal recognition phenomena and synaptogenesis, as it has been shown for the Con A-binding glycoprotein of the rat cerebellum (23); or 3) the UEA I-binding glycoprotein may also have a special function in the transmission and regulation of sensory impulses in a group of sensory nerves.

References

1. Müller,W., Klein,P.J., Newman,R.A., Uhlenbruck,G. 1980. Acta Neuropathol. 49, 101.
2. Caron,M., Denguier,M.A., Albe,X., Bisconte,J.C., Faure,A. 1981. Experientia 37, 1154.
3. Hori,T., Nishiyama,F., Matsutani,M., Teramoto,A., Takakura,K., Sano,K,. Hirano,H. 1981. Acta Neuropathol. 56, 67.
4. Schwechheimer,K., Schnabel,P., Möller,P. 1983. Acta Neuropathol. 61, 21.
5. Schwechheimer,K., Born,I.A., Kern,P., Weiss,G. These proceedings
6. Cotman,C.W., Taylor,D. 1974. J.Cell Biol. 62, 236.
7. Gurd,J.W., Fu,S.S. 1982. J. Neurochem. 39, 719.
8. Zanetta,J.P., Reeber,A., Dontewill,M., Vincedon,G. 1984. J.Neurochem. 42, 334.
9. Fischer,J., Klein, P.J. These proceedings
10. Knyihár-Csillik,E., Csillik,B. 1981. Progress in Histochem. Cytochem. V. 14. G. Fischer Stuttgart, New York, pp.1-137.
11. Csillik,B., Knyihár-Csillik,E. 1982. Neurosci. 7, 1149.
12. Price,J., Mudge,A.W. 1983. Nature 301, 241.
13. Hökfelt,T., Elde,R., Johansson,O., Luft,R., Nilsson,G., Arimura,A. 1976. Neurosci. 1, 131.
14. Nagy,J.I., Hunt,S.P. 1982. Neurosci. 7, 89.
15. Matsumoto,I., Osawa,T. 1969. Biochim.Biophys.Acta 194, 484.
16. Andres,K.H.1961. Z.Zellforsch. 55, 1.
17. Duce,I.R., Keen,P. 1977. Cell Tiss.Res. 185, 263.
18. Colmant,H.J. 1959. Arch.Psychiat.Nervkrankh. 199, 60.
19. Knyihár,E. 1971. Experientia 27, 1205.
20. Urich,H. 1976. In: Greenfield's Neuropathology (W.Blackwood and J.A.N. Corsellis, eds.). Edward Arnold Ltd.p. 694.
21. DiFiglia,M., Aronin,N., Leeman,S.E. 1982. Neurosci.7, 1127.
22. Stone,G.C., Hammerschlag,R. 1983. J.Neurochem. 40, 1124.
23. Groswald,D.E., Kelly,P.T. 1984. J. Neurochem. 42, 534.

LECTIN BINDING PATTERNS IN NORMAL HUMAN SALIVARY GLANDS

I. Antonio Born [1], Karl Schwechheimer [2], Heinz Maier [3], Peter Möller [1]

[1] Institut für Allgemeine Pathologie und pathologische Anatomie, [2] Institut für Neuropathologie, [3] Hals-Nasen-Ohrenklinik der Universität Heidelberg, D-6900 Heidelberg, GFR

Numerous reports are now appearing in the literature dealing with mapping of lectin binding to normal and pathological human and animal tissues. Thus the lectin binding to various human non-neoplastic tissues and the related tumors was the theme of a number of reports presented here (4,6-8,10,12,15-17). Earlier there has been but a few studies reported on lectin binding to salivary glands. Schulte and Spicer (18, 19) extensively reviewed the lectin target cells in rat and mouse salivary glands.

Here we want to report our systematic studies on the lectin binding to non-heoplastic human salivary glands.

Lectins, Vol. IV

Materials and Methods

The present paper wants to examine and to demonstrate lectin binding patterns in non-neoplastic human salivary glands. The tissue specimens of the parotid and submandibular gland were obtained 10 min after surgery. The tissue specimens were fixed in 10% buffered formaldehyde, Bouin's solution or 3% glutaraldehyde for 1-12 hours. Formaldehyde and Bouin fixed tissue were dehydrated overnight and embedded in paraffin. Lectin binding of 7 human parotid and 7 submandibular glands were examined with *Ulex europeaues* agglutinin (UEA I), peanut agglutinin (PNA), *Helix pomatia* agglutinin (PHA), *Glycine max* agglutinin (SBA) and Con A (cf. Appendix). All lectins were obtained from DAKO, Hamburg, FRG. An extended four step peroxidase (PaP) method (14) was used on 4 Bouin, 10 buffered formalin and 4 glutaraldehyde fixed, salivary glands. Con A binding was assessed according a modified procedure (1).

Results

Serous cells showed a fine or coarse vesicular intracytoplasmic binding with all lectins tested by UEA I in the submandibular rather than the parotid gland (Fig. 1). Mucinous cells in the submandibular gland showed diffuse intracytoplasmic binding with UEA I, Con A and HPA as well as a fine vesicular subnuclear reaction with PNA. The most characteristic feature was strong coarse vesicular basal binding with HPA (Fig. 2). Serous demilunes in submandibular glands showed the same patterns as serous cells qualitatively, but to a lesser extent (Fig. 3). Myoepithelial cells were always negative. Interlobular and striated ducts in parotid and submandibular glands exhibited a luminal, cytoplasmic and coarse vesicular binding with UEA I (Fig. 4), PNA, Con A, HPA (Fig. 5) and SBA (Fig. 6). HPA and SBA, however, gave a mosaic-like pattern. The most prominent result was the coarse vesicular reaction of Con A in interlobular duct cells (Fig. 7). Luminal surface in striated ducts showed a very strong reaction with UEA (Fig. 4) and SBA (Fig. 6) while cytoplasmic membrane was stained by UEA (Fig. 4) and HPA (Fig. 5). The most prominent finding was an intense intracytoplasmic binding of PNA in basal cells of intercalated,

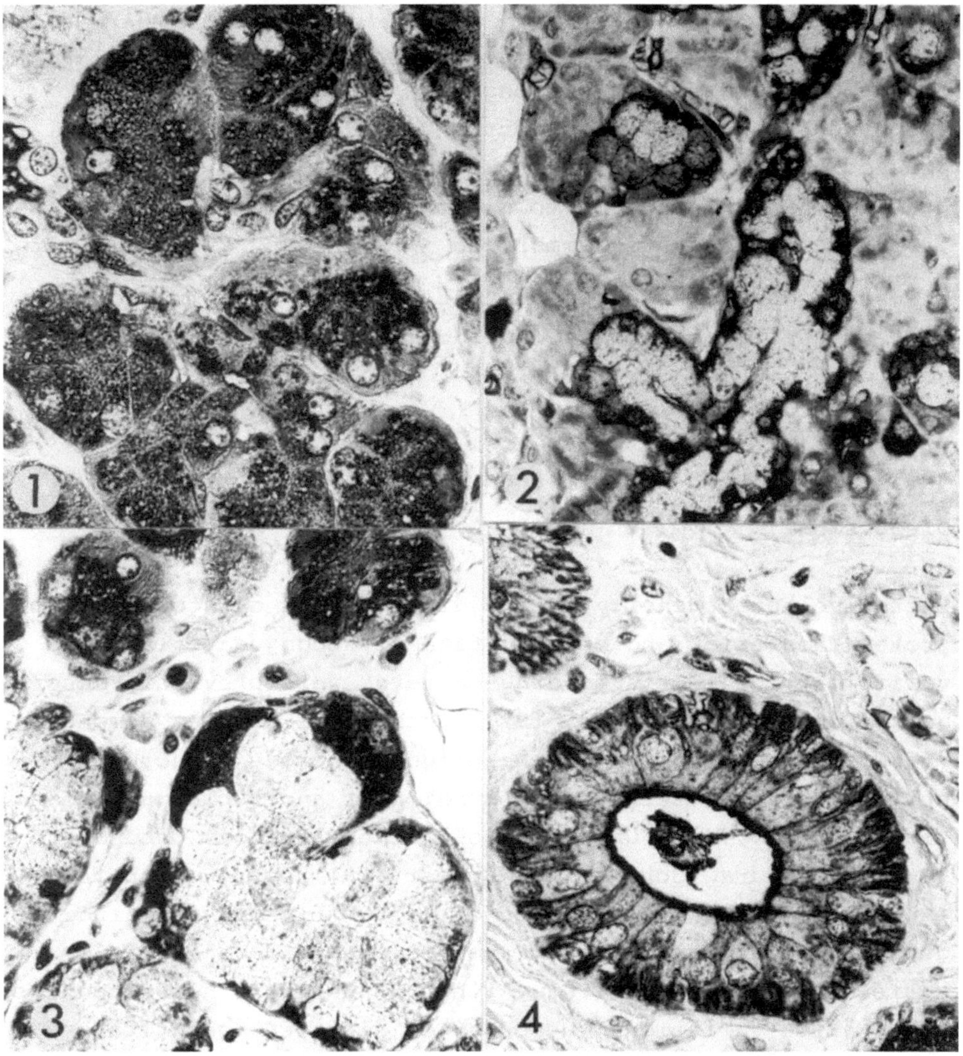

Fig. 1: *UEA I, parotid gland: Serous acini, x 480.*

Fig. 2: *HPA, submandibular gland: Mucinous acini, x 480.*

Fig. 3: *UEA I, submanidbular gland: Serous acini with serous demilunes, x 480.*

Fig. 4: *UEA I, parotid gland: Striated ducts, x 480.*

striated and interlobular ducts (Fig. 8-9), where only rare and weak staining with Con A (Fig. 10) and HPA (Fig. 5) was found. Sometimes, a few strongly PNA-positive basal cell were seen in the striated ducts. Vascular endothelium expressed high amounts of UEA I acceptor molecules. The binding of PNA, HPA and SBA to endothelial surface was facultative. The details are summarized in Table 1.

PAS-stain showed a very prominent reaction in acinus cells. Duct epithelial cells and basal cells were extremely weak.

Discussion

Extensive heterogeneity in the distribution of stored glycoprotein secretion and cell surface glycoconjugates and the variability of lectin binding patterns were previosly described in mouse and rat (18, 19). Biochemical studies demonstrated wide heterogeneity in the oligosaccharide structures of purified glycoproteins of submandibular and sublingual gland tissue (9, 11, 13). Unfortunately, these authors used different rat strains. Schulte and Spicer (19) report, with few exceptions, a high invariance of lectin binding from individual to individual within a species. These data generally correlate with our findings.Although there are not extensive qualitative variations, we observed different quantitative amounts of lectin acceptor molecules from gland to gland and from individual to individual. A comparison between different species is difficult. The discrepancies between the findings of the literature (18, 19) and our results might be species-related and are therefore not comparable.

The fact that the five lectins examined showed a variable reaction in all cell types of glandular tissue demonstrates

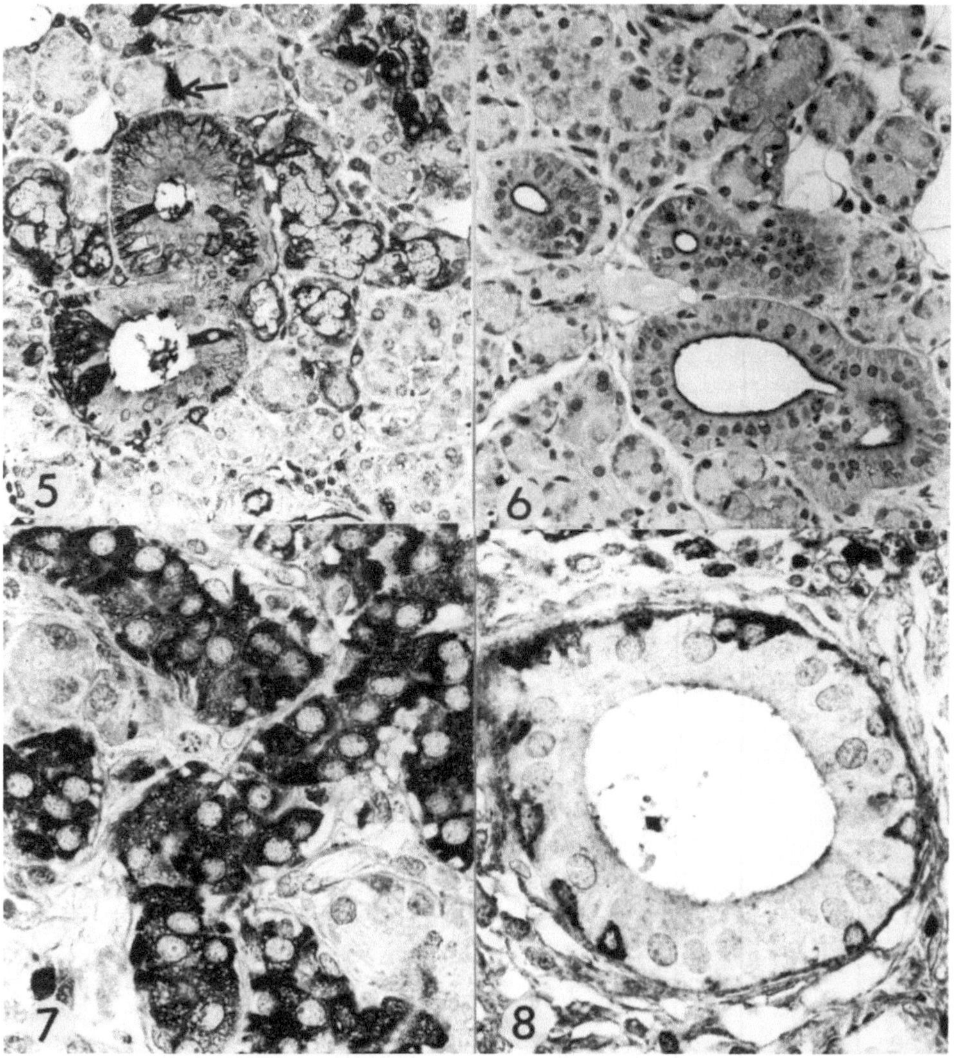

Fig. 5: *HPA, submandibular gland: Mosaic binding pattern in intercalated and striated ducts; note positive reaction in some serous acinus cells and single basal cells (arrows), x 180.*

Fig. 6: *SBA, submandibular gland: Apical linear and intracytoplasmic binding of striated ducts and moderate reaction of mucinous cells, x 180.*

Fig. 7: *Con A, parotid gland: Lectin binding in intercalated duct epithelium, x 180.*

Fig. 8: *PNA, parotid gland: Basal cells in striated ducts, x 480.*

Table 1

Lectin binding patterns of normal salivary glands

	UEA α-L-Fuc			PNA D-Gal-ß-(1-3)GalNac			Con A α-D-Glc α-D-Man		
PAROTID GLAND	1	2	3	1	2	3	1	2	3
Serous acini	+++	i	D	-/+	i	S	+	i	S
Myoepithelium	-	-	-	-	-	-	-	-	-
Intercalated duct	++	il	F	++/+++	l △	D	++++	il	D
Striated duct	+++	ilm	D	++++	l △	D	+	l △	S
Interlobular duct	+++	ilm	D	++++	l △	D	-/+	l △	S
Endothelium	+++	l	D	-	-	-	-	-	-
SUBMANDIBULAR GLAND	1	2	3	1	2	3	1	2	3
Serous acini	++++	i	D	-/+	i	S	++	i	S
Mucinous acini	+	i	D	-/+	i	S	-	-	-
Serous demilunes	++++	i	D	-/+	i	S	+	i	S
Intercalated duct	++	i	F	+++/++++	l △	D	++++	il	D
Striated duct	++++	il	D	++++	l △	D	+	l △	S
Interlobular duct	+++	ilm	D	++++	l △	-	-/+	l △	S
Endothelium	+++	l	D	-	-	-	-	-	-

	H P A -D-GalNac			S B A -D-GalNac/ D-Gal			P A S		
PAROTID GLAND	1	2	3	1	2	3	1	2	3
Serous acini	-/+	i	S	-/+	i	D	+++	i	D
Myoepithelium	-	-	-	-	-	-	-	-	-
Intercalated duct	+++	i	M	-/+	i	S	-/+	il	F
Striated duct	+++	im	M	+++	ilm	M	-/+	il △	M
Interlobular duct	+	im	M	+	ilm	M	-/+	i	D
Endothelium	++	l	Fac.	++	l	Fac.	-	-	-
SUBMANDIBULAR GLAND	1	2	3	1	2	3	1	2	3
Serous acini	++	i	S	-/+	i	S	++++	i	D
Mucinous acini	+++	i	D	-/+	i	D	++++	i	D
Serous demilunes	++	i	S	-/+	i	S	++++	i	D
Intercalated duct	++	i	M	-/+	i	M	-/+	il	F
Striated duct	++++	im	M	++++	ilm	M	-/+	il △	M
Interlobular duct	++	im	M	+	ilm	M	-/+	i	F
Endothelium	++	l	Fac.	++	l	Fac.	-	-	-

1. *Quantification of lectin binding:*
   - = negative, + = weak, ++ = moderate, +++ = strong, ++++ = extensive.
2. *Cellular distribution of lectin binding:*
   i = intracytoplasmic, l = luminal, m = membrane bound, △ *= basal cell.*
3. *Histological distribution of lectin binding:*
   S = single cell, F = focal, D =diffuse, M = mosaic pattern, Fac. = facultative.

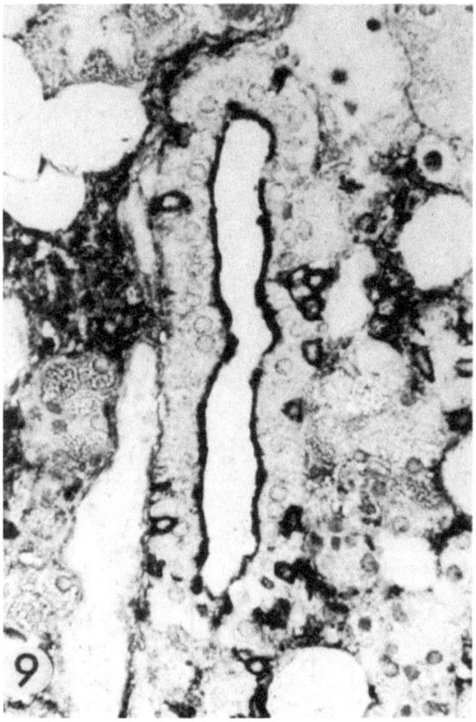

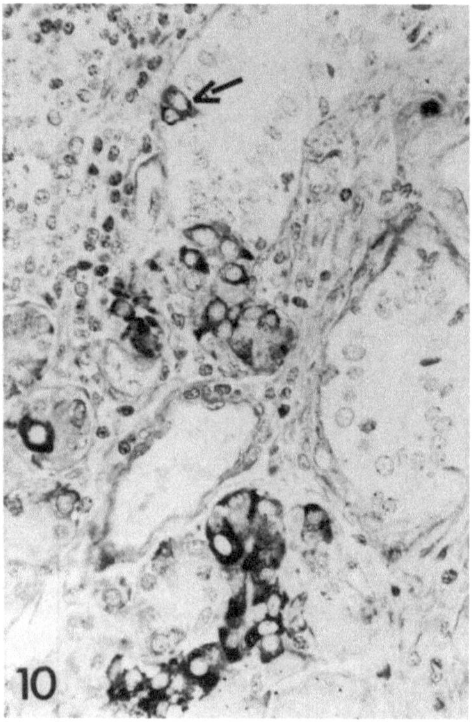

Fig. 9: PNL, submandibular gland: Basal cells in interlobular ducts, x 300.

Fig. 10: Con A, parotid gland: Strong reactivity in intercalated duct epithelium and single basal cells (arrows), x 300.

that the related carbohydrate residues (α-Fuc-UEA I, D-Gal-ß-(1-3)-GalNac - PNA, α-D-Man/α-D-Glc - Con A, α-D-GalNac/D-Gal - SBA, α-D-GalNac - HPA) are present in the intra- as well as the extracellular spaces of both gland types. The constant reaction of the various lectins, however, with specific glandular cells types favors the view that there are high amounts of α-L-fucose (UEA I) in serous acinus cell, α-D-mannose and α-D-glucose (Con A) in intercalated ducts, α-D-galactose Nac (HPA) and α-D-galactose (Nac/D-galactose (SBA) im mucous acinus cells as well as D-galactose-ß-(1-3)-GalNac in basal cells ("reserve cells") (5). Because of the

absolutely constant immunocytochemical reaction we propose UEA I, Con A and PNA as structural markers for serous acinus cells, intercalated ducts and basal cells, respectively. Even if HPA and SBA, which show moderate binding to mucinous cells markers, their mosaic binding pattern in diverse cell types of the salivary glands proves them of value as functional markers.
Further studies will be necessary to demonstrate the potential importance of lectin binding patterns for the identification of the cellular components in human salivary gland tumors. Moreover, immunoelectronmicroscopic techniques are required to visualize lectin acceptor structures at the ultrastructural level.

This work was supported by a grant of the Deutsche Forschungsgemeinschaft, Bonn-Bad Godesberg, FRG.

References

1. Avrameas, S., Karsenti, E., Bornens, M. In: Bittiger, H., Schnebli, H. P. (eds): Concanavalin A as a tool. John Wiley and Sons, London, New York, Sydney, Toronto, 85-93 (1976).

2. Born, I. A., Liewald, F., Möller, P.: Der Pathologe (abstract; in press).

3. Debray, H., Docount, P. C., Jenkins, K.: Carbohydr. Res. 87, 265-274 (1980).

4. Dus, D., Radzikowski, C., Debray, H., Christensen, B., Kieler, J. In: Bøg-Hansen, T. C., Breborowicz, J. (eds): Lectins. Biology, Biochemistry, Clinical Biochemistry, Vol. 4. De Gruyter, Berlin, New York 65-74 (1985)

5. Ferner, H., Gansler, H.: Zeitschr. Zellforsch. 55, 148-178 (1961).

6. Fischer, J., Klein, P.J..In: Bøg-Hansen, T. C., Breborowicz, J. (eds): Lectins. Biology, Biochemistry, Clinical Biochemistry, Vol. 4. De Gruyter, Berlin, New York 95-100 (1985)

7. Fischer, J., Klein, P. J.: In: Bøg-Hansen, T. C., Breborowicz, J. (eds): Lectins. Biology, Biochemistry, Clinical Biochemistry, Vol. 4. De Gruyter, Berlin, New York 101-108, (1985)

8. Fischer, J., Klein, P.-J., Csillik, B.. In: Bøg-Hansen, T. C., Breborowicz, J., (eds): Lectins. Biology, Biochemistry, Clinical Biochemistry, Vol. 4. De Gruyter, Berlin, New York, 117-126, (1985)

9. Herp, A., Wu, A. M., Moschera, J.: Mol. Cell. Biochem. 23, 27-44, (1979)

10. Holthöfer, H., Rapola, J.. In: Bøg-Hansen, T. C., Breborowicz, J. (eds): Lectins. Biology, Biochemistry, Clinical Biochemistry, Vol. 4. De Gruyter, Berlin, New York, 83-88, (1985)

11. Kenyen, G., Herman, G., Rorsignal, B.: Biochem. Biophys. Acta 297, 186-191, (1973)

12. Malmi, R., Söderström, K.-O.. In: Bøg-Hansen, T. C., Breborowicz, J. (eds): Lectins. Biology, Biochemistry, Clinical Biochemistry, Vol. 4. De Gruyter, Berlin, New York, 145-154, (1985)

13. Moschera, J., Pigmar, W.: Carbohydr. Res. 40, 53-67, (1975)

14. Möller, P.: Virchows Arch. (Pathol. Anat.) 396, 313-317, (1982)

15. Örntoft, T. F., Petersen, St. E.. In: Bøg-Hansen, T. C., Breborowics, J. (eds): Lectins. Biology, Biochemmistry, Clinical Biochemistry, Vol. 4. De Gruyter, Berlin, New York, 89-94, (1985)

16. Raedler, A., Böttcher, S., Stalling, F., Kaukel, E.. In: Bøg-Hansen, T. C. Breborowicz, J. (eds): Lectins. Biology, Biochemistry, Clinical Biochemistry, Vol. 4. De Gruyter, Berlin, New York, 137-144, (1985)

17. Rudolph, M., Samtleben, R.. In: Bøg-Hansen, T. C., Breborowicz, J. (eds): Lectins. Biology, Biochemistry, Clinical Biochemistry, Vol. 4. De Gruyter, Berlin, New York, 653-662, (1985)

18. Schulte, B. A., Spicer, S. S.: Histochem. J. 15, 1217-1218, (1983)

19. Schulte, B. A., Spicer, S. S.: Histochem. J. 16, 3-20, (1984)

20. Schwechheimer, K., Weiss, K., Möller, P.: Virchows Arch. (Pathol. Anat.) 402, 297-306, (1984)

LECTIN BINDING SITES ON BRONCHIAL CARCINOMA CELLS

A. Raedler, S. Böttcher, F. Stalling, E. Kaukel
Med. Klinik und Poliklinik, Universität Hamburg, FRG

Compared to their normal counterparts, carcinoma cells exhibit a diffent biological behaviour under in vitro conditions, such as a decreased contact inhibition and adherence and, moreover, an increased agglutinability by lectins (1, 2). This latter observation may be related to a further characteristic of transformed cells, i.e. a decreased synthesis of extracellular fibronectin (3). Starting from the observation of a better agglutinability of carcinoma cells with lectins, several epithelial carcinomas have been analysed with respect to the ability of their cell surface constituents to react with lectins of various specificities (see also: Örntoft and Petersen, Schwechheimer et al., Malmi and Söderström, Fischer and Klein, Rudolph and Samtleben, Duś et al., this volume). It was shown that glycoconjugates, especially those with terminal L-fucose, such as those displayed by carcinoma cells derived from kidney, pancreas and the colon, act as lectin binding sites (4-8). These glycoconjugates could not be detected on cells derived from normal tissue, at least not with the sensibility of the methods used (4-5).
In order to create conditions for an improvement of early diagnosis of bronchial carcinoma, experiments were designed to test whether lectins can be used for detection of determinants related to bronchial carcinomas. In detail these experiments give answer to the question whether:

Lectins, Vol. IV

1. The profile of lectin-binding glycoproteins differs between bronchial carcinoma cells and bronchial epithelium.
2. Carcinoma cells can be enriched by lectin-mediated panning.
3. Tumor-associated glycoproteins are shedded by bronchial carcinoma cells.

Material and Methods

The squamous bronchial carcinoma cell line (KNS 62) and the oat cell carcinoma cell line (SK-LC 17) were a kind gift from Dr. Fogh, Sloan Kettering Institut, N.Y. Cells from bronchial lavages were obtained by instillation of 50-100 ml 0.9% saline solution during bronchoscopy and subsequent density centrifugation. Cell lines, cells from bronchial lavages of patients with squamous cell carcinoma (n=7) and normal controls (n=9) as well as bronchial biopsies (n=2) were incubated in horseradish-peroxidase-(HRP)-coupled lectins (Peanut-Agglutinin,PNA; Wheat Germ-Agglutinin, WGA; Ulex-Europaeus-Agglutinin, UEA; Concanavalin A, ConA; Soy-Bean-Agglutinin, SBA; Ricinus-Communis-Agglutinin, RCA; Horseshoecrab-Lectin; Helix-Pomatia-Lectin, HPA; Lens-Culinaris-Lectin, LCA; Phaseolus-Vulgaris-Lectin, PHA; Medac, 0.5 mg/ml, 25´, RT), HRP-coupled carbohydrate residues (Lactosyl-, Fucosyl-, Chitobiosyl-residues, Medac, 0.5 mg/ml, 25´, RT) and OKT 9 (Ortho-mune, 1:10, 30´, RT, second antibody: goat-anti-mouse-IgG-HRP, Medac, 1:10, 30´, RT) and in 3´3-Diaminobenzidine (Serva, 15´, 4°). Specimens were then prepared for electron microscopy as decribed elsewhere (9). Controls were performed using additional inhibitory sugars specific for the lectins used (Medac, 0.1 mmol). In experiments with OKT 9 normal mouse serum was used instead of the monoclonal antibody.

Lectin mediated panning. Cells from the squamous carcinoma cell line KNS 62 and bronchial lavages were used. Panning was performed as described elsewhere (10). Briefly petri dishes (Ø=5.5 cm) were coated using 4 ml of a 0.001% SBA or UEA solution. Cells were incubated for 1h at room temperature. Adherent cells were recovered using a solution containing N-Acetyl-Galactosamine or L-Fucose (0.2 mmol, 10´, 4°C) and subsequently counted. Controls were performed assessing adhesion of cells in the presencs of the appropriate inhibitory sugar. Coating was controlled by incubation of Petri dishes coated with lectin/HRP in 3´3-Diaminobenzidine, resulting in a reduced translucence in case of effective binding.

Enzyme-linked-lectin-assays. Supernatants of the KNS 62 squamous cell line, the SK-LC 17 oat cell line and bronchial lavages were analysed using lectin-coated microplates (UEA, SBA, PNA, WGA, 100 µl, 0.01% lectin/per well). Surfaces were blocked using bovine albumine (Miles Sc., 400 µl, 1%, 1h, 37°C, 1h, 4°C). Wells were then incubated in HRP-coupled-lectin solutions (100 µl, 0.01%, RT) and O-Phenylenediamino (Sigma, 45', 37°C) subsequently. Extinction was measured at 492nm (Multiscan Photometer). Adjustment was performed using appropriate polycarbohydrate complexes (Medac). Controls were carried out analysing samples in the presence of inhibitory carbohydrates.

Results

Ultrastructural analysis of cell lines. Incubation of cells of the oat-cell carcinoma cell line with lectins and macromolecule-coupled carbohydrates revealed that between 81% and 100% of the carcinoma cells exhibit binding sites for UEA, WGA, PNA, RCA, SBA and LCA (Fig.1). All other lectins and all carbohydrates used, including ConA and PHA, do not bind to the cell surface of oat-cell carcinoma cells. OKT 9 reacts with

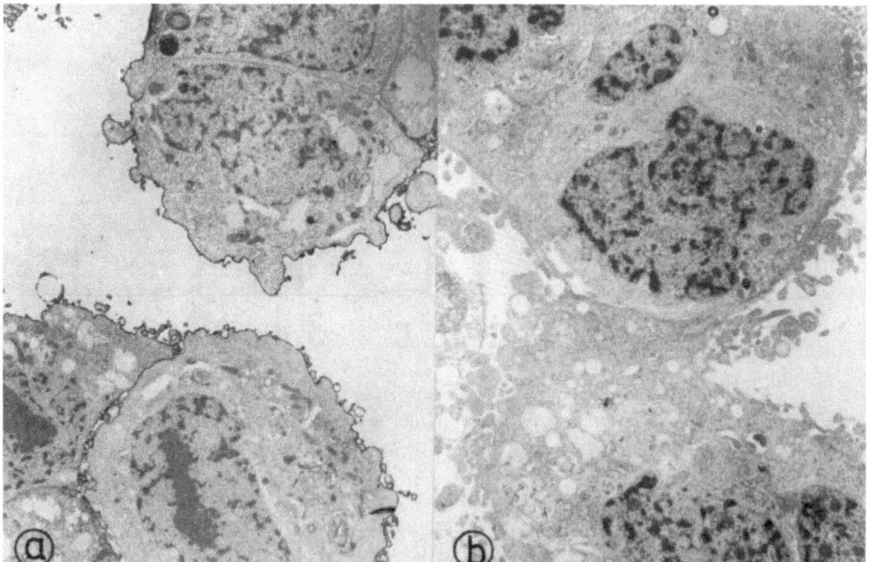

Fig. 1.: Oat cell carcinoma cells incubated in UEA (a, x 2700) and Con A (b, x 4160). The dark rim on the cell surface indicates lectin binding sites for UEA (a), whereas no binding is detectable in case of Con A (b).

cell membrane associated determinants only in 31% of SK-LC 17 cells. Most interestingly binding of lectins is significantly reduced on exposed cell membranes of aggregated cells compared to single cells.

Squamous carcinoma cells of the cell line KNS 62 showed a very similar labelling pattern. 71% to 100% of the cells reacted with UEA, WGA, RCA and SBA. There were no binding sites detectable for LCA, PNA, ConA and PHA.

Controls revealed that labelling of cell membranes was not detectable in the presence of inhibitory sugars. In experiments with WGA and N-Acetyl-Glucosamine we observed some binding of WGA which was inhibited by further addition of sialic acid. Controls using normal mouse serum instead of OKT 9 showed

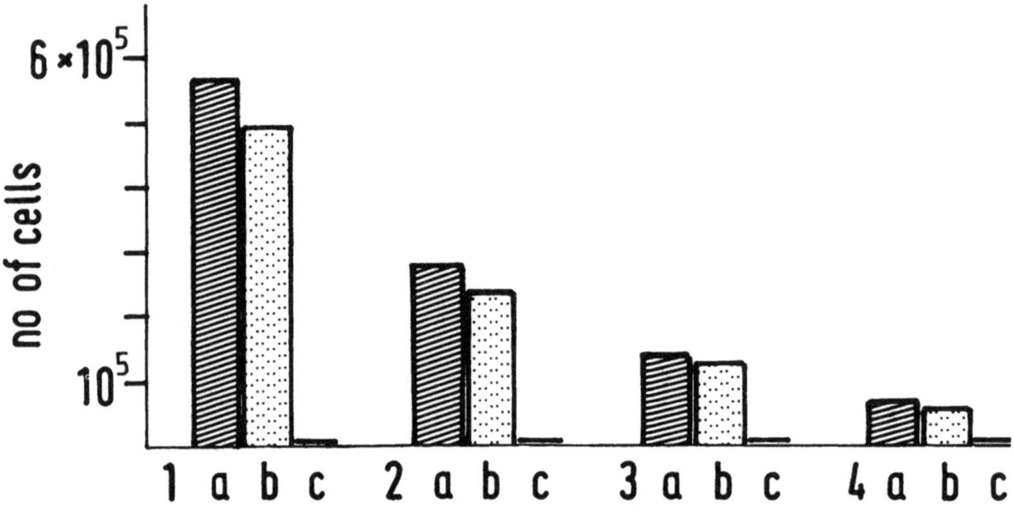

Fig. 2.: Number of adherent squamous carcinoma cells (KNS 62) after SBS and UEA panning. Almost 100% of the tumor cells can be recovered by SBA panning if no more than 2 x 10^5 cells were used. No adherence of tumor cells could be achieved if UEA was used for panning. (1a = 57 x 10^4 cells seaded, b = harvested after SBA panning, c = harvested after UEA panning; 2a = 28.5 x 10^4 cells seaded, b = harvested after SBA panning, c = harvested after UEA panning; 3a = 14.25 x 10^4 cells seaded, b = harvested after SBA panning, c = harvested after UEA panning; 4a = 7.125 x 10^4 cells seaded, b = harvested after SBA panning, c = harvested after UEA panning).

that an unspecitic reaction, e.g. via Fc-receptors, was not responsiblefor the labelling.
Ultrastructural analysis of bronchial lavage cells and bronchial biopsies. Analysis of bronchial lavage cells revealed a profile of lectin binding sites of oat cell and squamous carcinoma cells that is identical to that found on the cell surface of the cell lines examined. However, other cell types present in the lavage fluid i.e. lymphocytes, monocytes, macrophages and tissue fragments as well as parts of the basal membrane exhibit binding sites for lectins too. In contrast normal bronchial epithelial cells do not bind lectins in demonstrable amounts under the given conditions.

Lectin-mediated panning of squamous carcinoma cells. As shown in Fig.2 carcinoma cells adhere to SBA coated surfaces almost quantitatively, if the relation between the adherent cells and the surface used for panning is optimalized i.e. if the capacity of the latter is not exceeded. In contrast adhesion was not seen using UEA.

Lectin-mediated panning of squamous carcinoma cells in the presence of normal bronchial lavage cells. Lectin-mediated panning with SBA led to a quantitative recovery of carcinoma cells if the number was sufficiently low (Fig.3). It was found that through SBA panning, carcinoma cells could be enriched significantly. Varation of the number of carcinoma cells or the number of bronchial lavage cells showed that enrichment of the former is most impressive if low numbers of carcinoma cells are present. However carcinoma cells did not bind selectively and normal bronchial lavage cells could be found among the adherent cells, too (Fig.4).

Analysis of supernatants of bronchial carcinoma cells by an enzyme linked-lectin-assay. A panel of lectins have been tested to assess glycoconjugates in the supernatants of bronchial carcinoma cell lines and bronchial lavages of patients suffering from bronchial carcinoma. Although in control experiments the enzyme-linked lectin assay has proven to render reliable data for as little as 1-5 μg lectin or carbohydrate /100 μl, no increase in the concentration of glycoconjugates could be observed in the supernatants of carcinoma cells compared to the controls. However even the latter are characterized by a high backround activity, presumably due to the presence of lectin-binding glycoconjugates in the medium and bronchial fluid.

Discussion

Beside the significance of lectin binding sites for the understanding of carcinogenesis it is tempting to use these determinants for diagnostic purposis, in particular in case of bronchial carcinoma where the diagnosis is mostly achieved too late for curative therapy. On the other hand bronchial cells for cytological examination are easily won, e.g. through bronchial lavage.

The experiments presented in this study have shown that cytological analysis could probably be improved by introduction of lectins linked to an appropriate marker. They could help to differentiate between bronchial carcinoma cells and normal epithelium and moreover between squamous cell carcinomas and oat cell carcinomas. However lectins could not substitute light microscopical analysis since lectin binding sites are expessed also on other cell types such as lymphocytes and monocytes/macrophages that are present in the bronchial lavages.

A further approach to improve the diagnosis of bronchial carcinoma may be achieved by lectin-mediated panning. It was found that low numbers of carcinoma cells could significantly be enriched through panning procedures. Bronchial lavage cells adhere less frequently to the coated surface via spe-

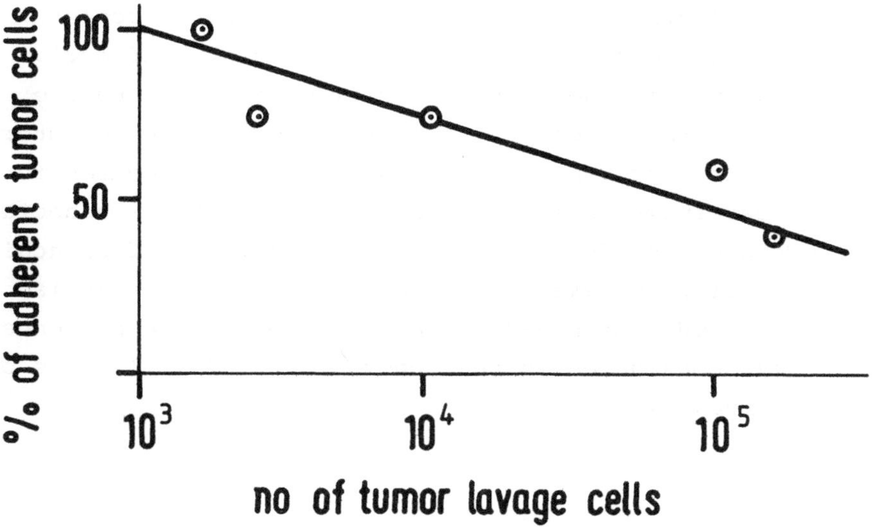

Fig. 3. SBA-mediated panning of squamous carcinoma cells (KNS 62) added to normal bronchial lavage cells. Increasing number of squamous carcinoma cells (x-axis) and constant number of normal bronchial lavage cells (10^6). If low numbers of carcinoma cells were added, complete recovery of carcinoma cells (ordinate) was achieved.

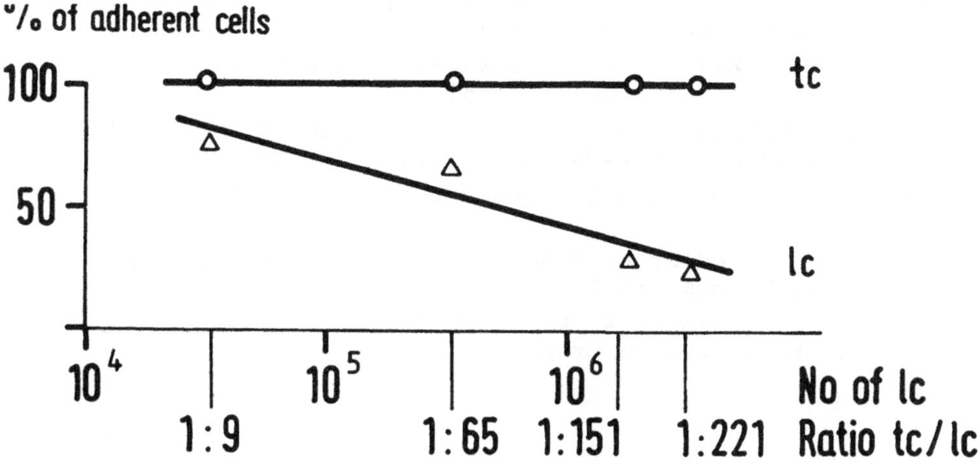

Fig. 4. SBA-mediated panning of a mixture of a constant number of squamous carcinoma cells (tc = KNS 62, 5 x 10^3) and an increasing number of normal bronchial lavage cells (lc). Reduced binding of lavage cells and complete adherence of carcinoma cells at high levels of cells.

cific (or unspecific) binding. Panning was not successful using UEA. This was shown not to be caused by ineffective coating, but may be due to peculiarities in regard to the distribution, exposition or affinity of the lectin receptor. A considerable improvement in the diagnostic procedures for bronchial carcinoma would be an assessment to tumor associated determinants shedded from the cell membranes of bronchial carcinoma cells. However up to now we failed to establish an enzyme-linked lectin assay that could prove the presence of such glycoconjugates in the supernatants of cell lines or bronchial lavages.

References

1. Burger MM (1969), Proc Natl Acad Sci USA 62:994
2. Inbar M, Sachs L (1969), Proc Natl Acad Sci USA 63:1418
3. Vaheri A, Mosher DF (1978), Biochem biophys Acta 516:1
4. Raedler A, Boehle A, Otto U, Raedler E (1982 b) J Urol 128:1109
5. Raedler A, Schmiegel WH, Raedler E, Arndt R, Thiele HG (1983 a), Exp Cell Biol 51:19
6. Miyauchi T, Muramatsu H, Ozawa M, Mizuta T, Suzuki T, Muramatsu T (1982), GANN 73:581
7. Miyauchi T, Yonezawa S, Takamura T, Chiba T, Tejima S, Ozawa M, Sato E, Muramatsu T (1982), Nature 229:168
8. Yonezawa S, Nakamura T, Tanaka S, Sato E (1982) JNCI 69:777
9. Raedler A, Raedler E, Scholz KU, Arndt R, Thiele HG (1983 b), Thymus 5:311
10. Raedler A, Fraenkel S, Klose G, Seyfarth K, Thiele HG Gastroenterology, in press

DISTRIBUTION OF GLYCOCONJUGATES IN NORMAL HUMAN TESTIS AND IN SEMINOMA

Raija Malmi, Karl-Ove Söderström

Departments of Anatomy and Pathology, University of Turku,
SF-20520 Turku 52, Finland

The glycoconjugates are found as structural elements of cell surfaces and intracellularly. Although sugars comprise only a small proportion of the weight of the cellular membrane, they are believed to be essential for cellular function. The changes in the carbohydrate chains of the glycocompounds are associated with cellular differentiation and malignant transformation (1).

Several studies on the interaction of lectins with sperm from various animal species have been conducted. Such studies have confirmed the segregation of surface molecules to specific morphological regions of the sperm and provided markers to follow developmental and maturational changes (2,3).

Changes that take place during differentiation of spermatogenic cells are less known. The early germinal cells, the spermatogonia of the seminiferous epithelium in adult animals proliferate, giving rise to cells which differentiate into spermatozoa. During the developmental process the germinal cells undergo two maturational divisions and transformation into spermatozoa which are secreted into the tubular lumen (4).

Molecular events associated with spermatogenesis are not well known. To define more precisely those alterations in cellular glycoconjugates that accompany germ cell differentiation, we

Lectins, Vol. IV

treated sections of human testis with lectins. In order to analyze the nature of glycoconjugates in malignant tissue, a seminoma was included in the study.

Materials and Methods

Both fluorescein isothiocyanate (FITC) and tetramethylrhodamine isothiocyanate (TRITC) conjugated lectins were purchased from Vector Laboratories (Burlingame, CA). The following lectins were used: Concanavalin A, Con A (D-Man, D-Glc); Lens culinaris, LCA (D-Man, D-Glc); Triticum vulgaris, WGA (D-GlcNAc-(1→4)-β-GlcNAc); Ricinus communis I, RCA I (D-Gal); Arachis hypogaea, PNA (D-Gal-β-(1→3)-D-GalNAc); Glycine maximum, SBA (D-GalNAc); Helix pomatia, HPA (D-GalNAc) and Ulex europaeus I, UEA I (L-Fuc). The terminal hexoses were purchased from Sigma chemical Co. (St. Louis, MO) and used in 0.2 M concentrations for the inhibition of lectin binding.

Normal human testis tissue was obtained at the Department of Forensic Medicine (University of Turku, Finland). Seminoma samples were obtained at the Department of Surgery (University Central Hospital, Turku, Finland). Histological study by routine stainings presented normal spermatogenesis. In a case of a seminoma, a fairly anaplastic variant of the tumour with degenerative regions and mitoses was revealed. Both tissue samples were fixed in Bouin's fluid and embedded in paraffin.

After deparaffination tissue sections were thoroughly washed in phosphate buffered saline, PBS, pH 7.2. The lectins were diluted in PBS supplemented with Ca^{++} and Mg^{++} (Dulbecco's solution, Orion, Finland) to contain 100 ug - 200 ug/ml, incubated on the sections for 30 min and washed thoroughly in PBS and distilled water. The specimens were mounted with Aquamount (Gurr, BDH Chemicals, England). Control sections were incubated with lectins that had been inactivated by their specific sugar inhibitors.

The slides were examined on a Leitz Dialux 20 fluorescent microscope equipped for epi-illumination and appropriate filter combinations for FITC and TRITC fluorescence. The specificity of stainings with each lectin was ascertained by comparing the intensity of fluorescence on sections stained with lectin plus sugar versus lectin alone.

A small piece (2x2x2 mm) of the seminoma tissue was prepared for electron microscopy. The tissue was fixed in 3 % glutaraldehyde in 0.1 M phosphate buffer (pH 7.4), postfixed in 1 % osmium tetroxide, dehydrated and embedded in Epon. Sections 70 nm thick were cut and stained with uranyl acetate and lead citrate and examined in a Jeol Tem 100 C electron microscope.

Results

Lectin binding during germ cell differentiation. On sections of Bouin-fixed human testis each lectin presented a characteristic binding pattern. LCA and Con A were bound to the cytoplasm of all germ cells, spermatid acrosomes, tubular walls and Leydig cells in interstitial tissue. A similar binding was observed with WGA. Differences in the binding intensity of the three lectins were noticed so that staining of the tubular wall by Con A was less pronounced (Fig. 1).
RCA I expressed a faint staining of germ cells and tubular walls and a moderate reaction in the cytoplasm of interstitial cells. However, the most intense fluorescence was observed in developing spermatids in their acrosomal regions (Fig. 2 a).
Restricted intratubular labeling by PNA was observed. Staining was faint in early germ cells and the main binding sites were located in spermatid acrosomes which were brightly fluorescent (Fig. 2 b). A similar, but less intense acrosomal staining was given by SBA (Fig. 2 c) and even a more selective staining of only some spermatid acrosomes was noticed by HPA (Fig. 2 d).
A particularly bright fluorescence of blood vessel walls was characteristic to UEA I labeling (Fig. 4 a).

Lectin binding in seminoma. In the seminoma Con A stained the cytoplasm of malignant cells in a granular manner (Fig. 3 b). The cytoplasm of seminoma cells was also brightly fluorescent with LCA and WGA though less granular (Fig. 3 c,d). RCA I was faintly bound to the cell surfaces of the malignant cells. Lymphocytes, the surroundings of blood vessels and degenerative regions of the malignant tissue were strongly reactive with these lectins.
PNA, SBA and HPA did not stain the seminoma tissue at all. UEA I labeled selectively vascular walls (Fig. 4 b).

Electron microscopy of seminoma. The electron microscopic exam-

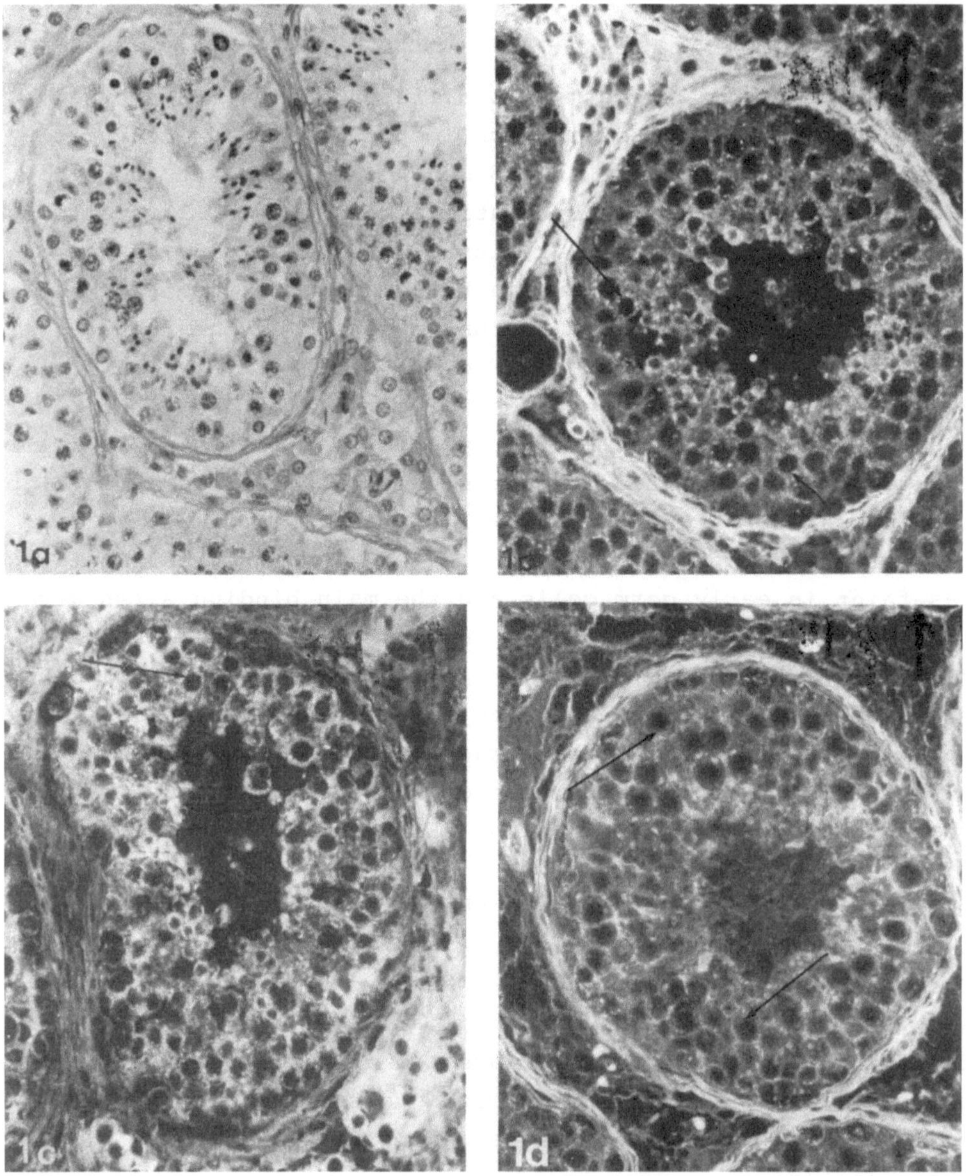

Fig. 1. (a) Histological section of human testis, showing seminiferous tubules with normal spermatogenesis (Hematoxylineosin staining). (b) TRITC-LCA, (c) FITC-Con A and (d) FITC-WGA stained the cytoplasm of all germ cells (arrows) and acrosomes of spermatids. The tubular walls and interstitial cells were also fluorescent. Magnifications: a x 175; b,c,d x 220.

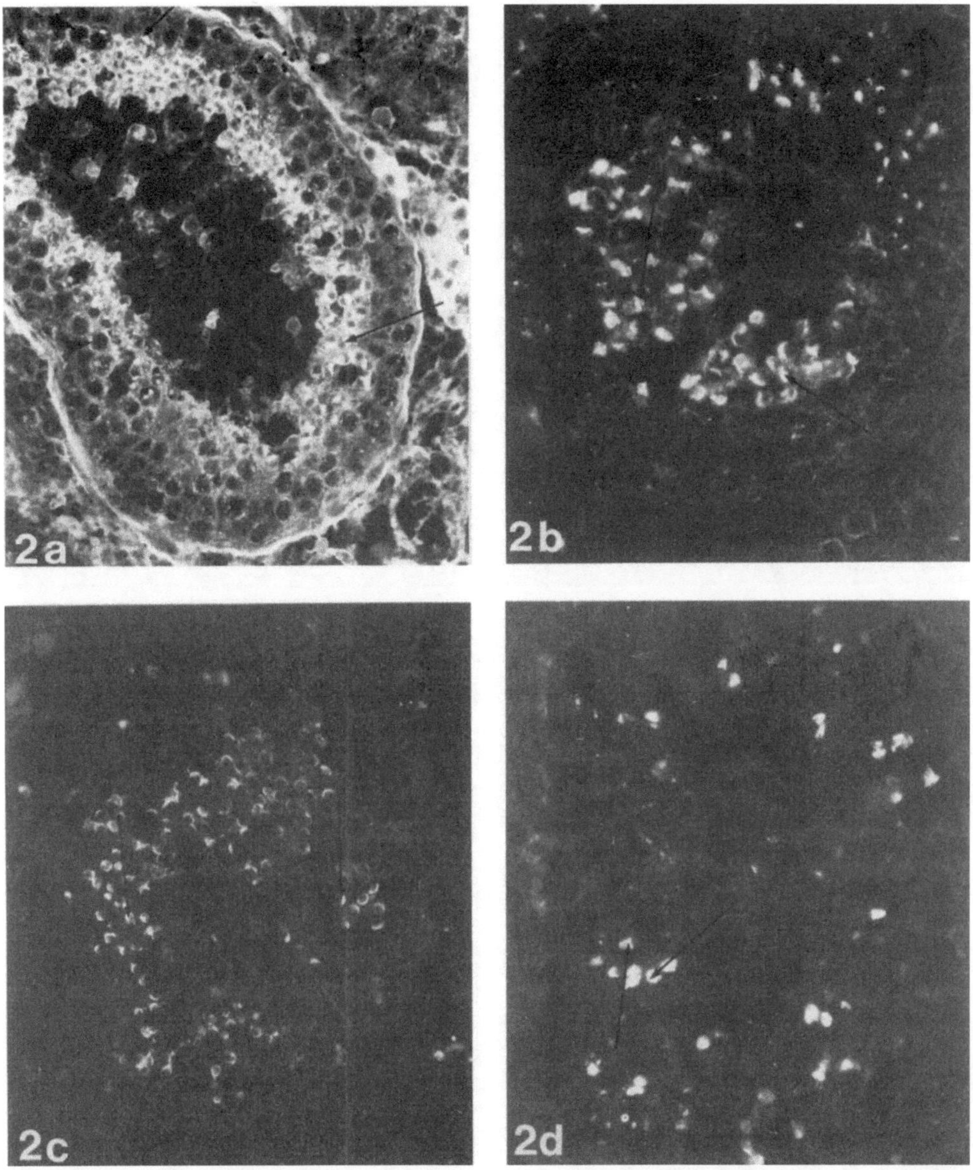

Fig. 2. (a) TRITC-RCA I bound to germ cell surfaces (arrowheads) and particularly to developing acrosomes (arrows), as well as to the tubular wall. (b) The main binding sites of FITC-PNA were in spermatid acrosomes (arrows) and (c) a similar but less intense acrosomal reaction was observed by FITC-SBA. (d) A selective staining of late spermatid acrosomes (arrows) was given by FITC-HPA. Magnifications: a,c x 220; b x 280; d x 380.

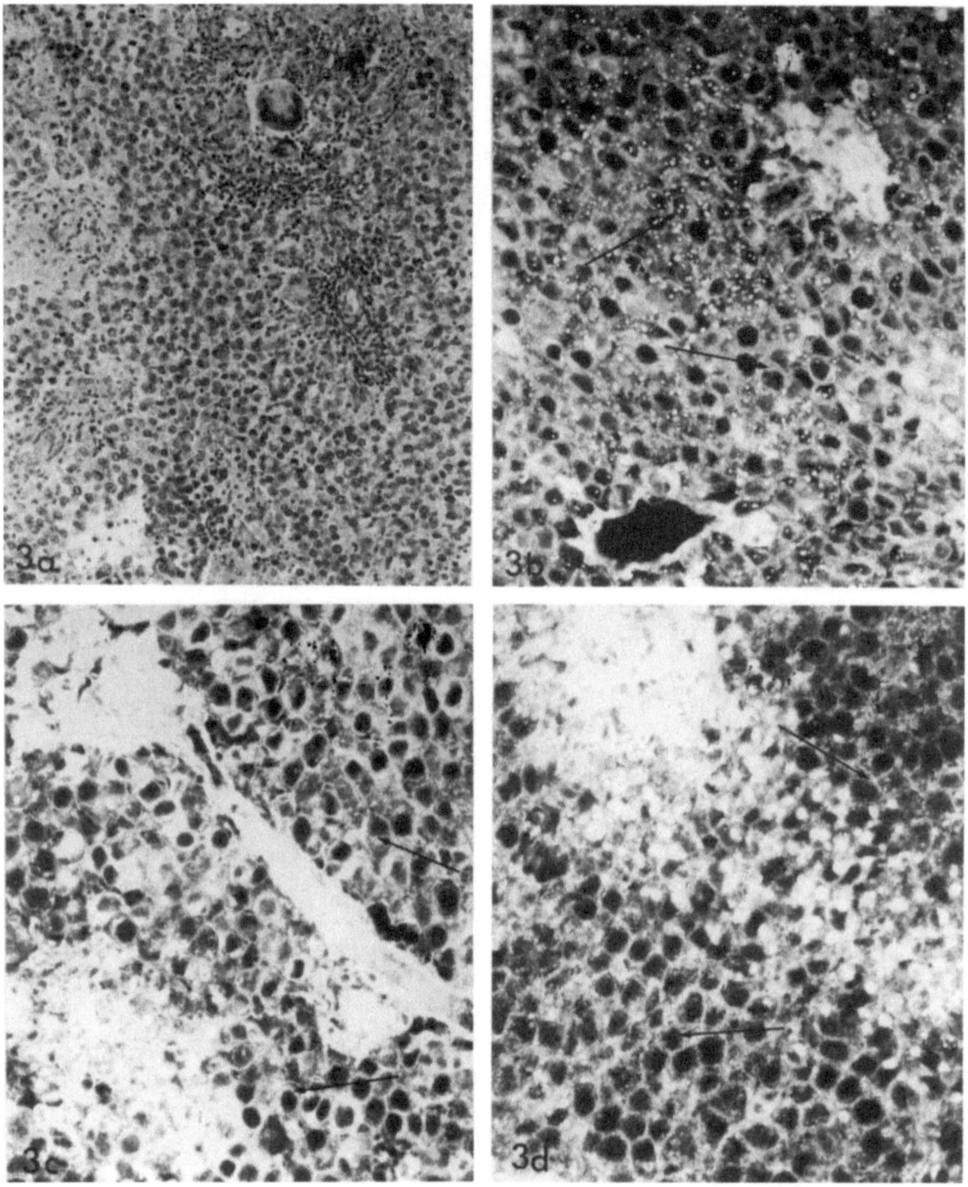

Fig. 3. (a) Histological section of seminoma (Weigert-van Gieson staining). (b) FITC-Con A stained the cytoplasm of all seminoma cells in a granular manner (arrows). (c) TRITC-LCA and (d) FITC-WGA bound also to the cytoplasm of malignant cells (arrows). Magnifications: a x 90; b x 175; c,d x 220.

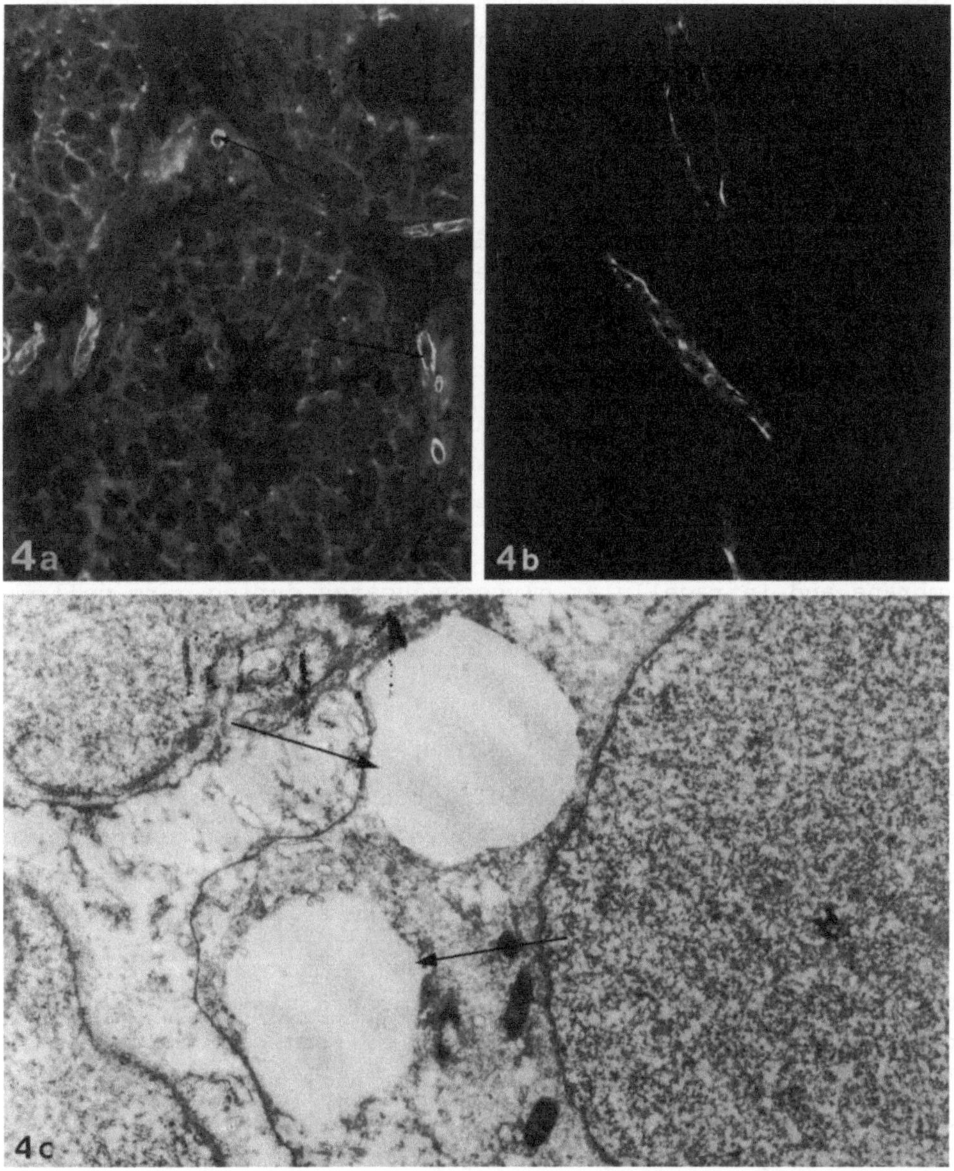

Fig. 4. FITC-UEA I was bound especially to the vascular walls (arrows) of human testis (a) and to those of the seminoma tissue (b). (c) An electron micrograph of seminoma showing cells with degenerative granules (arrows) in their cytoplasm. Magnifications: a,b x 220; c x 6000.

ination showed a typical seminoma with degenerative vacuoles in the cytoplasm of the malignant cells (Fig. 4 c).

Discussion

Several recent articles have been published on lectins as probes for the histochemical demonstration of glycoconjugates both in frozen and paraffin embedded sections of various tissues (5, 6, 7, 8, 9). Since paraffin embedding may be of value when considering the possible use of lectins in histopathological diagnosis, we have used paraffin sections in our study. Instead of formalin we have used Bouin's fluid as a fixative, as it caused enhancement of lectin fluorescence (9).

Our results show that there are considerable alterations in membrane and intracellular glycoconjugates during germ cell differentiation. The differences in carbohydrate composition in each region may reflect the function of the respective structure of human testis.

Con A and LCA which bind to mannosyl groups of the core region in the carbohydrate chains labeled the cytoplasm of both germ cells and Leydig cells. Either these cells may express a large amount of sugars detectable by the two lectins, or less numerous glycoconjugate molecules may expose several sugar residues detected by the two lectins used.

WGA interacts most strongly with oligomers of β-N-acetyl-glucosamine and its binding is enhanced by terminal sialic acid (10). Both interstitial and germ cells were shown to be rich in glycoconjugates detectable by this lectin. The same cells bound also RCA I, specifically bound by terminal galactosyl residues, but the reaction was less prominent.

Restricted binding was given by PNA, SBA and HPA. Though PNA receptors were faintly expressed already in early germ cells, the main reactive sites were localized in spermatid acrosomes.

Galactose specific PNA binds avidly D-Gal-β-(1→3)-D-GalNAc. Soybean lectin binds preferentially N-acetyl-D-galactosamine and less intensely D-galactose (10). The SBA reactive sites were also distributed in the acrosomes of spermatids. The similarities between the binding patterns of RCA I, PNA and SBA lectins may reflect the similar distribution of D-galactose and N-acetyl-galactosamine or a partial overlap of the lectin specificities. The most specific acrosomal reaction was observed by HPA which is specific for N-acetyl-galactosamine (10).

The acrosome is a highly modified lysosome-like structure and is known to contain several enzymes which are glycoproteins. Sperm capacitation and acrosome activation may be associated with the acrosomal polysaccharides. The periodic acid-Schiff reagent (PAS) has been used for visualization of these polysaccharides (11). The present work revealed a more detailed analysis of the carbohydrate composition of acrosomal substances which are abundant in terminal nonreducing N-acetyl-galactosamine and D-Gal-β-(1→3)-D-GalNAc, as judged from the specific labeling with PNA, SBA and HPA.

The tubular wall consists of basal lamina and connective tissue fibers, together with peritubular myoid cells. The basement membranes contain glycoproteins and proteoglycans. Our results suggest that in the tubular wall there are mannose containing glycoconjugates detectable by Con A and LCA, as well as some terminal nonreducing galactosyl residues detectable by RCA I, and sialic acid and N-acetyl-glucosamine residues specifically bound by WGA.

The seminoma tissue was positively stained by Con A, LCA and WGA and less prominently also by RCA I. The highly granular intracytoplasmic fluorescence observed with Con A may correspond to the degenerative granules seen in electron micrographs of the seminoma cells. The glycoconjugates of germ cells observed during normal spermatogenesis by these lectins may also be components of the malignant cells, or seminoma cells may synthesize new compounds of the same lectin specificity.The lack of PNA, SBA and HPA binding sites in the seminoma speaks in favor for the histogenetic origin of the tumour from early differentiation stages where normally no significant expression of these glycosubstances occurs.

The fucose specific UEA I bound to vascular walls both in normal and malignant testis tissue, and could be used as a marker for vascular endothelia as has been previously shown in other tissues (6).

In conclusion,alterations in cellular glycoconjugates during normal germ cell differentiation and in a seminoma have been described. Whether lectins will prove to be useful probes in the histopathological study of seminomas and other testicular disorders remains to be determined.

References

1. Nicolson, G.L., Robbins, J.C., Winkelhake, J.L.: Cellular Membranes and Tumor Cell Behavior, Williams and Wilkins, Baltimore (1975)
2. Koehler, J.K.: Arch.Androl. 6, 197 (1981).
3. Naaby-Hansen, S., Lihme, A., Larsen, K.P., Bøg-Hansen, T.C., Bjerrum, O.J.: The abstract book, Interlec 6, 78 (1984)
4. Leblond, C.P., Clermont, Y.: Ann.N.Y. Acad. Sci. 55, 548-573 (1952).
5. Faraggiana, T., Malchiodi, F., Prado, A., Churg,J.; J Histochem. Cytochem. 30, 451 (1982).
6. Holthöfer, H., Virtanen, I., Pettersson, E., Törnroth, T., Alfthan, O., Linder, E.: Lab. Invest. 45, 391 (1981).
7. Murata, F., Tsuyama, S., Suzuki, S., Hamada, H., Pzawa, M., Muramatsu, T.: J Histochem. Cytochem. 31, 139 (1983)
8. Nemanic, M.K., Whitehead, J.S., Elias, P.M.: J Histochem. Cytohem. 31, 887 (1983)
9. Söderström, K.O., Malmi, R., Karjalainen, K.: Histochemistry 80,475-579 (1984).
10. Goldstein, I.J.: Adv. Carbohydr. Chem. Biochem.,35,127(1978).
11. Clermont, Y.,Leblond, C.P.: Am.J. Anat. 96, 229 (1955).

DIAGNOSIS OF SOFT TISSUE TUMORS. ULEX EUROPAEUS LECTIN I IS A MARKER FOR BIPHASIC SYNOVIAL SARCOMA AND MESOTHELIOMA

Bettina Borisch
Institut für Pathologie, Abteilung Paidopathologie, der Universität Kiel, D-2300 Kiel

Udo Schumacher und Dieter Harms
Institut für Pathologie der Universität Kiel, D-2300 Kiel

Soft tissue can be defined as nonepithelial extraskeletal tissue of the body exclusive of the reticuloendothelial system, glia and supporting tissue of various parenchymal organs. By convention it also includes the peripheral nervous system because tumors arising from nerves appear as soft tissue masses and pose similar problems in diffential diagnosis and therapy. Soft tissue tumors are classified on a histogenetic basis according to the adult tissue they resemble (6). On a routine basis with good hematoxylin-eosin-stained sections from a well-sampled lesion only 15 - 20 % of soft tissue tumors remain not or not exactly classified according to their histogenesis (8, 16). Special stains and electron microscopy should be used in close conjunction to the routine methods. Recently immunohistochemical staining of soft tissue tumors for a variety of substances i. e. Myoglobin (4) and factor VIII-related antigen (2) using the peroxidase-antiperoxidase procedure of Sternberger (17) has been reported. Polyclonal antibodies and even monoclonal antibodies have been widely used. In order to visualize cell surface

Lectins, Vol. IV

glycoconjugates, lectins have been introduced into histochemistry both for light and electron microscopy (14). The lectin of Ulex europaeus I (UEA) has become a diagnostical aid for all tumors and and tumor like lesions of blood vessels by its capacity to bind to human endothelial cells (9). During our work on tumors of vascular origin we found a positivity for UEA I in endothelial cells but also in epithelial cells of biphasic synovial sarcoma. This finding prompted us to study tumors of serous membranes as well as tumors of epithelial origin in regard to their binding capacity for UEA I. In the present study we investigated 84 cases of soft tissue tumors for their binding capacity to the lectin UEA I.

Material and methods

Biopsy material of 33 tumors and tumorlike lesions of blood vessels, 22 synovial sarcomas, 21 mesotheliomas, 2 malignant schwannomas, 2 neuroblastomas, 2 rhabdomysarcomas was stained using a modified peroxidase-antiperoxidase method (1). After blocking the endogenous peroxidase first buffer-rinsing was followed an incubation with commercially available UEA I (E. Y. Lab. Inc., San Mateo, CA, USA, in West-Germany obtained from medac) at a dilution of 1:50 in buffer. Rabbit-anti-UEA I (also from medac) was used as a primary antibody, a swine-anti-rabbit and a rabbit PAP-complex (both from DAKO patts, Denmark) were used as linking and labelling antibody, respectively. Peroxidase reaction was visualized with 3,3-diaminobenzidine as chromogen. As controls 9 adenocarcinomas of colon and endometrium as well as 4 carcinomas of the breast were also included in this study. Lectin UEA I controls of the staining method were made by adding the competitive sugar L-fucose (0.1 M) to the lectin solution. All controls showed inhibition of specific lectin-staining.

Results

All endothelial cells in neoplastic and non-neoplastic tissues showed a strong diffuse homogenous intracytoplasmic positivity for UEA I. This staining pattern will be called the 'vascular pattern' (Fig. 1). Similarly to the 'vascular pattern' also a diffuse homogenous intracytoplasmic staining but with weaker intensity was found in epithelial components of biphasic synovial sarcoma and in epithelial components of mesotheliomas. The epithelial cells in biphasic synovial sarcoma, which are characterized by large, round or oval, vesicular nuclei and abundant pale staining cytoplasm with distinctly outlined cellular borders, are stained as described throughout the whole tumor (Fig. 2). So the cleft-like spaces lined by epithelial cells could be distinguished from cleft-like artifacts that are a result of shrinkage. The surrounding spindle cell or fibrous component doesn't show any staining for UEA I at all. However, with this staining method even very small amounts of epithelial differentiation in an otherwise monophasic fibrous synovial sarcoma can be detected (Fig. 2).

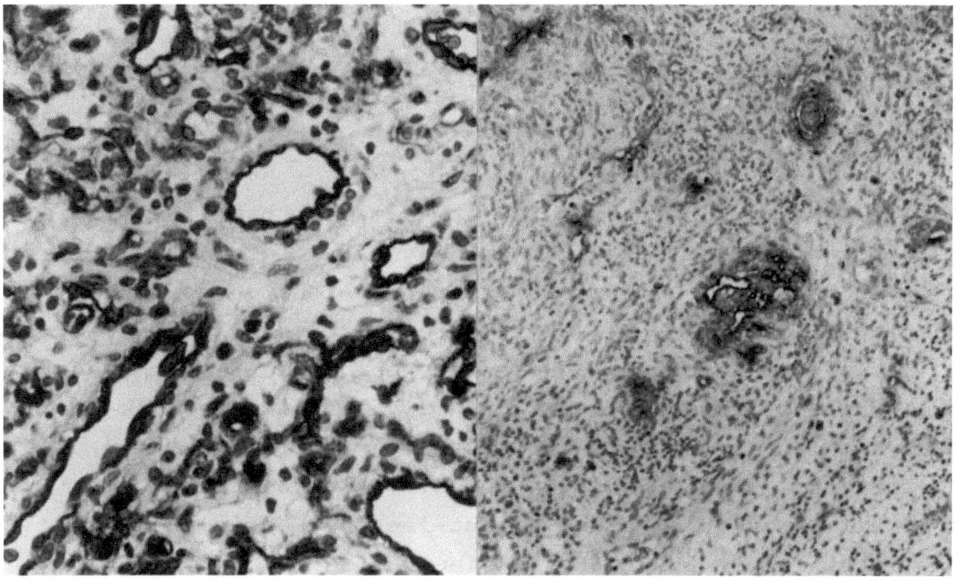

Fig. 1: Benign hemangioendothelioma, endothelial cells intensively stained with UEA I, "vascular pattern", 280 x.

Fig. 2. Small area of epithelial differentiation in a synovial sarcoma detected with UEA I. Epithelial cells show a weak, homogenous, diffuse intracytoplasmic staining for UEA I, 112 x.

As to the mesotheliomas most of them were of the epithelial type and could be readily identified by their characteristic tubulopapillary pattern (Fig. 3). This pattern consist of glandlike tubular or cystic spaces lined by cuboidal or flattened epithelial-like cells with uniform vesicular nuclei, one or two inconspicous nucleoli, and abundant eosinophilic cytoplasm with distinct cytoplasmic borders (Fig. 3). Like in biphasic synovial sarcoma these epithelial-like cells show a distinct homogenous intracytoplasmic UEA I-staining. Distinction of mesothelioma from primary or metastatic epithelial neoplasms may cause considerable difficulty. That is why we also investigated adenocarcinomas and breast carcinomas by means of UEA I-staining. Adenocarcinomas showed a quite different pattern: a granular intracytoplasmic positivity with a scattered distribution, compare Figs.3 and 4. In carcinomas of the breast the intensity of UEA I-staining also differed from cell to cell.

Discussion

Our results with UEA I demonstrate clearly that a lectin can be helpful in soft tissue tumor diagnosis. Lectins are widely used in the field of histochemistry. There are reports on lectin binding in malignant transformed tissue (5, 15, 12, 13, 10,7)as well as in normal human tissue (3). Most of these tumors are of epithelial origin, only very little work has been done on soft tissue tumors. Recently MIETTINEN and co-workers (11) demonstrated the biphasic character of synovial sarcomas by using peanut-lectin. Lectin-histochemistry is perhaps not a routine method but it provides further information as to the histogenesis of the tumors or tissues under examination. We did our work with UEA I; MIETTINEN and co-workers used peanut-lectin; other lectins well give new results. With a collection of the data on soft tissue we will come to a better understanding of the development of tumors arising from soft tissue. We also found a positivity for UEA I in tumors of epithelial origin. But the distribution pattern was different (Fig. 4). So the differential diagnosis between biphasic tumors and adenocarcinomas is possible even in difficult cases.

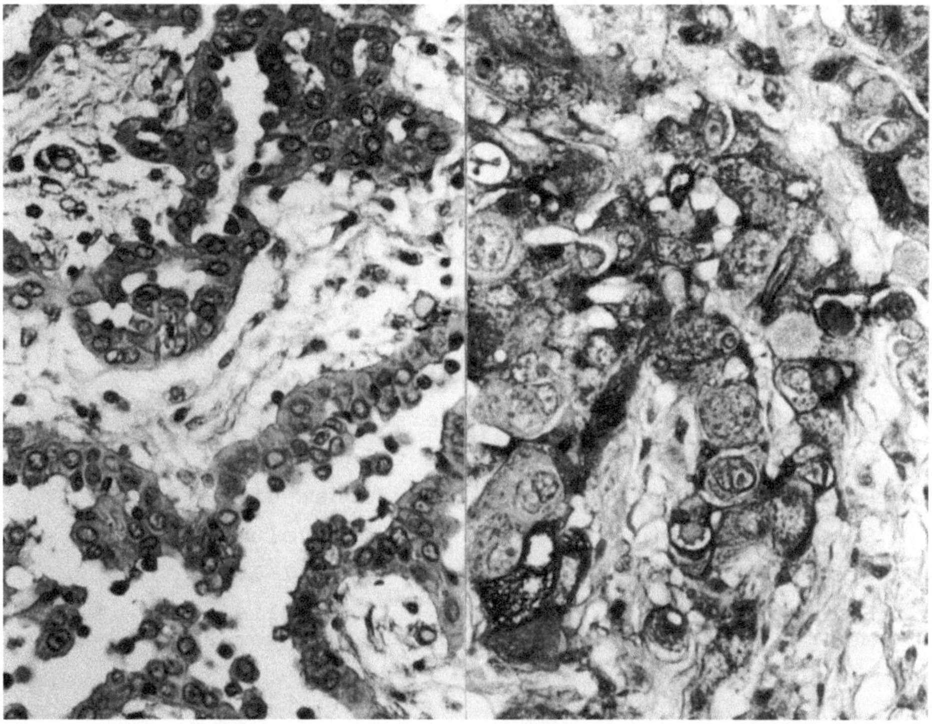

Fig. 3: Mesothelioma of peritoneum, characteristic tubulopapillary pattern; all epithelial cells of the tumor show a weak, homogenous, diffuse intracytoplasmic staining for UEA I, 280 x.

Fig. 4: Adenocarcinoma of the colon: granular, intracytoplasmic UEA I positively staining, intensity of the staining differs from cell to cell, 280 x.

Present adress: Bettina Borisch; Institut für Anatomie, Medizinische Hochschule Lübeck, Ratzeburger Allee 160, D-2400 Lübeck.

REFERENCES

1. Borisch, B., Möller, P., Harms, D. (1983). Pathologe 4, 241-243.
2. Burgdorf, W.H.C., Mukai, K., Rosai, J. (1981). Am J Clin Pathol 76, 197-200.
3. Born, I. A., et al., see this volume.
4. Corson, J. M., Pinkus, G. S. (1981). Am J Pathol 103, 384-389.
5. Dus, D., et al., see this volume.
6. Enzinger, F. M., Weiss, S. W. (1983). Soft tissue tumors, The C. V. Mosby Company, St. Louis.
7. Fischer, J., et al., see this volume.
8. Harms, D. (1982). Klin Pädiat 194, 275-280.
9. Hormia, M. (1982). Cell Biol Int Rep 6, 1123-1134.
10. Malmi, R., et al., see this volume.
11. Miettinen, M., Lehto, V.-P., Virtanen, I. (1983). Virchows Arch (Cell Pathol) 44, 187-199.
12. Orntoft, T. F. and S. E. Petersen, see this volume.
13. Raedler, A., et al., see this volume.
14. Roth, J. (1983). J Histochem Cytochem 31, 987-999.
15. Schwechheimer, K., et al., see this volume.
16. Stern, R. (1981). Human Pathol 12, 777-781.
17. Sternberger, L. A., Hardy, P. H., Cuculis, J. J., Meyer, H. G. (1970). J Histochem Cytochem 20, 829-831.

THE BINDING OF *DOLICHOS BIFLORUS* AGGLUTININ TO VASCULAR ENDOTHELIUM AND INTESTINAL EPITHELIUM OF MICE, A GENETIC DIMORPHISM.

H. Gepke Uiterdijk*, Bruce A.J. Ponder+

*Division of Tumour Biology, Subdivision of Genetics, Antoni van Leeuwenhoekhuis, Plesmanlaan 121, 1066 CX Amsterdam, Netherlands

Present address: Academic Hospital Leiden, Department of Rheumatology, geb 8, Rijnsburgerweg 10, 2333 AA Leiden, Netherlands

+Institute of Cancer Research, Haddow Laboratories, Clifton Avenue, Sutton, Surrey SM2 5PX, England

Dolichos biflorus agglutinin (DBA) is a plant lectin with specificity for terminal non-reducing N-acetyl galactosamine residues (1). In a search for polymorphisms suitable for use as histological markers for the study of mouse aggregation chimaeras, Ponder and Wilkinson (2) screened cryostat sections of tissues from 10 inbred strains of mice with a panel of 12 different lectin conjugates. They found only one polymorphism, for binding of DBA to vascular endothelium and intestinal epithelium. Localisation of binding sites to the endothelial cells was confirmed by electron microscopy (2). The application of this polymorphism to the study of chimaeric tissues has been reported elsewhere (2,3,4),

Here we review in more detail the binding pattern of DBA in mouse tissues and present new data which indicate that the polymorphism is due to alleles at a single locus which maps near *Re* on chromosome 11. The locus is named *Dlb*-1 (for

Lectins, Vol. IV

Dolichos lectin binding); the alleles are Dlb-1^{a} and Dlb-1^{b}(5).

Materials and Methods

Mice: The mice were bred and kept at the Institute of Cancer Research, (CBA/Ca), at the MRC Laboratories, Carshalton (DDK, RIII-ro, C57BL/6J, B10.A, C57BL/10ScSn and C57BL/6-re) and at the Mammalian Unit of the Antoni van Leeuwenhoek Research Laboratory, Amsterdam (BALB/cHeA, STS/A, GRS/A, LTS/Af, MAS, SM/Ja, 020/A, SJL/J, C57BL/10ByA and DBA/Af). All other mice from inbred strains screened were obtained from the MRC, Mill Hill, London. The mice used for screening for DBA binding patterns were aged from 3 weeks to 2 years : no age differences were detected. For the first single backcrosses 220 mice (110 ♀♀, 110♂♂) were used; for backcrosses to map the gene, 242 mice have been examined so far, aged 8-12 weeks. Patterns of DBA binding in embryos were examined in 4 C57BL/6J and 4 RIII-ro embryos at 7,8,9,11,13 and 18 days gestation.

Lectins: In addition to the 12 lectins used in the original screen for polymorphism (2), a further 11 lectins were screened to see if any either showed the same polymorphic binding as DBA, or a reciprocal polymorphic pattern. These lectins were those from : Jack fruit, Cotoneaster and *Mangifera indica* (peroxidase conjugates a gift from Dr. J.A. Forrester); *Limulus polyhemus* (LPA), *Wisteria floribunda* (WFA), *Helix pomatia* (HPA), *Vicia villosa* (VVA), *Codium fragile*, *Sofora japonica*, *Pisum sativum*, and *Maclura pomifera* (purified lectin or conjugates from Sigma). Peroxidase conjugates of DBA, and alkaline phosphatase conjugates of WFA, HPA, VVA, and *Codium fragile* a. were prepared at the Institute of Cancer Research. For methods, see ref (2).

Histology: Dissected tissues were prepared for cryostat sections by freezing on cork discs,or fixed in Carnoy's fixative or methacarn (Carnoy's with methanol substituted for ethanol). Fixed tissues were dehydrated and embedded in Histowax. Sections were cut at 5 µm. Frozen sections were fixed immediately before use in 10% formol saline (pH 7). Sections were incubated with lectin conjugates as previously described (2). Endogenous peroxidase was blocked with 0.1% phenylhydrazine HCl. After washing in PBS (phosphate buffered saline, pH 7.3), the sections were incubated with 100 µl each of lectin conjugate diluted in PBS containing 0.5% bovine serum albumin (BSA : Sigma). Dilutions were determined for each conjugate by experiment. Control sections were incubated in PBS-BSA without lectin conjugate or with the conjugate diluted in the presence of the appropriate inhibiting sugar (final concentration 2% W/V). Peroxidase was demonstrated using 3,3' diaminobenzidine as substrate, and alkaline phosphatase using Fast Red TR (2). Sections were counterstained with haemalum. Photographs were taken using Ilford Pan-F film and a blue-green filter (Ilford No. 307).

Genetics: We scored DBA binding to gut epithelium and vascular endothelium on sections of small intestine, adjacent mesentery and pancreas. The mesentery contains large vessels for scoring endothelial DBA binding, and the pancreatic tissue provided an internal control because pancreatic ducts stain with DBA peroxidase in all strains tested. The marker genes used for linkage were Hba, which induced differences in the tryptic peptides of the haemoglobin α-chain in mice and which was typed for by Dr. L. Skow, National Institutes of Health, Research Triangle Park, N. Carolina; Es-3, which is responsible for specific variations of an eserine-sensitive esterase in mouse erythrocytes and which was typed for by Dr. J. Hilgers, Antoni van Leeuwenhoek Research Laboratory, Amsterdam; and Re, which gives rise to a wavy coat and curly whiskers, which was typed for by Dr. M. Festing, MRC Experimental Embryology and Teratology Unit, Carshalton (6). Our preliminary mapping data were calculated according to Green (7). Full genetic results will be published elsewhere (8).

Results

The expression of DBA binding sites could be observed in fixed, paraffin embedded as well as in frozen tissues (cryostat sections). The optimum fixative was methacarn. Control sections (see Materials and Methods), and sections of known positive and negative tissues included in every experiment, always confirmed the specificity of the DBA-peroxidase binding. Of the 23 lectins screened, only DBA, HPA and WFA (all with specificity for terminal non-reducing N-acetyl galactosamine residues) showed the genetic dimorphism. Other lectins (VVA, Codium fragile, Soybean a.) with nominal N-acetyl galactosamine specificity did not show the dimorphism. No lectin gave a dimorphic pattern reciprocal to that seen with DBA.

Distribution of binding sites: All inbred strains of mice tested (Table 1) either showed DBA-HRP staining of vascular endothelium and not intestinal epithelium (E^+G^- phenotype), or expressed the reverse pattern (E^-G^+) (Fig. 1a,b). Lymphatic endothelium was negative in all mice (2).

In fetal mice up to about 11 days of gestation, the dimorphic patterns were not so distinct. All strains showed some DBA binding to both vascular endothelium and intestinal epithelium although stronger in whichever tissue would remain DBA-positive in the adult. A number of tissues were consistently DBA-positive in all strains (i.e. showed no dimorphism). These included yolk sac, developing otocyst, a cell population in embryonic thymus (9), an unidentified population of cells in lamina propria of small intestine, and the epithelium of stomach, pancreatic and biliary ducts, caecum and salivary gland ducts. There were consistent departures from the basic dimorphic pattern of expression of DBA binding sites in some adult tissues, notably in endothelium of E^+G^- mice, in which vessels in certain tissues lose DBA binding sites in late embryonic and early post-natal development (2), and in colonic epithelium, which was DBA-negative in some strains which were otherwise DBA-positive in gut epithelium (10).

Genetics: F1 hybrids between E^+G^-/E^-G^+phenotype parents were always E^+/G^+. DBA binding site expression is therefore co-dominant (Fig. 1c).

Backcrosses to the appropriate parent gave a 50-50 outcome indicating that DBA binding sites on intestinal epithelium and vascular endothelium are each determined by one gene. It is very likely that these genes are in fact alleles at a single locus for the following reasons. 21 recombinant inbred (RI) strains (14 BALB/c x STS from Dr. J. Hilgers, Antoni van Leeuwenhoek Research Laboratories and 7 C57L x SWR from Dr. B. Taylor, Jackson Laboratories) yielded 11 with E^+G^- phenotype and 10 with E^-G^+. If there were 2 separate loci, 4 phenotypes would be expected : (E^+G^-; E^+G^+; E^-G^+; E^-G^-). Assuming that the loci were unlinked and all phenotypes fully viable, the chance of producing only the 2 phenotypes which were observed would be $(1/2)^{21} = <10^{-6}$. Even if E^-G^- were lethal (we know

TABLE 1.

Classification of inbred strains of mice for DBA binding pattern.

Gene	Phenotype	Strains
<u>Dlb</u>-1^a	(E^+G^-) :	DDK, GRS/A, LTS.Af, MAS, SM/Ja, STS, SWR, RIII-<u>ro</u>
<u>Dlb</u>-1^b	(E^-G^+) :	AKR, A/Nimr, BALB/cHeA, C57L, C57BL/6J, C57BL/10ScSn, B10.A, C57BL/Ks, C57BL/Nimr, CBA/Ca, CBA/HN, C3H/Bi, C3H/He, DBA/1, DBA/2, DBA/Af, GE, HTG, IF, LP/Nimr, LT, NZB, 020/A, SEA/J, SJL/J, 101/H, 129/RrJ.

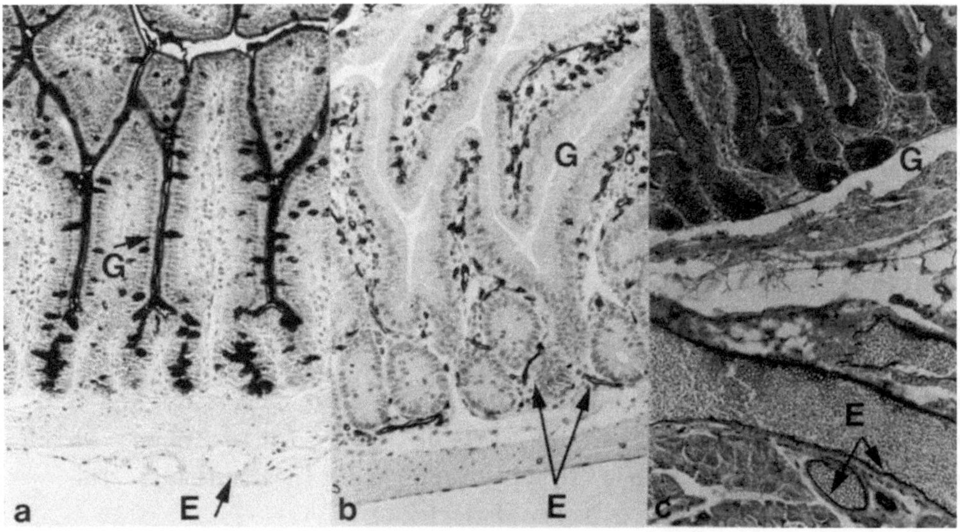

<u>Fig. 1</u>. Patterns of DBA-peroxidase binding to intestinal epithelium and endothelium. a C57BL/6 (x 160); b RIII-ro (x 160); c BALB/cHeA x STS/A (D1b-1^a x D1b-1^b) F_1 hybrid (x 130). Counterstained with haemalum. DBA-peroxidase binding sites appear black. E = endothelium; G = intestinal epithelium. Note that C57BL/6 is E^-/G^+; RIII-<u>ro</u> E^+/G^-; and the F_1 hybrid E^+G^+.

TABLE 2.

The genotype distribution pattern of Hba, Dlb-1 and Es-3 in the BALB/c x STS (CxS) RI series.

	Genotype of C x S RI strains													
Strain	1	2	3	4	5	6	7	8	9	10	11	12	13	14
Locus														
Hba	C	S	C	C	C	S	S	C	C	S	C	C	C	S
		X	X	X	X				X		X	X		
Dlb-1	C	C	S	S	S	S	S	C	S	S	S	S	C	S
		X		X		X								
Es-3	C	S	S	C	S	C	S	C	S	S	S	S	C	S

The genotype for Es-3 and Dlb-1 is concordant in 11 of the 14 RI strains (exceptions are 2,4 and 6). This concordance suggests that the genes may be linked.

from the F1 data that E^+G^+ is viable) the chances would be $(2/3)^{21} = 2 \times 10^{-4}$. A similar argument applies to the analysis of over 30 inbred strains, none of which had E^-G^- or E^+G^+ phenotypes. Inbred strains should be in linkage equilibrium, so the chance of this result if the two genes were unlinked would be at most $(2/3)^{30}$ or about 5×10^{-6}. The ratio of E^+G^- : E^+G^+ : E^-G^+ phenotypes in 78 F_2 animals (16:45:17) does not depart significantly ($p = 0.66$) from the 1:2:1 expected with a single locus.

We conclude that it is highly probable that the E^+G^- and E^-G^+ phenotypes are determined by alleles at a single locus, which we have designated Dlb-1 for Dolichos lectin binding. The alleles are Dlb-1^a (E^+G^-) and Dlb-1^b (E^-G^+).

Testing the 14 BALB/c x STS RI strains (11) for Dlb-1 suggested that the gene was on chromosome 11 (Table 2). Preliminary breeding data allow us to calculate (7) the gene to be lying 26.2 $\pm$ 3.8 cM from Es-3 and 38.9 $\pm$ 3.6 cM from Hba

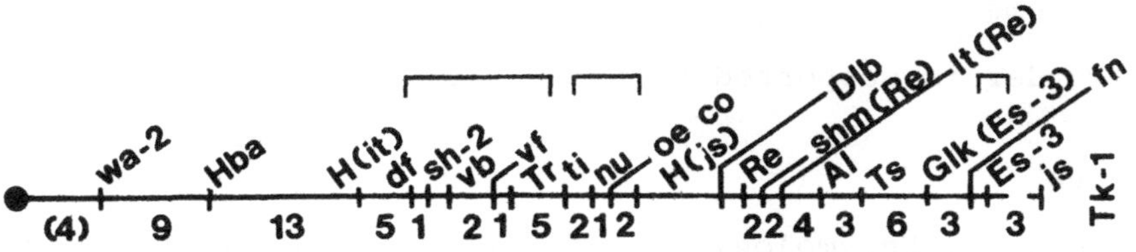

Fig. 2. Linkage map of chromosome 11, with provisional location of Dlb included, after Green (7).

and 4.0 ± 2.3 cM from Re on the linkage map of the mouse (Fig. 2). (Full genetic results will be published elsewhere (8)).

Discussion.

We have found a new marker gene, Dlb-1, which is recognisable on histological material. It maps to chromosome 11, 4.0 ± 2.3 cM proximal to Re. Because it is histologically demonstrable, the dimorphism has proven useful in the analysis of mosaic cell patterns in tissues of mouse aggregation chimaeras. A curious feature of the dimorphism is the reciprocal pattern of tissue distribution of DBA binding sites in the two dimorphic types. The mechanism by which this can be generated by an allelic difference at a single locus remains unclear. It seems possible the dimorphism involves a regulatory element which determines the tissue specific expression of a glycosyltransferase.

Apart from the dimorphism, DBA binding may provide a useful marker of vascular endothelium, for example in studies of the vascularisation of transplanted tumours, and of early neoplastic change in experimentally induced murine colonic cancer (unpublished observations). The finding of consistent tissue-specific differences in the expression of DBA binding sites on vascular endothelium is consistent with the possibility that endothelial cells may carry a 'tissue address' (2).

Acknowledgements

B.A.J. Ponder is supported by the Cancer Research Campaign.

References

1. Etzler, M.: Enzymology, Vol XXVIII, Academic Press, New York/London 1972.

2. Ponder, B.A.J., Wilkinson, M.M.:Dev.Biol.96, 535-541 (1983)

3. Schmidt, G.H., Garbutt, D.J., Wilkinson, M.M., Ponder, B. A.J.: J. Embryol. Exp. Morphol. (in press).

4. Schmidt, G.H., Wilkinson, M.M., Ponder, B.A.J.: Anat. Rec. (in press).

5. Uiterdijk, H.G., Ponder, B.A.J.: Research report in Mouse Newsletter, 71, 15 (1984).

6. Green, M.C.: Genetic variants and strains of the laboratory mouse, Gustav Fischer Verlag, Stuttgart/New York 1981.

7. Green, M.C.: Methodology in mammalian genetics, Holden Day Inc., San Francisco 1963.

8. Uiterdijk, H.G., Ponder, B.A.J., Hilgers, J., Skow, L., Festing, M., Van Nie, R. (in preparation).

9. Kasai, M., Takashi, T., Takahashi, T., Tokunaga, T.: J. Exp. Med. 159, 971-980 (1983).

10. Ponder, B.A.J., Festing, M., Wilkinson, M.M.: J. Embryol. Exp. Morphol. (submitted).

11. Taylor, B.A.: Origins of inbred mice, Academic Press, New York 1978.

RAT GASTROINTESTINAL GLYCOCONJUGATES. HISTOLOGICAL STUDIES WITH LECTINS AND MONOCLONAL ANTI-HUMAN BLOOD GROUP ANTIBODIES

Peschke P., Kuhlmann W.D.
German Cancer Research Center, D-6900 Heidelberg, FRG

Lectins are today the most specific molecular probes for the histological localisation of glycoconjugate glycosylation (1-5). Yet, clear-cut recommendations for reliable, optimal and highly sensitive histological staining schedules are still lacking. Therefore, the minimal amount of lectin necessary for histological detection was studied by markers such as fluorescein (FITC), peroxidase (HRP), and colloidal gold particle (CGP) labels in direct and indirect procedures. The test system involved binding of Griffonia simplicifolia (GS) agglutinin I to unfixed cryostat sections as well as to fixed and paraffin embedded tissue blocks of rat gastrointestinal mucosa.

For our studies on the nature of gastrointestinal glycoconjugates we use a panel of different lectins which are known to possess blood group reactivities. In order to correlate lectin binding profiles with those of monoclonal antibodies and conventional mucin stains, histological sections of rat stomach and duodenal mucosae were reacted with the respective lectins as well as with monoclonal antibodies against blood group substance A, B, and H. Stainings were also performed after chemical modification of sections, i.e. oxidation with periodic acid followed by reduction with sodium borohydride or not.

Lectins, Vol. IV

MATERIALS AND METHODS

Lectins. Lectins and lectin conjugates from Dolichos biflorus agglutinin (DBA), Helix aspersa agglutinin (HAA), Limulus polyphemus agglutinin (LPA), Lotus tetragonolobus agglutinin (LTA), Griffonia simplicifolia agglutinin I Isolectin B_4 (GS B_4), Arachis hypogaea agglutinin (PNA), and Ulex europaeus agglutinin I (UEA I) of controlled purity were obtained from E-Y Laboratories (San Mateo, USA).

Antibodies. Hyperimmune sera against lectins were prepared in rabbits and IgG fractions were isolated (6). For immunohistology and indirect lectin binding studies, second antibodies and their HRP and CGP conjugates were prepared as described (7). Mouse monoclonal antibodies against blood groups A, B and H were purchased from Cell Tech (Berks, UK) and DAKO (Santa Barbara, USA).

Tissue specimens. Corpus and duodenal mucosae from normal rats were fixed in 8-10% formaldehyde/12-24 h or in 99% ethanol - 1% acetic acid/12-18 h and embedded in paraffin. In parallel, cryostat sections were prepared from unfixed tissues from which some specimens were fixed in cold acetone.

Trypsinisation. 0.05-0.1% trypsin in 0.05 M Tris-HCl, pH 7.6 plus 0.1% $CaCl_2$, 5-10 min at 37° C was employed.

Chemical modification. Parallel sections were submitted to (a) periodate oxidation with 1% periodic acid in distilled water for 10 min at room temperature (8) and (b) periodate oxidation followed by reduction with 0.1% sodium borohydride in 1% Na_2HPO_4 for 30 min at room temperature (9). Inhibition of endogenous peroxidases was performed with 1% hydrogen peroxide in PBS for 1 h at room temperature (10).

Histological lectin staining. Minimal amounts of lectins, anti-lectin antibodies and conjugates necessary for staining were determined by serial dilution of the respective reagents. Incubation times of both labelled and unlabelled lectins were always 18 h at 4° C. In the case of indirect procedures, anti-lectin antibodies and labelling antibodies were always incubated for 30 min at room temperature. Unreacted lectins and antibodies were washed off by three successive washings. Peroxidase activity was revealed by incubation in 3,3'-diaminobenzidine (DAB) and H_2O_2 (11). Gold particles were visualized by a silver precipitation reaction (12).

Immunohistological blood group staining. In parallel to lectin stainings, sections were incubated with mouse monoclonal antibodies (isotype IgM) against blood group substances A, B and H, respectively. The antigen-antibody complex formed was visualized by HRP-labelled goat anti-mouse IgM antibody.

Double staining experiments. Slides for combined localisation of a given lectin binding site and the correlating blood group were sequentially treated: first, incubation in HRP labelled lectins developed with DAB/H_2O_2 and second, incubation with monoclonal antibodies against the respective blood group antigens, followed by HRP conjugated second antibodies and 4-chloro-1-naphthol/H_2O_2 substrate.

Controls. Specificity of lectin binding was checked on serial sections by incubation as above but supplemented with either of the respective sugar inhibitors (13). Specificity of immunohistological reactions was proved as described (14). For routine histological examination, sections were stained with hematoxylin-eosin, periodic acid-Schiff reaction, Alcian Blue at pH 2.5, and High iron diamine.

RESULTS AND DISCUSSION

Our observations on rat gastrointestinal glycoconjugates with lectins and monoclonal antibodies are summarized in Tables 1-3.

Unfixed and acetone fixed cryostat sections were compared with paraffin sections from both ethanol-acetic acid and formaldehyde fixed tissue blocks. Cellular details were better preserved in paraffin than in cryostat sections. Histological lectin binding was less sensitive after formaldehyde than after ethanol-acetic acid fixation inasmuch as higher concentrations of lectins were needed. This drawback could be overcome by trypsinisation of the sections. It is well known from immunohistological studies that fixation plays an important role in masking biomolecules (14) and partial reconstitution was obtained by treatment of rehydrated sections with proteolytic enzymes (15,16).

Under all experimental conditions, the indirect methods required lower concentrations of lectins than direct ones; indirect procedures increased sensitivity about 5-10 fold (Table 1). CGP labels were always of highest sensitivity when gold particles were further developed by a silver precipitation method. HRP was not as efficient in lectin localisation as CGP, but cytochemical staining was more convenient

Table 1

Minium amount of Griffonia simplicifolia isolectin I (B_4) necessary for the localisation of its binding to sections from differently processed tissues

Stainig method	Eth.-acet.acid fix. paraffin embedment microtomy, sections		Formaldehyde fixation paraffin embedment microtomy, sections		No fixation cyrostat sections air dried	No fixation cyrostat sections acetone fix.
	Rehydr.[a]	Tryps.	Rehydr.	Tryps.	Rehydr.	Rehydr.
Direct - FITC	500[b]	400	1200	900	800	800
Direct - HRP	300	100	300	200	700	700
Direct - CGP	60	20	100	40	-[c]	-
Indirect - HRP	50	5	70	20	-	-
Indirect - CGP	10	2	10	2	-	-

a No pretreatments, sections were passed after rehydration into the respective lectins

b Lectin concentration expressed as ng protein/ml

c Not done

Table 2

Histochemical carbohydrate stains applied to gastrointestinal mucosa, comparison of staining patterns with lectin binding sites

Histologic site	PAS	AB	HID	UEA I	GS I(B_4)	DBA	HAA	LPA	LTA	PNA
surface epithelium	+	+	-	+	-	-	-	-	+	+[a]
mucus neck cells	+	+	+	(+)	-	(+)	+	-	+	+
parietal cells	-	-	-	-	+	-	-	-	-	+
pyloric gland cells	-	+	+	(+)	-	+	+	-	-	-
columnar epithelium	-	-	-	-	+[a]	+[a]	-	+[b]	-	+[a]
brush border	+	+	-	+	-	+	-	-	+	+
villous goblet cells	+	+	+	+	-	+	-	-	-	-
crypt goblet cells	+	+	+	+	+	+	+	-	-	-
Brunner's glands	+	-	-	+	+	+	+	-	-	+

a supranuclear

b crypt columnar epithelium

Table 3

Histological staining with monoclonal antibodies and lectins of normal and chemically treated sections from gastrointestinal mucosa

Reagent	Localisation	Normal	HIO_4	HIO_4+$NaBH_4$
anti-H-antigen (monocl.ab)	surface epith.	+	-	-
	mucus neck cells	+	-	-
	pyl.gland cells	+	-	-
	goblet cells	+	-	-
UEA I	surface epith.	+	+	+
	mucus neck cells	+	+	-
	pyl.gland cells	+	+	-
	Brunner's gl.cells	+	+	-
	goblet cells	+	+	-
LTA	surface epith.	+	+	+
	mucus neck cells	+	+	-
	goblet cells	-	+	-
	brush order	+	+	-
PNA T-antigen	surface epith.[a]	+	+	+
	mucus neck cells	+	+	+
	parietal cells	+	-	-
	Brunner's gl.cells	+	+	+
	enterocytes[a]	+	-	-
	brush order	+	-	-
anti-A-antigen (monocl.ab)	mucus neck cells	+	+	+
	pyl.gland cells	+	-	-
	Brunner's gl.cells	+	+	-
DBA	mucus neck cells	+	-	-
	pyl.gland cells	+	+	+
	Brunner's gl.cells	+	-	-
HAA	mucus neck cells	+	+	+
	Brunner's gl.cells	+	+	+
anti-B-antigen (monocl.ab)	surface epith.	+	-	-
GS I (B_4)	parietal cells	+	-	-
	mucus neck cells	-	+	+
	Brunner's gl.cells	+	+	+
	crypt goblet cells	+	+	+

[a] supranuclear

in routine work. Direct FITC labellings proved to be of lowest sensitivity. Generally in our model, carbohydrate portions in glycoconjugates survived well, and excellent localisation of cellular lectin bindings was achieved. Moreover, controls with selected inhibitory sugars demonstrated that the observed lectin bindings depended on a specific lectin-sugar interaction.

Conventional mucin stains gave quite homogenous distribution within a given cell population while lectin binding was able to demonstrate the heterogenous nature of sugar residues in glycoconjugates (Table 2). In columnar epithelium of duodenum for example, lectins detected stages of glycosylation which were confined to different cell compartments. Comparable results have been observed in glycoconjugates of normal human salivary glands (Born et al., These Proceedings). Lectins also demonstrated changes in carbohydrate moieties in the course of goblet cell differentiation. Furthermore lectins bound to glycosylated non-mucus cell products as observed in parietal cells. Yet, the relationship of carbohydrate residues seen by the different lectin bindings was still unclear. Therefore, a systematic study with lectins of distinct specificities including normal and diseased gastrointestinal organs is necessary (17). Fischer and Klein as well as Örntoft and Peterson proposed the potential use of lectins as diagnostic reagents. Multiple approaches with other tissues are in progress (cf. These Proceedings).

Lectins used in our experiments are known to possess blood group reactivity against A, B, H and T (2). Hence, we suggested that the use of monoclonal anti-human blood group antibodies might detect comparable determinants. Yet, in histological tissues a certain lack of correlation was evident for localisation of blood group related antigen and lectin binding sites (Fig. 1,2). The latter represent distinct carbohydrate residues (18). This discrepancy was especially apparent in double staining experiments. Apart from identical staining patterns, lectins were able to detect other tissue components which were not revealed with monoclonal anti-blood group antibodies.

In the case of HRP as marker, endogenous tissue peroxidases must be irreversibly inhibited in order not to interfere with stains derived from the lectin labelling experiments. This step was routinely done by 1% hydrogen peroxide in PBS for 1h and proved useful for all types of histological immunoperoxidase reactions (14). While endogenous peroxidases became completely inhibited, subsequent lectin binding was not affected. In any case, the latter possibility must be recon-

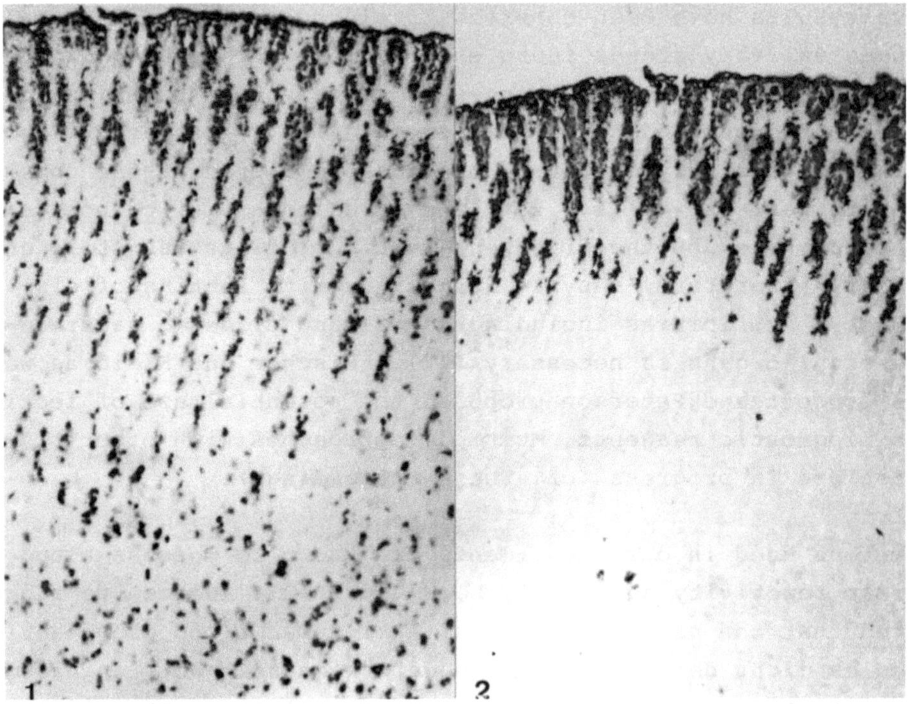

Fig. 1. Lectinhistology: UEA I-binding to rat gastric body mucosa. Magnification x130.
Fig. 2. Immunohistology: Monoclonal antibody against blood group substance H. Magnification x130.

ciled. For the inhibition of tissue peroxidases we did not employ schedules which involved periodic acid-borohydride treatments (19,20); these had deleterious effects on carbohydrates. While periodic acid considerably changed subsequent lectin binding, treatment of sections with periodic acid followed by sodium borohydride led to further impairments (Table 3). Thus, we observed: (a) loss of staining probably due to cleavage of vicinal glycols in terminal position of glycoprotein saccharide chains (4,8), (b) nonaffected labelling due to resistance of carbohydrate to oxidation or due to independence of lectin/antibody binding from vicinal diols, (c) new staining patterns perhaps caused by exposure of respective residues in a subterminal position.

In conclusion, lectins as specific molecular tools for the histological characterisation of glycoconjugates make necessary controlled tissue sampling and reagent preparation. In order to overcome descriptive phases of lectin staining patterns, it remains to be determined whether binding of a lectin is evidence of a defined entity with specific significance. This would require characterisation of the chemical structure. Moreover, it might be possible that digestion of sections with specific glycosidases and subsequent use of lectins will provide more knowledge of sugar sequences in glycoconjugates and, furthermore, prove specificity of lectin binding. From our studies we suggested that lectins detect qualitative and quantitative differences in carbohydrate composition of gastrointestinal glycoconjugates which are not necessarily inherent with serologically defined blood group types.

REFERENCES

1. Lis, H., Sharon, N.: Ann.Rev.Biochem.42, 541-574 (1973)

2. Goldstein, I.J., Hayes, C.E.: Adv. Carbohydr.Chem.Biochem. 35, 127-340 (1978)

3. Etzler, M.E., Branstrator, M.L.: J.Cell Biol. 62, 329-343 (1974)

4. Stoward, P.J., Spicer, S.S., Miller, R.L.: J.Histochem. Cytochem. 28, 979-990 (1980)

5. Sato, A., Spicer, S.S.: Histochem.J. 14, 125-138 (1982)

6. Goudswaard, J., Van der Donk, J.A., Noordzij, A., Van Dam, R.H., Vaernab, J.P.: Scand.J.Immunol. 8, 21-28 (1978)

7. Kuhlmann, W.D., Peschke, P.: Histochemistry 81, 265-272 (1984)

8. Suzuki, S., Tsuyama, S., Murata, F.: Histochemistry 73, 563-575 (1982)

9. Culling, C.F.A., Reid, P.E., Worth, A.W.: Meth. Achiev. Exp.Pathol. 10, 73-100 (1981)

10. Kuhlmann, W.D.: Histochemistry 44, 155-167 (1975)

11. Graham, R.C., Karnovsky, M.J.: J.Histochem.Cytochem. 14, 291-302 (1966)

12. Danscher, G.: Histochemistry 71, 81-88 (1981)

13. Kabat, E.A.: Structural concepts in immunology and immunochemistry. Holt, Rinehart and Winston, New York, Toronto, London 1976

14. Kuhlmann, W.D.: Immuno enzyme techniques in cytochemistry. Verlag Chemie, Weinheim 1984

15. Huang, S.H., Minassian, H., More, D.J.: Lab.Invest. 35, 383-390 (1976)

16. Denk, H., Radaszkiewicz, T., Weirich, E.: J.Immunol.Meth. 15, 163-167 (1977)

17. Kuhlmann, W.D., Peschke, P., Wurster, K.: Virchows Arch. (Pathol.Anat.) 398, 319-328 (1983)

18. Lemieux, R.U., Bock, K., Delbaere, L.T.J., Koto, S., Rao, V.S.: Can.J.Chem. 58, 631-653 (1979)

19. Isobe, Y., Chen, S.T., Nakane, P.K., Brown, W.R.: Acta Histochem. Cytochem. 10, 161-171 (1977)

20. Heyderman, E.: J.Clin.Pathol. 32, 971-978 (1979)

LECTIN-BINDING AND UPTAKE BY HUMAN MONONUCLEAR CELLS

Udo Schumacher
Pathologisches Institut der Universität Kiel
D-2300 Kiel

Ulrich Welsch
Anatomische Anstalt, Universität München
D-8000 München

Jürgen Barth, Kurt-Günther Ravens, Michael Derwahl
I. Medizinische Klinik der Universität Kiel
D-2300 Kiel

Phagocytosis can be divided basically into three steps. The adherence of the particles to be phagocysed on the outer cell membranes, the incorporation of the material in vesicles and the subsequent digestion. The binding and uptake process in phagocytes can be visualised by the use of electron dense markers such as ferritin in electron microscopy while the metabolic changes can be monitored by chemiluminescence, which is an equivalent of the release of super oxide anion and other oxygen species. (1,6). Human blood monocytes were used as a model system for phagocytosis. Following initial contact with lectins such as concanavalin A (Con A) or wheat germ agglutinin (WGA) oxidative metabolism is markedly enhanced in these cells (2, 3). To study the fate and further interaction of these lectins in monocytes ferritin labelled Con A and WGA were studied employing electron microscopic techniques.

Lectins, Vol. IV

Material and Methods

Heparinized (Vetren ^{R}) blood samples were drawn from healthy human donors. After dilution with equal amounts of PBS they were layered over a Ficoll gradient and centrifuged at 800 g for 10 mins. After washing isolated cells three times with Dulbecco's modified Eagle' medium (Gibco) the cell suspension was adjusted to 2000 cells/µl and 500 µl were used for each chemiluminescence assay. The assay was run on a Berthold biolumat LB 9505. Luminol at a final concentration of 20µg in 500µl cell suspension was used. Con A was used for assay in varying concentrations 10 - 200µg/ml, WGA 1 - 100µg/ml (Sigma) as final concentrations. For electron microscopy 100µg/ml ferritin Con A or 10µg/ml ferritin WGA (E.Y.-Laboratories) were used. For EM-preparations 500µl of a 6% glutaraldehyde solution was added at 3, 15 and 40 min after starting the assay. The cells were fixed for 2 hrs at 4°C and centrifuged down. After careful washing with phosphate buffer the cells were embedded in a 2% Agar Noble solution (Difco) and post-fixed with 2% OsO_4 for 2 hrs. After washing the cells again they were dehydrated and embedded in araldite. Ultrathin sections were contrasted with lead citrate and examined under a Zeiss EM 10. (5).

Results

Mononuclear cells respond to both lectins Con A and WGA with an initial increase of oxidant generation. The peak of this respiratory burst was seen 1 - 2 mins after starting measurement. However, 5 mins after starting measurement activity was almost reduced to the basal level. The effects of Con A and WGA were shown to be dose dependent (Fig. 1). It was also shown that Con A needs a higher concentration to evoke the burst of supernatant oxygen release than WGA, although the cells respond to a lower level of WGA. Using higher concentrations than those shown in Fig. 1 a suppression of chemiluminescence was seen, probably due to a toxic lectin effect on oxidative metabolism of cells. Comparing the chemiluminescence response of ferritin labelled lectins with those of native ones the peak maximum was lower even by using the same amounts of lectins.

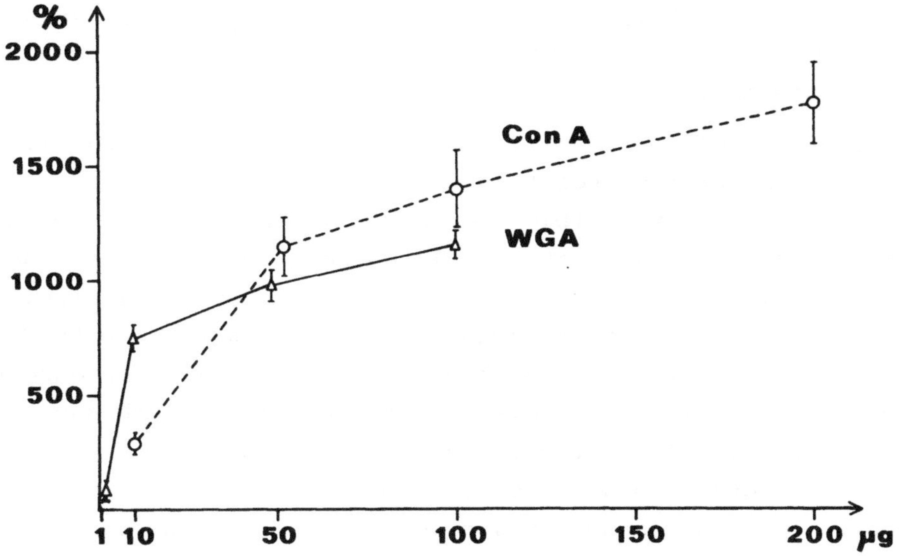

Fig. 1. Effect of different amounts of Con A and WGA to chemiluminescence. Maximal increase above base chamiluminescence is demonstrated in % ± SD. Peak levels were seen 90 - 120 sec after adding lectins.

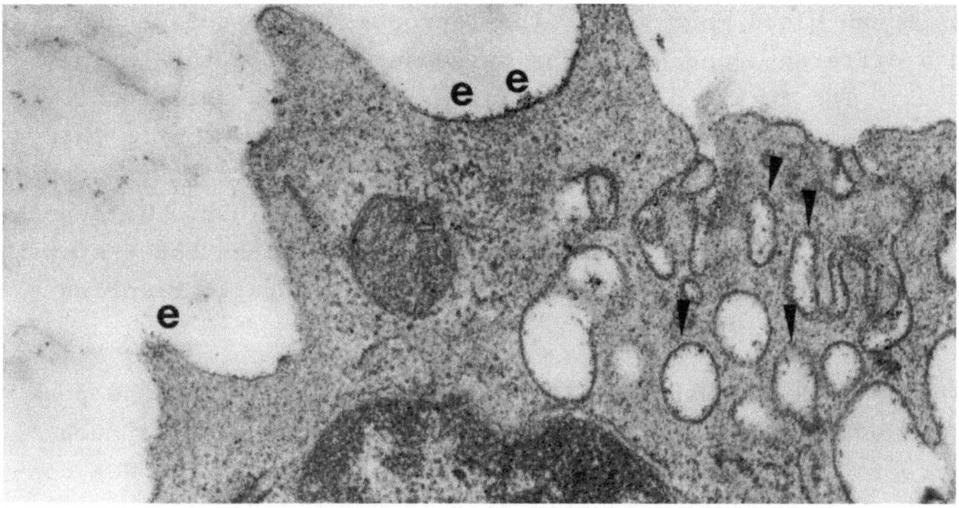

Fig. 2. Monocyte after labelling with WGA ferritin for 3 min. 39360 x. e = Extracellular space labelling of plasma membrane and subplasmalemal vacuoles can be seen (arrows).

After 3 mins monocyte plasma-membrane were strongly labelled with WGA (Fig. 2) which was also incorporated into subplasmalemal vacuoles.

However, labelling of cell membranes was also visible with Con A to a lower degree than with WGA and incorporation into vacuoles was only visible sporadically. After 15 min both lectins could be seen at the plasma membrane and in intracytoplasmic vacuoles and vesicles of various shapes and sizes. Binding of WGA was considerably stronger than that with Con A (Fig. 3). After 40 min prominent binding of WGA was observed in intracytoplasmic vacuoles and at the plasma membranes. (Fig. 4). The labelling of the plasma membrane and intracytoplasmic vacuoles and vesicles with Con A was still less prominent. In single monocytes, however, no labelling could be detected at all with Con A.

Discussion

Both lectins were able to induce a chemiluminescence response in human blood monocytes. The response was dose-dependent and different concentrations were used to evoke the maximum peak. The ferritin labelled lectins were less able to produce chemiluminescence response than the native ones. This may be due to the denaturation of the lectin during the coupling procedure or to the changed molecular weight. We suppose the second possibility is more likely since the response was greater with higher doses of ferritin labelled lectins (data not shown).

The EM data demonstrate a binding of the lectins at the peak maximum at 3 min at the cell membrane, the binding is much more intensive with WGA than with Con A although the WGA concentration used is less than that of Con A (10µg/ml or 100µg/ml, respectively). The subsequent uptake of WGA is also con-

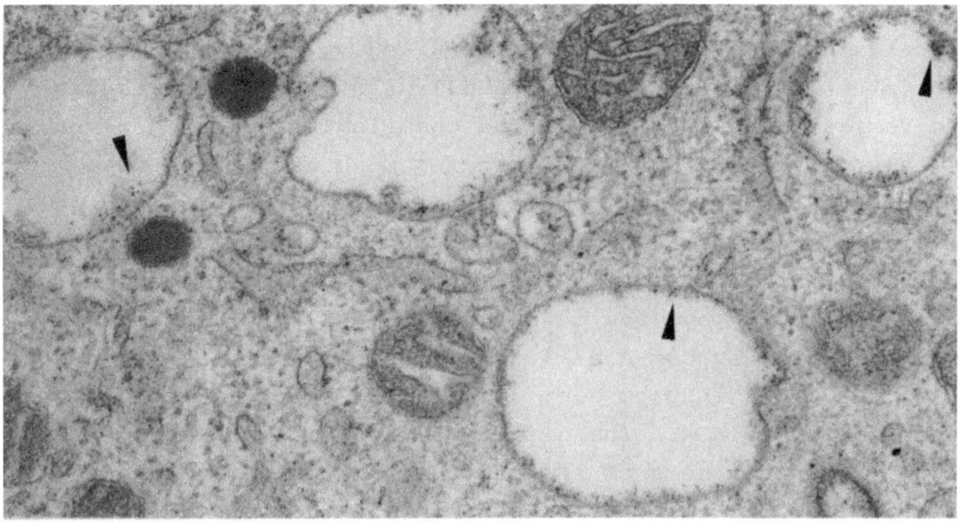

Fig. 3. Monocyte after labelling with Con A ferritin for 15 min. 64960x. (arrows).

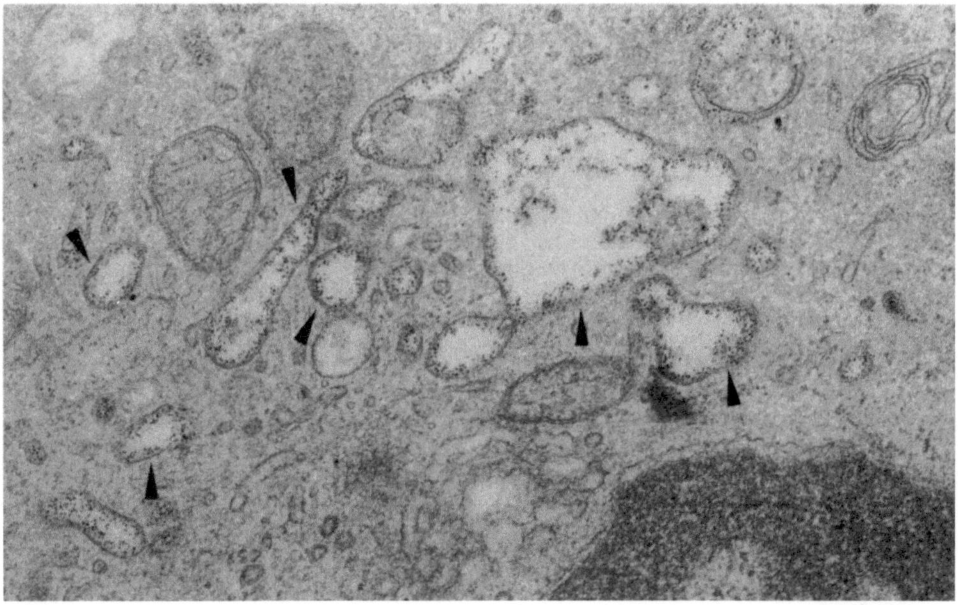

Fig. 4. Monocyte after labelling with WGA ferritin for 40 min. 38800x. Note the labelling of intracytoplasmic vacuoles and vesicles of various shape and size (arrows).

siderably stronger than that of Con A, while the cellular compartments are the same. Binding and uptake are still present after 40 min, while the labelling of the intracytoplasmic vesicles has increased. The chemiluminescence, however, is only present within the first few minutes. These data suggest that the initial contact of the lectins with the cell membrane is necessary to induce oxygen radical production. The receptors behave quite differently: the WGA receptor is labelled heavily by 10µg/ml WGA, while the labelling of the Con A receptor by Con A at 100µg/ml is less prominent, however, the Con A induced chemiluminescence peak is higher than that induced by WGA. The different labelling intensity may be due to the numbers of receptors on the cell membranes or different affinities of the lectins to the receptor.

References

1. Allen, R.C. and Loose, L.D.: Phagocytic activation of a luminol-dependent chemiluminescence in rabbit alveolar makrophages. Biochem. Biophys. Res. Comm. 69, 245 - 252, (1976).

2. Derwahl, M., Barth, J., Ravens, K.G. and Schumacher, U.: Mononuclear cell response to lectins: cAMP and chemiluminescence, see in these proceedings.

3. Barth, J., Ravens, K.G. and Schumacher, U.: Lectin induced modification of cell metabolic activities as measured by luminol enhanced chemiluminescence in human mononuclear cells. Europ. Journal of Cell Biology Supp. 4, 3, (1983).

4. Dalinden, C.L., Fehr, I. and Hugli, T.E. (1983): Role of cell surface contact in the kinetics of superoxide production by granulocytes. J. Clin. Invest. 72, 113 - 121.

5. Nagl, W.: Elektronenmikroskopische Laborpraxis. Springer, Berlin (1981).

6. Wrogemann, K., Weidemann, M.J., Peskar, B., Staudinger, H.J., Fischer, H.: Function and Structure of Immune System, Plenum Publishing Corporation, New York, 395 - 399 (1979).

RICIN BINDING IN RAT BRAIN

B.K. Groeger

Zoology Department, University College London, Gower Street, London WC1E 6BT

P.L. Debbage

Max Planck Gesellshchaft Klinische Forschungsgruppe für Multiple Sklerose, Würzburg, West Germany.

We have previously shown the presence of binding sites for the lectin ricin 120 in liver tissues (4,5). They are localized on the plasma membranes of the non-parenchymatous cells of the liver, and appear in tissue sections to be lining branching tube-like structures, apparently blood vessels. We also showed that ricin 120 binds mainly to a single protein in liver non-parenchymatous plasma membranes (5). We show here that ricin 120 also binds to blood vessels in the brain. However, the major ricin binding proteins in the brain vasculature and in liver plasma membranes are not the same, as demonstrated here by their different electrophoretic mobilities. Preliminary results reported here suggest that the ricin 120 binding proteins in rat brain may be of functional significance.

Materials and methods

Fluorescence histology. Frozen sections of rat brain, cerebral hemispheres and cerebellum were prepared and labelled with fluorescein-conjugated ricin (Sigma) as described previously (4), and preparations of rat brain microvessels dried onto slides were labelled in identical procedures.

Microvessel preparation. Four to six adult rats were killed by cervical dislocation, the brains removed and placed in Eagles medium buffered with HEPES, on ice. The meninges and choroid plexuses were removed by dissection at a binocular microscope. The cerebella were processed separately. The brains were homogenised in 1 ml/g cold medium in a loose-fitting Dounce, using 30 strokes. The homogenate was mixed with four volumes of cold Lymphoprep (Nyegaard & Co, Norway), sieved through metal mesh of 400 µm mesh size, then centrifuged at 1,500g_{av}, 4°C, for 15 min. The myelin pad was discarded, and the pellet (microvessels, cell nuclei, and a few intact cells) was resuspended in 10 ml cold medium. The suspension was placed on a 10 cm column of 400 µm glass beads and rinsed with 80ml cold medium. The lower 9 cm of beads was discarded, and the microvessels released from the top 1 cm by gentle agitation in 10 ml cold medium, adapted from previous methods (3,6). Endothelial cells and microvessels were separated (8).

Electron microscopy. Microvessels were fixed in 3% glutaraldehyde for 20 minutes, rinsed, and incubated in 0.15 M lysine for 60 min, then in 5×10^{-7} M ricin 120-peroxidase conjugate (Sigma) for 60 min. After thorough rinsing in PBS the microvessels were incubated in 0.1% diaminobenzidine (DAB) + 0.01% hydrogen peroxide in for 20 min, rinsed, dehydrated in alcohols, and embedded in Araldite. Ultrathin sections were contrasted with uranium and lead salts, and viewed in a Zeiss EM. Vibratome sections (50 µm) were cut from the brain of a rat previously perfused with 5% glutaraldehyde, incubated with ricin 120 peroxidase, developed as above, and processed for electron microscopy.

Plasma membrane fractions. All procedures were carried out at 4°C. Tissue from 10 rats was homogenized in a Dounce homogenizer, in 1 mM bicarbonate, allowed to stand on ice for 3 min, re-homogenized briefly, then pelleted at 3,000 g_{av} (3 min). The resulting pellet was resuspended in 10 volumes fresh bicarbonate and loaded onto a 45% sucrose pad and spun at 4,000 g_{av} for 60 min. The interface material was collected, resuspended in PBS and pelleted at 3,000 g_{av} (3 min). The pellets contained large sheets of plasma membranes (11), and were stored at -20°C.

Gel electrophoresis and electroblotting. Total protein concentrations of plasma membrane fractions were determined by the method of Lowry et al, (1951). SDS-PAGE was carried out with 50-100 µg of protein on 10% separating gels (7). Gels were either stained with Kenacid blue R or blotted (see below) then stained. Low molecular weight markers (Pharmacia, 14,000-94,000) were run as standards on each gel. Electroblotting from gels to nitrocellulose sheets was carried out at 30 V and 100 mA for 16 hrs (9). Standards were stained with 0.5% Amido black.

Ricin labelling of blots. Membrane sample and microvessel blots were incubated with ricin 120-fluorescein for 30 min at 5×10^{-7} M or with ricin 120-peroxidase (preincubated in 1% bovine serum albumin) at 1×10^{-8} M, rinsed and developed in DAB as above.

Ricin 120-peroxidase perfusion of rats. Rats of approximately 200g weight were injected via the tail vein with 0.6 ml ricin 120 peroxidase at a concentration calculated to give a blood concentration of 3×10^{-7} M (assuming a blood volume of 8ml). 1 hour later, 1% Evans blue dye was injected. After 5 minutes, the rats were etherised, perfused with 10 ml PBS then sacrificed. Brains and livers were carefully removed and either snap-frozen for frozen sections or fixed in 5% glutaraldehyde, then incubated with DAB as above.

Glutaraldehyde perfusion of rats. Rats were perfused with 5% glutaraldehyde prior to sectioning. Tissue was then incubated with ricin peroxidase and developed with DAB and processed for electron microscopy as above.

Results

In brain sections ricin 120 binds to blood vessels and to myelin, (Figure 1). In the cerebellum there are additional binding sites, as yet not accurately localized. Microvessels isolated from brain bind ricin 120 strongly (Figure 2), mainly in the basal membranes (results not shown).

Electrophoretic and blot analysis of brain microvessels prepared from forebrain and from cerebellum reveals the

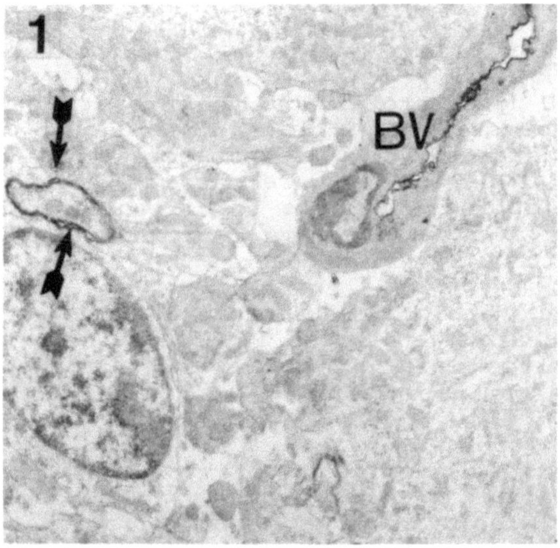

Fig. 1. Electron micrograph of glutaraldehyde perfused rat cerebellum labelled with ricin 120-peroxidase. Reaction product is seen in blood vessels, BV, and disrupted myelin (arrows). x 6,000.

presence of prominent ricin 120 binding glycoproteins of MW 74,000 and 114,000 (Figure 3a). These could not be detected in similar analyses of brain plasma membrane fractions (Figure 3b,3c). However, in plasma membrane fractions from some brain regions (cerebellar hemispheres, and, possibly, the basal ganglia), a ricin 120-binding glycoprotein of MW 31,000 occurs. (It appears that the heavy ricin binding bands at the end of the tracks with the 31,000 MW glycoprotein are degradation products of this band (unpublished data)).

Figure 3 compares these observations with results obtained from liver and spleen plasma membrane fractions, in which a

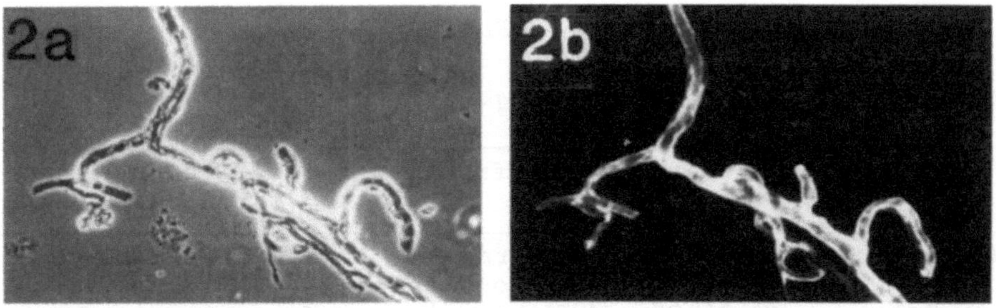

Fig. 2. (a) Phase image of isolated rat forebrain microvessels. (b) Corresponding fluorescent image after labelling with ricin 120-fluorescein conjugate. x 140.

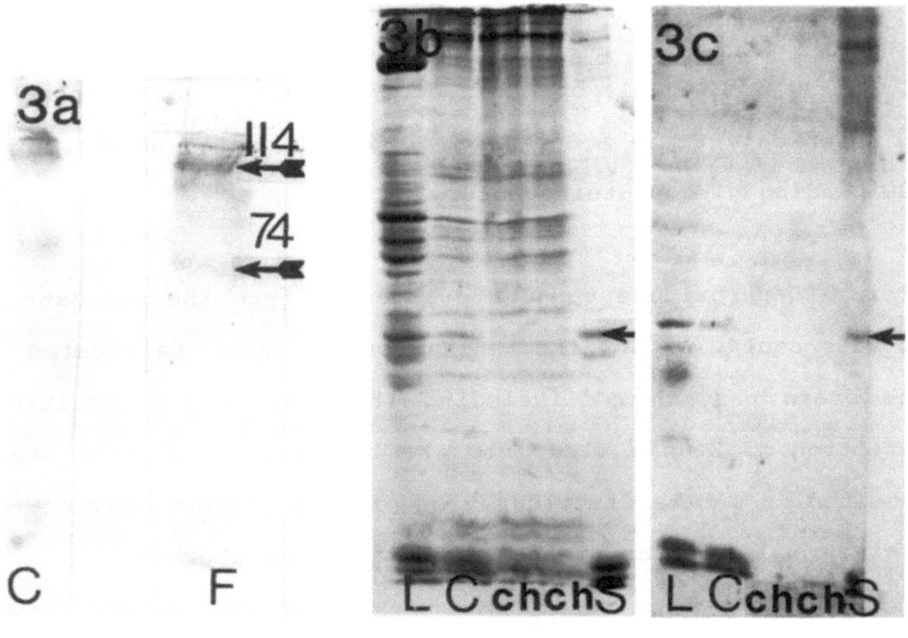

Fig. 3. (a) Blot of cerebellar (c) and forebrain (F) microvessels, labelled with ricin 120-peroxidase, the 114,000 and 74,000 molecular weight bands are arrowed (gel not shown).

(b) SDS-PAGE of plasma membrane enriched fractions lane: (L) rat liver; (C) cerebellum; (ch) & (ch) cerebral hemispheres; (S) spleen. The arrow idicate the 31,000 band.

(c) The corresponding blot labelled with ricin 120-peroxidase, labelled as for gel in 3(b).

TABLE 1

Relationships of ricin binding.

	Plasma membrane fraction	Blood vessel isolates
Rat Liver	MW 31,000 (HMW trace) (non-parenchymatous cells)	Not determined
Rat Spleen	MW 31,000 (HMW trace)	Not determined
Forebrain	None	HMW (prominent) (in basal membranes)
Cerebellar Hemispheres	MW 31,000	HMW (prominent) (in basal membranes)

HMS = High molecular weight band.

ricin 120 binding glycoprotein of MW 31,000 is prominent, but in which only trace amounts of higher molecular weight ricin 120 binding glycoproteins occur.

Ricin 120-peroxidase conjugate injected into the bloodstream of rats could, after circulating for one hour, be located in the brain parenchyma, including the myelin. In addition, injection of Evan's Blue into the bloodstream at this stage resulted in a blue discolouration of some regions only of the brain, in particular of the cerebellar hemispheres.

Discussion

Our data indicate differences in the ricin-binding proteins associated with the vasculature and the parenchyma of brain, liver and spleen. The glycoprotein of MW 31.000 identified in liver and spleen (4,5) may be the same as that of similar electrophoretic mobility which is present in plasma membranes from some brain areas, particularly the cerebellar hemispheres. The vasculature extends widely throughout the brain, and the restricted distribution of this protein, together with its absence from brain microvessels, show that it is not associated with brain blood vessels. Our electrophoretic data do not provide an analysis of the myelin glycolipids, some of which (the galactolipids) probably account for the histochemical localization of ricin 120 in myelin, and we have not yet localized this protein histochemically. Ricin-binding sites have been reported in synaptosomes (1) and in microglia *in vitro* (2), and these are possible sites in which to search for this 31,000 MW protein. A group of higher molecular weight glycoproteins does however, appear to be associated exclusively with the vasculature, and with prominent ricin 120 binding bands of brain microvessels. They are also evident in small amounts in some liver plasma membrane preparations, consistent with the presence of numerous blood vessels in the liver. We have not isolated hepatic blood vessels for a formal comparison however, (it is not possible to isolate microvessels from liver).

Both sets of ricin 120-binding proteins may be extracellular, the 31,000 MW protein being localized at the surface membranes of non-parenchymatous liver cells (11), and the higher molecular weight ones being associated mainly with the vascular basal membranes.

The discovery of two prominent ricin-binding proteins in the cerebral microvasculature prompted us to examine possible functional rôles for them. The rapid entry of ricin-peroxidase conjugate into the brain indicates there is a functional rôle for ricin binding proteins in the brain vasculature, and the permeablity of the vasculature to Evan's Blue after circulation of ricin 120 in the bloodstream suggests that ricin disturbs the so-called "blood-brain barrier" (10). Our preliminary evidence suggests that this disturbance is most severe in those brain regions, the cerebellar hemispheres, which contain the 31,000 MW ricin-binding protein in addition to those associated with the vasculature.

References

1. Bittiger H, Schnebli HP: Nature 249, 370 (1974)
2. Debbage PL, O'Dell DS, James DW, In: Lectins, Biology, Biochemistry, Clinical Biochemistry Vol 2, 273 (ed) TC Bøg-Hansen (1982)
3. Goldstein JW, Wolinsky JS, Csejtey J, Diamond I: J Neurochem 25, 715 (1975)
4. Groeger BK, Williams LG, Pigott J, Ziska P, O'Dell DS, Williams, DJ, Franz H, Debbage PL In: Lectins, Biology, Biochemistry, Clinical Biochemistry, Vol 3, 179 (eds) TC Bøg-Hansen & GA Spengler (1983)
5. Groeger BK, Williams DJ, Franz H, O'Dell DS: Eur J Cell Biol Suppl 6, vol 33, 9 (1984)
6. Joó F, Karnushina I.: Cytobios (1973)
7. Laemmli UK: Nature 227, 680 (1970)
8. Lidinsky WA, Drewes LR: J Neurochem 41, 1341 (1983)
9. Meadow P, Rowe PN, Wells P: J Gen Microbiol 130, 631 (1984)
10. Westergaard E: Acta Neuropath (Berlin) 39, 181 (1977)
11. Williams LG, Williams DJ, O'Dell DS, Debbage PL: Bioch Soc Trans 10, 38 (1982)

LECTIN PROBES FOR VISUALIZATION OF CELL DIFFERENTIATION MARKERS ON EMBRYONIC CELLS

Anne-Marie Duprat[(1)], Pierre Rougé[(2)], Sylvie Flottes[(2)] and Mylène Bleys[(2)]

(1) UA CNRS n° 04675, Laboratoire de Biologie générale, Université Paul Sabatier, 118 route de Narbonne, 31 062 Toulouse, France.

(2) Laboratoire de Biologie cellulaire, Faculté des Sciences Pharmaceutiques, Université Paul Sabatier, 35 chemin des Maraîchers, 31 062 Toulouse, France.

The process by which embryonic cells become differentiated depends on complex tissue and cell interactions and implies intracellular molecular mechanisms leading to the expression of genes specific for each differentiated cell-type. The plasma membrane of these cells plays a crucial role in various events such as recognition, interaction and functional specialization occurring during cell differentiation. Among the major plasma membrane constituents the glycoconjugates, molecules involved in the determination and differentiation processes (1-6), can be identified and visualized easily *in situ* on intact cell surfaces using labelled lectin probes. The affinity of these marker molecules for special carbohydrate residues allows not only to visualize a static plasmalemma phenotype but also to observe changes in the membrane components and thus approach the dynamics of the system.

In the present work, we describe studies with fluorescent lectins on the relationship between neural determination and changes in cell surface carbohydrates of vertebrate (amphibian) neuronal precursor cells grown and differentiated *in vitro*.

MATERIAL AND METHODS

1. Dissociated cell cultures:

Pleurodeles waltl gastrulae and early neurulae (stages 8 and 13, respectively (7), were used in this study. Several different lays of eggs were used. After removal of the jelly coat

Lectins, Vol. IV

and vitelline membrane, the competent neural target tissue (presumptive ectoderm) from gastrulae and the neural primordium (neural plate and neural fold) from neurulae, were excised with platinium threads mounted on stalk. The tissues were then transfered in Ca2+ and Mg2+ -free Barth solution (88mM NaCl, 2.4mM $NaHCO_3$, 2.0mM Na_2HPO_4, 0.1M KH_2PO_4, 0.5mM EDTA) and dissociated by gentle pipetting. Cell counts were performed with a hemocytometer monitored under a phase-contrast microscope and the Trypan blue dye exclusion test was used as the criterion for viable cells. Individual cultures contained 3.75 x 10^4 cells. The isolated cells were cultured as described (1, 2,8) in Barth balanced solution (9) supplemented with 1 mg/ml bovine serum albumin (Sigma), 100 UI/ml penicillin and 100 µg/ ml streptomycin, in modified 35 mm Petri dishes (10) on plastic or on UV irradiated dried film of rat tail tendon collagen (11) After a 48 h culture in air atmosphere at 20°C, the medium was removed and replaced by Barth saline solution without bovine serum albumin. Cultured cells, issued from an embryonic layer of identical precursor cells, differentiate in two different ways according to whether they have been induced or not (Figure 1). Non-induced cells are engaged in an epidermal differentiation while induced cells differentiate into different cell-types (neural, pigment cells, etc...).

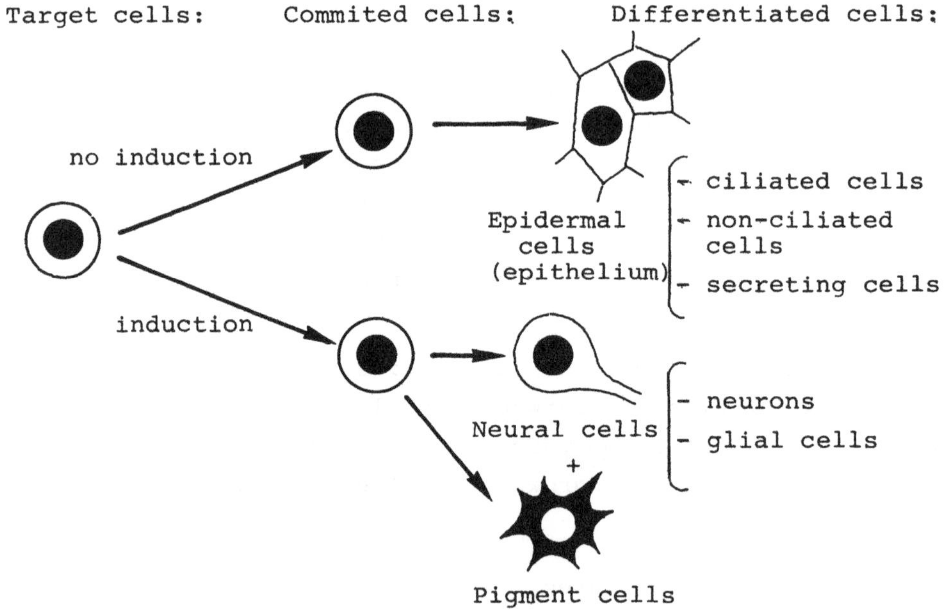

Fig. 1. Differentiation of presumptive ectodermal cells (target cells) in vitro with and without neural induction.

In this study, the observations have been only focused on two cell-types which can be easily identified phenotipically in culture conditions: neurons and epidermal cells, cultured under identical experimental conditions. At the beginning of the culture all the cells filled by yolk platelets are round and without any phenotypical features. After 48 h at 20°C, they attach and spread out over the substrate of the culture. The time-course of appearance of morphological differentiation is 3-4 days. All the cells progressively differentiate in primary culture over 2-3 weeks.

2. Lectins:

The lectins used were pea (PSA), lentil (LCA) and soybean (SBA) lectins.

Isolation of lectins: Pea (*Pisum sativum* L., petit provençal nain) and lentil (*Lens culinaris* Medik., verte du Puy) lectins were isolated from seed meal extracts essentially as in (12). Soybean (*Glycine max* (L.) Merr., cv Hodgson) lectin was obtained by affinity chromatography on Sepharose-N-caproylgalactosamine (13). The purity of the lectins was checked by polyacrylamide gel electrophoresis either at basic (14) or acidic (15) pH. Protein fractions were stained as described in (16).

Labelling of lectins: Fluorescent labelled lectins, FITC-PSA (fluorescein isothiocyanate labelled PSA) and FITC-LCA (fluorescein isothiocyanate labelled LCA), were prepared according to (17). Labelling of LCA and SBA by TRITC (tetramethylrhodamine isothiocyanate) was performed under the same conditions as FITC labelling, but using a fluorochrome solution of twice the concentration. The fluorochromes were dissolved in N,N'-dimethylformamide prior to coupling and 20-40 µg per mg of protein were added to lectin solubilized in 0.1M Na_2CO_3 (pH 9.0) under continuous stirring. The reaction was continued in the dark at 4°C for 4 h. In all cases, the labelled lectins were separated from unbound fluorochrome by affinity chromatography.

Lectin treatments: Lectin solutions (50 µg/ml of saline Barth solution), were prepared immediatly before used and cultured cells at various stages of their determination/differentiation were incubated in lectin solutions, in the dark, during 5 min. Cells were then carefully rinsed with Barth solution and examined for labelling. Our observations have been conducted on 5 different culture series and on 5 cultures in each serie.

3. Fluorescent microscopy:

A Leitz Dialux epifluorescence microscope with an incident source HBO 50 orthomat with filter I_2 (BP 450-490, LP 515) was used to examine cells for FITC labelling and with filter M_2 (BP 546/14, LP 580) in order to examine cells for TRITC labelling. Fluorescence micrographs were made on Ektachrome 160 ASA film with an orthomat Leitz camera.

RESULTS

At the earliest stage studied (competent neural target cells, st. 8), the dissociated cells all bound each lectin tested (PSA, LCA, BSA). However, in agreement with a previous quantitative study performed on the non-dissociated target tissue (2) differences in the fluorescence intensity were observed and judged on an arbitrary scale of 0 to +++ : +++ for SBA, ++ for PSA and + for LCA. Preincubation of lectins with their hapten sugars (0.1M α-methylmannoside for PSA and LCA and 0.1M galactose for SBA) prevented their binding to cells.

Immediatly after neural induction (neural plate and neural fold cells, st. 13), all the round isolated cells presented the same fluorescent pattern and bound all the tested labelled lectins. At this developmental stage, changes were visualized in the labelling of the cell surfaces of induced cells in regard with previous observations on non-induced cells (st. 8).

When the cells were maintened in primary cultures for several days, they differentiated and stricking differences appeared in lectin-binding molecules on the cell surface according to their differentiating pathway:

- after 3 days in culture at 20°C, as early as differentiation proceeded, disappearance or increase of fluorescence were expressed.
- up to 15 days in culture, phenotypical differentiation progressed in all cultured cell-types and important changes in binding pattern were noted according to the cell-type differentiation (Table 1): non-induced cells rapidly reaggregated in culture *in vitro* (after 24 h), forming an epithelial sheet and differentiated into various epidermal cells (ciliated, non-ciliated and secreting cells). No fluorescence was observed on epithelial cells with PSA and LCA whereas SBA showed positive labelling. No fluorescence was observed on secreting

cells after PSA or LCA incubation; in contrast, a very extensive labelling was visualized after SBA treatment. Ciliated cells did not display any fluorescence with PSA, LCA and SBA.
We concluded that the epidermal cells did not present α-D-mannose, α-D-glucose and derivative sugars in their plasmalemma surface but, by contrast, these cells had α-D-galactose and galactose derivatives (except ciliated cells which did not present any of the carbohydrate residues studied here).

In contrast with non-induced cells, the binding pattern persisted on differentiating neurons incubated with PSA and LCA, whereas this cell-type showed a totally negative pattern using SBA (Table 1). We concluded that neuroblasts did not present α-D-galactose and galactose derivatives on their plasmalemma but had α-D-mannose, α-D-glucose and their derivatives (Table 1).

Table 1

Fluorescent patterns of induced and non-induced cells observed with various fluorescent lectins. Fluorescence intensity was judged on an arbitrary scale of 0 (no fluorescence) to +++ (very high fluorescence) (cf 2).

Fluorescent lectins	Induced cells: neurons	Non-induced cells: epidermal cells		
		non-ciliated cells	ciliated cells	secreting cells
PSA (glucose, mannose and derivatives)	++	0	0	0
LCA (glucose, mannose and derivatives)	+	0	0	0
SBA (galactose and derivatives)	0	++	0	+++

DISCUSSION

In this work, we demonstrated specific changes of cell surface carbohydrates during the *in vitro* differentiation of competent precursor cells into either neurons (neural induced cells) or epidermal cells of different types (non-induced cells).

The prevention of the lectin binding to cells by preincubation with their hapten sugars lends support to the idea that the described lectin-binding pattern from cells was not the result of unspecific adsorption on cell surface.

With regard to the events responsible for neural determination, we could not detect any changes in the plasma membrane glycoconjugates of the induced but yet undifferentiated precursor cells, immediatly after their commitment. Very early in ontogeny neural commitment did not result in a modification of cell surface polysaccharides detectable with labelled lectin probes. Possibly, a modification does not occur or is too subtle to be detected by the technique used.

Thus, it is the transition from an initial phenotypical undifferentiated state to a differentiated state that involves detectable changes or rearrangements of the plasmalemma components. That PSA and LCA do not bind to epidermal cells suggests that mannose and glucose sites are not present or accessible on the new cell surface. In the same way, that SBA does not bind to neurons reflects a loss or a "masking" of the plasma membrane galactose sites (18). These data on changes in molecular structure of the membrane are to compare with previous observations on changes in the adhesive properties of non-induced and induced neural cells (3, Duprat et al., submitted). Isolated non-induced cells strongly reaggregate *in vitro* forming a tissue whereas induced cells remain isolated or form clusters.

On the basis of our experiments, PSA, LCA and SBA binding sites can be considered as interesting markers of non-induced and induced neural cells. Our experiments, in agreement with those on pigment cells, adrenergic and cholinergic neurons from Sieber-Blum and Cohen (19), Braun et al. (20), Pfenninger and Maylié-Pfenninger (21), Schwab and Landis (22), Zurn (23) and Reano et al. (24), clearly show that the expression of certain major membrane glycoconjugates by cultured embryonic cells is correlated with their ability to differentiate into various pathways. Similar changes in the expression of plasma membrane glycoconjugates have been revealed between normal and transformed cells; this particularity could be useful for diagnostic purposes (25-33, these proceedings).

Labelled lectins which easily permit to visualize and follow the changes of glycoconjugates on the cell surface constitute an excellent and promissing tool for studying quantitative or qualitative differentiation at the cell membrane level.

Moreover, the correlation between glycoconjugates and differentiation state makes these glycoconjugates attractive candidates for the mediation of further intracellular communication or functional relationship, specific for different kinds of cells and tissues. In this respect, these changes in the biochemical composition of the cell surface is likely to be important in cell-cell and cell-environment interactions for the further morphogenesis according to the ontogenic development.

REFERENCES

1. Duprat, A.M., Gualandris, L., and Rougé, P.: J. Embryol. exp. Morph. 70, 171-187 (1982).
2. Gualandris, L., Rougé, P., and Duprat, A.M.: J. Embryol. exp. Morph. 77, 183-200 (1983).
3. Gualandris, L., Duprat, A.M., and Rougé, P.: Eur. J. Cell. Biol., suppl. 4, 88 (1983).
4. Duprat, A.M., Gualandris, L., Kan, P., and Foulquier, F.: in The Role of Cell Interactions in Early Neurogenesis, (Duprat, A.M., Kata, A.C., and Weber, M. ed.), Plenum Publ. Co., New York, NATO ASI Series, 3-20 (1984).
5. Takata, K., Yamamoto, K.Y., and Ozawa, R.: Wilhelm Roux Arch. Dev. Biol. 190, 92-96 (1981).
6. Takata, K., Yamamoto, K.Y., Ishii, I., and Takahashi, N.: Cell Differentiation 14, 25-31 (1984).
7. Gallien, L., and Durocher, M.: Bull. biol. Fr. Belg. 91, 97-114 (1957).
8. Duprat, A.M., Zalta, J.P., and Beetschen, J.C.: Exptl Cell Res. 43, 358-366 (1966).
9. Barth, L.G., and Barth, L.J.: J. Embryol. exp. Morph. 7, 210-222 (1959).
10. Hawrot, E., and Patterson, P.H.: Methods Enzymol. 58, 574-584 (1979).
11. Bornstein, M.B.: Lab. Invest. 7, 134 (1958).
12. Agrawal, B.B.L., and Goldstein, I.J.: Biochim. Biophys. Acta 147, 262-271 (1967).
13. Allen, A.K., and Neuberger, A.: FEBS Lett. 50, 362-364 (1975).
14. Davis, B.J.: Ann. N. Y. Acad. Sci. 121, 404-427 (1964).
15. Reisfeld, R.A., Lewis, U.J., and Williams, D.E.: Nature 195, 281-283 (1962).
16. Chrambach, A., Reisfeld, R.A., Wyckoff, M., and Zaccari,J.: Anal. Biochem. 20, 1-15 (1967).
17. The, T.M., and Feltkamp, T.E.W.: Immunology 18, 865-873 (1970).
18. Moscona, A.A.: Science 171, 905-907 (1971).
19. Sieber-Blum, M., and Cohen, A.M.: J. Cell Biol. 76, 628-638 (1978).
20. Braun, S.J., Sweadner, K.J., and Patterson, P.H.: J. Neurosci. 1, 1397-1406 (1981).

21. Pfenninger, K.H., and Maylié-Pfenninger, M.F.: J. Cell Biol. 89, 536-546 (1981).

22. Schwab, M., and Landis, S.: Dev. Biol. 84, 67-78 (1981).

23. Zurn, A.D.: Dev. Biol. 94, 483-498 (1982).

24. Reano, A., Faure, M., Jacques, Y., Reichert, U., Schaefer, H., and Thivolet, J.: Differentiation 22, 205-210 (1982).

25. Orntoft, T.F., and Petersen, S.E., (these proceedings).

26. Holthöfer, H., and Rapola, J., (these proceedings).

27. Born, I.A., Schwechheimer, K., Maier, H., and Möller, P., (these proceedings).

28. Schwechheimer, K., Born, I.A., Kern, P., and Weiss, G., (these proceedings).

29. Raedler, A., Böttcher, S., Stalling, F., and Kaukel, E., (these proceedings).

30. Malmi, R., and Söderström, K.O., (these proceedings).

31. Fischer, J., and Klein, P.J., (these proceedings).

32. Fischer, J., Klein, P.J., and Csillik, B., (these proceedings).

33. Dua, D., Radzikowski, C., Debray, H., Christensen, B., and Kieler, J., (these proceedings).

LECTINS AS HISTOCHEMICAL MARKERS OF HELMINTHIC ANTIGENS

Miodrag Movsesijan, Dragica Borojević and Margita Čuperlović
Institute of Endocrinology, Immunology and Nutrition - INEP
11080 Zemun, Yugoslavia

The search for host-protective antigens of parasitic helminths resulted in ample evidence that antigenic properties are mostly connected with glycosylated molecules bound to helminth surface structures or released by worms as products of their metabolism (1-5). Therefore, glycoproteins and glycolipids of helminths are now more systematically investigated by various techniques of chromatography, radiolabelling, electrophoresis, etc. (6-8). A wider application of lectins, with their unique properties to distinguish particular carbohydrate residues, has not yet been found in the examination of parasite antigens. There are only a few examples of lectin application in parasitology, such as the determination of surface sugar residues of *S. mansoni* in various developmental stages (9-13), lectin-histochemistry of *Ascaris suum* (14), the identification of surface glycoproteins of *T. cruzi* (8) or the purification of surface glycoprotein of *Trypanosoma congolense* (23).

Lectins, Vol. IV

The use of lectins as probes for glycosylated molecules of parasitic helminths, however, offers possibilities not only for topological information about their distribution and molecular composition, but for subsequent separation of components which are expected to have antigenic properties as well (15, 16). Lectin histochemistry can be used to select the appropriate lectin for a chosen antigen, which can thus be fractionated with great specificity. In this work lectins of kidney beans (PHA) and wheat germ (WGA) were conjugated with fluoresceinisothiocyanate (FITC) and used to study histological sections of *Fasciola hepatica*, *Echinococcus granulosus* protoscolexes, encapsulated larvae of *Trichinella spiralis* and of *Ascaris suum*.

Material and Methods

Adult parasites of *F. hepatica* and *A. suum* as well as hydatid cysts of *E. granulosus* were collected from naturally iinfected swine. *T. spiralis* was obtained from an artificially infected rat. Worms and protoscolexes isolated from hydatid cysts were washed with saline and fixed in cold 95% ethanol. Tonque tissue of the infected rat was also fixed in cold ethanol. All samples were then cleared in xylol, embedded in paraffin at 50°-55°C and kept at 4°C until use.

Phytohaemagglutinin (PHA) was isolated from red-spotted beans (17) and wheat germ agglutinin (WGA) was prepared from crude wheat germ by affinity chromatography (18). Lectins were conjugated with FITC (19).

Dewaxed paraffin sections of the investigated tissues (5 μ thick) were incubated for 30 minutes at room temperature in solutions of lectin-FITC conjugates (1 mg/ml), then rinsed in phosphate-buffered saline and mounted in buffered glycerol saline. In control experiments each lectin conjugate was pre-incubated with equal volumes of 0.25 M solution of inhibitory sugar (N-acetyl-D-galactosamine for PHA and N-acetyl-D-glucosamine for WGA).

Sections were examined with a Fluoval 2 (Zeiss, Jena) fluorescent microscope.

Results

The morphology of adult *F. hepatica*, studied by histological methods, revealed the existence of cuticular epithelium, parenchyma, gum with its single-cell epithelial layer, yolk glands, genital organs, muscular layer, myoblasts and excretory tubules (20). Of all these structures, only the external membrane of the cuticular epithelium was intensively stained by both applied lectin-FITC conjugates (as shown in Fig. 1 for WGA-FITC staining and Fig. 2 for PHA-FITC). The spines which were inserted in this layer remained unstained. The subepithelial muscle layer and epithelial cells were also dark. The vitelline cells in yolk glands were fluorescent after staining with WGA-FITC conjugate (Fig. 1), while PHA-FITC was not bound to this tissue either, labelling only the cuticular epithelium (Fig. 2).

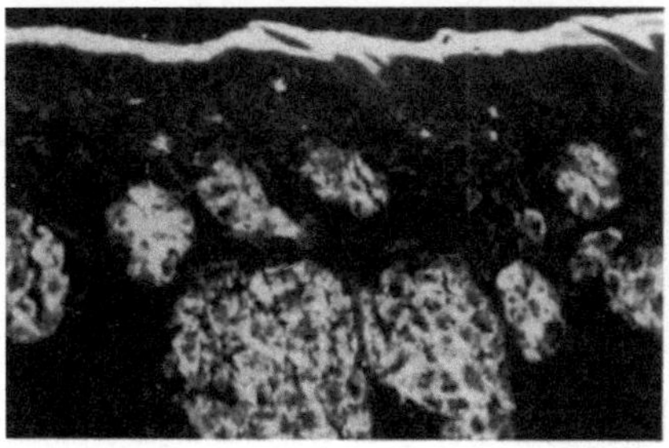

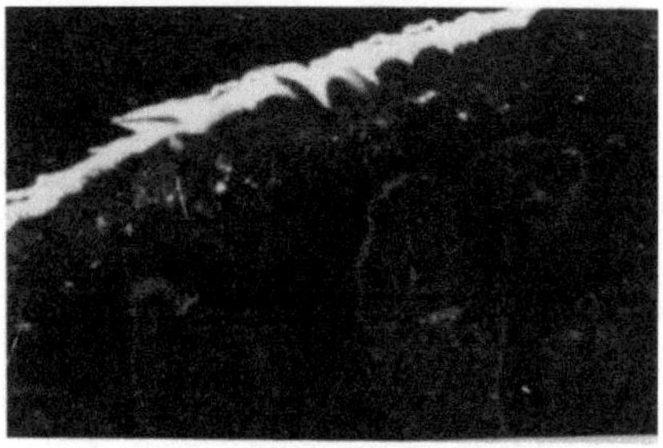

Fig. 1 and 2. Sections of F. hepatica treated with WGA-FITC (left) and PHA-FITC (right).

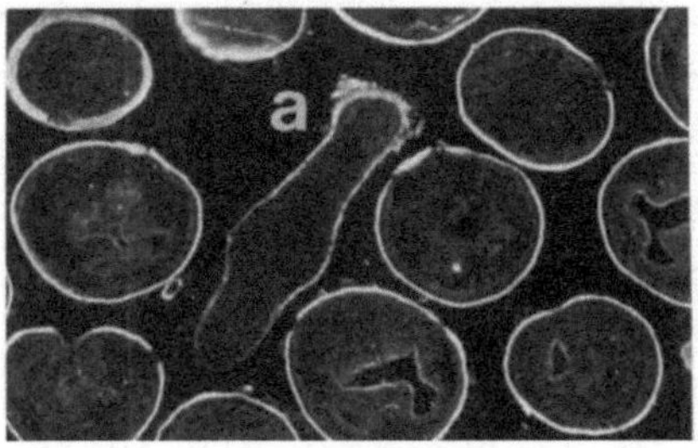

Fig. 3. Sections of invaginated and evaginated (a) E. granulosus stained by WGA-FITC.

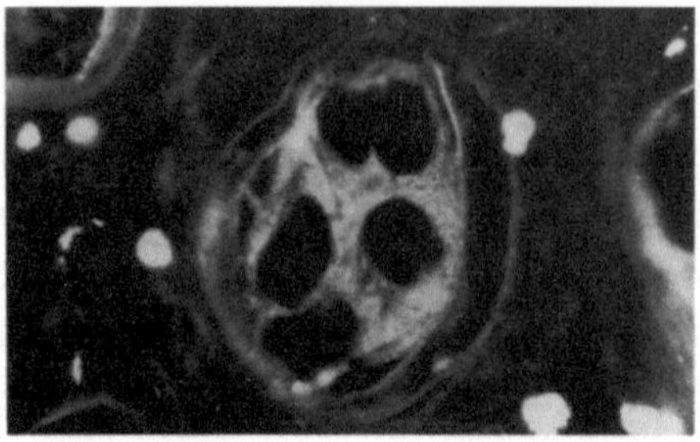

Fig. 4. Cross section of encapsulated T. spiralis muscle larvae treated with PHA-FITC.

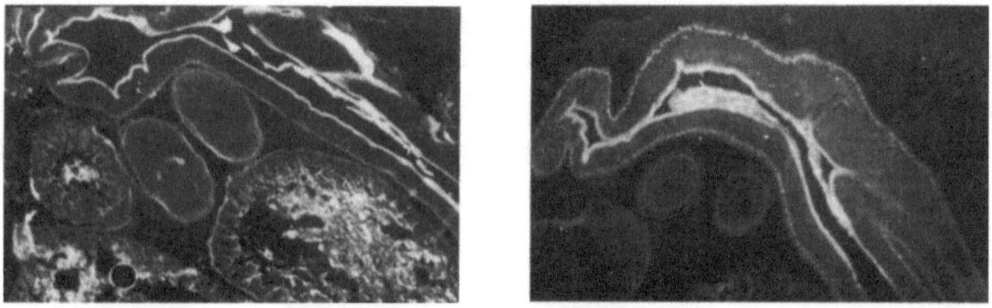

Fig. 5 and 6. Sections of A. suum treated with WGA-FITC (left) and PHA-FITC (right). The surface of intestine and uterus is stained with WGA-FITC. The surface of intestine, intestinal content and nuclei of epithelial cells are stained with PHA-FITC.

E. granulosus in its developmental stage of protoscolex can appear in invaginated or evaginated form. Both forms are visible in Fig. 3. It is possible to distinguish cuticulae, well developed muscles belonging to the rostellum and suckers, as well as excretory systems, on the histological sections of both forms. The surface of the invaginated protoscolex was uniformly and clearly stained by WGA-FITC conjugate, as well as the surface of the rostelum and the apical region of the evaginated protoscolex (A). PHA-FITC conjugate did not react with *E. granulosus* protoscolex sections.

The infective stage of *T. spiralis* are larvae which are encysted in host striated muscles. Histological studies of this stage exhibit a well developed capsule enclosing the larvae

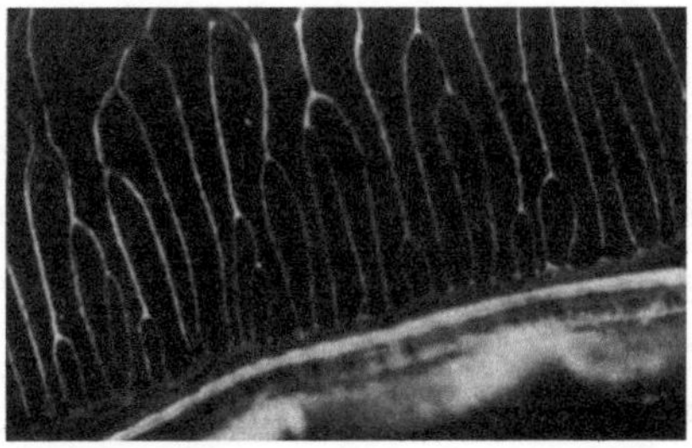

Fig. 7. Section of A. suum treated with WGA-FITC. Several layers of cuticle and surface of muscle cells show bright fluorescence.

composed of multilayered cuticulae, digestive and reproductive organs and stychocytes. As shown in Fig. 4 the material surrounding the parasite is intensively stained by PHA-FITC conjugate. The parasites themselves were not stained by the lectin conjugate and appeared in Fig. 4 as dark regions surrounded by bright fluorescence. Bright spots were also seen outside the encapsulated larvae and corresponded to the host cells stained by lectin. The fluorescent staining was not observed when these sections were treated with WGA-FITC conjugate.
The adult for of *A. suum* is the largest of all the investigated parasites, with very well developed organs and systems. Histological sections of *A. suum* reveal the cuticula, muscle, intestine and reproductive organs of the worm stained by lectin-FITC conjugates (Fig. 5-7). The well developed multilayer cuticula of *A. suum* was very brightly stained with WGA-FITC. Several layers of the cuticula were distinct: the cortex, matrix and fibre layers, which are further subdivided (Fig. 5). This lectin was also bound to the inner and outer surface of the uterus and to its contents, while the ovaria and hypodermis were stained slightly (Fig. 6). Both membranes of the cylindrical epithelial cells of the intestine reacted strongly with WGA-FITC.

Treatment with PHA-FITC provided the visualization of many intestinal structures (Fig. 7). There was strong fluorescent

staining at the surface of the cylindrical epithelial cells. The nuclei of these cells were also stained but less intensively. The content in the intestinum, when present, was very brightly stained. Other examined tissues: the cuticula, hypodermis, muscles and reproductive organs were not visualized by staining with PHA-FITC conjugate.
Control sections treated with lectin-FITC conjugates which were preincubated with the appropriate inhibitory sugars confirmed that the reactions of lectins with parasite tissues were specific, reflecting the distribution of the respective glycosylated molecules.

Discussion

Lectin histochemistry of the examined helminths revealed that sugar moieties reacting with PHA or WGA were mostly distributed on the epithelial surface of the cuticula and other peripheral tissues. The possibility to thus label and then separate glycosylated molecules of these regions is of particular importance, since the surface of the parasites appears to be a major determinant of the host - parasite interaction and the source of antigenc materials. The helminth cuticula is apparently composed of secreted collagen combined with hyaluronic acid and chondroitin sulphate-containing acid mucopolysaccharides (21). This explains the high affinity for lectins, characteristic for this tissue. Some variations in the composition of the cuticulae of the examined species were also revealed by lectin histochemistry.

Superficial layers of *F. hepatica*, *A. suum* and *E. granulosus* protoscolex exhibit a high affinity for WGA, but not for PHA. The specificity of WGA for N-acetyl-glucosamine, or better for β /1-n/ linked repeated sequences of this unit, suggests that a similar structure is a constituent of the cuticula and adhering mucus. It is known that the mucopolysaccharides coating the epidermis of *F. hepatica* or *A. suum* play a role in protecting the parasite from the digestive enzymes of the host (21). This layer is also the site of worm enzymes and the source of materials antigenic to the host. The high affinity for WGA now offers possibilities for further investigations of these molecules using lectin affinity chromatography or some other method based on the lectin - sugar interaction. Encapsulated *T. spiralis* is also covered by material reacting with lectin, but here it was PHA. The surface of the parasite undergoes a series of changes during the life cycle, which are the manifestations of protection and adaptation to the host environment. So, in the three developmental stages of *T. spiralis* completely different surface antigens were found (5), but they were all glycosylated molecules possessing an affinity for Concanavalin A. Lectin histochemistry revealed that they also shared an affinity for PHA, at least in the third developmental stage. Surprisingly, internal tissues of the encapsulated parasite did not react with either of the two applied lectins.

The overall distribution of lectin conjugates on the helminth histological sections corresponds to the localization of antigen - antibody complexes when similar sections were treated by anti - parasite antibodies (22). It can be thus assumed that some of the components which act as antigens also react with lectins, which can be the basis for their further investigation.

References

1. Soulsby, E. J. L.: Ann. N. Y. Acad. Sci. 113, 492-509 (1963).
2. Clegg, J. A., Smith, M. A.: Adv. Parasitol. 16, 165-218 (1978).
3. Philipp, M., Rumjaneck, F. D.: Mol. Biochem. Parasitol. 10, 245-268 (1984).
4. Philipp, M., Parkhouse, R. M. E., Ogilvie, B. M.: Nature 287, 538-540 (1980).

5. Parkhouse, R. M. E., Clark, N. W. T.: Mol. Biochem. Parasitol. 9, 319-327 (1983).

6. Samuelson, J. C., Caulfield, J. P.: J. Cell. Biol. 94, 363-369 (1982).

7. Rumjanek, F. D., Curiel, M.: Mol. Biochem. Parasitol. 7, 183-195 (1983).

8. Anders, R. F., Howard, R. J., Mitchell, G. F.: Immunology of Parasitic Infections. EDS. S. Cohen, K. S. Warren. Blackwell Sci. Publ. Second. Ed. Oxford, 29-73 (1982).

9. Yoshino, T.P., Cheng, T. C., Renwrantz, L. R.: J. Parasitol. 63, 818-824 (1977).

10. Murrell, K. D., Taylor, D. W., Vannier, W. E., Dean, D.A.: Exp. Parasitol. 46, 247-255 (1978).

11. Bennett, J. L., Seed, J. L.: J. Parasitol. 63, 250-258 (1977).

12. Simpson, A. J. G., Smithers, S. R.: Parasitol. 81, 1-15 (1980).

13. Simpson, A. J. G., Correa-Oliveira, R., Smithers, S. R., Sher, A.: Mol. Biochem. Parasitol. 8, 191-205 (1983).

14. Movsesijan, M., Borojević, D., Čuperlović, M.: Acta Veterinaria, 34, 31-40 (1984).

15. Pelley, R. P., Pelley, R. J., Hamburger, J., Peters, P. A. Warren, K. S.: J. Immunology 117, 1553-1555 (1976).

16. Snary, D., Hudson, L.: FEBS Lett. 100, 166-170 (1979).

17. Čuperlović, M., Movsesijan, M., Djordjević, D.: Jugoslav. Physiol. Pharmacol. Acta 17, 211-218 (1981).

18. Vretblad, P.: Biochim. Biophys. Acta 434, 169-176 (1976).

19. The, T. H., Feltkamp, T. E. W.: Immunology 18, 865-873 (1970).

20. Pantelouris, E.M.: The Common Liver Fluke. Pergamon Press, London (1965).

21. Lee, D.L.: Adv. Parasitol. 4, 187-254 (1966).

22. Stankov, M., Movesisjan, M., Kovanzic, D.: Acta Parasitol. Jugoslav. 6, 31-36 (1975).

23. Rautenberg, P., Reinwald, E., Kaufmann, E., Risse, H.J. In: Lectins, Biology, Biochemistry, Clinical Biochemistry Vol 2. (ed. T.C. Bøg-Hansen) W. de Gruyter, Berlin, 1982 pp. 619-632.

Part III

Lectin Tools

EFFECT OF pH AND METAL IONS ON THE INTERACTIONS BETWEEN CONCANAVALIN A AND CARBOHYDRATE.

An Approach to Standardization of Affinity Electrophoresis for Con A - Carbohydrate Interaction - Part II.

Kazusuke Takeo, Masanori Fujimoto, Ryosuke Suzuno, Tatehiko Tanaka, Kazuyuki Nakamura, and Akira Kuwahara
First Department of Biochemistry, Yamaguchi University School of Medicine
J-755 Ube, Japan

Polyacrylamide gel disc electrophoresis of concanavalin A (Con A) has been carried out generally in an acidic medium, mainly using the discontinuous buffer system described by Reisfeld et al(1). While a discontinuous buffer system is superior to a continuous buffer system to gain sharp band, the choice of pH value is quite limited. Furthermore, it is impossible to perform electrophoresis at any constant concentration of metal ions. We have reported previously a continuous buffer system for polyacrylamide gel disc electrophoresis in the broad pH range(2). Using this buffer system, we have investigated the effects of pH and metal ions on the interactions between Con A or succinylated Con A (s-Con A) and dextran.

Lectins, Vol. IV

Methods

Polyacrylamide gel disc electrophoresis was carried out at 20°C with a thermostatic electrophoresis apparatus (3) using the continuous buffer system. Table I shows the stock buffer solutions for the continuous buffer system for polyacrylamide gel disc electrophoresis. The stock buffer solutions were diluted 20 times to the final concentration and used as the working buffer for the electrode vessel and the electrophoresis gel buffers. Ca(Ac)2 was added to the working buffer solutions. Electrophoresis gels, 5 cm in height, were prepared using 5.1% polyacrylamide gel for Con A and 6.9% polyacrylamide gel for s-Con A by photopolymerization (2). For determination of dissociation constants (K) (4), one set of electrophoreses was run with 12 tubes, each composed of duplicated gels with 6 different concentrations of dextran (0, 0.04, 0.08, 0.12, 0.16, and 0.20%). The relative migration distance (R_o or r) was expressed as the ratio of the migrating distance of the protein to that of the indicator dye, methylene blue or bromophenol blue. K values were calculated according to the following equation (Eq. 1) (4):

$$1/r = (1/R_o)[1+(c/K)] \quad \ldots\ldots\ldots\ldots\ldots\ldots\ldots (1),$$

where R_o and r are the relative migration distances of Con

Table 1. Stock buffer solutions of continuous buffer system for polyacrylamide gel disc electrophoresis. Working buffer solutions were prepared by dilution of the stock buffer solutions on 20-fold, except of the histidine buffers(*). The latter was diluted 4-fold.

pH	HCl (2M)	HAc (2M)	β-alanine (2M)	Histidine (0.2M)	Imidazole (2M)	Tris (2M)	Diethanolamine (2M)	H_2O
	ml	ml	ml	ml	ml	ml	ml	ml
3.5	100		200					300
3.6	100		250					250
3.8	100		300					200
4.3		300				100		200
4.6		200				100		300
4.9		150				100		350
5.6		140		1600				1260*
6.0		100		2000				900*
6.8		100			150			350
7.0		100			200			300
7.2		100			250			250
8.0		100				150		350
8.2		100				200		300
8.6		100				300		200
9.1		100					200	300
9.3		100					300	200

A or s-Con A in the absence and presence of dextran in the gels and c is the concentration of dextran in the gels. Demetallized Con A (apo-Con A) and demetallized s-Con A (apo-s-Con A) were prepared by preparative affinity electrophoresis (5). Commercial Con A was dialyzed overnight against 20 mM EDTA at pH 5.0 and then applied to the preparative electrophoresis (6) with 5.0% polyacrylamide gel containing 0.5% dextran, using the buffer system described by Reisfeld et al.(1). Con A was separated into 3 bands depending on the affinity to dextran. The fastest band was extracted from the gel. The fraction showed no affinity to dextran. This fraction was used as apo-Con A. Apo-s-Con A was prepared from commercial preparation of s-Con A using the same method as apo-Con A.

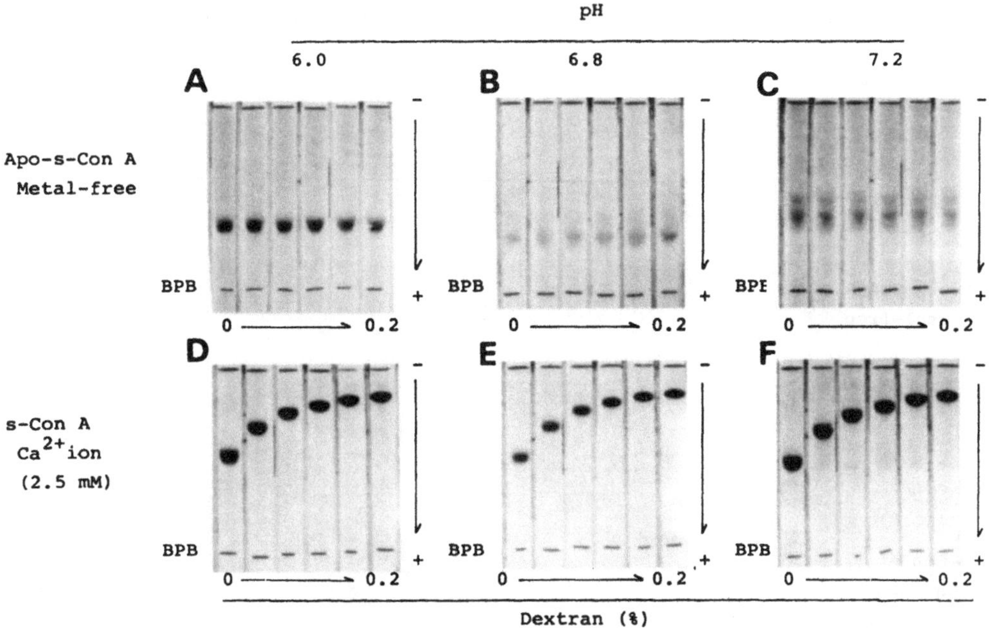

Fig. 1. Affinity patterns of apo-s-Con A and metallized s-Con A at neutral pH. Electrophoresis was carried out using 6.9% polyacrylamide gel at 20°C at pH 6.0 (A and D), pH 6.8 (B and E), and pH 7.2 (C and F), and in the absence of metal ions (A, B, and C) and the presence of 2.5 mM calcium ion. One set of electrophreses was run with 12 tubes, each composed of duplicated gels of 6 different concentrations of dextran (0, 0.04, 0.08, 0.16, and 0.20%). The protein sample migrated from cathode to anode. Bromophenol blue was used as the reference. Its position (BPB) was shown by the copper wire stuck in. The protein was stained by Coomasee BBR-250.

Results and Discussion

1) Apo-s-Con A migrates as a single band in the pH range between 4.3 and 6.0 as seen in Fig. 1A. At pH 6.8 (Fig. 1B), it is dispersed into several bands and, at higher pH, the intensity of the dispersion is increased, as seen in Figs. 1 and 2. At pH 8.6 (Fig. 2B) and 9.1 (Fig. 2C), it is separated into one major and four minor bands. While apo-s-Con A regains full binding activity, migrating as a sharp single band in the presence of 2.5 mM Ca^{2+} ion below pH 6.8 (Figs. 1D and E), the minor bands appear at the same position as seen on the electrophoresis pattern obtained in the absence of Ca^{2+} ion above pH 7.2 (Fig. 1F or Figs. 2D, E, and F). Those minor bands have no binding affinity to

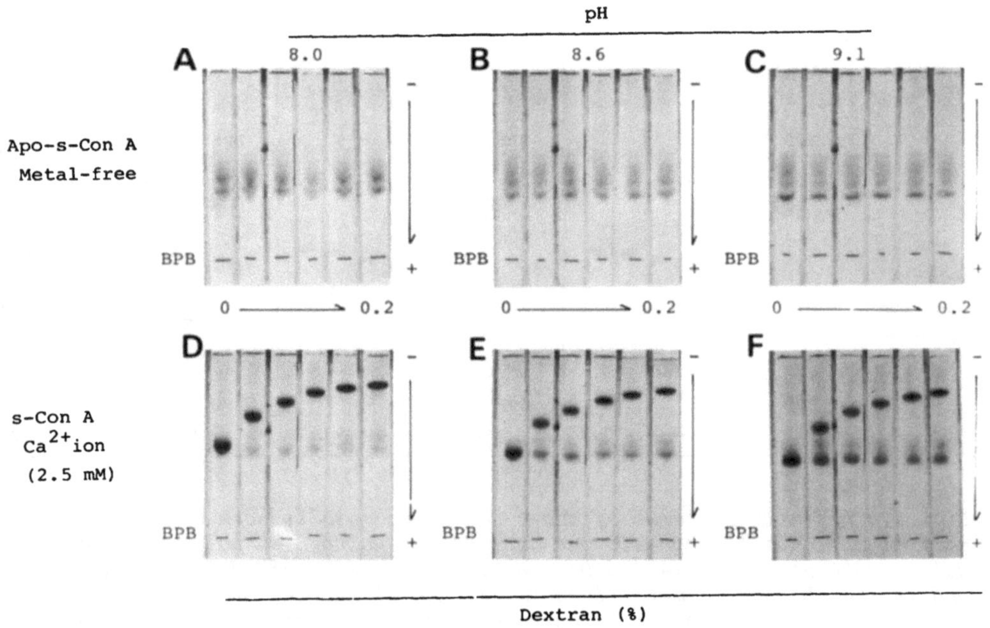

Fig. 2. Affinity patterns of apo-s-Con A and metallized s-Con A at slightly alkaline pH. Electrophoresis was carried out at pH 8.0 (A and D), pH 8.6 (B and E), and pH 9.1 (C and F). The other conditions of electrophoresis were the same as in Fig. 1.

dextran. The dissociation constant of the active s-Con A for dextran remains constant in the pH range between 4.3 and 9.1, estimated to be 0.24 to 0.27 mM.

2) Apo-Con A migrates as a single band in the pH range between 4.3 and 9.1, as seen in Figs. 3 and 4. At pH 4.3 (Fig. 3A) and 5.6 (Fig. 3B), it migrates from anode to cathode, while at pH 6.8 (Fig. 3C), it migrates from cathode to anode. Thus, its pI value is estimated to be 6.1. It does not regain the binding activity to dextran, even in the presence of 2.5 mM Ca^{2+} ion at pH 4.3 (Fig. 3D) and 4.9.

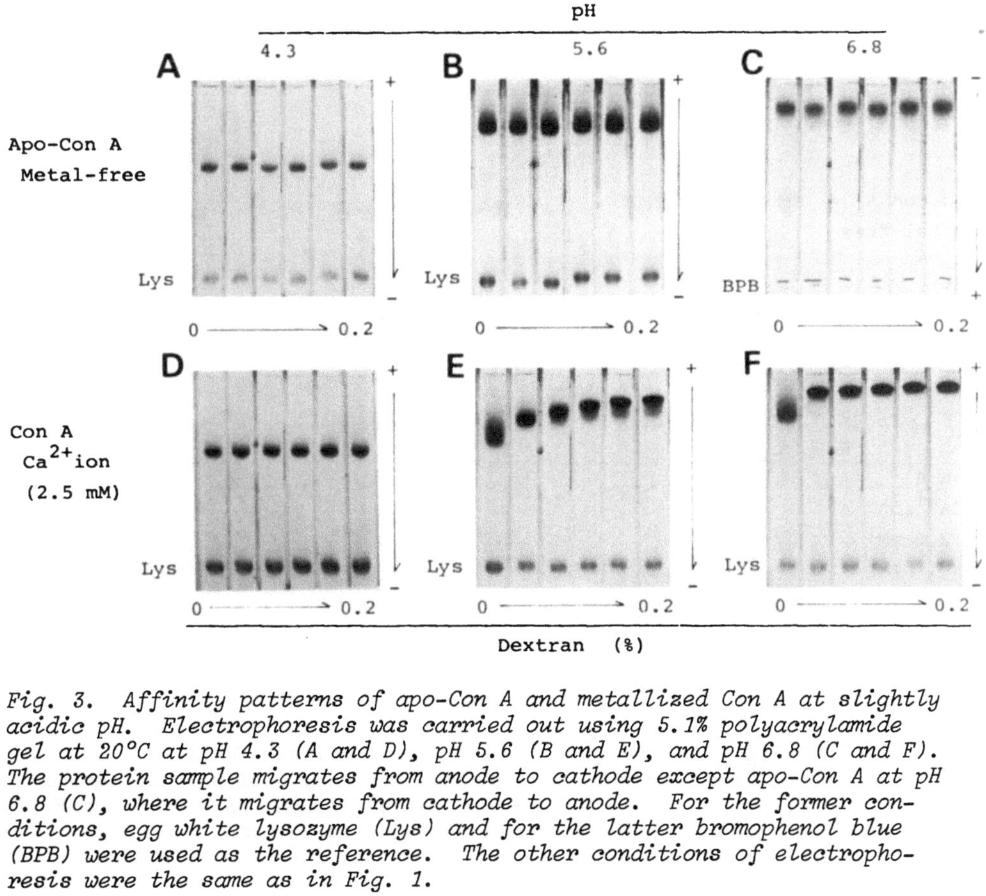

Fig. 3. Affinity patterns of apo-Con A and metallized Con A at slightly acidic pH. Electrophoresis was carried out using 5.1% polyacrylamide gel at 20°C at pH 4.3 (A and D), pH 5.6 (B and E), and pH 6.8 (C and F). The protein sample migrates from anode to cathode except apo-Con A at pH 6.8 (C), where it migrates from cathode to anode. For the former conditions, egg white lysozyme (Lys) and for the latter bromophenol blue (BPB) were used as the reference. The other conditions of electrophoresis were the same as in Fig. 1.

When 0.01 mM Co^{2+} ion is added together with Ca^{2+} ion, it regains the binding activity separating into two bands, the fully metallized and the partially metallized Con A (5). In the presence of 2.5 mM Ca^{2+} ion and 0.1 mM Co^{2+} ion at pH 4.3, it was fully metallized and migrated as a single band(5). At pH 5.6 in the presence of 2.5 mM Ca^{2+} ion (Fig. 3E), it regains the binding activity, separating into two bands (I and II) in the same manner as seen at pH 4.3 in the presence of 2.5 mM Ca^{2+} and 0.01 mM Co^{2+} ions. While apo-Con A migrates from cathode to anode at pH 6.8 (Fig. 3C), the metallized Con A migrates from anode to cathode as a single band, having a higher binding activity in the presence of

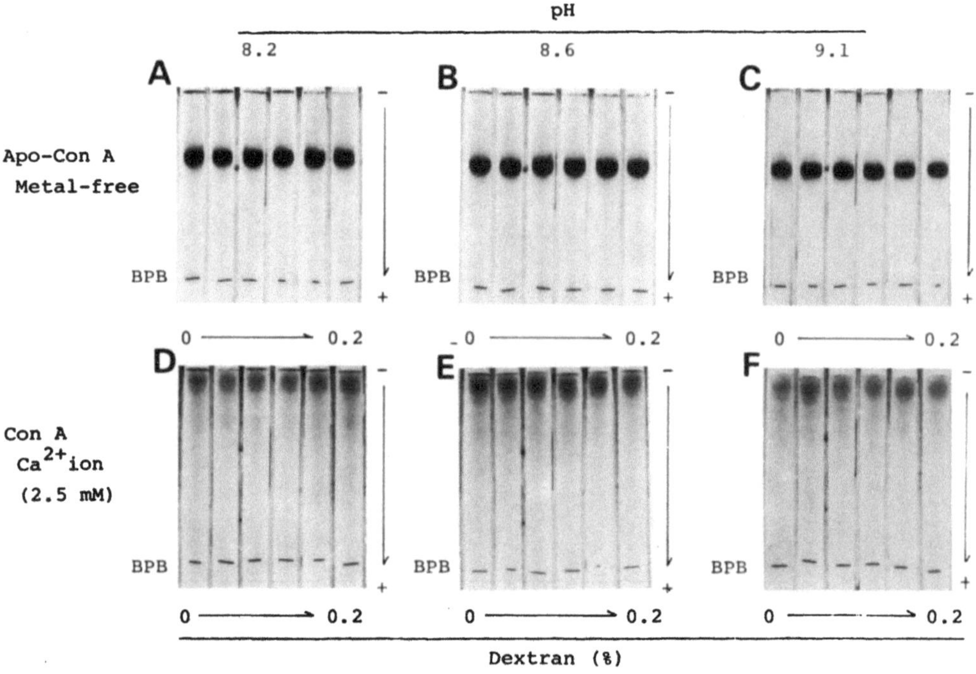

Fig. 4. Affinity patterns of apo-Con A and metallized Con A at slightly alkaline pH. Electrophoresis was carried out using 5.1% polyacrylamide gel at 20°C at pH 8.2 (A and D), pH 8.6 (B and E), and pH 9.1 (C and F). The protein sample migrates from cathode to anode. Bromophenol blue (BPB) was used as the references. The other conditions of electrophoresis were the same as in Fig. 1.

2.5 mM Ca^{2+} ion at pH 6.8 (Fig. 3F). In the pH range between 8.2 and 9.1, it remains at the origin of the separating gel as a diffuse band in the presence of 2.5 mM Ca^{2+} ion, as seen in Figs. 4D, E and F. While the pI value for the metallized Con A could not be estimated exactly, it moves to more alkaline pH level, 1.5 to 2.0 pH units higher than those for apo-Con A. This shift of pI appears to be a result of the binding of metal ion to apo-Con A.

3) Con A is a homogeneous protein which exists as an equilibrated mixture of various molecular forms. It exists as a tetrameric structure at neutral pH and predominately as a dimeric structure at pH level below 5.6 (7). In the presence of metal ions, it exists electrophoretically in 3 forms: Type I, the native or the fully metallized form which has full binding activity, Type II, the partially metallized form which has partial binding activity, and Type III, the demetallized form, or apo-Con A, which has no binding activity (7). Distribution of those 3 forms depends on pH and concentration of metal ions (5). Thus, in the presence of 0.34 mM Ca^{2+} ion at pH 8.6, as seen in Fig 5, Con A was not fully metallized and a small fraction of Con A (Fig. 5B)

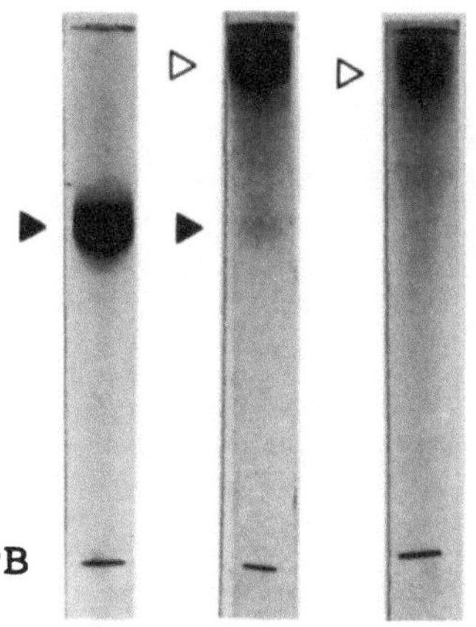

BPB

A B C

Fig. 5. Effect of calcium ion concentration on metallization of apo-Con A. Electrophoresis was carried out using 5.1 % polyacrylamide gel containing 0.2% dextran at 20°C at pH 8.6. Patterns A, B, and C were obtained in the absence and the presence of 0.34 mM and 2.5 mM calcium ion, respectively. ▶: Apo-Con A and ▷: Metallized Con A. The other conditions of electrophoresis were the same as in Fig. 4. See patterns B and E of Fig. 4.

appears at the same position where apo-Con A appears (Fig. 5A or Fig. 4B). In this respect, Con A interaction with carbohydrates or glycoproteins should be performed in the presence of a sufficient concentration of Ca^{2+} ion (more than 1 mM), preferrentially adding a second metal ion such as Co^{2+} ion (0.1 mM).

References

1. Reisfeld, R. A., Lewis, U. J., and Williams, D. E.: Nature 201, 281-283 (1962).
2. Takeo, K., Fujimoto, M., and Kuwahara, A.: Lectins 3, 397-404 (1983).
3. Takeo, K.: Lectins 2, 583-594 (1982).
4. Takeo, K. and Nakamura, S.: Arch. Biochem. Biophys. 153, 1-7 (1972).
5. Kuwahara, A., Fujimoto, M., Suzuno, R., and Takeo, K.: Electrophoresis '83, Walter de Gruyter, Berlin-New York, 643-646.
6. Takeo, K., Suzuno, R., Fujimoto, M., Kuwahara, A., Ogata, H., and Nakamura, S.: Bull. Yamaguchi Med. Sch., 23, 165-183 (1976).
7. Kalb, A. J. and Lustig, A.: Biochim. Biophys. Acta 168, 366-367 (1968).

CROSSED AFFINOIMMUNOELECTROPHORESIS WITH GALACTOSE SPECIFIC LECTINS IN THE FIRST DIMENSION

J.A. Kint and J.G. Leroy
Childrens Hospital, Ghent State University Medical School,
Ghent, Belgium.

The technique of two dimensional affinity electrophoresis in agarose gel with a free lectin in the first dimension was introduced by Bøg-Hansen and Brogren (1). The technique has since been applied with great success (2-4). The combination of quantitative immunoelectrophoresis and affinity electrophoresis with lectins has yielded unexpected findings such as the polymorphism of some serum proteins (e.g. alpha-1-acid glycoprotein, alpha-feto-protein, C-reactive protein). The method has been applied mostly with the more popular lectins such as concanavalin A and wheat germ lectin (2). Although almost any lectin can be used in this technique, the major class of galactose specific lectins with high affinity for the agarose itself such as ricinus communis agglutinin (RCA), peanut agglutinin or crotalaria lectin are not suitable. Binding to agarose prevents any interaction between the lectin and the galactoproteins. Therefore we have looked into finding an alternative to the agarose as the medium for the first dimension electrophoresis. Cellulose acetate membranes seemed an attractive choice as they do not interact specifically with lectins.

Lectins, Vol. IV

Materials and Methods

RCA was purified by affinity chromatography according to Tomita et al. (6). Cellulose acetate membranes (Gelman, Ann Arbor, Michigan, USA) were used in a L.K.B. Multiphor electrophoresis apparatus. The membranes were supported by glass plates (60 x 60 x 1 mm) after being cut to the appropriate length. After soaking for a few minutes in the electrophoresis buffer (0.05 M Veronal, pH = 8.6 with or without lectin) the membranes were carefully laid at the edge of these glass plates. The excess of buffer liquid was removed by gentle blotting with filter paper. Eight such plates could be run simultaneously. Contact between the first and second row of plates was made by paper wicks. The first dimension electrophoresis was carried out under the following conditions: 0.05 M Veronal buffer pH = 8.6, 200 Volts, 45 minutes, 2 milliampères per membrane - cooling temperature 15°C. Following the first electrophoresis the plates were removed from the tank and 3 milliliter of a preheated (60°C) solution of 1% agarose in electrophoresis buffer together with 20-40 microliter of a specificity antibody (Dako, Denmark) was applied to each glass plate. By holding a plastic bar on top of the cellulose acetate membrane, the agarose solution was prevented from covering completely the membrane. The agarose was only allowed to make contact for about 2-3 mm along the total length. After solidification the plastic bar was removed, the plates were put back in the tank and the second electrophoresis was started (40 volts - 18 hours, 2 milliampères per plate). The drying, washing and staining of the plates was carried out as described by Kröll (7). Sample application (0,5 µl) was carried out with a Hamilton syringe.

Results and discussion

During our attempts to characterize the interaction between galactose specific lectins and human serum proteins, we were hindered by the high affinity between RCA and agarose. Although it was possible in a few cases to detect some interaction, most of the time a highly stained background was the only result. A gel of polyacrylamide seemed at first worthwhile but resulted only in diffuse and weak immuneprecipitates presumably because of insufficient diffusion of the antigen and the antibodies in such a gel. Therefore we turned to the combination of a cellulose acetate membrane and agarose gel. In order to demonstrate the resolving power of crossed affinoimmunoelectrophoresis with cellulose acetate membrane in the first dimension, the separation of the orosomucoid fractions by electrophoresis in the presence of Con A was carried out in duplo (Fig. 1). First agarose gel was used in both dimensions (Fig. 1A). A second time a cellulose acetate membrane instead of agarose was used in the first dimension (Fig. 1B). In both cases the first dimension medium contained 1 milligram of Con A per milliliter and agarose with anti orosomucoid antibodies in the second dimension. As can be concluded from Fig. 1, both separations yielded about the same resolution. Cellulose acetate has only minor drawback: Coomassie Blue is difficult to remove from the membrane and this results in a permanent dark blue background. However, this dark background does not interfere with the image of the peaks in the second dimension. The clear separation of the orosomucoid fractions allowed us to use then the same technique but with RCA instead of Con A.

Ziska and Franz (9) could show that α-2-macroglobin, haptoglobin, IgM and β-lipoprotein were the only proteins from human serum which were bound to the galactose specific Viscum album lectin. Therefore we tested our technique by analysis of two of these proteins, α-2-macroglobin and haptoglobin.

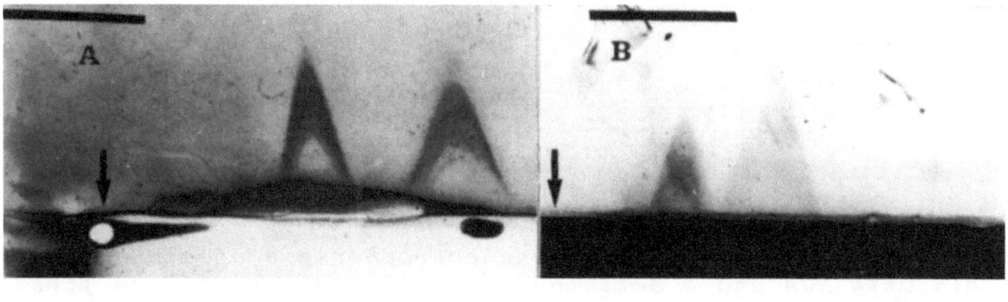

Figure 1. Affino-immunoelectrophoresis of orosomucoid from human serum. First dimension: 1% agarose (A) or cellulose acetate membrane (B), both with 1 mg/ml Con A. Second dimension (A and B): 3 ml of 1% agarose with 40 µl of anti orosomucoid. The arrow indicates the origin. Cathode is to the right for the first dimension and to top for the second dimension. The bar indicates 1 cm. Notice the dark blue background of the cellulose acetate membrane and the dark "rocket" of precipitated glycoproteins at the origin of A.

Alpha-2-macroglobin. Crossed affinoimmunoelectrophoresis of human serum with increasing amounts of RCA in the first dimension and anti α-2-macroglobin in the second dimension resulted in the following observations:

1. High concentrations of RCA resulted in a smaller area of the α-2-macroglobin peak, with a complete disappearance at RCA concentrations of 10 mg/ml (Fig. 2).

One should notice that although the height of peak a is smaller than in peak b, c or d, the surface area is not. The smaller area in peaks e, f, g and h indicates the progressive binding of the α-2-macroglobin protein to the lectin. The complete disappearance suggests that all α-2-macroglobin molecules contain galactose in a terminal position.

2. At low RCA concentration the smaller α-2-macroglobin peak showed an electrophoretic mobility which was proportional to the

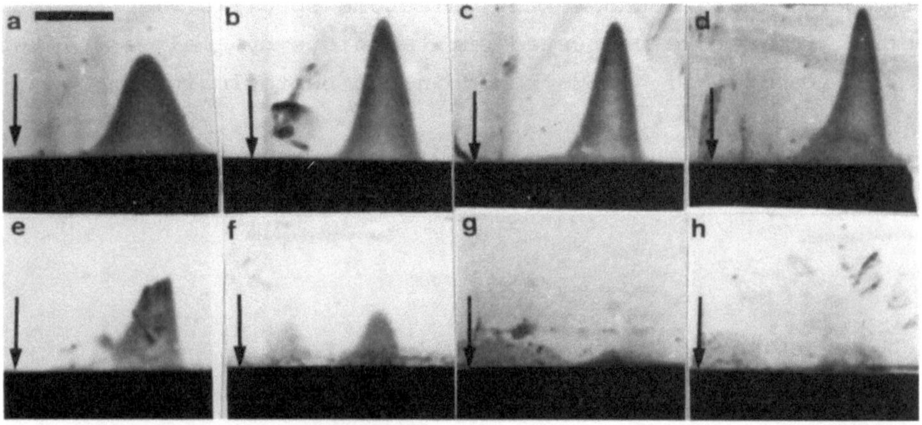

Figure 2. Crossed Affino-immunelectrophoresis of alpha-2-macroglobin with RCA and a cellulose acetate membrane in the first dimension. The membranes were soaked in buffer with RCA at the following concentrations:
a = 0,00 mg/ml; b = 0,15 mg/ml; c = 0,30 mg/ml; d = 0,60; e = 1,25 mg/ml; f = 2,50 mg/ml; g = 5,00 mg/ml; h = 10,00. One microliter of human serum was applied. The second dimension contained 20 μl of anti alpha-2-macroglobin, per ml of agarose gel.
The meaning of the arrow and the bar and the position of the cathodes are the same as in figure 1.

RCA concentration (Fig. 3). We assume that in the membrane the soluble lectin-glycoprotein complex shows a higher mobility than the free α-2-macroglobin does.

3. It was impossible to separate the α-2-macroglobin peak in more one fraction. Whatever concentration of RCA was used, no second peak was seen. This finding constrasts strongly with the effect of affinoimmunoelectrophoresis with Con A which at high concentration results in multiple peaks of several serum proteins (1).

Haptoglobin. Mixtures of human serum and of different amounts of hemoglobin were prepared and subjected to crossed affinoimmunoelectrophoresis with RCA (10 mg/ml) in the first dimension and with anti haptoglobin antibodies in the second dimension. The peak of haptoglobin in normal serum disappears completely

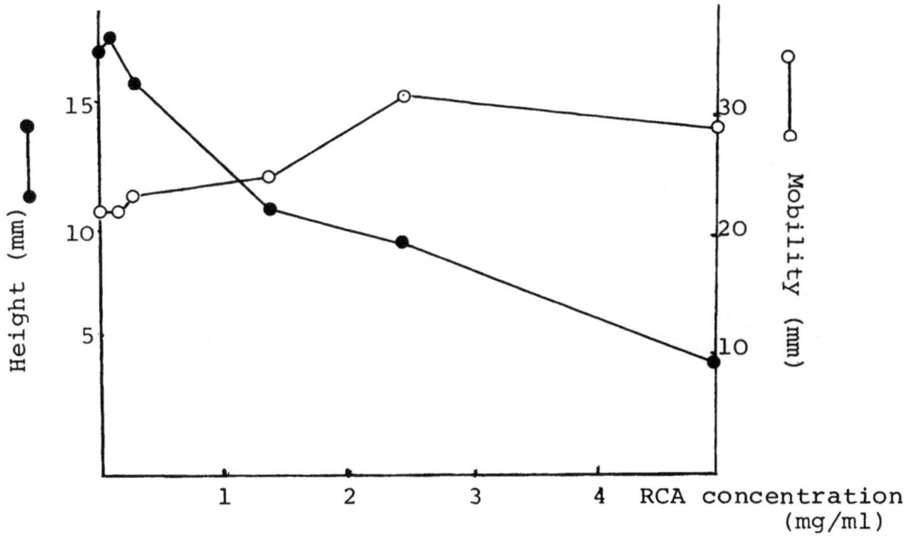

Figure 3. Mobility and height of peak of alpha-2-macroglobin in a RCA-containing cellulose acetate membrane.
Open circles: mobility of peak; closed circles: height of the peak. Abscissa: concentration of RCA in the membrane.

in the RCA-containing membrane. However, after mixing increasing amounts of hemoglobin with the serum, the haptoglobin peak reappears, suggesting that the binding of hemoglobin to haptoglobin has some consequence for the affinity of RCA to the exposed galactose residues of the haptoglobin. We assume that in the hemoglobin-haptoglobin complex, galactose is no longer available for binding to RCA. This fits well with studies of Con A binding, cf. Katnik et al. (These proceedings).

Conclusion. A technique is described which allows two dimensional affinity electrophoresis with galactose specific lectins in the first dimension. By using cellulose acetate membrane for the first dimension, binding of the galactose lectin to the agarose support is prevented.
Experiments with haptoglobin and alpha-2-macroglobin could not demonstrate any polymorphism in either of the two galactoproteins. Following its complex formation with hemoglobin, haptoglobin could no longer be bound to RCA.

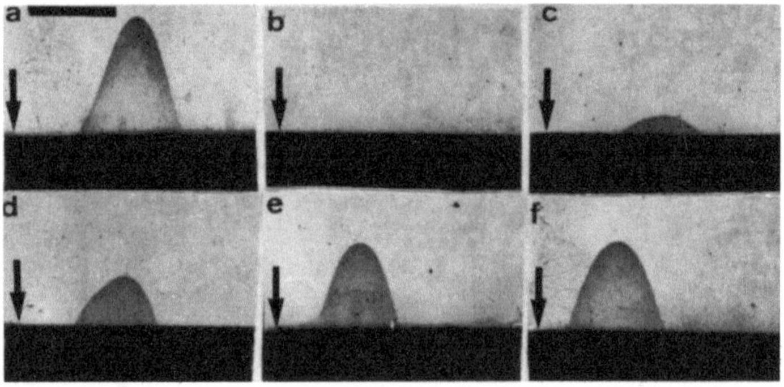

Figure 4. Crossed affino-immunoelectrophoresis of haptglobin from human serum. First dimension: cellulose acetate membrane. Second dimension: agarose gel with 20 µl of antihaptoglobin per milliliter of gel.
a = control experiment without RCA.
b-f = RCA (5 mg/ml) in the first dimension.
Before analysis the serum was mixed with hemoglobin to a concentration of:
a = 0,00 mg/ml; b = 0,00 mg/ml; c = 1,87 mg/ml; d = 3,75 mg/ml; e = 7,50 mg/ml; f = 15,00 mg/ml.

References

1. Bøg-Hansen T.C. and Brogren C.H. Scand. J. Immunol. 4, suppl. 2, 135-139 (1975).

2. Bøg-Hansen T.C. and Takeo T. Electrophoresis 1, 67-71 (1980)

3. Bøg-Hansen T.C. in Solid phase biochemistry. Analytical Synthetic Aspects, 223-251, Editor William H. Scouten, John Wiley and Sons, New York (1983).

4. Bjerrum O.J. and Bøg-Hansen T.C. Biochemical Analysis of membranes, Editor A.H. Maddy, Chapman and Hall, London p. 378 (1981).

5. Horejsi V., Ticha M and Kocourek J. T.I.B.S. 4,1,N6(1979)

6. Tomita M., Kurokarva T., Onzozakik M., Ichiki N., Osawa T. and Ukita T. Experientia 28, 84-85, 1972.

7. Kröll J. Scand. J. Clin. Lab. Invest. 22, 112-114 (1968)

8. Jayle M.F. Les haptoglobines, pp 1-112, Masson, Paris (1962).

9. Ziska P. and Franz H. Acta Biol. Med. Germ. 38, 677-700 (1979).

NEITHER SERUM FERRITIN NOR ITS CON A-BINDING FRACTION ARE TUMOR MARKERS IN RENAL CARCINOMA

Mette M. Andersen[1)] A. Lihme[2)] Susanne Noack[1)] T.C. Bøg-Hansen[2)]

[1)] Laboratory of Medical Immunology, Rigshospitalet and The Finsen Laboratory, The Finsen Institute, Copenhagen (present address)

[2)] The Protein Laboratory, University of Copenhagen, Denmark

The concentration of serum ferritin in normal healthy state is regulated by the liver according to the iron stores of the organism (19). In patients with malignant tumors, however, the serum ferritin concentration is abnormally increased (11, 13, 15, 17, 20, 21). Tissue ferritin concentrations are often very high in malignant tumors (1, 22) and the question of whether tumor tissue contributes to serum ferritin has often been raised. Since patients with renal cancer in low clinical stages have their tumors removed completely by operation, patients with this tumor type were chosen for the present study. Having studied the effect of tumor removal on the serum ferritin concentration, we turned to look at the possible contribution by the leucocytes.

Ferritin in serum contains a con A-binding fraction (23), whereas ferritin from organs and leucocytes does not bind to con A by immunoelectrophoretic techniques (4). We therefore also studied the effect of tumor removal on the con A-binding serum ferritin fraction.

Finally, we continued our work comparing different ferritins in order to find characteristics which have to be defined in

Lectins, Vol. IV

order to provide ferritin standards for immunoelectrophoretic work with ferritin.

Materials and Methods

Patients: All patients had renal carcinomas in stage I or II/III, (staging was performed by the Copenhagen Study Group of Renal Carcinoma), i.e. without distant metastases and all had a radical nephrectomy. Eighteen patients were used for comparative studies of tumor ferritin and total and con A-binding serum ferritin. Serum was obtained the day before operation and again, from ten of the patients, at the first post-operative clinical checkup 1 - 12 weeks later. Sera from 20 volunteers were used for comparison. Five other patients were used for studies of the immediate effect of tumor removal on the concentration of plasma and leucocyte-associated ferritin. From these patients about 20 ml blood was drawn in heparinized tubes on the morning of operation and again three to eleven days later. The leucocytes were isolated and washed 6 times in Hank's buffered salt solution and each wash supernatant was harvested. Blood cells from three regular blood donors and two healthy volunteers were treated similarly for comparison.

Tumors: The renal carcinoma tumors were obtained at operation and frozen ($-20\ ^{\circ}C$) until further analysis. At the day of analysis the tumors were thawed, cut by scissors and homogenized in a Potter-Elverhjelm blender in diemal buffer 1:14 (w/v). In three cases the tumors were large enough for sampling of two pieces. Kidney tissue, obtained at autopsy from two accident victims, was treated similarly.

Immunoelectrophoretic techniques: The concentration of serum and tumor ferritin was quantitated by radioimmunoelectrophoresis using as references a serum pool and the pooled kidney tissue, respectively (5). Ferritin in renal tumor tissue was compared with ferritin in liver against anti renal tumor and anti liver ferritin antibodies in rocket immunoelectrophoresis (16). The anti liver ferritin antibodies were purchased from DAKOpatts, Copenhagen and the anti tumor ferritin antibodies prepared as described in (1). The purified liver ferritin was a gift from Dr. Strandberg Pedersen. The con A-binding ferritin was quantitated by crossed lectin affinity radioimmunoelectrophoresis (4). The following amounts and concentrations were used routinely: 10 - 30 µl serum in the application well; 350 µg con A/cm^2 in the first dimension gel; $0.5 - 1 \times 10^{-3}$ µl/cm^2 of unlabelled anti liver ferritin and $0.5 - 1 \times 10^{5}$ CPM ^{125}I-iodinated anti liver ferritin in the antibody containing gel. 2% PEG 6000 plus 2% Ficoll was also added to this gel to improve immunoprecipitation. Finally 1.5% α-methyl-D-glucopyranoside was included to dissolve the con A affinity precipitate formed during the first dimension electrophoresis which was run at 5 V/cm. The second dimension was run at 2 V/cm over night. After electrophoresis the plates were pressed for 24 hours, dried and subjected to autoradiography for 3 - 5 days at -20 °C. The relative areas of the immunoprecipitates of the con A-binding and non-binding ferritin were determined by planimetry. The percentage of con A-binding was calculated as

$$\frac{\text{con A-bd.}}{\text{con A-bd.} + \text{non-bd.}} \times 100\%.$$

The concentration of ferritin in plasma and in supernatants from leucocyte washings were quantitated by Sample tube immunoelectrophoresis (3). The electrophoresis was run overnight at 4 V/cm against anti liver ferritin antibodies with a serum pool ferritin as reference. Furthermore, IgG was quantitated in the same washing supernatants, as described in (2).

Results and discussion

Ferritin is not a tumormarker.

The concentration of total ferritin in the 18 preoperative sera varied from 30% to 600% of the reference serum pool and the concentration of ferritin in the respective tumor tissues varied from 10% to 9600% of the reference kidney tissue, Fig. 1. There appears to be some correlation between the serum ferritin concentrations and the log transformed values for tumor ferritin ($r = 0{,}5$; $p \leq 0{,}05$). When two samples were analysed from the same tumor these differed up to two fold in ferritin concentration. One double sample is included in the material illustrated in Fig. 1 (double circles).

Neither the tumor ferritin concentration nor the preoperative serum ferritin concentration were correlated to the clinical tumor stage of the patients (I or II/III).

Con A-binding ferritin in normal control sera represented 87.0% ($\pm$4.4 S.D.) and in preoperative tumor patient sera 77.8% ($\pm$12.1 S.D.). However, the lower percentage of con A-binding ferritin was exclusively found in patient sera with high total serum ferritin, Fig. 2. In serum taken postopera-

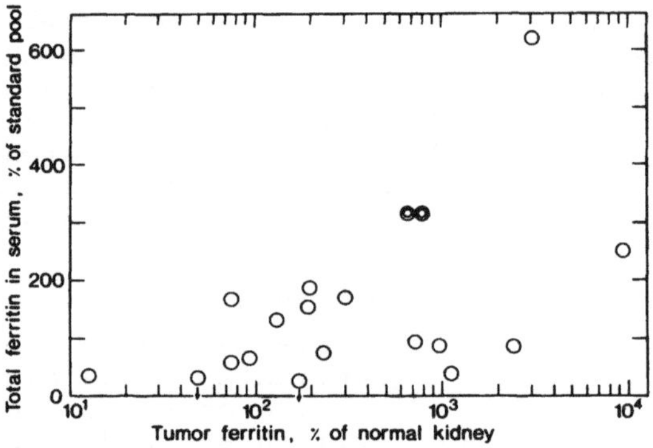

Fig. 1. Concentration of ferritin in serum against that in the respective tumor tissue from 18 renal cancer patients. From one patient two pieces of tumor tissue were analysed (double circles). In two patients the serum ferritin was less than 30% of serum pool (arrows pointing downwards from the circles).

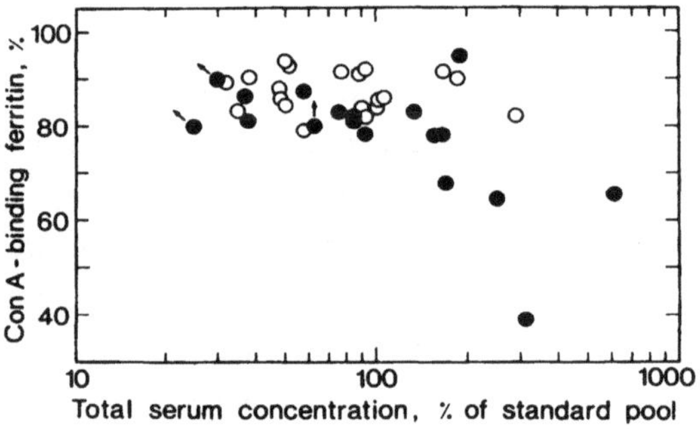

Fig. 2. Percentage of Con A-binding ferritin against total ferritin in sera from 18 preoperative patients (black circles) and 20 controls (open circles). In three patients the results were less precise due to the precipitates being rather loose, and the values shown for these three are minimum values (arrows).

tively the con A-binding ferritin was 80.4% ($\pm$6.5 S.D.), and again the lowest values were found in sera with the highest serum ferritin, Table I.

A low con A-binding fraction of ferritin has been seen in patients suffering from chronic liver diseases and iron overload (8, 9) and cannot be regarded as specific for tumorbearing patients. Only in sera from patients suffering from acute infections do the two ferritin fractions increase in parallel (6).

The increase in non-binding ferritin could be explained by direct contribution by any organ or cell type since no tissue-ferritin, including renal carcinoma and leucocyte ferritin, bind to con A (4). However, other possibilities exist since it is known that con A-binding ferritin has a different turnover and a different mechanism of breakdown (12, 24).

Fig. 3 illustrates the timecourse of the ferritin concentration after the operation. The ratio of pre- to post-operative ferritin is lower than 1 in the first weeks after operation. This demonstrates that the impact of operation in terms of raising serum ferritin is greater than that of the tumor. About day 50 the ratio increases above 1. This result is in accordance with results by Birgegård et al. (7) demonstrating that serum ferritin increases as acute phase proteins in general do, but that it remains increased for several weeks.

We thus conclude that renal carcinoma tumors do not contribute significantly to serum ferritin. Consequently, neither ferritin nor its con A-binding fraction can be regarded as tumor markers for renal carcinoma.

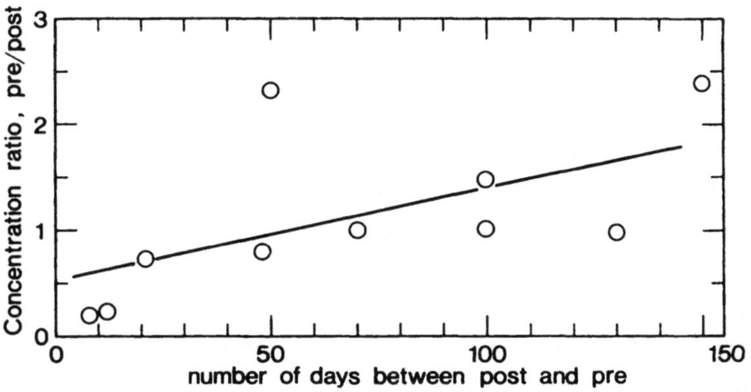

Fig. 3. Concentration ratio of preoperative to postoperative serum ferritin in 10 patients against the time interval between the respective samples. The least squares regression equation is $y = 0.0088x + 0.516$ and the correlation coefficient is $r = 0.61$. The slope is different from 0 ($p = 0.06$).

Table I

Patient no	Total ferritin in serum(*) pre	post	Percentage con A-binding ferritin pre	post
1	38	112	81.2	77.1
2	25	92	80(+)	85.3
3	37	53	86.5	84.4
4	170	210	67.7	67.3
5	252	108	64.4	88.1
6	30	30	90(+)	80(+)
7	314	214	39.3	72.5
8	65	65	80(+)	80(+)
9	75	75	82.8	88.9
10	58	25	87.4	80(+)
Mean values	106.4	98.4	75.9	80.4
±1 S.D.	±97.6	±63.2	±14.5	±6.5

(*) percentage of reference serum pool.

(+) the precipitate of the con A-binding ferritin was too diffuse for precise quantitation.

Leucocyte-associated ferritin.

The possible contribution by the leucocytes to the increased serum ferritin concentration was investigated by quantitation of the ferritin content in plasma and in 6 successive washing supernatants from leucocyte washings (2). Fig. 4 shows the patterns of such quantitations of leucocyte-associated ferritin in 2 patients typical of all 5. The results demonstrated that there is no difference in either the 'loosely' associated ferritin (i.e. the steep part of the curve) or the 'strongly' associated ferritin (i.e. the flat part of the curve) before and after operation. When the concentration of ferritin in the washed leucocytes was converted to concentration per milliliter of packed cells - assuming a cell volume of 200 μm^3

- we found this to be about 500 times that of the plasma in each case. Thus the leucocyte and plasma ferritin concentrations are probably correlated. The concentration of ferritin in leucocyte washing supernatants and cell pellets was about 10 times lower in the control group, but the respective IgG concentrations were in the same range as for the patients. Our results suggest that leucocyte ferritin synthesis is increased both as a result of the presence of the tumor in the patients and as a result of operation. Lipschitz et al. (18) have also demonstrated that leucocyte ferritin increases after surgery, and in vitro studies have shown that ferritin synthesis in monocytes increases several fold as a result of inflammatory stimuli (25).

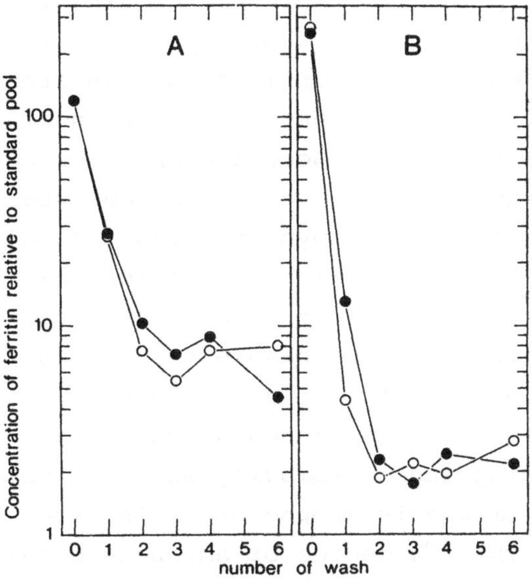

Fig. 4. Demonstration of identical leucocyte-associated ferritin before and after operation in 2 patients (A and B). Concentration of ferritin in plasma (0) and 6 successive wash supernatants before (black circles) and after operation (open circles). Patients A and B had cells taken 3 and 11 days respectively after operation. The values expressed as relative concentration /10^8 cells washed in 10 ml medium per wash

In conclusion the increased serum ferritin concentration in tumor patients probably result from stimuli which are of a more general inflammatory nature. Whether these stimuli react directly on liver cells, leucocytes or both remain to be answered.

On the quantitation of ferritin.

Though it is doubtful whether ferritin is of any value as a tumormarker there is still considerable interest in measurements of this protein.

Serum ferritin contains at least two different fractions in terms of antibody reactivity: the con A-binding i.e. the carbohydrate containing and the non-binding i.e. without carbohydrate. The precipitate of the con A-binding fraction is less dense i.e. contains fewer antibody molecules per square millimeter, than the non-binding ferritin as seen from the autoradiograph, Fig. 5. This phenomenon was seen in all sera, normal as well as pathological. Some precipitates could not be estimated precisely (Table I), and this could be due to individual differences in the carbohydrate structure of the ferritins. The fewer ^{125}I-antiferritin molecules reacting with con A-binding ferritin may lead to overestimation of this fraction in techniques which measure the area size and underestimation in other techniques (e.g. competitive radioimmunoassays). In earlier studies we found that the sum of the two areas (con A-binding and non-binding when quantitated separately by rocket immunoelectrophoresis) is greater than the sum area (4, 5). The low degree of antibody binding could be due to shielding of antigenic sites by the carbohydrate, but this ferritin, the 'G'-subunit (10), could possibly circulate partly as a monovalent, uncomplexed subunit ferritin in serum.

In our previous study we showed - in accordance with the well-known difference in antigenic reactivity between acidic and basic ferritins (14) - that the acidic ferritins of heart and Hela cells are not precipitated by anti liver and anti kidney ferritin antibodies (4).

In the present study we compared renal carcinoma and liver ferritins which both can be precipitated by the same antibodies (1,4) to find that both the slope and the shape of the standard curves could differ. The difference between the slopes was found with some antisera but not with others, demonstrating that some rabbits distinguished between the two ferritin types, whereas others did not. The difference between the shape of the curves was caused by differences in the degree of purification of the respective ferritins. The purified liver ferritin had a straight curve with linear correlation between the ferritin concentration and the rocket height. Both unpurified renal carcinoma and unpurified liver ferritin had bended curves possibly due to ferritin complexes or aggregated polymers with a slow electrophoretic mobility. In immunoelectrophoretic work minor antigenic differences may be disclosed only when working in diluted antigen/antibody

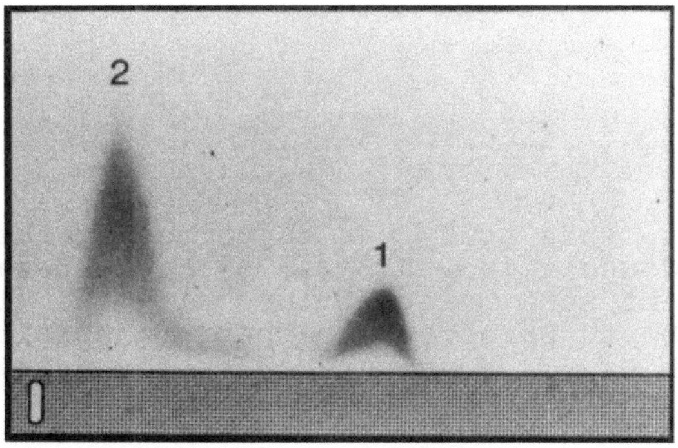

Fig. 5. Crossed lectin-affinity-radioimmuno-electrophoresis of serum ferritin (autoradiograph). Numbers 1 and 2 denotes the non-binding and Con A-binding fractions respectively. Anode to the right in the first dimension electrophoresis. The percentage of Con A-binding serum ferritin was calculated as $(\frac{2}{1+2} \times 100\%)$.

systems. In accordance we found that although all renal tumors had sharp, symmetric precipitates in the large (visible) scale in crossed immunoelectrophoresis, this changed as they were gradually diluted.

Using anti liver ferritin antibodies the anodic leg of the precipitates became gradually more diffuse and 'lifted' after which it disappeared completely, when diluted about 1:100 in the radioimmunoelectrophoretic system. This demonstrated that also renal carcinoma contains acidic ferritins.

In conclusion, the following points have to be taken into consideration in order to define an international standard for ferritin:

1) the carbohydrate composition,
2) the composition in terms of acidic or basic subunits,
3) the heterogeneic molecular composition of ferritins in biological material (free subunits, aggregated complexes etc.).

Acknowledgements. This study was supported by the Danish Medical Research Council (No. 512-10170). The renal carcinoma tumors and patient sera were obtained with the kind help from Dr. Witt, Dept. Surgery D, Bispebjerg Hospital, Copenhagen. We thank Dr. J. Konnerup-Madsen for drawing the figures, and Ms Jytte Lentz for secretarial assistance.

References

1. Andersen, M.M.: Prot. Biol. Fluids 27, 347-350 (1979).
2. Andersen, M.M.: Scand. J. Immunol. 15, 399-407 (1982).
3. Andersen, M.M., Krøll, J.: Scand. J. Immunol. 17, Suppl. 10, 217-223 (1983).
4. Andersen, M.M., Lihme, A., Bøg-Hansen, T.C.: Lectins-Biology, Biochemistry, Clinical Biochemistry (Bøg-Hansen, T.C., ed.) Vol. 2, pp.475-485, Walther de Gruyter, Berlin - New York 1982.
5. Andersen, M.M., Lihme, A., Bøg-Hansen, T.C.: Ibid. pp. 487-496.
6. Birgegård, G.: Clin. Sci. 59, 385-387 (1980).
7. Birgegård, G., Hällgren, R., Killander, A., Strömberg, A., Venge, P., Wide, L.: Scand. J. Haematol. 21, 333-340 (1978).
8. Chapman, R.W., Bassendine, M.F., Laulicht, M., Gorman, A., Thomas, H.C., Sherlock, S., Hoffbrand, A.V.: Digst. Dis. Sci. 27, 111-116 (1982).
9. Chapman, R.W.G., Gorman, A., Laulicht, M., Hussain, M.A., Sherlock, S., Hoffbrand, A.V.: J. Clin. Pathol. 35, 481-486 (1982).
10. Cragg, S.J., Wagstaff, M., Worwood, M.: Biochem. J. 199, 565-571 (1981).
11. Grail, A., Bates, G., Ward, A.M., Jones, W.G., Hancock, B.W.: Eur. J. Cancer Clin. Oncol. 18, 261-269 (1982).
12. Halliday, J.W., Mack, U., Powell, L.W.: Br. J. Haematol. 42, 535-546 (1979).
13. Ito, H., Takagi, Y., Ando, Y., Kubo, A., Hashimoto, S., Tsutsui, F., Kurihara, S.: Obstet. Gynecol. 55, 358-362 (1980).
14. Jones, B.M., Worwood, M., Jacobs, A.: Clin. Chim. Acta 106, 203-214 (1980).
15. Kew, M.C., Torrance, J.D., Derman, D., Simon, M., Macnab, G.M., Charlton, R.W., Bothwell, T.H.: Gut 19, 294-299 (1978).
16. Laurell, C-B.: Anal. Biochem. 15, 45-52 (1966).
17. Linder, M.C., Wright, K., Madara, J.: Enzyme 27, 189-198 (1982).
18. Lipschitz, D.A., Cook, J.D., Finch, C.A.: Proc. Soc. exp. Biol. Med. 148, 358-364 (1975).
19. Mack, U., Cooksley, W.G.E., Ferris, R.A., Powell, L.W., Halliday, J.W.: Br. J. Haematol. 47, 403-412 (1981).

20. Marcus, D.M., Zinberg, N.: J. Natl. Cancer Inst. 55, 791-795 (1975).

21. Nitti, D., Fabris, C., Del Favero G., Farini, R., Grassi, F., Farini, A., Baccaglini, U., Pedrazzoli, S., Piccoli, A., Lise, M., Naccarato, R.: Digest. 25, 258-262 (1982).

22. Weinstein, R.E., Bond, B., Silberberg, B.K.: Cancer 50, 2406-2409 (1982).

23. Worwood, M., Cragg, S.J., Wagstaff, M., Jacobs, A.: Clin. Sci. 56, 83-87 (1979).

24. Worwood, M., Cragg, S.J., Williams, A.M., Wagstaff, M., Jacobs, A.: Blood 60, 827-833 (1982).

25. Worwood, M., Hourahane, D., Jones, B.M.: Br.J.Haematol. 56, 31-43 (1984).

LECTIN BLOTTING OF NORMAL AND DERIVATISED MEMBRANE PROTEINS

Søren Naaby-Hansen, Allan O.F.Lihme, Thorkild C.Bøg-Hansen, and Ole J. Bjerrum

The Protein Laboratory, University of Copenhagen,
34, Sigurdsgade, DK-2200 Copenhagen N, Denmark.

Since the early seventies, lectins have been used to characterize and quantify specific carbohydrates on membrane surfaces. Until now, the biochemical characterization of membrane lectin receptors has mainly depended on gel immunoprecipitating techniques (1) or complex procedures involving e.g. affinity chromatography together with analytical gel electrophoresis (2).

The principle of immobilizing electrophoretically separated macromolecules to synthetic membranes, e.g. nitrocellulose, has been a refinement for their subsequent biochemical characterization (3, 4). In comparison to conventional gels, the immobilized proteins are concentrated in a thin layer on the surface and thus accessible to high-molecular-weight ligands e.g. lectins. This gives increased sensitivity and shorter

Lectins, Vol. IV

processing time (4). By electroblotting it is easy to transfer proteins separated in gels to the membranes (3). In some cases several replicas can be produced of the same gel and stored sheets can be reassayed (5). By the combination of this technique with the sensitive monitoring system from ELISA i.e. detection with enzyme conjugated ligands, the powerful tools of immunoblotting (3) and lectin blotting (6, 7) have appeared.

The aim of the present work was to exploit the use of conjugated lectins for characterization of membrane glycoproteins by lectin blotting. Besides determination of the specificity and sensitivity of the technique, the possibility of performing enzymatic or chemical modifications of the nitrocellulose bound glycoproteins before the lectin binding was examined.

Materials and Methods

Reagents. 3-amino-9-ethyl carbazole, dithiothreitol, phenylmethylsulphonyl fluoride, Na-p-tosyl-L-lysine chloromethyl ketone, sodium periodate and the monosaccharides D-galactose, D-glucose and N-acetyl-glucosamin were obtained from Sigma. Amido-black, pyronin G and the detergents sodium dodecylsulfate (SDS), Triton X-100 and Tween 20 were purchased from Merck. Acrylamide and bisacrylamide were from Serva. Aprotinin was from Novo. Neuraminidase (V. cholera) and human transferrin were purchased from Behringwerke. Nitrocellulose sheets (GSWP, 304 FO, pore size 0.22 m) were from Millipore. Ampholytes (pH 4-8), horse-radish peroxidase (HRP conjugates of Con A, LCA, WGA, PNA and RCA-1, biotinylated Con A, LCA, WGA, STA, PHA-L and VFA and HRP conjugated avidin were obtained from KEM-EN-TEC, Lemchesvej 11, 2900 Hellerup, Denmark. The conjugates were delivered in 0.1 M NaCl and 15 mM NaN_3 at a protein concentration of 0.1 - 1 mg per ml.

Membrane materials. Semen from (3-8) normal human donors were pooled and the spermatozoa isolated by centrifugation (1000 x g, 15 min). The cells were counted and washed x 3 in 10 vols of Dulbecco's buffer, once in phosphate buffered saline, pH 7.4

(PBS) and finally suspended to 7 x 10^8 cells/ml in Tris/HCl buffer, pH 8.6 containing 5 mM EDTA, 2 mM PMSF, 10 mM tosyllysinechloromethyl ketone and 500 U/ml of aprotinin (final concentrations) (8), and stored at -20°C. Solubilization of 7x10^8 cells/ml with SDS to a final concentration of 1% (w/v) was performed at 95°C for 2 min. DNA and insoluble material were removed by centrifugation at 40,000 x g for 40 min at 4°C. A similar solubilization in 1% (v/v) Triton X-100 took place under gentle rotation for 30 min at 4°C, followed by centrifugation 40,000 x g for 40 min at 4°C. For isoelectric focusing, 7 x 10^8 freshly prepared cells (pelleted at 1000 x g, 15 min) were solubilized directly in a lysis buffer containing 8 M urea, 2% (v/v) Triton X-100, 5% (w/v) ampholytes and 5% (w/v) dithiothreitol under gentle rotation for 30 min at 4°C and centrifuged as above. Human erythrocyte membrane was isolated from freshly drawn blood as described in (9). The protein concentration was spectrophotometrically determined after addition of 2% (w/v) SDS:

$$\frac{A_{280}^{1cm} - A_{310}^{1cm}}{1.3}$$ directly gives the protein concentration in mg/ml.

Electrophoretic procedures. SDS-PAGE was performed as described by Laemmli (10) in a linear gradient gel, T: 7.5-20%, C: 0.5-1.5% (bis), using a 5% stacking gel (C: 0.52% (bis)). Two-dimensional electrophoresis was performed as described by O'Farrell (11). Electrotransfer to nitrocellulose was performed as described by Kyhse-Andersen (12).

Lectin binding. Excess binding sites on the nitrocellulose was blocked by incubation for 2 min at 20°C in PBS containing 2% (v/v) Tween 20 followed by twice washing in PBS. The transfer was checked by staining of a nitrocellulose strip with Amido black (13). The other strips containing the transferred polypeptides from one lane of the SDS-PAGE were incubated with 1-5 µg of the respective HRP-conjugated or biotinylated lectins in 10 ml PBS containing 0.05% (v/v) Tween 20. In case of Con A incubation 1 mM $CaCl_2$, 1 mM $MnCl_2$ and 1 mM $MgCl_2$ were added. The incubation took place at 20°C under gentle rocking for 1-48 h (normally 16 h), whereafter surplus of lectins were removed by rinsing in PBS (5 times). Binding of HRP-conjugated lectins to the immobilized glycoproteins was visualized by staining with 3-amino-9-ethyl carbazole (8, 13). Biotin-conjugated lectins were incubated with 1.0 µg HRP-conjugated avidin in 10 ml PBS for 2 h at 20°C and developed as above. Blots of 2-dimensional polyacrylamide gels were incubated with 5 µg HRP-conjugated lectins in 150 ml PBS (plus ions) for 16 h at 20°C. Neuraminidase treatment of proteins bound to nitrocellulose was performed with 0.01 U neuraminidase in 20 ml 0.05 M acetate buffer, pH 5.5 containing 1 mM $CaCl_2$ by incubation for 45 min at 37°C. After 2 rinsings with PBS, the lectin incubations were performed as described above. Periodat oxidation of electroblots was performed for 30 min at 20°C in

20 ml PBS containing 20 mM $NaIO_4$, whereafter the strips were washed twice in PBS and incubated with lectins.

The sensitivity of the lectin blotting was examined with human transferrin and glycophorin of the erythrocyte membranes (14) as model proteins. For SDS-PAGE 30 µl of 3-fold dilutions of stocks (1 mg/ml) was applied. Lectin blotting was performed for 16 h as described above with HRP-Con A, HRP-LCA and HRP-WGA. Specific inhibition of the lectin binding (PNA, RCA-1, WGA and Con A) was performed by adding beforehand 0.2-0.4 mM of the monosaccharides (D-galactose, D-galactose, N-acetylglucosamine and D-glucose) to the lectin incubation buffer.

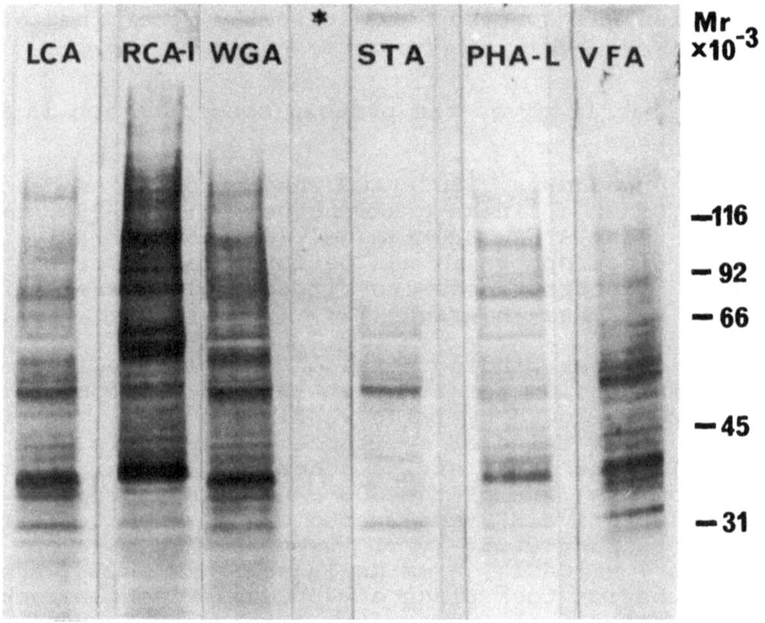

Fig. 1. Lectin blotting of human spermatozoal glycoproteins. 0.1 mg of Triton X-100 solubilized proteins was separated by SDS-PAGE, electroblotted to nitrocellulose membranes and incubated with the lectins indicated. Positions of molecular weights markers are indicated to the right. In (A), (B) and (C) 0.5 µg/ml of HRP conjugated lectins and in (D), (E) and (F) 0.5 µg/ml of biotinylated lectins were applied. After primary incubation of the nitrocellulose strips for 16 h, lanes (D-F) were further incubated with 0.15 µg/ml HRP-avidin. Peroxidase activity was developed with H_2O_2 and 3-amino-9-ethyl carbazol as coupling salt. Stars indicate the position of pyronin G for indication of the position of the application wells.

Results

Two model systems were investigated viz. membrane glycoproteins of human spermatozoa and human erythrocytes. Fig. 1 shows the binding of 3 HRP-conjugated lectins (WGA, LCA, and RCA-1) and 3 biotinylated lectins (STA, PHA-L and VFA) to spermatozoal proteins separated in SDS-PAGE and electroblotted to nitrocellulose. Similar experiments for HRP-Con A and HRP-PNA are shown in Figs. 2A and 2C. Each lectin gave reproducible patterns with different staining patterns for the various lectins. Furthermore, application of biotinylated Con A, PNA and WGA resulted in patterns similar to those obtained with the same HRP-conjugated lectins. The resolution of the separation on SDS-PAGE was not high enough to exclude the existence of additional lectin-binding species within the same molecular weight range. Compare for example Fig. 2A where only 11 lectin-binding bands could be discerned with Fig. 2B where the spermatozoal proteins were separated by combining isoelectric focusing with a second dimension SDS-PAGE (2D-gel electrophoresis) before the electroblotting. In this case more than 50 separate spots were reproducibly visualized with HRP-Con A (Fig. 2B). With HRP-PNA only 7 major spots were detected (Fig. 2D) of which 7 were the same as the Con A spots (Fig. 2B, arrows).

The specificity of the observed binding was tested by inhibition experiments with the relevant sugars. Thus the binding of HRP-conjugated PNA, WGA, RCA-1 and Con A in concentrations of

1-10 μg/10 ml for 16 h was inhibited by incubation in presence of 0.2 M D-galactose, 0.4 M N-acetylglucosamine, 0.2 M galactose and 0.3 M D-glucose, respectively (cf. Fig. 3, lanes C, G and J). Incubation with irrelevant sugars in the same concentrations did not inhibit the lectin binding. With incubation times of more than 16 h higher concentrations of monosaccharides were necessary to inhibit the lectin binding. However, after 24 h of lectin incubation extra non-inhibitable staining was observed.

The detection limits for glycoproteins electroblotted to nitrocellulose after SDS-PAGE were determined for transferrin and glycophorin with HRP-conjugated LCA and Con A, and WGA, respectively. Under the conditions employed as little as 100 ng of the glycoproteins could be detected.

Derivation of glycoproteins on the nitrocellulose membrane. The erythrocyte and spermatozoa membrane proteins were desialylated and oxidized with $NaIO_4$ and the possibility of performing lectin binding was examined. As shown in Fig. 3, lane A, PNA binding could be demonstrated to erythrocyte glycoproteins after these had been treated with neuraminidase while immobilized to nitrocellulose. Before the neuraminidase treatment no PNA binding was observed (Fig. 3, lane B). The specificity of the demonstrated PNA binding to desialylated protein was supported by inhibition with 0.2 M D-galactose (Fig. 3, lane C). No changes of the WGA-binding pattern to desialylated erythrocyte proteins were observed (Fig. 3, lanes D and E). The bind-

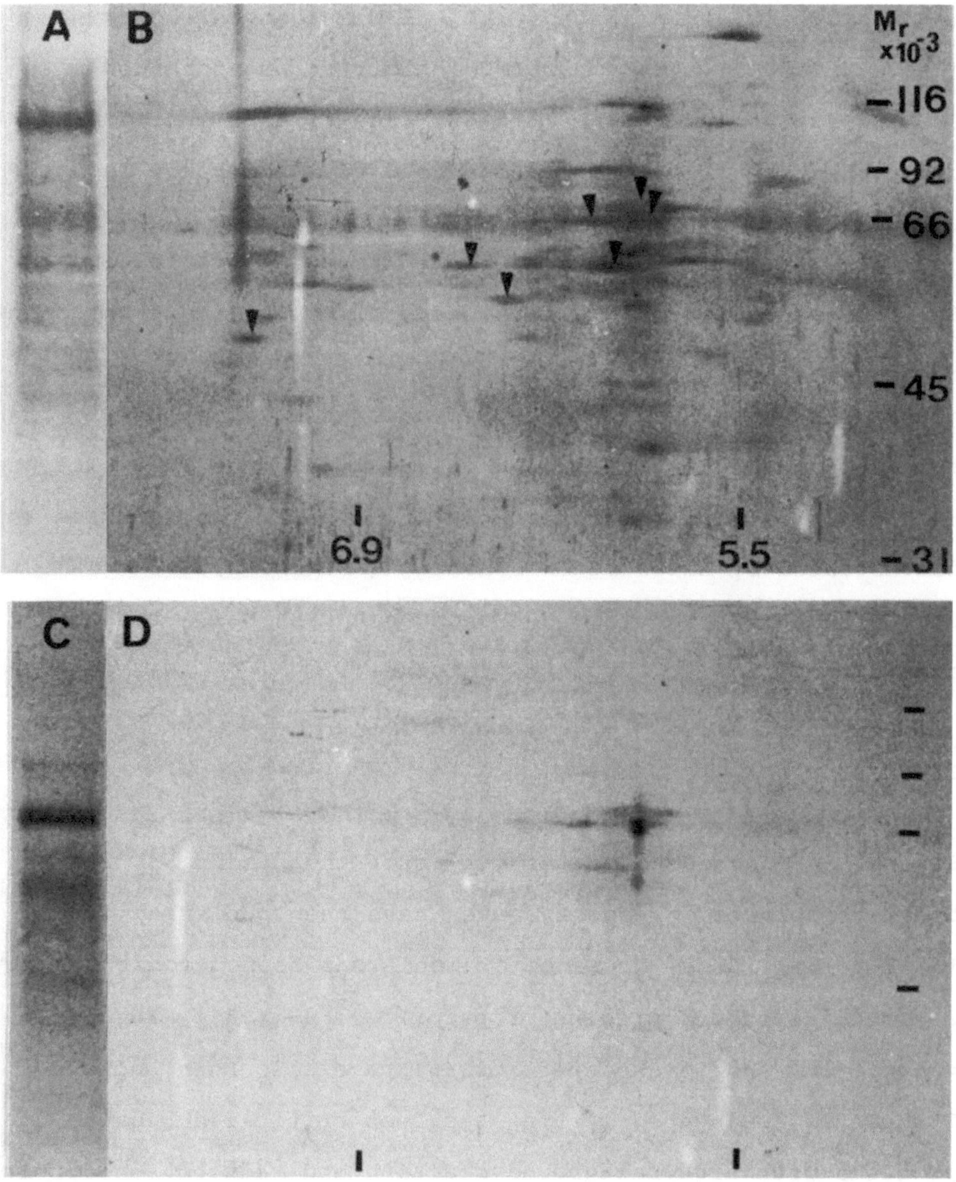

Fig. 2. Lectin blotting of human spermatozoal proteins (∼ 0.2 mg) separated with SDS-PAGE in (A) and (C) and by 2D-gel analysis in (B) and (D). After electroblotting to nitrocellulose incubations have been performed with 1 µg/ml HRP-Con A in (A) and (B) and HRP-PNA in (C) and (D). Conditions and other designations were as in Fig. 1. Arrows point to proteins binding both lectins. Molecular weight markers are indicated to the right and the pH-gradient for the isoelectric focusing of the first dimension at the bottom.

ing was otherwise inhibited with 0.4 M N-acetylglucosamine (Fig. 3, lane G). Only partial inhibition was obtained with 0.2 M N-acetylglucosamine (Fig. 3, lane F). Neither did desialylated spermatozoal proteins show any difference in the PNA and WGA binding.

By $NaIO_4$-oxidation of electroblotted glycoproteins the binding of lectins could be abolished. Lane H of Fig. 3 shows the neutralized binding of PNA to spermatozoal glycoproteins. Lanes I and J show the normal and inhibited binding of PNA. The protein core was still present on the nitrocellulose membrane as demonstrated by Amido black staining (not shown). The same results were obtained for WGA and Con A with $NaIO_4$-treated spermatozoal and erythrocyte proteins.

Discussion

Several approaches to demonstration of the binding of lectins to electroblotted glycoconjugates have been published (6, 7, 15, 16, 18). Thus, presence of Con A has been demonstrated by a direct secondary binding of peroxidase (7, 15). Autoradiography (16) and immunodetection (6) have also been employed. The use of conjugated lectins has been criticized because of problems with high background staining and reduced sensitivity (6). However, we find that blotting with conjugated lectins is a fast, simple and reliable procedure for the study of carbohydrate moities in complex mixtures of glycoproteins. As shown in the present communication, it works well with a number

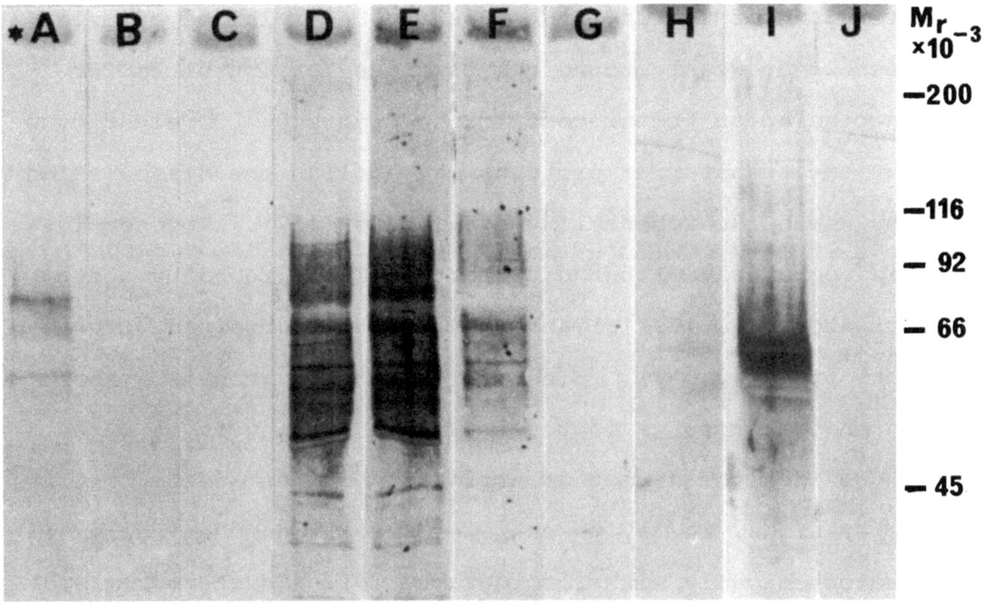

Fig. 3. Lectin blotting of derivatized proteins. Analysis of SDS-PAGE separated proteins from (A)-(C) of human erythrocyte membranes (20 μg) and (D)-(J) of human spermatozoas (∿ 0.1 mg) blotted to nitrocellulose. First lane (A) and (D) have been incubated with neuraminidase 0.0005 U/ml for 45 min at 37°C and lane (H) with 1 mM $NaIO_4$ for 30 min at 20°C. Then binding of HRP-PNA to lanes (A-C) and (H-J) and HRP-WGA to lanes (E-G) was performed. Specific inhibition of the lectin binding was performed by addition to the lectin solution in (C) and (J) of 0.2 M D-galactose and (F) and (G) of 0.2 M and 0.4 M N-acetylglucosamine, respectively. Conditions and designations were otherwise as for Fig. 1.

of lectins. The reactivity and specificity are retained for both the peroxidase-conjugated and the avidin-conjugated lectins as their binding to electroblotted glycoproteins were inhibited by the presence of 0.2-0.4 M of the respective inhibiting monosaccharides.

The use of Tween 20 as blocking agent (13) completely removed background staining. The extra non-inhibitable binding observed after 24 h of lectin incubation may be explained by release of Tween 20 or weakly bound polypeptides leaving sites for lectin binding on the nitrocellulose membrane. Covalent coupling of the transferred proteins may abolish the dissociation e.g. by use of diazobenzyloxymethyl paper (17). The sensitivity is comparable with the detection limits found by other investigators (7, 15, 16) and equal to that obtained with immunoblotting (3, 13). Electroblots of 2-D gels is a short cut to characterize glycoprotein without having it in pure form. Besides the direct determination of the lectin binding characteristics, it may be characterized immunochemically (6) and radiochemically with fluorography (8). Furthermore, as shown in the present communication, it is possible to perform enzymatic modifications of the immobilized glycoproteins whereby a closer analysis of the molecular carbohydrate structure can be performed. Thus it should be possible to follow the changes in glycosylation of individual proteins on the spermatozoes carried out by the epidymal glycosyl transferases. Until now the glycoproteins have only been observed by fluorescent lectins (18) and surface radiolabelling (19, 20).

Lectin blotting should also be useful for the planning and evaluation of purification procedures e.g. lectin affinity chromatography. In this case incubations for short periods should be applied which theoretically would correspond best to the chromatographic binding because low affinity binding is avoided.

In conclusion, lectin blotting is a specific and sensitive technique for identification of electrophoretically separated glycoproteins. Furthermore, it is possibly to perform chemical and enzymatic modification of the glycoproteins *in situ* on the nitrocellulose membrane.

Acknowledgements

Mrs. Jane Tullberg is thanked for skilful technical assistance. Financial support has been provided by the Danish Medical Research Council (Grants Nos. 12-4305, 12-4336), the Harboe Foundation and Gerda and Aage Haensch's Foundation. During this work SNH held a scholarship from NOVO Research Division.

References

1. Bøg-Hansen, T.C.: Solid-Phase Biochemistry (Scouten, W.H., ed.), pp. 223-251, J. Wiley & Sons, New York, 1983.

2. Causse, H. and Rougé, P.: Lectins - Biology, Biochemistry, Clinical Biochemistry (Bøg-Hansen, T.C., ed.) Vol. 3, pp. 559-572, Walter de Gruyter, Berlin - New York, 1983.

3. Gershoni, J.M. and Palade, G.E.: Anal. Biochem. 131, 1-15 (1983).

4. Bio-RAD Laboratories, Bulletin 1080EG, 1-37 (1981).

5. Manabe, T., and Okuyama, T.: Electrophoresis '84 (Neuhoff, V., ed.), pp. 387-399, Verlag Chemie,

6. Glass, W.F., Briggs, R.C., and Hnilica, L.S.: Anal. Biochem. 115, 219-224 (1981).

7. Clegg, J.C.S.: Anal. Biochem. 127, 389-394 (1982.

8. Naaby-Hansen, S., and Bjerrum, O.J.: J. Reprod. Immunol. (1983), in press.

9. Bjerrum, O.J., and Bøg-Hansen, T.C.: Biochim. Biophys. Acta 455, 66-89 (1976).

10. Laemli, U.K.: Nature 227, 680-685 (1970).

11. O'Farrell, P.H.: J. Biol. Chem. 250, 4007-4011 (1975).

12. Khyse-Andersen, J.: J. Biochem. Biophys. Methods (1983), in press.

13. Bjerrum, O.J., Larsen, K.P., and Wilken, M.: Modern Methods in Protein Chemistry (Tschesche, H., ed.), pp. 79-124, Walter de Gruyter, Berlin, 1983.

14. Marchesi, V.T.: Methods Enzymol. 28, 252-254 (1972).

15. Hawkes, R.: Anal. Biochem. 123, 143-146 (1982).

16. Bartles, J.R., and Hubbard, A.L.: Anal. Biochem. 140, 284-292 (1984).

17. Renart, J., Reiser, J., and Stark, G.R.: Proc. Natl. Acad. Sci. U.S.A. 76, 3116-3120 (1979).

18. Virtanen, I., Badley, R.A., Paasivuo, R., and Lehto, V.-P.: J. Cell. Biol. 99, 1083-1091 (1984).

19. Voglmayr, J.K., Fairbanks, G., and Lewis, R.G.: Biol. Reprod. 29, 767-775 (1983).

20. Brown, C.R., von Gloss, K.I., and Jones, R.: J. Cell Biol. 96, 256-264 (1983).

ONLY THE CON A-BINDING FRACTION OF THE β CHAIN OF HAPTOGLOBIN IS BIOLOGICALLY ACTIVE

Iwona Katnik, Tadeusz Guszczynski, Wanda Dobryszycka

Department of Pharmaceutical Biochemistry, Medical Academy, Szewska 38/39; 50-139 Wroclaw, Poland

Human haptoglobin is a serum glycoprotein and exists as three main genetic types: 1-1, 2-2, and 2-1. Haptoglobin 1-1 consists of two identical β (heavy) chains and two identical α (light) chains linked by disulphide bridges. Haptoglobins 2-2 and 2-1 are composed of polymers of increasing molecular weight owing to the presence of the α^2 chain (17.6 kDa) which is almost a duplicate of the α^2 chain (9.1 kDa). The β chain (40 dKa), containing carbohydrate is identical in the three genetic haptoglobin types. Haptoglobin is known to form a complex with hemoglobin, showing peroxidase activity (1). In our previous papers (2,3) the interaction of haptoglobin with Con A as well as fractionation of haptoglobin on Con A-Sepharose, were studied. INdependently of the genetic types, Con A-non binding and Con A-binding variants were obtained. There is good evidence that the main antigenic determinants, the hemoglobin-binding site and the Con A-binding site are located on the β chain (2-4). In the present report we have further characterized the Con A-binding by the β chain.

Lectins, Vol. IV

Materials and Methods

Human haptoglobin, type 2-1 was prepared from a cancer ascites fluid (5). The β chain was further purified according to (6). Anti-haptoglobin antiserum was raised in rabbits (4), and the immunoglobulin fraction was precipitated by $(NH_4)SO_4$. Affinity chromatography was carried out as described earlier (3). The preparation of the haptoglobin β chain in 0.005 M HEPES buffer, pH 6.9, was fractionated on a column of Con A-Sepharose equilibrated with the same buffer and next with the same buffer containing 215 mM methyl-α-D-glucopyranoside.

The haptoglobin concentration was determined by the peroxidase method of Jayle (7). Sialic acid was assayed by the method of Jourdian et al. (8), hexosamine according to Rondle and Morgan (9), and neutral carbohydrates by the orcinol method of Hewitt (10), respectively.

Double immunodiffusion was carried out by the technique of Ouchterlony (11). Crossed immunoelectrophoresis was performed by the method of Weeke (12) in 1% agarose gel containing 0.02 M Tris/veronal buffer, pH 8.6. SDS polyacrylamide gel electrophoresis was carried out by the method of Laemmli (13).

Results

Affinity chromatography on Con A-Sepharose. Haptoglobin β chain was isolated from human ovarian cancer ascitiv fluid and submitted to affinity chromatography on a Con A-Sepharose column. Two peaks corresponding to a non-binding fraction and a binding fraction were specifically eluted. The non-binding and binding fractions were collected and used for further experiments. They accounted for approximately 15 and 85%, respectively.

Molecular mass determinations. Polyacrylamide gel electrophoresis of the fractions obtained at the affinity ohromatography of the β chain on Con A-Sepharose showed that the non-binding and the binding fractions of the β chain each consisted of one band of M_r 40 kDa, identical to that of the unfrcationated preparation of β chain.

Carbohydrate composition. As can be seen in Table 1, the composition of neutral carbohydrates, hexosamine, and sialic acid in the Con A-fractionated variants was similar to the respective content in the unfractionated β chain. The Con A-binding fraction contained slightly amounts of the examined sugars.

Peroxidase activity with hemoglobin. When we measured the peroxidase activities with hemoglobin of the Con A-non binding and the binding fractions of the β chain, we found that the ability to form an active complex with hemoglobin was exhibited only be the Con A-binding fraction. Moreover, this fraction showed higher peroxidase activity than the chain not fractionated through Con A-Sepharose (3.3 units versus 1.2 units). The non-binding fraction did not show any peroxidase activity.

Double immunodiffusion in agar gel. As can be seen in Fig. 1, the Con A-binding fraction of the β chain precipitated anti-haptoglobin antibodies revealing immunological identity with

Table 1.

Carbohydrate composition of the haptglobin β chain fractions obtained by Con A-affinity chromatography.

Preparation	Neutral carbohydrates %	Hexosamine %	Sialic acid %
β chain	9.8	6.7	6.48
Con A-non binding fraction	8.27	7.0	6.98
Con A-binding fraction	8.25	6.63	5.87

Neutral carbohydrates were measured according to Hewitt (10), hexosamine to Rondle and Morgan (9), and sialic acid to Jourdian et al. (8), respectively.

unfractionated haptoglobin 2-1 in the same way as the Con A-binding fraction of haptoglobin 2-1. Partial identity was obtained with the unfractionated β chain, indicating a deficiency in antigenic determinants on the β chain. The Con A-non binding fraction of the β chain did not form any precipitate with anti--haptoglobin antibodies.

These results were substantiated by crossed immunoelectrophoresis, where we found that the peaks of the non-binding fraction of the β chain and of unfractionated β chain were less intensive than the peak of the Con A-binding fraction in spite of identical amounts of the preparations taken for the experiments.

Discussion

The experiments described in the present paper indicate that by means of affinity chromatography on Con A-Sepharose, two fractions (variants) of the β (heavy) chain of human haptoglobin may be separated: an active (the Con A-binding) and an inactive fraction (the Con A-non binding). The question arises whether the Con A-binding and the non-binding fractions of the β chain represent an in vivo molecular microheterogeneity.

The Con A-binding fraction of the β chain exhibits higher peroxidase activity when complexed with hemoglobin than the unfractionated β chain. The conclusion of our experiments might be that the Con A-non binding fraction has somehow lost specific sites for binding of hemoglobin and antibody, therefore affinity chromatography on Con A-Sepharose may be used for pre paration of the highly active haptoglobin β chain.

It should be stressed that the β chain used in our experiments was isolated from haptoglobin derived from ascitic fluids of women with cancer of the ovaries. Carbohydrate changes are well known to occur in pathological situations. Thus, acid α_1-glycoprotein in circulating blood of cancer patients showed a different ratio of Con A-binding and non-binding forms in comparison

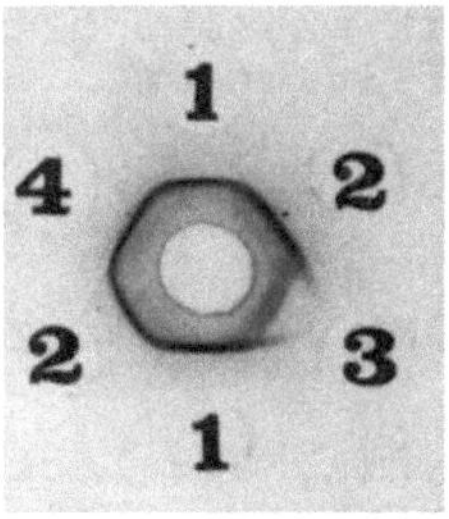

Fig. 1. Double immunodiffusion of Con A-binding fractions of haptoglobin 2-1 and the β chain.
The center well contained rabbit anti-haptoglobin immunoglobulins. Outer wells: 1: Haptoglobin 2-1. 2: The Con A-binding fraction of the β chain. 3: The β chain. 4: The Con A-binding fraction of Hp 2-1 (0.1% solutions).

with the preparation from normal blood (14,15). The same was shown with serum haptoglobin of cancer patients (unpublished) and those with toxemia of pregnancy (16). With respect to circulating haptoglobin its microheterogeneity has been related to differential sialic acid contents and in part to intrinsic carbohydrate differences (17).

The carbohydrate content of the products of the β chain fractionation by the use of Con A-Sepharose did not show apparent difference (Table 1). Con A-binding and Con A-non binding fractions exhibited also identical electrophoretic mobility and molecular mass. Therefore, the formation of two molecular variants might result from different structural forms of glycan chains.

Triantennary carbohydrate structures are known to be present in Con A-non binding fractions of glycoproteins (18). On the other hand in pathological, the concentration of biantennary structures (Con A-reactive) was found to be higher than in normal sera (19). Our results cannot be directly related to the normal situation. Determination of glycan chains per molecule of haptoglobin β chain isolated from normal and pathological sera could answer the problem.

Acknowledgement. This work was supported by grant No.II.1.2.3. from the Polish Academy of Sciences.

References

1. Bowman, B.H. and Kurosky, A.: Adv. Hum. Genet. 12, 189-261 (1982).

2. Dobryszycka, W. and Katnik, I.: Lectins - Biology - Biochemistry - Clinical Biochemistry, Vol.II. Walter de Gruyter, Berlin. New York. 382-392 (1982).

3. Katnik, I.: Biochim. Biophys. Acta - in press.

4. Katnik, I. and Dobryszycka, W.: Biochim. Biophys. Acta 670, 17-24 (1918).

5. Dobryszycka, W. and Lisowska, E.: Biochim. Biophys. Acta 121, 42-49 (1966).

6. Bernini, L.F. and Borri-Voltattorni, C.: Biochim. Biophys. Acta 200, 203-209 (1970).

7. Jayle, M.F.: Bull. Soo. Chim. Biol. 33, 876-880 (1951).

8. Jourdian, G.N., Dean, L. and Roseman, S.: J. Biol. Chem. 246, 430-435 (1971).

9. Rondle, C.J.M. and Morgan, W.T.J.: Biochem. J. 61, 586-592 (1955).

10. Hewitt, L.F.: Biochem.J. 31, 360-366 (1937).

11. Ouchterlony, O.: Handbook of Experimental Immunology, Blackwell Sci. Publ., Oxford and Edinburgh, 655-706 (1967).

12. Weeke, B. Scand. J. Immunol., 2, suppl. 1, 47-56 (1973).

13. Laemmli, U.K.: Nature 227, 680-685 (1970).

14. Hansen, J.E.S., Larsen, U.A. and Bøg-Hansen, T.C.: Clin. Chim. Acta 138, 41-47 (1984).

15. Katnik, I., Bøg-Hansen, T.C., Gerber, J. and Hansen, J.E.S. - in preparation.

16. Dobryszycka, W., Katnik, I., Wozniak, M., Dobryszycka, E., Dopierala, E., Agopsowicz, G. and Sward, J.: Le Pharmacien Biologiste 18, 27-32 (1984).

17. Dobryszycka, W. and Krawczyk, E.: Comp. Biochem. Physiol. 62B, 111-113 (1979).

18. Krusius, T., Finne, J. and Rauvala, H.: FEBS Letters 71 117-120 (1976).

19. Nicollet, I., Lebreton, J.P., Fontaine, M. and Hiron M.: Biochim. Biophys. Acta 668, 235-245 (1981).

POST-SYNTHETIC MODIFICATIONS OF PROTEINS CHANGE DURING DEVELOPMENT.
ACID PHOSPHATASES FROM DIFFERENT GRASS TISSUES

Irena Lorenc-Kubis and Bronslawa Morawiecka
Institute of Biochemistry, University of Wroclaw, Poland

In addition to the histochemical use of lectins (see other papers, these proceedings), lectins can be useful for structural characterization of glycoproteins (1-5). Earlier we have shown that acid phosphatases of grass seeds are glycoproteins which interact with lectins, and lectin affinity immunoelectrophoretic studies have proven Con A to be a suitable lectin for affinity chromatography of these enzymes (6).

The aim of the present study was to describe changes in the post-synthetic modifications of proteins, during development and tissue differentiation by isolating the acid phosphatases from different tissues of 3 species of grass by affinity chromatography on Con A-Sepharose and chatecterizing the purified isoenzymes by their interaction with lectins and by immunological analyzes.

Lectins, Vol. IV

Materials and Methods

Surface-sterilized caryopses of Hordeum sativum, Poa pratensis and Triticum vulgare were germinated in moist saw dust at 25°C with daily light. Leaves and coleoptiles of 14 days seedlings were harvested, frozen and then stored at - 30°C. Ground caryopses, samples of coleoptiles and leaves were homogenized with 6 volumes (w/v) of 0.1 M acetate buffer, pH 5.1 at 2-4°C and centrifuged. The supernatant was used for purification of acid phosphatases as described in (7). Affinity chromatography was performed with Con A-Sepharose (Pharmacia Fine Chemicals) from which the bound proteins were displaced with 15 mM- and 300 mM alpha-methyl-D-mannopyranoside. Gel filtration was performed on Sephadex G-200, pH 5.0. Hydrophobic interaction chromatography on phenyl-Sepharose (Pharmacia Fine Chemicals) was carried out according to (8). Acid phosphatase activity was measured spectrophotometrically (9) and developed histochemically as described in (6). Protein was determined by the method of Lowry et al. (10). Double affinity diffusion with free lectins: Concanavalin A (Con A), Lentil lectin (LCA), Soy bean agglutinin (SBA), Wheat germ agglutinin (WGA), Helix pomatia lectin (HPA) was performed in 1% agarose gel, pH 8.6. The lectins (10 µl of 5% Con C) were placed in the outsidewells, and the isoenzymes (10 µl of 0.5% Con C) in the centre well. The diffusion took place 48 h. Rocket affinity electrophoresis was performed according to (11) and immunodiffusion according to Ouchterlony (12). Quantitative immunoelectrophoresis including modification with Con A was carried out as described in (2,13, 14). Antibodies were purified from rabbit antiserum (15) raised against purified acid phosphatase B_1 from Poa pratensis caryopses.

Results

Acid phosphatases were isolated from caryopses, coleoptiles and leaves of Hordeum sativum, Poa pratensis, and Triticum vulgaris. The enzymes from the tissues were separated by affinity chromatography on Con A-Sepharose into two fractions: the non-binding (AcPase A) and the Con A-binding fraction (AcPase B). The Con A-binding AcPase B could be eluted with 10 mM alpha-methyl-D-mannopyranoside. The enzyme was further purified by gel filtration on Biogel P-100 and hydrophobic interaction chromatography on phenyl-Sepharose. Active fractions were eluted from the phenyl-Sepharose column with 0.8 M ammonium sulphate pH 8.7 (AcPase) B_1) and with Tris-glycine buffer pH 8.7 (AcPase B_2). The purified isoenzymes (AcPase B_1 and AcPase B_2) from all tissues were analyzed by double immunodiffusion using antibodies against AcPase B_1 from Poa pratensis caryopses. Fig. 1. shows the results of the test.

In all cases a cross-reaction was observed indicating that there is an immunological relationship among the polymorphic forms of the acid phosphatase in the grass group. Therefore we tested the isoenzymes carbohydrate relation using different lectins (Fig. 2).

Double diffusion revealed that AcPase B_1 and B_2 from all investigated tissues interact with free Con A except for both isoenzymes from coleoptiles and leaves of Hordeum sativum, which did not form any precipitate. Among the other lectins examined, LCA gave a grace affinity precipitate with isoenzymes of Poa prapratensis caryopses, coleoptiles and leaves while HPA and SBA interacted with the isoenzymes of Triticum vulgare. The interaction of Triticum vulgare AcPases with SBA and HPA indicate that the carbohydrate part of these isoenzymes contain exposed not only mannose but also N-acetylgalactosamine units. In control experiments (without Con A) the enzymes moved out of the plate. The enzymes activity of most grass tissues was retained in the gels with Con A and formed affinity precipitates.

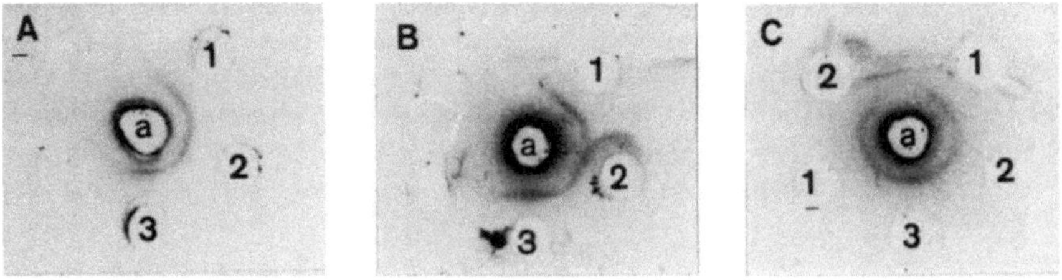

Fig. 1. Ouchterlony immunodiffusion of AcPase B_1 from: Caryopses (1), coleoptiles (2) and leaves (3) of Poa pratensis (A), Triticum vulgare (B) and Hordeum sativum (C) against anti-seeds AcPase B_1 of Poa pratensis antiserum (a).

The rocket affinity precipitates of AcPase B_1 and B_2 showed significant differences in their morphology and staining intensity (Fig. 3 A,B,C). Purified AcPase B_1 and B_2 from caryopses of all grassess both isoenzymes of Triticum vulgare leaves, AcPase B_1 from Poa pratensis and Triticum vulgare coleoptiles, AcPase B_2 from Poa pratensis leaves interact with free Con A and form a distinct affinity precipitate. In the case of AcPase B_1 from Poa pratensis leaves and AcPase B_2 from coleoptiles of Poa pratensis and Triticum vulgare, only a trace affinity pre-

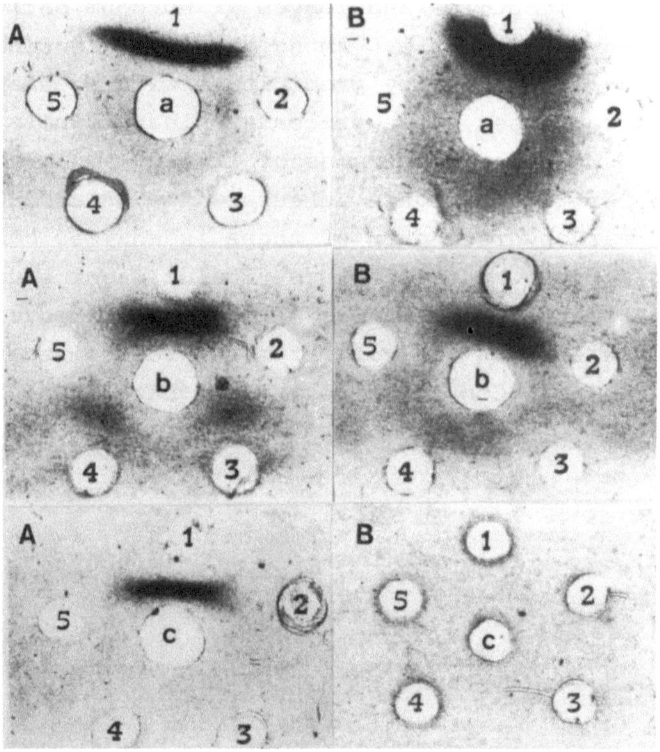

Fig. 2. Double diffusion in agarose gel. Lectins are directed against AcPase B_1 from caryopses (A), and leaves (B). 10 μg of AcPase B_1 of Poa pratensis (a), Triticum vulgare (b) and Hordeum sativum (c) are placed in the centre well. Lectins (100 μg) in outside wells: 1. Con A, 2. LCA, 3: SBA, 4. HPA and 5. WGA.

cipitate was formed (Fig. 3 B,C). The isoenzymes from Hordeum sativum leaves and coleoptiles did not form any affinity precipitate with free Con A.

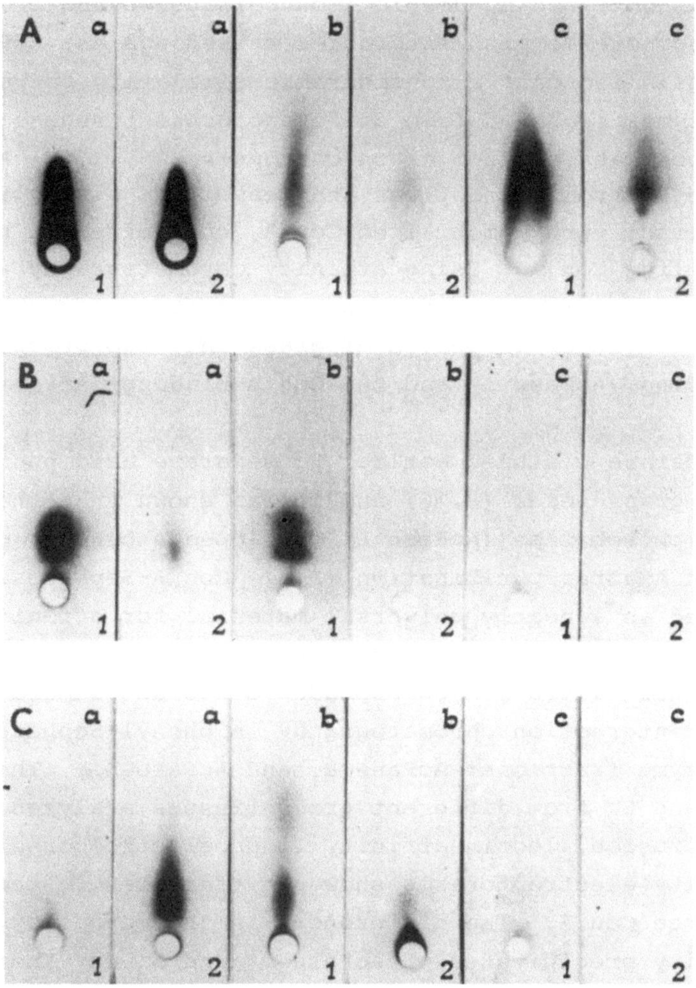

Fig. 3. Rocket-affinity electrophoresis of 7 µg AcPase B_1 (1) and AcPase B_2 (2) of caryopses (A), coleoptiles (B) and leaves (C) from Poa pratensis (a), Triticum vulgare (b) and Hordeum sativum (c) into agarose gel containing Con A (75 µg/cm^2).

In spite of their binding to immobilized Con A, the enzymes exhibited differences in their precipitation with free Con A, which could be connected with the number of binding sites to this lectin.

Discussion

The results obtained by double immunodiffusion and crossed immunoelectrophoresis with antibodies raised against AcPase B_1 of Poa pratensis indicate a considerable similarity between the all isoenzymes isolated from different grass tissues. Similar to acid phosphatases from grass caryopses (16,17,18) AcPase B_1 and AcPase B_2 from coleoptiles and leaves are also glycoproteins which interact with immobilized Con A, but differ in precipitating with free Con A. Using affinity chromatography with Con A-Sepharose we separated acid phosphatases from caryopses, coleoptiles and leaves of 3 species of grass into two groups: the Con A-non binding (AcPase A) and the Con A-binding (AcPase B).

Con A-Sepharose was used earlier to separate acid phosphatases from some grass seeds (6,16) and it was shown that the non-binding acid phosphatase (AcPase A) developed a binding capacity toward Con A after purification (17). Con A-Sepharose can be recommended as a nearly universal material for affinity chromatography of grass acid phosphatases. AcPase B was further purified by gel filtration on Sephadex G-200 and separated by hydrophobic interaction chromatography on phenyl-Sepharose into two isoenzyme fractions: AcPase B_1 and AcPase B_2. The isoenzymes B_1 and B_2 from different grass tissues analyzed by lectin double diffusion, lectin affinity immunoelectrophoresis and rocket affinity electrophoresis showed differences in their affinity to free Con A. The differences in the morphology of rocket affinity precipitates of AcPase B_1 and B_2 may indicate structural differences between the isoenzymes in their sugar component (11). AcPase B_1 and AcPase B_2 from coleoptiles and leaves of Hordeum sativum interact with immobilized Con A but do not precipitate with free Con A. This may indicate that the coleoptile and leave isoenzymes have only one binding site to Con A per molecule with mannose in the carbohydrate part, while the caryopses isoenzymes posses two or more binding sites to this lectin (1).

The distinct differences between the acid phosphatases of caryopses and coleoptiles or leaves AcPases of Hordeum sativum in their capacity to bind to free Con A may reflect changes in the sugar component during development as a result of the post-synthetic modifications.

Acknowledgements. This work was supported by the Polish Academy of Sciences - grant M.R. II 1.2.6.

References

1. Bøg-Hansen, T.C., Brogren, C.-H.: Scand. J. Immunol. 4, Suppl. 2, 135-139 (1975).

2. Bøg-Hansen, T.C., Bjerrum, O.J., Ramlau, J.: Scand. J. Immunol. 4, Suppl. 2, 141-147 (1975).

3. Kornfeld, R., Ferris, C.: J. Biol. Chem. 250, 2614-2619 (1975).

4. Debray, H., Pierce-Crétel, A., Spik, G., Montreuil, J. In: Lectins, Biology, Biochemistry, Clinical Biochemistry, T.C. Bøg-Hansen and G.A. Spengler (Ed.) Vol. 3, Walter de Gruyter, Berlin, New York, 1983, pp. 335-350.

5. Dall'Olio, F., Serafini-Cessi, F., Scannavini, M., Campadelli-Fiume, G. In: Lectins, Biology, Biochemistry, Clinical Biochemistry, T.C. Bøg-Hansen and G.A. Spengler (Ed.) Vol. 3, Walter de Gruyter, Berlin, New York, 1983, pp. 351-359.

6. Lorenc-Kubis, I., Bøg-Hansen, T.C. In: Lectins, Biology, Biochemistry, Clinical Biochemistry, T.C. Bøg-Hansen (Ed.) Vol. 1, Walter de Gruyter, Berlin, N w York, 1981, pp. 157-166.

7. Lorenc-Kubis, I., Morawiecka, B.: Acta Soc. Bot. Polon. 42, 369-377 (1973).

8. Bjerrum, O.J., Blirup-Jensen, S., Larsen, P.: Prot. Biol. Fluids 27, 775-780 (1980).

9. Yoshida, H., Tamiya, N.: J. Biochem. 69, 525-534 (1971).

10. Lowry, O.H., Resenbrough, N.J., Farr, A.L., Randall, R.J.: J. Biol. Chem. 193, 265-275 (1951).

11. Bøg-Hansen, T.C., Bjerrum, O.J., Brogren, C.-H.: Anal. Biochem. 81 78-87 (1977).

12. Ouchterlony, O. In: Handbook of Experimental Immunology Weir, D.M. (Ed.) Blackwell Sci. Publ. Oxford and Edinburgh, pp. 655-706 (1967).

13. Bøg-Hansen, T.C., Brogren, C.H., Murrow, Mc. I.: Inst. Brew. 80, 443-446 (1974).

14. Weeke, B.: Scand. I. Immunol. 2, Suppl 1, 47-56 (1973).

15. Harboe, N., Ingild, A.: Scand. J. Immunol. 2, Suppl. 1, 161-164 (1973).

16. Lorenc-Kubis, I., Morawiecka, B., Bøg-Hansen, T.C. In: Lectins, Biology, Biochemistry, Clinical Biochemistry, T.C. Bøg-Hansen (Ed.) Vol. 2, Walter de Gruyter, Berlin, New York, 1982, pp. 509-515.

17. Lorenc-Kubis, I. In: Lectins, Biology, Biochemistry, Clinical Biochemistry, T.C. Bøg-Hanse, and G.A. Spengler (Ed.) Vol. 3, Walter de Gruyter, Berlin, New York, 1983, pp. 379-386.

18. Lorenc-Kubis, I., Morawiecka, B.: Acta. Biochim. Polon. 27, 345-352 (1980).

GLYCOCONJUGATE INTERACTIONS IN SOYBEAN ROOT NODULES

Robert B. Mellor, Dietrich Werner
Fachbereich Biologie, Botanisches Institut der Philipps-Universität, D-3550 Marburg

Following the infection of soybean (*Glycine max*) root cells by *Rhizobium japonicum* both eukaryotic and prokaryotic cells are transformed. In the plant cell development of the central vacuole ("tonoplast") ceases and the cell becomes enlarged and filled with peribacteroid membrane-enclosed bacteria (1). Although the peribacteroid membrane is believed to be provided by the plant, (2, 3, 4, 5) the bacteria can influence the properties of the membrane by an unknown mechanism (6). Thus by using a variety of bacterial mutants we may influence the processes by which membranes are made.

Methods

The time-dependent infection of phytotron-grown *Glycine max* var Mandarin seedlings by *Rhizobium japonicum* strains 61-A-101, 61-A-24 and RH-31 Marburg was as before (3).

Cellular subfractionation using sucrose density gradient centrifugation was as described (3). Marker enzyme distribution through such gradients have been given previously (3, 4) as have electron microscopical checks of purity (5). Endoplasmic reticulum (ER), mitochondria and peribacteroid membrane fractions were repurified on subsequent gradients.

Electron microscopy of whole nodules was performed as in (7).

The enzymes galactosyltransferase and N-acetylgalactosaminyltransferase were assayed as in (8). Briefly, a routine assay contained 0.3 mg asialoagalactofetuin (prepared according to (9)), 0.28 KBq (^{14}C) nucleotide sugar and 100 µg enzyme in 1 ml of 30 mM Tris HCl buffer, pH 7.5 with 10 mM $MgCl_2$, 2 mM ATP and 10 mM DTT. Assays were stopped after 1 h at 28°C by the addition of TCA (to 6 % final concentration) and filtered onto a glass fibre disk. This disk was washed sequentially with chloroform:methanol (2:1), chloroform:methanol: water (3:48:47) and 8 % TCA before the radioactive content was determined.

Electrophoresis was achieved in buffers with SDS as before (10) and, where appropriate, electroblotted onto cellulose nitrate sheets (Millipore) at 10 V. cm^{-1} for 30 min in a 0,2 mM Tris glycine buffer, pH 8.0, containing 20 % methanol. Blots were removed and blocked with 2 % Tween 20 for 15 min. After removal of excess Tween blots were stained with lectin horseradish peroxidase conjugate (500 mU HRP per blot made to 3 mg/ml lectin with free lectin) whereupon the washed blot was coloured in 0.05 mM acetate buffer, pH 5.5 containing 0.01 % H_2O_2 and 1 % ethylaminocarbozol. Dried blots could be used immediately for autoradiography.

"Dot-blots" were obtained after a similar fashion to electroblots using protein samples, applied as 5 µl spots directly onto a cellulose nitrate strip in a range from 0.008 µg to 20 µg protein. Binding was estimated by eye on a scale strong (++++) to trace (+).

Protein was estimated from TCA precipitates (11) and sucrose concentration by refractometry.

Materials

Isotopes, UDP (^{14}C) galactose sp.act. 2 GBq $mmol^{-1}$ and UDP (^{14}C) N acetylgalactosamine sp.act. 2 GBq $mmol^{-1}$ were from Amersham Buchler. Lectin conjugate was purchased either from E-Y Labs (via Medac, Hamburg) or was a kind gift of Kem-En-Tec, (Hellerup, Denmark). Unless otherwise specified, all other materials were from either Sigma or Serva.

Results and Discussion

1) Tissue. The progress of the symbioses between *Glycine max* and 3 of the *R. japonicum* strains under study are shown in figure 1. The strain 61-A-101 is an effective strain and the other 2 (61-A-24 and RH-31 Marburg) are called ineffective due to their lack of ability to fix atmospheric N_2. 21 days post-infection the effective strain has induced the plant to form the new membrane type, the peribacteroid membrane (Fig. 1a). In nodules formed with strain 61-A-24 however, such membranes as are formed are unstable and rapidly break down (12) leaving the naked bacteroid in the host cell cytoplasm

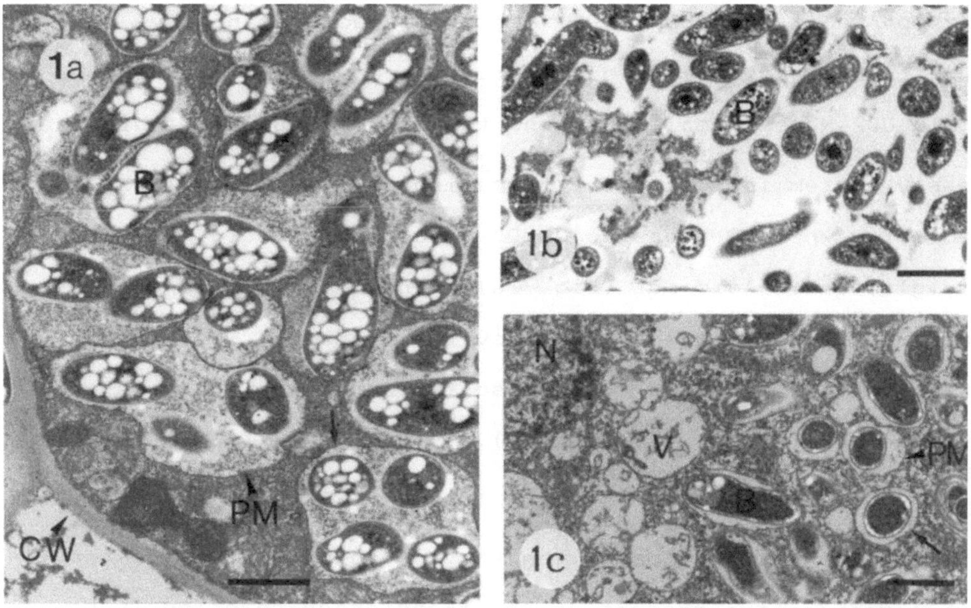

Fig. 1. Thin sections through Glycine max nodules infected with R. japonicum strains a) 61-A-101, b) 61-A-24 and c) RH 31. B: bacteroid, CW: cell wall, N: nucelus, PM: peribacteroid membrane and V: vacuole. bar = 1 μm

(Fig. 1b). Fig. 1c shows that with the mutant RH-31 Marburg the bacteroids inhabit peribacteroid structures which look normal, but close to the host cell nucleus, large compartments are observed at a very early age, were bacteroids are lyzed (13). Thus the bacterial genome can direct plant membrane flow in at least in three directions: 1. stability, 2. lysis, 3. fusion.

2.1) Characterization of the enzymes. Nodule homogenates catalysed the formation of a TCA-insoluble (^{14}C) product from UDP-(^{14}C)-galactose and asialoagalactofetuin at a rate proportional to enzyme concentration up to 400 µg of acceptor concentration (Fig. 2b). Asialoagalactofetuin could also accept sugar from UDP-(^{14}C)-N-acetylgalactosamine, but the two transferase were two activities because the addition of cold UDP-N-acetylgalactosamine to the aminosugar transferase assay system inhibited the formation of TCA-insoluble radioactivity wheras excess cold UDP-galactose or cold UDP-N-acetylglucosamine did not (Fig. 2a). Inhibition followed a similar pattern with the galactosyltransferase. This agrees with earlier findings that these two sugar transferases are in different donor classes (14).

2.2) Characterization of the enzyme product. The TCA-insoluble radioactive product of the transferase reaction is believed to be a re-glycosylated form of asialoagalactofetuin, because asialoagalactofetuin, after polyacrylamide gel elec-

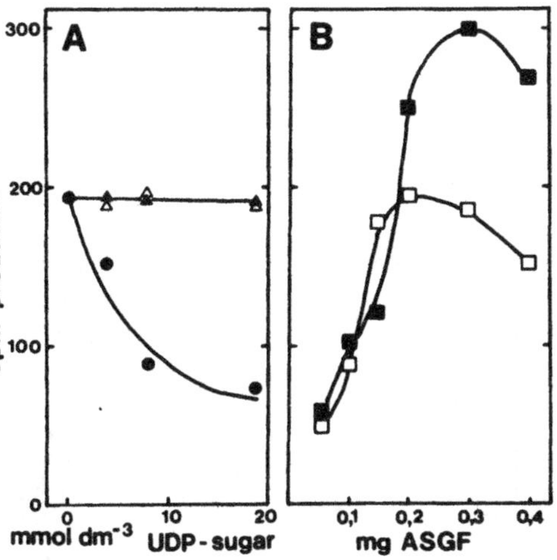

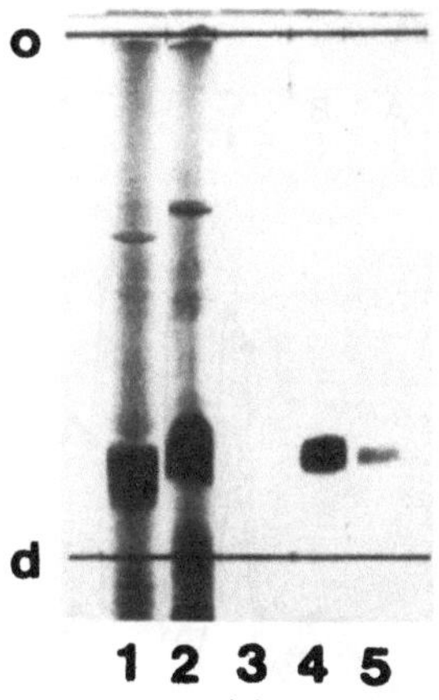

Fig. 2A. Effect of excess cold UDP-galactose (△), UDP-N-acetylglucosamine (▲) and UDP-N-acetylgalactosamine (●) on the incorporation of radioactivity into a TCA-insoluble form from UDP (^{14}C) N-acetylgalactosamine by homogenates of effective root nodules.

Fig. 2B. Effect of exogenous ASGF on the incorporation of galactose (■) and N-acetylgalactosamine (□) from the respective nucelotide-sugar catalyzed by homogenates of effective root nodules into TCA-insoluble product.

Fig. 3. Lane 1: Coomassie blue stained gel of 80 µg fetuin (Sigma type III); Lane 2: As above, but with 80 µg ASGF; Lane 3: PNA-stained electroblot from a gel similar to lane 2; Lane 4: PNA-stained electroblot of a reaction mixture (total protein 120 µg) containing 80 µg ASGF and 0.28 KBq UDP-(^{14}C) N-acetylgalactosamine; Lane 5: Autoradiograph of lane 4.

trophoresis and electroblotting, did not react with peanut agglutinin-horseradish peroxidase conjugate (Fig. 3). However, reaction mixtures of asialoagalactofetuin and UDP(^{14}C)-N-acetylgalactosamine contained one major radioactive protein of approximately the same mobility as asialoagalactofetuin which was also the major PNA positive protein in the preparation (Fig. 3).

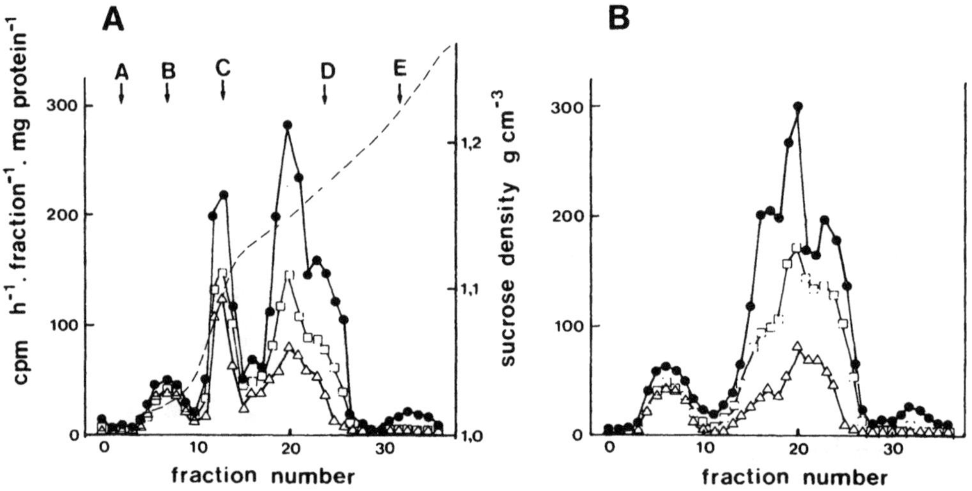

Fig. 4. Distribution of (A) UDP-ASGF galactosyltransferase and (B) UDP-ASGF N-acetylgalactosamine transferase amongst collected fractions after sucrose density gradient centrifugation of nodule homogenates. Arrows A-E represent the peaks of tonoplast, cytoplasm, ER, mitochondria and bacteroids both with and without peribacteriod membranes respectively. Nodules were infected with bacterial strains 61-A-101 (●), RH 31 Marburg (□) or 61-A-24 (△). The dotted line indicates sucrose density.

3.1) Subcellular distribution of the enzymes. After sucrose density gradient centrifugation, enzymic activity was widely distributed throughout the collected gradient fractions (Fig. 4a). The origin of the activity in the soluble (cytoplasmic) fraction is not clear, but since some glycosyltransferases are not tightly bound to membrane systems (15), it is possible that this soluble activity represents molecules lost from membranes during homogenation. Galactosyltransferase is often used as a marker enzyme for Golgi fractions (16, 17, 18) but it has also been found in ER in systems where rapid membrane proliferation takes place (8). Two peaks of enzyme

Table 1

The specific activities of UDP-ASGF galactosyltransferase (Gal) and UDP-ASGF N-acetylgalactosamine transferase (GalNAc) in nodule homogenates of Glycine max with differing Rhizobium japonicum strains 17 days post infection.

Rhizobium	Gal	GalNAc
61-A-101	1260	1647
RH-31 Marburg	712	1377
61-A-24	643	631

Activity is given as cpm radioactive product formed per hour per mg protein. Activities of both enzymes in uninfected mature root tissue was variable but always much lower than the above values.

activity were prominent in our gradients. One cosedimented with ER and the second was a broader peak which, on the basis of density, lipid labelling and marker enzyme activity (19), we have previously designated as Golgi (4). The finding of galactosyltransferase in this region of the gradient tends therefore to support the idea that these fractions contain Golgi membranes.

Fig. 4b shows that the N-acetylgalactosaminyltransferase activity was overwhelmingly confined to the Golgi-containing fractions. Although this was the case in all the symbioses studied, the difference between the strains in terms of aminosugar transferase was much greater than that for the galactosyltransferase. Around 17 days post-infection the

Table 2

Reactivity of lectins towards subcellular fractions obtained from an effective symbiosis (with strain 61-A-101): PBM2 indicates peribacteroid membranes from the ineffective symbiosis formed by strain 31-Marburg.

Lectin	"Tonoplast"	Cytoplasm	ER	"Golgi"	Mitochondria	PBM	PBM2
CON A	+	++	+++	++	+	+	+
PSA	+	+	+++	+	++	++	+
LCA	+	++	+++	+	+	++	+
WGA	+	+	+	+	+	+	+
GSI	+	negative	+	+	+	+	+
SBA	+	negative	+	++	+	++	+
PNA	+	negative	negative	++	negative	+++	++

Negative *indicates no binding was apparent using up to 20 μg of protein per sample.*

metabolic activity of the nodules peak (see 12, 7, 13) and at this time the differences in transferase activity between the 3 types of symbiosis were even clearer (Table 1).

3.3) Distribution of products. Whilst an exogenous acceptor asialoagalactofetuin, was converted to a PNA positive form through the action of the transferase enzymes, the subcellular distribution of PNA-positive material was not great. As expected, peak Golgi-containing fractions were PNA-positive (Table 2), as were "tonoplast" membranes, which are also thought to be derived directly from Golgi structures (20). Fractions such as ER and mitochondria, which are not made from Golgi and are therefore not influenced by the amino sugar transferase, were all PNA negative, although their reactivity towards other lectins indicate that they contain glycoconjugates. Peribacteroid membranes reacted not only with lectins specific for sugars found in normal N-glycosylated glycoproteins, but also contained the highest concentration of PNA-positive material of all the fractions tested. Peribacteroid membrane fractions from the symbioses making lesser quantities of peribacteroid membrane had lower concentrations of PNA-positive material (Table 2). This indicates that peribacteroid material has been previously processed by Golgi apparatus. We speculate that PNA-positive sugar conformations on glycoconjugates may be important in the segregation or protection of material within the Golgi and that those bacterial strains lacking the inducer of the correct glycosyltransferase can build only partial and/or ineffective symbioses.

This paper provides a background for the use of lectins as subcellular markers.

Acknowledgements.
Glycine max var Mandarin was supplied by K. Behm GmbH, Hamburg and *Rhizobium japonicum* was from the Nitragin Company of Milwaukee. Dr. Mörschel provided previously unpublished electronmicrographs. We thank also Dr. Bøg-Hansen for demonstrating many of the methods used and the Deutsche Forschungsgemeinschaft for financial support.

References

1. Sutton, W.D., Pankhurst, C.E., Craig, A.S.: Internat. Rev. Cytol. Suppl. 13, 149-177 (1981).

2. Robertson, J.G., Warburton, M.P., Lyttleton, P., Fordyce A.M., Bullivant, S.: J. Cell Sci. 30, 151-174 (1978).

3. Mellor, R.B., Dittrich, W., Werner, D.: Physiol. Plant Pathol. 24, 61-70 (1984).

4. Mellor, R.B., Christensen, T., Bassarab, S., Werner, D.: Z. Naturforsch. (in press)

5. Mellor, R.B., Mörschel, E., Werner, D.: Z. Naturforsch. 39c, 123-125 (1984).

6. Sutton, W.D., Paterson, A.D.: Plant Sci. Lett. 30, 33-41 (1983).

7. Werner, D., Mörschel, E.: Planta 141, 169-177 (1978).

8. Mellor, R.B., Lord, J.M.: Planta 147, 89-96 (1979).

9. Kim, Y.S., Perdomo, J., Nordberg, J.: J. Biol. Chem. 246, 5466-5476 (1971).

10. Laemmli, U.K.: Nature 227, 680-685 (1970).

11. Lowry, O.H., Rosebrough, N.J., Farr, A.L., Randall, R.J.: J. Biol. Chem. 193, 265-275 (1951).

12. Werner, D., Mörschel, R. Stripf, R., Winchenbach, B.: Planta 147, 320-329 (1980).

13. Werner, D., Mörschel, E., Mellor, R.B., Bassarab, S.: Planta (in press)

14. Pierce, M., Turley, E.A., Roth, S.: Internat. Rev. Cytol. 65, 1-47 (1980).

15. Mellor, R.B., Roberts, L.M., Lord, J.M.: Biochem. J. 182, 629-631 (1979).

16. Fleischer, B., Fleischer, S. (1970): Biochim. Biophys. Acta 219, 301-319 (1970).

17. Howell, K.E., Ito, A., Palade, G.E.: J. Cell Biol. 79, 581-589 (1978).

18. Appelkvist, E.L., Bergman, A., Dallner, G.: Biochim. Biophys. Acta 512, 111-122 (1978).

19. Ray, P.M., Shininger, T.L., Ray, M.M. (1969): Proc. Natl. Acad. Sci. USA 64, 605-611 (1969).

20. Robertson, J.G., Lyttleton, P., Bullivant, S., Grayston, G.F.: J. Cell Sci. 30, 129-149 (1978).

A HUMAN BRONCHIAL BLOOD GROUP A ACTIVE GLYCOPROTEIN. STRUCTURAL CHARACTERISTICS STUDIED BY A QUANTITATIVE LECTIN PRECIPITATION ASSAY

Albert M. Wu
Department of Veterinary Pathology, College of Veterinary Medicine, Texas A&M University, College Station, Texas 77843 U.S.A.

Carol E. Synder, Anthony Herp
Department of Biochemistry, New York Medical College
Valhalla, New York 10595 U.S.A.

Rhong-Phone Hsieh
School of Medical Technology, College of Medicine, National Taiwan University Taipei, Taiwan

Human bronchial mucus glycoproteins, constituents of the secretions found in the respiratory tract, play an important protective role in the normal functioning of the airway (1-3). These glycoproteins are high molecular weight compounds composed of a protein core attached to numerous oligosaccharide side chains through O-glycosidically linked N-acetylgalactosamines (4-8). Encountered in addition to N-acetylgalactosamine are N-acetylglucosamine, galactose, fucose and sialic acid, which comprise from 60 to 80% of the molecule (5,6,9,10). The bronchial glycoprotein represent a very heterogeneous group with respect to composition and molecular size of their carbohydrate side chains (5,7,8,11-13). Lectins have been widely used to probe the structural and functional role of cell surface carbohydrates (14,15) and as a tool for detecting, isolating and characterizing glycoproteins such as blood group substances (16,17) and glycolipids (18). By examining their

Lectins, Vol. IV

behavior toward well-defined lectins (19-23), information about specific sugars, anomeric linkage, and position (non-reducing or internal) in glycoproteins can be deduced (19). In recent years, lectins have also been used for detecting glycoconjugate in glandular cells of human bronchial mucosa (24,25) and fractions of human bronchial mucus glycoprotein from sputum of patients suffering from chronic bronchitis (26). In the present report, the chemical structure of lectin receptors of a healthy human bronchial glycoprotein and its binding characteristics were investigated by a quantitative precipitin assay. From the information obtained, structural features of the oligosaccharides at the terminal non-reducing end were proposed.

Materials and Methods

Five ml of tracheobronchial secretions were aspirated and collected from a single individual. The glycoprotein was prepared as described by Synder et.al.(27). Purity was ascertained by SDS-slab gel electrophoresis on 6.25% acrylamide gels (29), and by analytical ultracentrifugation in the presence and absence of denaturing solvents. The mild acid hydrolyzed glycoprotein was prepared by hydrolysis of the native glycoprotein at 80 C for two hours at pH 2.0, dialysis against distilled water and lyophilization.

The protein content was measured by the procedure of Lowry et. al. (30) with crystalline bovine serum albumin as the standard. Sialic acid was determined by the resorcinol (31) and the thiobarbituric acid methods, using N-acetylneuraminic acid as the reference (32). Total hexosamines were estimated by the Elson-Morgan method (33). Differential hexosamine analysis was performed on a Beckman 120C Amino Acid Analyzer using a citrate buffer (0.35M with respect to Na^+), pH 5.28 under the same conditions of hydrolysis as those used for total hexosamine analysis (34). For the analysis of fucose, the cysteine-sulfuric acid reaction (35) was used with a heating period of ten minutes. Galactose was analyzed by the method of Dische and Danilchenko (36). In some instances, sugars were quantitated by gas-liquid chromatography as their alditol acetate derivatives according to Niedermeier (37) after hydrolysis of the glycoprotein in 2M trifluoracetic acid at 120 C for ninety minutes. A Supelco SP-2340 column (3 ft) and Perkin Elmer Model 801 gas chromatograph equipped with a flame ionization detector was used with nitrogen as the carrier gas. Blood group activity of the glycoprotein was determined by a hemagglutination-inhibition assay according to Kabat and Mayer (38).

Seeds used for the preparation of lectins were purchased from F. W. Schmacher Co., Sandwich, Mass. The Maclura pomifera (MPA), Sophora japonica (SJA), Lotus tetragonolobus (LTA), Ulex europaeus II (UEA-II), Helix pomatia (HPA), Dolichos biflorus, (DBA) and Wistaria floribunda (WFA) lectins were purified by adsorption to insoluble polyleucyl hog gastric (A+H) mucin (39-41), and eluted by melibiose (23), DGal (22), LFuc (42), 35% ethylene glycol (43), DGalNAc (44,45), and lactose (46), respectively. Bauhinia purpurea alba (BPA) lectins were isolated by affinity chromatography on acid-treated Sepharose 6B using DGal as eluant (20). Ulex europaeus I (UEA) lectin was purified by affinity chromatography on a 2,4-fucosyl-O-polyacrylamide gel (47) using fucose as the eluent (42). Geodia cydonium (GCA) was provided by Dr. H. Bretting (Zoologisches Institut and Museum Universitat, Hamburg). Canavalia ensiformis agglutinin (Con A), peanut lectin (PNA) Ricinus communis agglutinin (RCA) and ricin were purchased from Boehringer Mannheim Biochemical, Indianapolis, IN and wheat germ lectin (WGA) from Sigma Chemical Co., St. Louis, MO. Their carbohydrate specificities are listed in the Appendix of this book.

Quantitative precipitation assay was performed by a microprecipitin technique (38). A fixed amount of lectin was added to varying amounts of glycoprotein. The lectin amount was chosen in the range of 40 to 60 µg protein or 4.9 to 7.1 µg N (Nitrogen). Total volume was 250 µl. The Total nitrogen and protein content of the washed precipitate was estimated by the ninhydrin method (48) and by the Lowry method (30), using bovine serum albumin as the standard. The precipitated N was expressed in % of the lectin N, disregarding the glycoprotein N contribution.

Results

The fraction of bronchial glycoprotein used for this study was purified from secretions of a healthy individual and exhibited blood group A activity (27). It contained 23% protein and 77% carbohydrate, with molar ratios of DGalNAc:DGlcNAc:DGal:LFuc:SA of 1.0:1.4:1.4:2.7:0.3. No mannose or glucose was detected. Seven amino acids asp, thr, ser, pro, glu, ala, and gly were the major amino acid constituents and composed 85% of the total protein present. Only one band was seen at the top of the gel on SDS-slab gel polyacrylamide electrophoresis. One symmetrical, hyperfine peak was observed on analytical study, the interaction of various lectins with the purified glycoprotein was used to secure chemical information about the occurrence of specific sugars, the configuration of anomeric linkages and their possible position (terminal non-reducing or internal). However, two questions must be considered for the present approach: **(A)** Is the specificity of lectins sufficient to elucidate the chemical structures of complex carbohydrates? and **(B)** Are the negative results caused by solubility characteristics of the lectin-glycoconjugate complexes? Such problems can be resolved by interacting the glycoprotein with several lectins sharing a common specificity.

HPA (44) and DBA (45) were reported to be specific for blood group A determinant (DGalNAcα1→3DGalβ1→3DGlcNAc, and DGalNAc. According to current findings, the Forssman pentasaccharide is about 23 and 8 times as active as blood group A determinant for DBA and HPA lectins, respectiely (50). The bronchial glycoprotein reacted strongly with HPA and DBA indicating that blood group A determinant sequence (DGalNAcα1→3DGalβ1→3DGlcNAc), and/or Forssman specific oligosaccharides play an important role in its binding activities. The mild acid hydrolyzed glycoprotein reacted well with WFA demonstrating that the above determinants are sterically masked by other long chain carbohydrates and/or are substituted by sialic acid and/or LFuc (50).

By using the lectin-horseradish peroxidase conjugate technique to examine human tracheal glands (24), it was also found that 90% of the mucous tubule cells in two blood group A specimens showed strong affinity for DBA, moderate in blood group AB individuals, while only about 30% of these cells exhibited moderate staining in blood group O specimens. Combining our present finding, it can be concluded that a majority of the oligosaccharide chains in tracheal mucus cells of blood group A individuals contain DGalNAcα1 linked residues or Forssman also contains several of the above oligosaccharides substituted by LFuc and/or sialic acid (20).

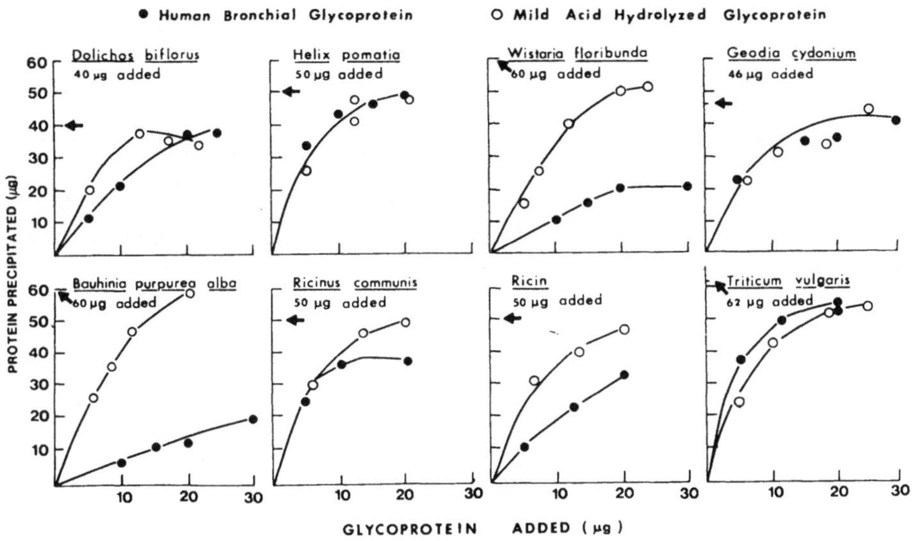

Fig. 1. Quantitative Precipitation Curves of Bronchial Blood Group A Active Glycoprotein and its Mild Acid Hydrolyzed Product with Various Lectins "●" Glycoprotein "O" Mild acid hydrolyzed product.

The purified glycoprotein precipitated only about 35% of each of the WFA and BPA (Fig. 1), but the mild acid hydrolyzed material reacted with over 85% of each of these lectins. From this, we may infer that its structure may be complex. It also suggests that the non-reducing carbohydrate moieties contain acid-labile sugars such as sialic acid and fucose, and explains the weak interaction of the native material with these lectins. The high activity of the intact and mild acid degraded material with wheat germ indicate the presence of DGalNAc residues at the non-reducing end or at interior positions in the sugar

TABLE I

Comparative precipitating activity of human bronchial glycoprotein and its mild hydrolyzed glycoprotein with various lectins

Lectin	Amount of lectin µg N	µg protein	Maximal Amount of Lectin (N) in the Precipitate (%) BG	H	Amount of glycoprotein required for 50% Precipitation (µg) BG	H
HPA	5.3	50	95	95	5	5
DBA	6.6	40	95	95	10	5
WFA	6.9	60	40	85	-	12
GCA	6.3	46	85	95	10	10
BPA	6.2	60	32	100	-	7
RCA	4.9	50	75	100	5	5
Ricin	4.9	50	75	95	15	5
PNA	4.9	53	11	17	-	-
MPA	5.9	46	11	16	-	-
SJA	7.1	52	30	40	-	-
WGA	5.6	62	88	86	5	7
Con A	5.0	56	8	20	-	-
LTA	5.4	48	20	25	-	-
UEA-I	5.6	59	30	4	-	-
UEA-II	5.1	59	14	15	-	-

Abbreviations: BG = bronchial glycoprotein, H=mild acid hydrolyzed product.
The maximal amount of N in the precipitate was calculated as % of lectin nitrogen added, thus disregarding the contribution of N in the glycoprotein (see Materials and Mehods).

ultracentrifugation in 6M urea (27). After mild acid treatment, about 10% of GalNAc, 90% of sialic acid, and 70% of LFuc of bronchial glycoprotein were removed.

The lectin precipitating properties of the purified native bronchial glycoprotein and of its mild acid hydrolyzed product are shown in Fig. 1 and Table I. Since the contribution of protein content or nitrogen content of the bronchial glycoprotein is small, the contribution of the protein part derived from the glycoprotein in the precipitate of lectin-glycoprotein complex is even smaller. Therefore, the contribution of the protein part of the glycoprotein or glycopeptide was not considered in the calculation (see Table 1). When both the ninhydrin and Lowry methods were used to estimate the precipitate, it was found that about the same precipitation profiles were obtained. The bronchial glycoprotein reacted equally well with HPA and DBA, and precipitated over 95% of each of these lectins; approximately 5 μg of glycoprotein was required for 50% precipitation of the lectin. These results indicate that the native glycoprotein probably has carbohydrate determinants specific for both HPA and DBA. No change in reactivity was seen with these lectins after mild acid hydrolysis.

The intact glycoprotein also exhibited a high affinity for GCA, RCA and ricin, precipitating about 75% of each of these lectins (46μg, 50μg, and 50μg added). After mild acid hydrolysis, a further increase in reactivity was observed with these lectins. These results indicate that the carbohydrate moiety of the bronchial glycoprotein as well as its mild acid hydrolyzed product contained several of the following carbohydrate determinants at the non-reducing ends: Forssman-specific oligosaccharides (DGalNAcα1→3DGalNAcβ1→3DGalα1→4DGal β1→4DGlc;, DGalNAcα1→3DGalNAc), blood group A active specific oligosaccharides (DGalNAcα1→3DGal; DGalNAcα1→3DGalα1→3DGlcNAc), anti I Ma determinants (DGalβ1→4DGlcNAc β1→6DGal; DGal β1→4DGlcNAc) and/or DGalNAc. The original glycoprotein chains (49). Both the native and mild acid treated glycoprotein reacted poorly with peanut agglutinin and MPA which are specific lectins for the DGalβ1→3DGalNAc sequence (20-23).

Although the glycoprotein contains substantial amounts of LFuc, it precipitated less than 30% of the LFuc-specific lectins, nor did its mild acid hydrolyzed product affect the precipitability of LTA and UEA-II. However, its ability to precipitate UEA-I decreased significantly, after treatment with mild acid.

The bronchial glycoprotein reacted weakly with Con A, and the mild acid treated product increased only slightly in reactivity, which is consistent with the absence of mannose in our glycoprotein, and is probably due to exposure of DGlcNAc residues (49).

Discussion

Although human bronchial glycoproteins have been extensively investigated for many years, little is known about the sequence of the complex carbohydrate side chains or their biological structure-function relationships (8). With respect to the chemical composition, the bronchial glycoprotein used for this study, in general, resembles those reported previously for pulmonary secretory glycoproteins isolated from healthy subjects (5), patients suffering from cystic fibrosis (7), chronic bronchitis (13), or asthma (6,9,10). The glycoproteins are comparable in the following aspects: **(A)** hexosamine (DGalNAc & DGalNAc), DGal, LFuc and sialic acid comprises from 60-80% of the molecules, **(B)** none or less than 2% of DMan, DGlc or sulfate were present, and **(C)** asp, ser, thr, glu, pro, gly, and ala were the major amino acids constituting about 75% of total amino acids. However, the molar ratio of carbohydrates is found to vary among preparations obtained from different healthy subjects. In our specific oligosaccharides (50). Characterization of these carbohydrate side chain in bronchial mucus glycoprotein is still hampered by the amount available. However, the presence of a Forssman specific oligosaccharide may be proven by antiserum or monoclonal antibody.

RCA, ricin and GCA share the same specificity for DGalβ1→4(3)DGlcNAc containing oligosaccharides. Recently, it was found that the I Ma determinant (DGalβ1→4DGlcNAcβ1→6DGal) was the best inhibitor for RCA. This inhibitor was about 1.4 and 2.2 times better than lactosamine and lactose, respectively while the former (DGalβ1→4DGlcNAc) was about two times more active than its β1→3 linked isomers (Wu and Kabat, in preparation). Both the native and the mild acid hydrolyzed glycoprotein reacted well with RCA, ricin, and GCA. Moreover, the mild acid hydrolyzed product of the bronchial glycoprotein exhibited a substantially increased reactivity toward BPA. It can therefore be assumed that the carbohydrate side chains of the bronchial glycoprotein are as complicated as the human blood group A_1 active glycoprotein purified from human ovarian cyst fluid (21) and also contain DGalβ1→4DGlcNAcβ1→6DGal and/or DGalβ1→3DGlcNAc (DGalNAc) at their terminal nonreducing end and furthermore that some of these oligosaccharides are substituted by DGalNAc, LFuc and/or sialic acid (20).

While LFuc is an important terminal sugar and present in high amounts in the native glycoprotein, the glycoprotein precipitated only about 30% of the LFuc specific lectins (Table I), which is possibly due to steric effects as blood group H activity was blocked by either DGalNAc or DGal in blood group A or B individuals. DGalβ1→3DGalNAc residues have been found in bronchial glycoproteins prepared from individuals suffering from asthma (6) and chronic bronchitis (13). However, both the native and mild acid treated glycoprotein precipitated poorly with peanut agglutinin and MPA. These results may be due to long chains of carbohydrate residues which are covering or sterically hindering access of DGal β1→3DGalNAc residues in the interior of the oligosaccharide chains to the combining sites of the lectins. This was further confirmed by histochemical findings (24).

From the data herein, some information regarding structural aspects of the oligosaccharides at the terminal non-reducing ends is obtained. However, studies on the combining sites of monoclonal antibodies of human bronchial glycoprotein will provide additional information.

Acknowledgements

This research is partially supported by grants from the Texas Agricultural Experiment Station (TAES H6194), US Department of Agriculture (USDA/SEA/AR58-519B-0880 and USDA 82-CSRS-2-1035), and USDA/SEA Formula Animal Health Funds (Project 6648). The authors would like to thank Aurora Bibb for her secretarial assistance in preparing this manuscript.

References

1. Lopez-Vidriero, M.T., Das, I.: Respiratory Defense Mechanisms, Part I (Brain, J.D. et al., eds.) New York 5, 289-356 (1977).
2. Sachdev, G.P., Myers, F.J., Horton, F.O., Fox, O.F., Wen, G., Rogers, R.M., Carubelli, R.: Biochem. Med. 24, 82-94 (1980).
3. Vishwanath, S. Ramphal, R.: Infection and Immunity 45,197-202 (1984).
4. Boat, T.F., Cheng, P.W.: Fed. Proc. 39, 3067-3074 (1980).
5. Lafitte, J.J., Lamblin, G., Lhermitte, M., Humbert, P., Degand, P., Roussel, P.: Carbohydr. Res. 56, 383-389 (1977).
6. Feldhoff, P.A., Bhavanandan, V.P., Davidson, E.A.: Biochemistry 18, 2430-2436 (1979).
7. Roussel, P., Lamblin, G., Degand, P., Walker-Nasir, E., Jeanloz, R.W.: J. Biol. Chem. 250, 2114-2122 (1975).
8. Van Halbeck, H., Dorland, L., Vliegenthart, J.F.G., Hall, W.E., Lamblin, G., Lhermitte, M., Boersma, A., Roussel, P.: Eur. J. Biochem. 127, 7-20 (1982).
9. Bhaskar, K.R., Reid, L.,: J. Biol. Chem. 256, 7583-7589 (1981).
10. Creeth, J.M., Bhaskar, K.R., Horton, J.R.: Biochem. J. 167, 557-569 (1977).
11. Lamblin, G., Lhermitte, M., Boersma, A., Roussel, P., Reinhold, V.: J. Biol. Chem. 255, 4595-4598 (1980).

12. Hook, G.E.R., Bell, D.Y., Gilmore, L.B., Nadeau, D., Reasor, M.J., Talley, F.A.: Lab. Invest. 39:342-356 (1978).
13. Lamblin, G., Humbert, P., Degand, P., Roussel, P.: Clin. Chim. Acta. 79, 425-436 (1977).
14. Pereira, M.E., Kabat, E.A.: Crit. Rev. Immunol. 1, 33-78 (1979).
15. Lis, H., Sharon, N.: The Antigens (Sela, M., ed.), vol IV, 429-529, Academic Press, New York (1977).
16. Watkins, W.M.: The Glycoproteins (Gottschalk, A., ed.), 2nd ed, part B, 830-891, Elsevier, Amsterdam (1972).
17. Kristiansen, T.: Methods in Enzymology (Jakoby, W.B., Wilchok, M. eds), vol 34, part B, 331-341, Academic Press, New York (1974).
18. Richards, R.L., Moss, J., Alving, C.R., Fishman, P.H., Brady, K.O.: Proc. Natl, Acad. Sci. U.S.A. 76, 1673-1676 (1979).
19. Wood, C., Kabat, E. A., Ebisu, S., Goldstein, I.J.: Ann. Immunol. Inst. Pasteur 129C, 143-158 (1978).
20. Wu, A.M., Kabat, E.A., Gruezo, F.G., Allen, H.J.: Arch. Biochem. Biophys. 204, 622-639 (1980).
21. Wu, A.M., Kabat, E.A., Pereira, M.E., Gruezo, F.G., Liao, J.: Arch. Biochem. Biophys. 215, 390-404 (1982).
22. Wu, A.M., Kabat, E.A., Gruezo, F.G., Poretz, R.D.: Arch. Biochem. Biophys. 209, 191-203 (1981).
23. Sarker, M., Wu, A.M., Kabat, E.A.: Arch Biochem. Biophys. 209:204-218 (1981).
24. Schulte, B.A., Spicer, S.S.: J. Histochem. Cytochem. 31, 391-403 (1983).
25. Mazzuca, M., Lhermitte, M., Lafitte, J.J., Roussel, P.: J. Histochem. Cytochem. 30, 956-966 (1982).
26. Lhermitte, M., Lamblin, G., Lafitte, J.J., Degand, P., Roussel, P., Mazzuca, M.: Carbohydr. Res. 92, 333- (1981).
27. Synder, C.E., Nadziejko, C.E., Herp, A.: Intl. J. Biochem. 14, 895-898 (1982b).
28. Le Treut, A., Lamblin, G., Houdret, N., Degand, P., Roussel, P.: Biochimie 63, 425-434 (1981).
29. Weber, K., Osborn, M.: J. Biol. Chem. 244:4406-4412 (1969).
30. Lowry, O.H., Rosebrough, N.J., Farr, A.L., Randall, R.J.: J. Biol, Chem. 193, 265-275 (1951).
31. Svennerholm, L.: Biochim. Biophys. Acta 24:604-611 (1957).
32. Aminoff, D.: Biochem. J. 81:384-390 (1961).
33. Boas, N.P.: J. Biol. Chem. 204:553-563 (1953).
34. Moschera, J., Pigman, W.: Carbohydr. Res. 40:53-67 (1975).
35. Dische, Z., Shettles, L.B.: J.Biol.Chem 175:595-603 (1948).
36. Dische, Z., Danilchenko, A.: Anal. Biochem. 21:119-124 (1967).
37. Niedermeier, W.: Anal. Biochem. 40:465-475 (1971).
38. Kabat, E.A.: Kabat and Mayer's Experimental Immunochemistry, 2nd ed, 127-128, Charles C. Thomas, Springfield, IL (1961).
39. Kabat, E.A.: Blood Group Substances: Their Chemistry and Immunochemistry, 2nd ed, Academic Press, New York (1956).
40. Tsuyuki, H., Von Kley, H., Stahmann, M.A.: J. Amer. Chem.

Soc. 78:764-767 (1956).
41. Kaplan, M.E., Kabat, E.A.: J. Exp. Med. 123:1061-1081 (1966).
42. Pereira, M.E., Kabat, E.A.: Biochemistry 13:3184-3192 (1974).
43. Pereira, M.E., Gruezo, F., Kabat, E.A.: Arch. Biochem. Biophys. 194, 511-525 (1979).
44. Hammarstrom, S., Kabat, E.A.: Biochemistry 8:2696-2705 (1969).
45. Etzler, M.E., Kabat E.A.: Biochemistry 9:869-877 (1970).
46. Sugii, S., Kabat, E.A.: Biochemistry 19:1192-1199 (1980).
47. Horejsi, V., Kocourok, J.: Biochim. Biophys. Acta. 336: 329-337 (1974).
48. Shciffman, G., Kabat, E.A., Thompson, W.: Biochemistry 3, 113-120 (1964).
49. Goldstein, I.J., Hayes, C.E.: Adv. Carbohydr. Chem. Biochem. 35, 127-340 (1978).
50. Baker, D.A., Sugii, S., Kabat, E.A., Ratcliffe, R.M., Hermenthin, P., Lemieux, R.V.: Biochemistry, 22, 2741-2750 (1983).
51. Bretting, H., Phillips, S.G., Klumpart, H.J., Kabat, E.A.: J. Immunol. 127, 1652-1658 (1981).
52. Pereira, M.E., Kabat, E.A., Lotan, R., Sharon, N.: Carbohydrate Res. 51, 107-118 (1976).
53. Peters, B.P., Ebisu, S., Goldstein, I.J., Flashner, M.: Biochemisty 18,5505-5511 (1979).
54. Pereira, M.E., Kabat, E.A.: Biochemistry 13:3184-3192 (1974).
55. Pereira, M.E., Kisailus, E.C., Gruezo, F.G., Kabat, E.A.: Arch. Biochem. Biophys. 185, 108-115 (1978).
56. Pereira, M.E., Gruezo, F.G., Kabat, E.A.: Arch. Biochem. Biophys. 194, 511-525 (1979).

LECTIN AFFINITY OF HUMAN PLACENTAL OPIATE RECEPTORS

Annie Valette[(1)], Pierre Rougé[(2)], Jean Cros[(1)] and Eric J. Simon[(3)]

(1) Laboratoire de Pharmacologie et Toxicologie Fondamentales du CNRS, 205 route de Narbonne, 31 400 Toulouse, France.
(2) Laboratoire de Biologie cellulaire, Faculté des Sciences Pharmaceutiques, Université Paul Sabatier, 35 chemin des Maraîchers, 31 062 Toulouse, France.
(3) Departments of Pharmacology and Psychiatry, New York University Medical Center, New York, N.Y. 10 016, USA.

Three distinct classes of opioid binding sites have been characterized by their pharmacological properties and their differential localization. These classes, refered as mu, delta and kappa, display a better affinity for morphine, enkephalines and benzomorphan, respectively. However, little is known about the chemical composition of opiate receptors except that they contain protein (1) and phospholipid (2). A recent report of Gioannini et al. (3) shows that opioid binding sites solubilized from the brain of several species are retained on insolubilized wheat germ agglutinin (WGA). This finding constitutes the first indication that opiate receptors may be glycoproteins, like many other membrane receptors. Since the results reported in (3) have been obtained on preparations containing a mixture of mu, delta and kappa binding sites, the aim of the present study was to investigate if different types of opiate receptors could be discriminated on the basis of their carbohydrate moiety. We have previously demonstrated the presence of opiate binding sites on the syncitiotrophoblast brush border membrane (4) and, subsequently, we have provided evidence that they are mainly of the kappa-type (5). In the present work, placental brush border membranes have been used as a source of kappa receptors and their interaction with various immobilized lectins was investigated. The results have been compared with those obtained with mu receptors.

Lectins, Vol. IV

MATERIAL AND METHODS

Isolation of lectins: Con-A was isolated from *Canavalia ensiformis* DC. seed meal (Serva), according to the procedure described by Agrawal and Goldstein (6). The three other α-D-mannose/α-D-glucose specific lectins (PSA from pea: *Pisum sativum* L., petit provençal nain; LCA from lentil: *Lens culinaris* Medik., verte du Puy and LaTA from *Lathyrus tingitanus* L.), were obtained by affinity chromatography on Sephadex G-100, as previously reported (7). Soybean seed lectin (SBA) was prepared as described by Allen and Neuberger (8) and wheat germ agglutinin (WGA) was isolated according to Bouchard et al. (9). The purity of these lectins was thoroughly checked by polyacrylamide gel electrophoresis either at basic (10) or acidic pH (11).

Immobilized lectins: The above mentioned lectins were covalently bound to CNBr-activated Sepharose 4B (Pharmacia), according to the manufacturer's instructions. In all cases, the amount of insolubilized lectin was 5 mg/g wet gel.

Membrane preparation: Human term placenta were collected on ice immediatly after cesarian section. Brush border membranes were prepared essentially as previously described (5) and used either immediatly or kept frozen at -70°C until needed. Crude membrane fractions were prepared from rat whole brain and thalamus, and cerebellum from rabbit and guinea pig. The tissues were homogenized in 10 vol of 50 mM Tris-HCl (pH 7.4) and centrifuged at 100,000 x g for 30 min. The resulting pellets were washed once and spun again for an additional period of 30 min at 100,000 x g. The membranes obtained were kept frozen at -70°C until used.

Radioactive soluble fraction derived from ^{3}H-etorphine labelled membrane: Membrane fractions (3 mg protein/ml) were incubated for 30 min at 37°C in 50 mM Tris-HCl (pH 7.4) with 1.5 nM ^{3}H-etorphine with and without 10^{-5} M levorphanol. Aliquots were removed and assayed to determine the extent of membrane binding. The remaining prebound membranes were diluted 1:1 with 1% digitonine in 50 mM Tris-HCl (pH 7.4) and solubilized by vigourous shaking at 0°C for 30 min. A 1% digitonine concentration has been choosen because it insures a good solubilization of active receptors (12). The supernatant after centrifugation at 100,000 x g for 30 min at 4°C was applied to a lectin column and eluted as described below. Aliquots were assayed for prebound solubilized receptors present prior to lectin chromatography.

Lectin chromatography: Lectin affinity columns were stored in 50 mM Tris-HCl (pH 7.4) containing 1% sodium azide. The buffer used for elution (buffer A) was 50 mM Tris-HCl (pH 7.4) for LCA, LaTA, PSA, SBA and WGA and buffer A containing 1 mM $CaCl_2$

and 1 mM $MgCl_2$ for Con-A. Before each use, columns (1 ml of gel) were extensively washed with at least 300 column volumes of buffer A. Solubilized receptor (3 ml diluted with buffer A 1:3.3) was then applied to the lectin column and the flow was stopped for 45 min, with frequent shaking in order to resuspend the gel. Unbound material was collected in the flow-through (fraction 1). The column was washed with 10 ml buffer A containing 0.05% digitonine (fraction 2) and again with 10 ml buffer (fraction 3). A 0.05% digitonine was used at this stage to prevent receptor aggregation. Displacer (5 ml buffer A containing 0.1M inhibiting sugar and 0.05% digitonine) was applied and column flow was stopped for 45 min, after which the eluate was collected (fraction 4). The column was rinsed with 5 ml buffer A containing 0.05% digitonine (fraction 5). The entire procedure was carried out at 4°C. Fractions 1-3 and 4-5 correspond respectively to the unretained and retained fractions.

Assay of lectin column fractions: ^{3}H-etorphine bound to solubilized opioid receptors was separated from the ligand by precipitation with 15% polyethylene glycol and subsequently assayed.

Protein analysis: The sensitive modified Lowry method of Bensadoun and Weinstein (13) was used, with bovine serum albumin (Sigma) as standard.

RESULTS

Human placental kappa opiate receptors were prebound with ^{3}H-etorphine and solubilized with digitonine. These prebound solubilized receptors were incubated on various lectin columns of different specificities and eluted using the corresponding displacing sugars. As shown in Table 1, the WGA-column retained a significant proportion of kappa receptors, whereas other immobilized lectins such as Con-A, LCA, LaTA, PSA and SBA did not interact with solubilized opiate receptors from human placenta. Total recovery of applied receptors from the different columns ranged from 76 to 95% (Table 1).

Approximately 23% of ^{3}H-etorphine specifically bound to human placental kappa receptors was eluted from the WGA-column by N-acetylglucosamine. When prebinding was performed in the presence of excess levorphanol, no radioactivity was retained on the WGA-column indicating that only the kappa receptors speci-

Table 1

Recovery of solubilized human placental kappa receptors from various lectin columns. The percentage recovered is calculated from the amount applied on the column. Data are given with s.e.m.

Lectin column	Inhibitory sugar used as eluent	n° of exp.	Recovery of solubilized opioid receptors %	
			unretained by the lectin	retained by the lectin
WGA	N-acetylglucosamine	8	67.5 ± 3.0	23.0 ± 1.5
Con-A	α-methylmannoside	6	76.5 ± 3.5	7.0 ± 1.5
PSA	α-methylmannoside	2	87.0	3.0
LaTA	α-methylmannoside	2	76.0	6.5
LCA	α-methylmannoside	2	70.0	6.5
SBA	galactose	2	90.5	2.5

fically labelled with ^{3}H-etorphine interact with WGA. The percentage of solubilized kappa receptors retained on WGA was not modified by recycling the flow-through during 2 h or by increasing the column volume 3-fold. In addition, when the radioactivity present in the flow-through fractions was reapplied to another WGA-sepharose column, 11% was retained and eluted by N-acetylglucosamine (Table 2). The simultaneous addition of solubilized kappa receptors and N-acetylglucosamine to the WGA-column resulted in the complete inhibition of the receptor-lectin interaction.

When we repeated the WGA experiment with receptors solubilized from human placenta and rat brain under the same experimental conditions, we found that the degree of retention of brain opiate receptors on WGA was twice as high as that measured with receptors from human placenta (Table 3). These results were in agreement with those previously reported for rat brain

Table 2

Specific interaction of solubilized human placental kappa receptors from WGA-column under various experimental conditions. The percentage recovered is calculated from the amount applied on the column. Data are given with s.e.m.

Experimental conditions	Specific binding %	
	retained on WGA	retained on WGA in control experiment
WGA-column volume increased 3-fold	18.0	20.0
Preparation recycled during 2 h on the WGA-column	18.5	17.5
Fraction 1 (flow-through) reapplied on a 2nd WGA-column	11.5 ± 3.0	23.5 ± 1.5

by Gioannini et al. (3).

The rat brain contains a mixture of mu, delta and kappa opiate receptors. In order to determine if the extent to which opiate receptors interact with immobilized WGA depends upon the nature of the receptors, we have compared the results obtained with preparations enriched in mu receptors such as rat thalamus or rabbit cerebellum with those obtained with kappa receptors solubilized from human placenta or guinea pig cerebellum. Data (Table 3) show that mu receptors interact equally well with insolubilized WGA whatever their tissue origin. In addition, the measured retention of mu receptors on the WGA-column is 2-times higher than that measured with placental kappa opiate receptors. Kappa receptors from guinea pig cerebellum though retained somewhat better than placental receptors (36%) are retained at a level significantly below that seen for mu receptors from the two tissues studied (52% for rabbit cerebellum and 54.5% for rat thalamus)

DISCUSSION

Our results confirm that opiate receptors of kappa- and mu-ty-

Table 3

Interaction between opioid receptors solubilized from different tissues and immobilized WGA. The percentage is calculated from the amount applied on the column. Data are given with s.e.m.

Tissues	Opioid receptor type %	n° of exp.	Recovery of solubilized opioid receptors (%)	
			unretained by WGA	retained by WGA
Human placenta	kappa 100 %	8	67.0 ± 3.0	23.0 ± 1.5
Rat brain	mu + delta + kappa	4	28.5 ± 3.0	49.5 ± 4.0
Rat thalamus	mu mainly	3	28.0 ± 3.0	54.5 ± 12.0
Rabbit cerebellum	mu mainly	3	33.0 ± 8.0	52.0 ± 6.0
Guinea pig cerebellum	kappa 85 %	4	55.0 ± 6.0	36.0 ± 1.5

pe contain carbohydrates and thus are, at least in part, glycoproteins. They also show that different types of opiate receptors do not interact equally well with immobilized WGA. By using human placenta as a source of opiate receptors which are essentially of the kappa-type, it has been shown that kappa receptors are retained on WGA but not on any other immobilized lectins tried (Con-A, PSA, LCA, LaTA and SBA). However, only 23% of the kappa receptors applied to the WGA-column were eluted by N-acetylglucosamine and thus retained by the lectin. Such a result could be possibly explained on the basis of a certain microheterogeneity of the carbohydrate moiety of the kappa-type receptors. The percentage of placental receptor retained on WGA-column is 2-times lesser than that obtained under the same conditions with receptors solubilized from rat brain, which is a mixture of mu, delta and kappa receptors. We have conducted studies to try to understand the reason for the observed difference in percent retention on WGA for receptors from human placenta compared to those solubilized from rat brain. These percentages are not significantly modified either by recycling the flow-through or by increasing the column volume, suggesting that differences in retention measured with various receptors preparations are independant of the WGA-column and chromatographic conditions. One can hypothesize that such differences in retention reflect in fact differences in the carbohydrate moieties of the three types of opioid binding sites. To test the validity of this hypothesis we have compared retention of receptors from human placenta with receptors from central nervous system known to contain predominantly mu receptors (rat thalamus and rabbit cerebellum). We have also studied the retention of another preparation containing 85% of kappa receptors, the guinea pig cerebellum. This experiment could not be done with delta receptors since any region of the rat brain contains predominantly this type of opiate receptors. The results indicate that the percentage of mu receptors either from rat thalamus or rabbit cerebellum that specifically interact with WGA is always 2-times higher than that measured with placental kappa receptors. Kappa receptors from guinea pig cerebellum are retained better than placental receptors but at a level significantly below that seen for mu receptors. The discrepancies observed between kappa receptors whatever their tissue origin could be explained by the fact that guinea pig cerebellum contains only 85% kappa receptors and the remaining sites (mu and delta) could account for the elevated interaction with WGA. In addition, the nature of membranes in which these receptors are inserted are structurally and functionnally different and an influence of the membrane environment could therefore be responsible for the observed differences. Nevertheless, the present results indicate that mu and kappa receptors do not interact with the same extent with WGA.

Although WGA is generally accepted as rather specific for N-acetylglucosamine and its 1-4 linked oligomers, it has also been shown to interact with sialic acid and sialoglycoproteins (14). Investigations are running on to determine if the differences in retention on insolubilized WGA represent a difference in the carbohydrate moieties of the different types of opiate receptors, i.e. if either N-acetylglucosamine (monomers or oligomers) or sialic acid are involved in the opiate receptor-WGA interaction.

REFERENCES

1. Simon, E.J., Hiller, J.M., and Edelman, I.: Proc. Nat. Acad. Sci. USA 70, 1947-1949 (1973).
2. Lin, H.K., and Simon, E.J.: Nature 271, 383-384 (1978).
3. Gioannini, T., Foucaud, B., Hiller, J.M., Hatten, M.E., and Simon, E.J.: Biochem. Biophys. Res. Commun. 105, 1128-1134 (1982).
4. Porthe, G., Valette, A., and Cros, J.: Biochem. Biophys. Res. Commun. 101, 1-6 (1981).
5. Porthe, G., Valette, A., Moisand, A., Tafani, M., and Cros, J.: Life Sci. 31, 2647-2654 (1982).
6. Agrawal, B.B.L., and Goldstein, I.J.: Biochim. Biophys. Acta 147, 262-271 (1967).
7. Rougé, P., and Père, D.: in Lectins, Biology, Biochemistry, Clinical Biochemistry, (Bøg-Hansen, T.C. ed.), De Gruyter, Berlin, 2, 137-150 (1982).
8. Allen, A.K., and Neuberger, A.: FEBS Lett. 50, 362-364 (1975).
9. Bouchard, P., Moroux, Y., Tixier, R., Privat, J.P., and Monsigny, M.: Biochimie 58, 1247-1253 (1976).
10. Davis, B.J.: Ann. N. Y. Acad. Sci. 121, 404-427 (1964).
11. Reisfeld, R.A., Lewis, U.J., and Williams, D.E.: Nature 195, 281-283 (1962).
12. Rueg, U., Cuenob, T.S., Hiller, J., Gioannini, T., Howells, R.D., and Simon, E.J.: Proc. Natl. Acad. Sci. USA 78, 4635-4638 (1981).
13. Bensadoun, A., and Weinstein, D.: Anal. Biochem. 70, 241-250 (1976).
14. Bhavanandan, V.R., and Katlic, A.W.: J. Biol. Chem. 254, 4000-4008 (1979).

QUANTITATIVE INHIBITION OF LECTIN BINDING TO IMMOBILIZED CARBOHYDRATES

Václav Hořejší, Vladimír Matoušek
Institute of Molecular Genetics, Czechoslovak Academy of Sciences
142 20 Praha 4, Czechoslovakia

Marie Tichá and Tomáš Trnka
Department of Biochemistry and Department of Organic Chemistry, Charles University
128 40 Praha 2, Czechoslovakia

Inhibition of some typical reactions of lectins by a series of carbohydrates is often used to characterize their binding specificity. Such reactions inhibitable by specific carbohydrates are usually precipitation of polysaccharides or agglutination of cells; these, and in fact all other lectin activities, are primarily based on the interaction of the lectin with a soluble or cell surface bound carbohydrate. Usually, the concentrations of the soluble carbohydrates are estimated which inhibit the test reaction e.g. by 50%. The lower is this $\underline{C}_{50}$ concentration, the stronger is the lectin-inhibitor interaction. $\underline{C}_{50}$ is often assumed to reflect directly the value of dissociation constant of the lectin-inhibitor complex.

In the present study we have examined in more detail this method with the aim to define the conditions under which it can really yield the dissociation constants.

We have analyzed the simplest system: a reference carbohydrate is immobilized on a solid phase (small beads, fixed cells, plastic surface) so that its final concentration in the sys-

Lectins, Vol. IV

tem is $\underline{A}$. After addition of a labeled lectin (concentration $\underline{P}$) and equilibration, the fractions of bound ($\underline{b}$) and free (soluble, $\underline{s}$) lectin are measured. The addition of a soluble carbohydrate (concentration $\underline{C}$) into this system will result in an inhibition of binding of the lectin to the immobilized carbohydrate, so that the $\underline{s}/\underline{b}$ ratio will increase with increasing $\underline{C}$. An analysis of this concentration dependence should yield the sought dissociation constant $\underline{K}$. Our results indicate that this method is simple and rather generally applicable for quantitative studies on the lectin-carbohydrate and other protein-ligand interactions. Major potentially complicating factors identified by theoretical analysis are: 1) multiple interactions of lectin molecules with immobilized carbohydrates and 2) too high lectin concentrations. Detailed theoretical analysis of this system and additional experimental data concenrning also the application of this method to the study of the monoclonal antibody-antigen complexes will be published elsewhere (1).

Yet another application of this method is its use for identification of microgram amounts of monosaccharides. This can be done by elution of the sugar spot from the chromatogram and testing its ability to inhibit the binding of selected labeled lectins to carbohydrates immobilized on the polystyrene microculture plate wells. This method can distinguish between the enantiomers of common monosaccharides, i.e. to assign them to the D- or L-series (2).

Theory

The model analyzed is schematically shown in Fig. 1. Immobilized carbohydrate (open circles) bound to a solid phase, soluble carbohydrate (full circles) and a lectin (depicted as bivalent, but generally $\underline{n}$-valent) are present in the solution. After equilibration various complexes between the lectin and the carbohydrates will be present (No. 1-4), as well as uncomplexed lectin (No. 5) and carbohydrate molecules. The formation of the type 6 complex is supposed to be negligible due to unfavourable sterical effects. This system can be quantitatively described by a set of equilibrium equations (defining all the equilibria involved) and a set of mass balance equations relating the equilibrium concentrations of all types of particles present to the known values of total lectin concentration ($\underline{P}$), total soluble carbohydrate concentration ($\underline{C}$) and total immobilized carbohydrate concentration ($\underline{A}$), which can be determined separately. Combining all these equations leads after rather laborious algebraic manipulations (given in detail in ref. 1) to equation 1:

$$\frac{(A - b).C}{ns(A - b) - bL} - \frac{n - 1}{n}\,\frac{b}{s} = \frac{K}{L}\,\frac{A - b}{b} + 1 \qquad \text{Eqn. 1}$$

($\underline{A}$ - total immobilized carbohydrate concentration, $\underline{L}$ - intrinsic dissociation constant of the lectin-immobilized carbohydrate complex, $\underline{K}$ - intrinsic dissociation constant of the lectin-soluble carbohydrate complex , $\underline{b}$ - equilibrium concentration of all bound forms of the lectin, $\underline{s}$ - equilibrium concentration of all soluble forms of the lectin ($\underline{s} + \underline{b} = \underline{P}$, i.e. total lectin concentration), $\underline{n}$ - the number of carbohydrate binding sites in the lectin molecule).

Fig. 1. Schematic representation of the system lectin with immobilized carbohydrate and soluble carbohydrate.

This equation describes completely the system and could be in principle used for estimation of K from the experimentally found dependence of s and b on C (at constant A and P), however it is too complicated for practical use. It is considerably simplified if the equilibrium concentrations of both immobilized and soluble carbohydrates are much higher than equilibrium concentrations of the complexes, i.e. practically if $P \ll A, C$. Then equation 1 is reduced to equation 2:

$$s/b = L.C/nA.K + L/nA \qquad \text{Eqn. 2}$$

which is already very well suited for the determination of K: for each value of C (starting from C = 0) the value of s/b is estimated and the linear plot s/b vs. C yields K as its intercept with the abcissa axis. It should be noted that it is not necessary to know in advance any other parameters, such as L, P, n or A. It is only necessary to immobilize a suitable carbohydrate so that after addition of a trace amount of labeled lectin a measurable s/b ratio is established after equilibration in the absence of any soluble carbohydrate and then to measure its dependence on C. Either s or b can be measured (in cpm) depending on the arrangement used, as total radioactivity added is known. Equation 2 is analogous to that used for similar purpose in affinity electrophoresis (3).

Materials and Methods

The lectins concanavalin A (Con A), pea seed lectin (PSA), Ulex europaeus lectin I (UEA) and Ricinus communis agglutinin (RCA) were purified by affinity chromatography on O-glycosyl polyacrylamide gels (4), iodination was performed by the chloramine T method (5).
Natural or synthetic glycoconjugates were immobilized by adsorption onto the wells of polystyrene microculture plates

(Falcon), in some cases pretreated by 1% glutaraldehyde solution. Optimal coating concentrations (0.2 - 5 mg/ml) were found for each substance empirically such to produce a system in which less than 70% of the labeled lectin subsequnetly added was specifically bound after equilibration in the absence of any soluble carbohydrate. To the coated wells 10 µl volumes of solutions containing trace amounts of ^{125}I-labeled lectin (10^{-9} - 10^{-10} M, approximately 1 - 10 x 10^4 cpm in the total volume) and the soluble carbohydrate in phosphate buffered saline pH 7.2 containing 1% periodate-oxidized serum albumin were added. After a 48 - 72 h incubation at 4° C in a humid chamber the wells were rapidly washed and radioactivity bound to individual wells was measured; alternatively, radioactivity in the supernatants was estimated. The losses of specifically bound radioactivity during washing were negligible. The results were corrected for nonspecific adsorption and the values of s/b plotted against C.

Paper chromatography of standard sugars was performed on Whatman No. 1 paper in the butanol - acetic acid - water (10 : 1 : 3) solvent system. Individual spots contained 10 - 100 µg of the sugar. After chromatography, the paper was dried and a leading strip was developed (6) to detect the positions of the separated sugars. The spots in untreated part of the chromatogram were cut out and extracted in 3 ml water. The eluates were lyophilized, dissolved in 20 - 100 µl solution containing 0.5 M NaCl, 10 mM phosphate buffer pH 7.2 and 1% periodate-oxidized serum albumin, mixed with the stock solutions of ^{125}I-labeled lectins and pipetted into the glycoconjugate-coated polystyrene wells as described above. After an overnight incubation and rapid washing the radioactivity associated with the wells was measured and compared to the values obtained in positive and negative control wells (eluates from the blank parts of the chromatogram and 1 M solutions of the best inhibitors, respectively).

TABLE 1

Dissociation constants of the lectincarbohydrate complexes as determined by the binding inhibition method.

Lectin	Immobilized carbohydrate	Soluble carbohydrate	K (mol/l)
Con A	gelatin or ovalbumin	D-mannose	1.2×10^{-4}
		D-glucose	1.1×10^{-3}
		transferrin (human)	6.0×10^{-6}
PSA	α-D-mannosyl copolymer	D-mannose	3.4×10^{-4}
		D-glucose	1.5×10^{-3}
UEA	hog gastric peptone	L-fucose	4.2×10^{-4}
RCA	hog gastric peptone or α-D-galactosyl copolymer	D-galactose	5.0×10^{-4}
		lactose	1.5×10^{-4}

Experimental Results

Dissociation constants of the lectin - carbohydrate complexes obtained by plotting $\underline{s}/\underline{b}$ vs. $\underline{C}$ are given in Tab. 1. These values are close to the values obtained by other methods (3, 7 - 12).

To ascertain that the condition of validity of the binding-inhibition method, i.e. $\underline{P} \ll \underline{A}$, was met, the values of $\underline{A}$ were determined by Scatchard analysis of the dependence of $\underline{s}/\underline{b}$ vs. $\underline{P}$. In all cases the $\underline{A}$ values were at least 1 - 2 orders higher than the values of $\underline{P}$ used in our experiments (not shown).

TABLE 2

Inhibition of ^{125}I-lectin binding to the glycoconjugate-coated polystyrene wells (coated with gelatin UEA) by microgram amount of sugars eluted from paper chromatograms.

Sugar	Lectin		
	Con A	RCA	UEA
D-glucose	+	-	-
L-glucose	-	-	-
D-mannose	+	-	-
L-mannose	-	+	-
D-galactose	-	+	-
L-galactose	-	-	+
D-fucose	-	+	-
L-fucose	-	-	+
D-arabinose	-	-	-
L-arabinose	-	±	-
D-rhamnose	n.t.	n.t.	n.t.
L-rhamnose	-	+	-

+ strong inhibition (>80%), ± weak inhibition (40-50%) and - lack of inhibition (less than 10%). D-rhamnose was not tested.

The inhibition of labeled lectin binding to immobilized carbohydrates was also used for discrimination between the enantiomers of common monosaccharides after paper chromatography (Tab. 2). Clearly, the enantiomers of glucose, mannose, fucose and arabinose can be distinguished on the basis of their reactivity with Con A, UEA and RCA. D-rhamnose could not be tested but it would certainly behave entirely differently from L-rhamnose. No lectin inhibited well by D- or L-xylose and D- or L-ribose could be found. Similarly, monosaccharides separated and identified by paper chromatography of hydrolysates of some glycoproteins could be assigned to the D- or L- series (2).

Discussion

The binding inhibition method of dissociation constant estimation appears to be attractively simple and general. The results obtained in this way compare well with those obtained by other methods (1, 3, 7 - 12). It should be applicable for many ligand-binding proteins. Two basic conditions must be met in order to obtain meaningful results by means of simple plots (equation 2): First, bi- or multivalent interaction of the protein with immobilized ligand must be avoided. Second, protein concentrations used must be much lower than the ligand concentrations. The first condition is automatically met for monovalent proteins, which are therefore ideally suited for this method. For multivalent proteins this condition will be probably best fulfilled if the ligand is immobilized on a relatively rigid carrier (plastic surface, fixed cells) at relatively low surface density, so that the protein molecule cannot span the distance between two neighboring ligands. Even if this mean distance between the immobilized ligand molecules is relatively small, the multiple interactions are rather unlikely due to sterical factors on such rigid carriers (13). Multiple interactions are expected to occur in greater extent in the case of flexible supports, such as hydrophilic gels or unfixed cells (14, 15).
The second condition ($\underline{P} \ll \underline{A}, \underline{C}$) can be usually fulfilled by using minimal concentrations of proteins labeled to high specific activities. Increased protein concentration (comparable to $\underline{A}$ or $\underline{C}$) causes more or less severe nonlinearities of the $\underline{s}/\underline{b}$ vs. $\underline{C}$ plots which lead to underestimation of the apparent strength of the interaction studied (1).
Our method appears to have several advantages for $\underline{K}$ estima-

tion as compared with other methods (equilibrium dialysis, affinity chromatography, affinity electrophoresis, differential spectroscopy etc.). It requires labeled lectin and not labeled carbohydrate, so that it should be useful especially in the comparative studies on a series of carbohydrates interacting with a single lectin. Carbohydrates widely differing in molecular weight, charge and other properties can be studied in parallel. For efficient labeling of lectins a number of simple and mild procedures exist.

Adsorptive immobilization of macromolecular carbohydrate derivatives on polystyrene surface as used in the present study is convenient and simple, but covalent attachment to other types of supports may be better in some situations, as higher capacity and better stability can be achieved in this way (13).

From our analysis it is also obvious that the $\underline{C}_{50}$ values (see Introduction) are proportional to $\underline{K}$ only if the above mentioned conditions (the lack of multiple interactions with immobilized carbohydrate and $\underline{P} \ll \underline{A}, \underline{C}$) are met.

Also, the possibility to discriminate between the enantiomers of sugars by this method may be of importance. So far this could be done only with relatively large amounts of sugars by polarimetry or by chromatographic separation of glycoside derivatives containing optically active aglycons (16, 17). As compared with those methods, our method is simple and requires only microgram amounts of the sugars tested.

Acknowledgements. The authors are indebted to Professor G. Uhlenbruck and Dr. G. Gachelin for some of the standard sugars used; to Mr. J. Josek for his technical assistance; and to Mrs. L. Vyhnálková for the careful typing of the manuscript.

References

1. Hořejší, V., Matoušek, V.: Mol. Immunol., in press.
2. Hořejší, V., Tichá, M., Trnka, T.: Anal. Biochem. 137, 170-174 (1984).
3. Hořejší, V., Tichá, M., Kocourek, J.: Biochim. Biophys. Acta 499, 290-300 (1977).
4. Hořejší, V., Kocourek, J.: Biochim. Biophys. Acta 538, 299-315 (1978).
5. Greenwood, F.C., Hunter, W.H., Glover, J.S.: Biochem. J. 89, 114-123 (1963).
6. Kocourek, J., Tichá, M., Koštíř, J.: J. Chromatogr. 24, 117-124 (1966).
7. So, L.L., Goldstein, I.J.: Biochim. Biophys. Acta 165, 398-404 (1968).
8. Young, N.M., Leon, M.A.: Biochim. Biophys. Acta 365, 418-424 (1974).
9. Bøg-Hansen, T.C., Takeo, K.: Electrophoresis 1, 67-71 (1980).
10. Trowbridge, I.S.: J. Biol. Chem. 249, 6004-6012 (1974).
11. Podder, A., Surolia, A., Bachhawat, B.K.: Eur. J. Biochem. 44, 151-160 (1974).
12. Olsnes, S., Saltvedt, E., Pihl, A.: J. Biol. Chem. 249, 803-810 (1974).
13. Kennel, S.J.: J. Immunol. Meth. 55, 1-12 (1982).
14. Eilat, D., Chaiken, I.M.: Biochemistry 18, 790-795 (1979).
15. Mason, D.W., Williams, A.F.: Biochem. J. 187, 1-20 (1980).
16. Gerwig, G.J., Kamerling, J.P., Vliegenthart, J.F.G.: Carbohyd. Res. 62, 349-357 (1978).
17. Gerwig, G.J., Kamerling, J.P., Vliegenthart, J.F.G.: Carbohyd. Res. 77, 1-7 (1979).

QUANTITATIVE STUDY OF LECTIN-MEDIATED CELL AGGREGATION

M. Duszyk, M., K. Kawalec and J. Doroszewski

Medical Center of Postgraduate Education, Department of Bhophysics and Biomathematics, 01-813 Warsaw, Marymoncka 99, Poland.

A characteristic property of lectins is their ability to promote cell aggregation. This phenomenon is the subject of numerous investigations (1-3). There are several hypotheses concerning lectin-mediated cell aggregation (for reviews see 4,5). However, none is sufficiently general to take various types of cells or various lectins into acount.

In the aggregation process two basic steps can be distinguished. The first is bringing cells together to produce contact between their surfaces. This contact may be coincidental (e.g. due to shaking) or elicited in a controlled way, e.g. in Couett's or Poiseuille's (9). In the latter case it is possible to evaluate the number of collisions between the cells in a definite time inverval. The second step consists in formation of bonds between cell surfaces. The probability of bond formation depends on several factors such as the amount of lectin adsorbed, the association strength and the diffusion of lectin receptors in the cell membrane.

Lectins, Vol. IV

Theory

Analysis of the aggregation process in Poiseuille's flow allows to determine changes in the concentration of the suspension (number of cells and aggregates per unit volume) as a function of parameters characterising flow and bond formation (5,7). In particular it is possible to establish the diffusion coefficient for molecules in the plane of the cell membrane (9).

When the lectin concentration in the medium is so high that a further increase does not influence cell aggregation, the change in concentration of the suspension can be expressed by equation (1)

$$N/N_o = 4\int_o^1 x(1 - x^2)\exp(-\frac{kxz}{1-x^2})dx \qquad (1)$$

Where x is the dimensionless integral variable, k is a constant, N is the suspension concentration at distance z from the beginning of the tube, and N_o is the initial suspension concentration. By fitting the results of aggregation measurements the constant k can be calculated (7).

In the case of low lectin concentrations where is a small change in concentration causes a large modification in aggregations, the change in cell concentration is described by equation (2).

$$N/N_o = 4\int_o^{x_k} x(1 - x^2)\exp(-\frac{kxz}{1-x^2})dx + (1 - x_k^2)^2 \qquad (2)$$

where $x_k = \sqrt{l \cdot D}$, where l is a constant and D is the diffusion constant of receptors in the membrane. This equation allows to calculate the relative change of the diffusion constant of receptors in the cell membrane plane (7).

The aggregation of control cells and of those subjected to the action of fixation in formaline is measured. The measurements are made at low concentration of lectin the medium. Equation (2) is used for calculation of the results. Comparison of the

x_k values calculated with and without formalin allows to determine the relative change of the receptor diffusion constant in the plane of the cell membrane.

Experimental

We performed experiments to determine the number of receptor sites for Ricinus communis, RCA I, at the surface of rat thymocytes as well as the relative change of the diffusion constant for receptors in the membrane of living and of formalin-fixed cells. For determination of the number of receptor sites on the cell surface, 2×10^6 thymocytes in 1 ml were incubated at 37^oC for 30 min with various concentrations of ^{125}I-labelled RCA I which was re-chromagotraphed after labelling. Radioactivity bound with the cells was estimated. As the control served cells incubated with ^{125}I-RCA I in the presence of 0.05 M lactose. The difference in cell-bound radioactivity was taken as a measure of the specific lectin binding. To find the number of lecti-binding sites on the cell surface, the results of the experiments were plotted by the method of Steck and Wallach (8) according equation (3)

$$\frac{1}{L_b} = \frac{1}{KncL} + \frac{1}{nc} \qquad (3)$$

where L_b is the amount of RCA I bound to the cells (μg/ml), K is the affinity constant of lectin (M^{-1}), n is the number of lectin-binding sites per cell, c is the number of cells (Fig. 1). On the basis of the molecular mass of 120 000 of RCA I, n is 2.65×10^6 per cell and K is 1.25×10^7 M^{-1}. The number of receptor sites at the cell surface is necessary to know when the absolute value of the diffusion constant for the receptors on the membrane is being determined (7).

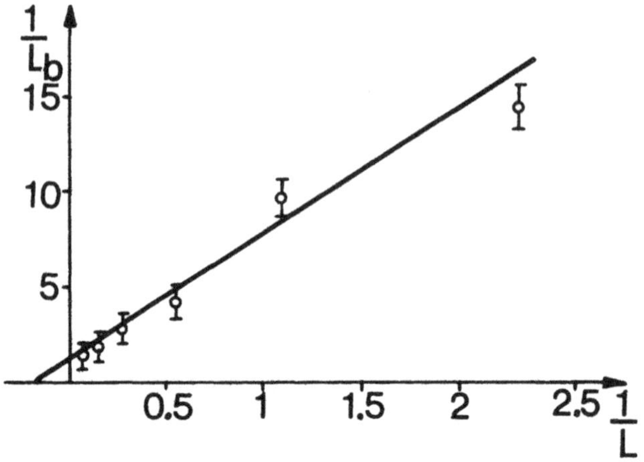

Fig. 1. Steck and Wallach plot of binding of ^{125}*I RCA I to thymocytes. L = the lectin concentration (µg/ml).* L_b *= the lectin bound to the cells (µg).*

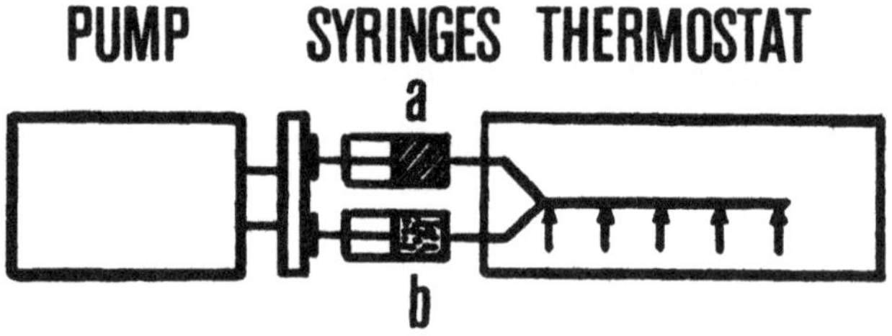

Fig. 2. Experimental device. a = cell suspension. b = RCA I. The cell suspension (4000000 per ml) and the lectin solution were supplied from separate 5 ml syringes to the polyethylene tube (660 mikrometre in diameter and 5.5 m in length). The tube was coiled and placed in a thermostat chamber. The arrow indicates where aliquots were taken at a distance of z meter.

Cell aggregation was measured by means of the device shown in Fig. 2. The cell suspension and RCA I solution were supplied from separate syringes to the polyethylene tube by pushing the syringe at constant speed. An aliquot was taken every 1.1 meter from the inlet and the concentration was determined by means

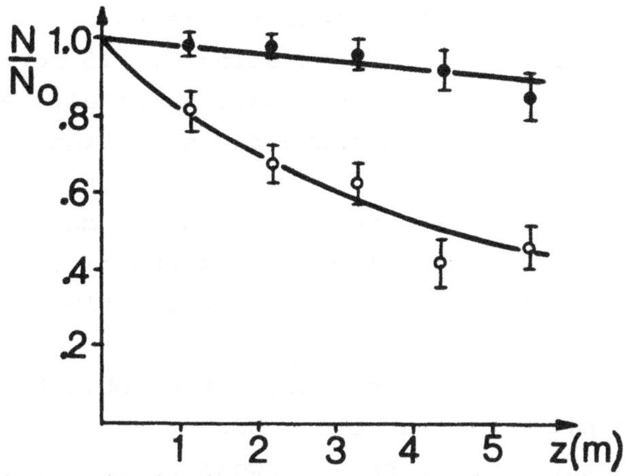

Fig. 3. Change in living (o) and fixed cells concentration (o) under the same flow conditions and lectin concentration (0.5 μg/ml).

Table 1.

Kinetics of cell aggregation

z (m)	1.1	2.2	3.3	4.4	5.5
N/N_o	0.78	0.68	0.59	0.52	0.30

of a cell counter (Picoscale 4, Medicor, Hungary). The results of the experiments are presented in the Table I and in Fig. 3. The table shows the results of cell aggregation measurements at a lectin concentration of 1 μg/ml.

The cange in cell and aggregate concentration is dependent on the length of the flow channel. It is, however, independent of the flow rate. The results of aggregation measurements at concentrations of lectin higher than 1 μg/ml were within the limits of accuracy equal with those shown in the Table 1. Fitting of the results presented in Table 1 to equation (1) allowed to determine the constant k as 0.39 M^{-1}.

Fig. 3. shows the results of aggregation measurement of living and fixed cells at a lectin concentration of 0.5 µg/ml. Fitting to the results shown in Fig. 3 to equation (2) allowed us to determine the x_k values and to compare the diffusion constants for receptors in the membrane. $D_1 = 0.12\ D_2$ was obtained, where D_1 is the diffusion constant for receptors in the membrane of living ones. Thus, cell fixation decreases the diffusion constant for receptors in the membrane plane about ten times.

This method may be useful for comparing the mobility of lectin receptors in the plane of the cell membrane and for determining the influence of the cytoskeleton as well as of environmental factors as temperature and pH.

References

1. Nicolson, G.L., Poste, G.: Bioch. Biophys. Acta 554, 520-531 (1979).
2. Loor, F.: Exp. Cell Res. 82, 415-425 (1973).
3. Jones, N.M., Perry, R.: Exp. Cell Res. 128, 41-48 (1980).
4. Lis. H., Sharon, H.: Annu. Rev. Biochem. 43, 541-574 (1973).
5. Nicolson, G.L.: Int. Rev. Cytol. 39, 99-190 (1974).
6. Duszyk, M.: Thesis, University of Warsaw, (1984).
7. Duszyk, M., Doroszewski, J.: Submitted for publication to J. Cell. Biol.
8. Steck, T.L., Wallach, D.: Biochim. Biophys. Acta 87, 510-522 (1965).
9. Happel, J., Brenner, H.: Low Reynolds Number Hydrodynamics, Noordhoff International Publishing, Leyden (1973).

PROLIFERATIVE RESPONSE OF HUMAN T LYMPHOCYTES TO A MIXTURE OF TWO LECTINS

K.E. Wiktorowicz, W. Niedbala, A. Mackiewicz
Department of Immunology and Rheumatology, University School of Medicine, ul. Szkolna 8/12, 60-967 Poznan, Poland

Mitogenic lectins are widely used in studies on the mechanism of lymphocyte activation and in the evaluation of the immunological competence of lymphocytes. Target cells for activation are T lymphocytes, but the effect varies with the kind of lectin (9,10), the genetic background of the lymphocyte donor (3, 14,23), and of the lymphocyte subset engaged in the response (1,13,15,21,26). Few trials deal with to stimulation of lymphocytes by a mixture of mitogens in order to analyse the cell triggering, the cellular collaboration or the lymphocyte subsets involved in the blastogenic response (11,22,24). In this paper, the response of purified T lymphocytes to a mixture of two lectins has been studied.

Lectins, Vol. IV

Materials and Methods

Preparation of mononuclear cells: freshly drawn samples of heparinized venous blood of 10 normal donors were centrifuged for 10 min at 350 g. The buffy boats of leucocytes were removed, diluted 1:4 with Hanks solution and centrifuged for 30 min at 400 g at 4°C on a Ficoll Uropolinum gradient (d = 1.077 g/cm^3) according to (2). Interphase cells were collected, washed twice with Hanks solution and resuspended in Eagle's medium.

Depletion of macrophages: cell suspension (5 x 10^6/ml) in Eagle's medium supplemented with 20% of autologoous plasma was incubated on a sterile glass Petri dishes for 2 h at 37°C in a humidified atomosphere containing 5% CO_2. Next, nonadherent cells were washed out from Petri dishes with Eagle's medium and incubated for 45 min at 37°C with carbonyl iron particles with sequential magnet treatment to deplete the phagocytes. Glass adherent cells, scraped off with a rubber policeman, contained over 60% of monocytes indentified on the basis of ANAE staining (12) and were used as a source of monocytes in some experiments.

Purification of T lymphocytes: 50 ml glass tubes were filled with lymphocyte suspension (4 x 10^6/ml) and SRPB (3% solution in Eagle's medium), preincubated for 10 min at 37°C, centrifuged for 5 min at 200 g and incubated for 2 h in ice. Next, half of the supernatant was taken out, calf serum (heat inactivated and absorbed with SRBC) was added to a concentration of 30% and the pellet was gently resuspended and underlayed with a Ficoll Uropolinum gradient. Tubes were centrifuged for 30 min at 400 g at 4°C and the pellet of cells from the bottom (SRBC-binding T lymphocytes) were collected, washed 3 times with cold Hanks' solution and treated with hypotonic solution (0.83% NH_4CL Tris buffer). This population, used in experiments as T lymphocytes, contained at least 95% of T lymphocytes, checked on the basis of ANAE staining (17).

Proliferation assay: 1 x 10^5 cells per well were cultured on flat-bottomed microplates (Falcon, USA) in a total volume of 0.2 ml of Eagle's medium supplemented with 10% of autologous plasma. Streptomycin (100 µg/ml) and pneicillin (100 IU/ml) were added to the culture medium. To avoid the influence of differences in proliferation kinetics of different lymphocyte subsets, the amethopterin block was used in all experiments. Amethopterin (Methotrexat, Lederle) was added to the culture medium at a final concentration of 2.2 x 10^{-7} M and blocked endogeneous thymidine synthesis to cause accumulation of proliferating cells at the G_1/S boundary (23). The block was reversed by addition of ^{3}H-thymidine. All cell cultures were performed in triplicate. Phytohaemagglutinin (PHA, HA 17, Wellcome), concanavalin A (Con A, Pharmacia), pokeweed mitogen (PWM, Sigma) and wheat germ agglutinin (WGA, Sigma) were used as mitogens.

Cells were cultured for 3 days at 37°C in a humidified atmosphere containing 5% Co_2. Two hours before harvest, 1 μCi (37 kBq) ^{3}H-thymidine (^{3}H-Tdr, UVVVR, CSSR) per well was added. Cells were collected on glass fibre filters and washed 3 times with water. Radioactivity was estimated in a liquid scintillation counter.

Results

We stimulated lymphocytes with a mixture of two mitogens. When unseparated lymphocytes were stimulated by PHA and PWM together, the response was higher than compared to lymphocytes stimulated by PHA or PWM alone. This was hot observed when total lymphocytes were stimulated by a mixture of PHA and Con, A Table 1.

The response of purified T lymphocytes to PHA, PWM and Con A was found to be much lower in comparison to the response of unseparated lymphocytes. T Lymphocytes responded higher to mixtures of Con A and PHA, Con A and PWM, and Con A and WGA than

Table 1.

Unseparated lymphocytes stimulated by two mitogens.

PHA	μg/ml		0.3	0.06	0.012	0
Control	μg/ml		65	38	8	3
PWM		10	75	64	44	43
		2	67	49	30	37
		0.4	66	46	23	26
Con A	μg/ml	5	21	21	19	11
		1	56	26	20	23
		0.2	64	21	7	4

Incorporation of ^{3}H-Tdr in thousands of cpm.

to any one of the lectins used alone, Table 2. In some ranges of lectin concentration, the response was synergistic.

Addition of glass adherent cells (monocytes) to the population of T lymphocytes increased the response to the lectins, but abrogated the synergistic effect observed in response to two mitogens, Table 2.

Discussion

The response of human peripheral blood mononuclear cells to the mixture of PHA and PWM was usually a little higher than to each of the mitogens used alone. The augmentation of the response, similar to that observed by Steffen et al. (24), could be caused by activation of different cell subpopulations. PHA is a T cell mitogen, and PWM activates both T and B lymphocytes (1, 13,2126), ^{3}H-Tdr incorporation in lymphocyte cultures stimulated by PHA and Con A together was usually lower than in the cultures stimulated with PHA alone. This is in agreement with the data of Steffen et al. (29) and Schmidtke and Najarian (22). The subpopulation of human peripheral blood lymphocytes responsive to Con A overlap with the one reactive to PHA (18,26).

Thus, an additive response or at least a somewhat higher response should be expected in the case of stimulation by the mixture of PHA and Con A. Such a response to PHA plus Con A was observed in the cultures of mouse splenocytes (27), where two populations of cells differing in their reactivity to PHA and Con A were found (7). The decrease of ^{3}H-Tdr incorporation by lymphocytes stimulated with PHA and Con A together could reflect some regulatory processes during lymphocyte activation.

Monocyte depleted T lymphocytes did not respond to any of the mitogens, what is in agreement with data of the other authors (8,13,16,19,20). The mixtures of two lectins (Con A plus PHA, Con A plus PWM, Con A plus WGA) were able to induce a prolife-

Table 2.

Mixed stimulation with two lectins. Incorporation of ^{3}H-Tdr in thousands cpm by cultures of purified T lymphocytes and cultures of T lymphocytes supplemented with 20% of autologous glass adherent cells.

lymphocytes		purified				supplemented			
Con A	μg/ml	5	1	0.2	0	5	1	0.2	0
PHA (μg/ml)	0	1	0	1	1	50	3	0	0
	0.3	8	2	2	2	70	36	21	26
	0.06	1	1	0	0	53	4	1	3
	0.012	5	3	0	0	1	2	2	1
PWM (μg/ml)	0	2	1	0	0	45	3	1	0
	10	4	2	1	1	12	25	16	12
	2	8	2	1	0	8	14	7	4
	0.4	5	4	2	0	3	3	2	3
WGA (μg/ml)	0	1	1	1	0	18	5	1	0
	20	10	4	1	1	7	3	1	0

rative response of purified T lymphocytes. Similar results were obtained by Favero et al. working on mouse thymocytes stimulated by a mixture of two lectins (4).

The proliferation of T lymphocytes induced by Con A and WGA together is of special interest. Some authors consider WGA as a nonmitogenic lectin, a potent inhibitor of PHA and Con A induced blastogenesis (6,10); others consider WGA a weak T specific mitogen (5). Our results seem to confirm that WGA is a specific T mitogen.

Favero et al. (4) have provent that the enhanced response of mouse thymocytes to two lectins is caused by a cooperation of two cell populations. This could also be the reason for the synergistic T lymphocyte response here, but it should be borne in mindt that a human peripheral blood T lymphocyte population is more heterogenous than a thymocyte population. In further investigations, a more detailed analysis of the T subsets used in the experiments will be necessary.

Acknowledgements. Authors are indepted to prof. S. Mackiewicz, prof. J. Steffen, dr. J. Favero, dr. M. Kotecki and dr. M.S. Stolzmann for stimulating discussions. The skilful technical assistance of Ms. A. Walczak is greatly appreciated. Mrs. Reni Storgaard is thanked for typing the manuscript.

References

1. Bedger, M.P., McGiven, A.R., Fitzgerald, P.H.: Immunology 37, 785-792 (1979).
2. Bøyum, A.: Scand. J. Immunol., 5, suppl. 5, 9-15 (1976).
3. Buckley, P.J., Wedmer, H.J.: J. Immunol., 119, 9-18 (977).
4. Favero, J., Marti, J., Dormand, J., Boonafous, J.C. These proceedings.
5. Gordon, L.K., Hammill, B., Parker, C.W.: J. Immunol., 125, 814-819 (1980).
6. Greene, W.G., Waldemann, T.A.: J. Immunol., 124, 2979-2987 (1980).
7. Hävry, P., Roberts, P.J., Ranki, A., Weber, T.: Cell. Immunol. 25, 121-128 (1976).
8. Issekutz, T., Jebra, H., Geha, R.S.: Clin. Immunol. Immunopathol., 23, 634-647 (1982).
9. Kondracki, E., Milgrom, F.: J. Immunol., 118, 381-387 (1977).
10. Larsson, E.L., Coutinho, A.: Eur. J. Immunol., 10, 93-99 (1980).
11. Nespoli, L., Vitiello, A., Maccario, R., Lanzavecchia, A., and Ugazio, A.U.: Scand. J. Immunol., 12, 165-170 (1980).
12. Niedbala, W., Tabaczewski, P., Wiktorowicz, K.: Immunol. Letters, 6, 327-329 (1983).
13. van Oers, M.H.J., Pinkster, P., Zeijlemaker, W.P.: Int. Arch. Allergy appl. Immunol., 58, 53-66 (1979).
14. Osa, S.R., Weksler, M.E.: Cell. Immunol., 32, 391-399 (1977).
15. Peterman, G.M., Altman, L.C., Corey, L.: J. Immunol., 126, 1547-1552 (1981).

16. Potter, M.R., Moore, M.: Clin. Exp. Immunol., 27, 159-164 (1977).

17. Ranki, A., Tötterman, T.H., Hävry, P.: Scand. J. Immunol., 5,1129-1138 (1976).

18. Rawson, A.J., Daniele, R.P.: Res. (NY), 20, 421-427 (1976).

19. Rosenberg, S.A., Lipsky, P.E.: J. Immunol., 126, 1341-1345 (1981).

20. Rosenstreich, D.L., Farrar, J.J., Daugherty, S.: J. Immunol., 116, 131-139 (1976).

21. Schechter, B.: Lymphocyte stimulation, Castellani, A. (ed.), Plenum Press, New York and London, 1-13 (1980).

22. Schmidtke, J.R., Najarian, J.: Immunol., 114, 742-746 (1975).

23. Steffen, J.A., Stolzmann, W.M.: Exp. Cell REs., 56, 453-460 (1969).

24. Steffen, J.A., Swierkowska, K., Michalowski, A., Kling, E., Nowakowska, A.: Mutagen-induced chromosome damage in Man, Evans, H.J. and Lloyd, O.C. (eds.), University Press, Edinburgh, 89-107 (1978).

25. Stiffel, C., Liacopoulos Briot, M., Decreusefond, C., Lambert, F.: Eur. J. Immunol., 7, 291-297 (1977).

26. Touraine, J.L., Touraine, F., Hadden, J.W., Hadden, E.M., Good, R.A.: Int. Arch. Allergy appl. Immunol., 52, 105-117 (1976).

27. Waterfield, J.D., Ekstedt, R.D., Möller, G.: Scand. J. Immunol., 6, 615-623 (1977).

ANTI H GROUPING WITH THE *SAMBUCUS EBULUS* L. LECTIN.

M. Janicot, M. Lalaurie, J. Berlan and F. Miquel.

Laboratoire d'Anatomie et Physiologie Humaines U.E.R. des Sciences Pharmaceutiques 15, avenue Charles Flahault, F-34060 Montpellier Cedex, France

Sambucus ebulus L. (daneworth) is a perennial with annually herbaceous growth, about 1,50 m high, with a radicating stock. The shoots appear as early as february; it is in full bloom from June to July; the ripe fruit in late August are berries with a black juice which contains three nutlets. *Sambucus ebulus* L. contains an anti O (H) lectin particularly active in the fruit and seeds (1).

It is inhibited by D-galactose and related sugars (2-3). Because of its ability to fix on H membrane sites the lectin was used for the differentiation of A_1O, A_2 and BO phenotypes in adult and new-born babies.

Lectins, Vol. IV

Materials and Methods

Gathering was done under specific soil and climatic conditions, taking into account the optimum vegetative periods, blossoming and fructification (1). The plants were dried in the shade in well ventialted rooms. The extracts were prepared according to the usual technique (1/10 wight in isotonic medium). After being ground in the liquid, then macerated for 12 hrs at 20°C, the final extract was separated through a long centrifugation.

Titration was done in a saline or albuminous medium (20% ox-albumin). The titre was given by the last positive agglutination, after the extract had been diluted in a ratio 2 geometrical progression. Different phenotypes of ABO blood cells (all of them Rh+) were used; they came from donors' bloods at the Transfu-

Table 1

Distribution of lectin Sambucus edulus.

	O(H)	A	A	B
Radicating stock	256	512	256	256
Roots of rootlets	0	2	0	0
Young shoots in earth	32	64	32	64
Green stems	0	0	0	0
Buds	32	8	8	8
Flowers	64	2	2	2
Gynaeceum + calyx	64	16	8	8
Unripe fruits green	256	16	256	16
Unripe fruits, red	1024	32	1024	64
Ripe fruits, black	1024	64	512	128
Seeds	512	128	256	256

The results are given as titre of extracts of each plant part. The titre was given by the last positive agglutination, after the extract had been diluted in a ratio 2 geometrical progression.

sion Center in Montpellier and were grouped through standard test serums, among which was the eel serum for O(H).

Results

Table 1 shows the distribution in hte plant of the agglutinating activity. The stock is active but not specific. The leaves and specially the fruit show a marked anti O(H) activity. In the seeds the agglutinin can be found in the teguments and the proteic layer, whereas in the stock it is in the wood and bast system. In the leaves, sunshine increases the synthesis of the agglutinating substance as well as its anti H specificity, which is even stronger in the fruit.

Agglutination experiments showed the lectin was inhibited by D-galactose, D-lactose and D-fucose (see Table 2). In electropho-

Table 2

Inhibition of lectins from eel-serum and Sambucus ebulus fruit by sugars.

	X	Y	Sugars (240 mM)	Eel serum	Sambucus edulus
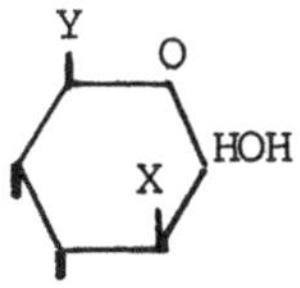	OH	CH_3	L-fucose	+	-
	H	CH_3	D-digitoxose	+	-
	OH	H	D-arabinose	+	-
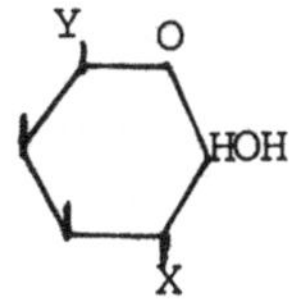	OH	CH_3	D-fucose	-	+
	OH	CH_2OH	D-galactose	-	+
	NH_2	CH_2OH	D-galactosamine	-	+
			lactose	-	+

+ = inhibition, - = no inhibition.

The Sambucus edulus lectin was extracted from the fruit.

Table 3

Frequency of the phenotypes obtained for 5047 bloods.

Sambucus ebulus	FRANCE	EUROPE	
Phenotypes %	Phenotype %	Phenotypes %	Genotypes Calculated
O = 44,64	O = 44	O = 2,9	OO = 45
A_1 = 38,38	A_1 = 38,35	A_1 = 31	A_1A_1 = 3,5
			A_1A_2 = 2,6
			A_1O = 25
A_2 = 6,7	A_2 = 6,7	A_2 = 9,6	A_2O = 9,2
			A_2A_2 = 0,5
A_1B = 2,13	A_1B = 0,75	A_2B = 1,1	
B = 7,70	B = 8	B = 10	BB = 0,7
			BO = 9,3

The Sambucus edulus lectin was extracted from the fruit.

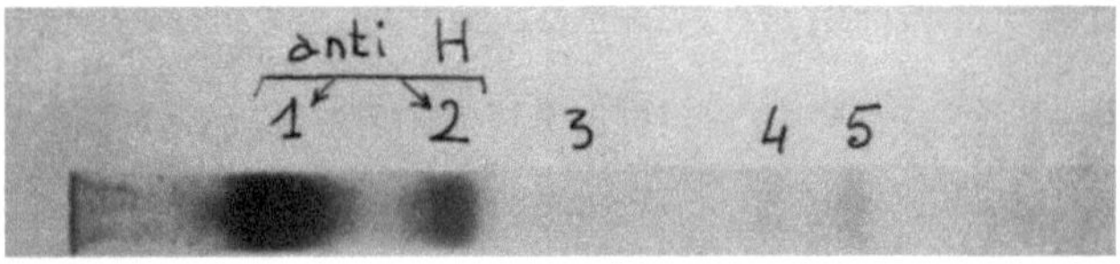

Fig. 1. Polyacrylamide gel electrophoresis of the seed lectin. Conditions were 5% polyacrylamide gel, 120 V, 24 mA, TRIS buffer pH 8.2, migration 1 h. Clearcut bands 1 and 2: anti H activity on 0 red cells. Clearcut bands 3, 4 and 5: no activity, weak concentration.

resis experiments with 100 µl of the aqueous extract we obtained 5 bands (Fig. 1). In separate electrophoresis experiments we isolated the first two bands, both with anti H activity.

Table 4

Differences in the A_1O and BO phenotypes found with Sambucus edulus and eel.

Sambucus %	Eel %	Genotypes calculated
A_1O_{Se} = 10,7	A_1O_{eel} = 1,82	A_1O = 25
BO_{Se} = 5,03	BO_{Se} = 2,35	BO = 9,3

A_1O_{Se} = A_1 and BO_{Se} = B revealed by the fruit extract of the Sambucus ebulus.
- A_1O revealed by the eel-serum.

Applications

The fruit-extract obtained through a dilution aqueous of 1/120, was used for differentiation of A_1O, A_2 and BO phenotypes in 5047 adult bloods. The results are shown in Table 3 and reveal that the percentages obtained with our extract from *Sambucus ebulus* were closer to the genotypes than those obtained with the eel-serum commonly used (Table 4). We found that the phenotypes defined by *Sambucus edulus* as A_1O_{se} and BO_{se} contained a lot of H membrane sites.

In grouping of the red cells of newborn babies containing H antigens on the membrane, 200 reacted with the extract of the *Sambucus ebulus*-fruit, whereas groupings were more difficult with standard test-serums (3).

Another test extract from the seeds was prepared; it has been used for more than 2 years in routine groupings at the Blood Transfusion Center in Montpellier through an automated system, the "Groupamatic" (4). This test extract, an aqueous saline extract of seeds at 1/10 to which AB inhibiting substances can be added, is used at a dilution of 1/120. The extract can be stored frozen in small amounts to be diluted when needed.

Discussion

The anti O(H) lectin of the *Sambucus ebulus* L. can be found in all the plant parts; it is highly specific in the fruit, but its practical use is hindered by the presence of anthocyanic pigments. This lectin inhibited by D-galactose and D-fucose is different from the eel-serum, L-fucose specific, and currently used as anti O(H). The lectin of the *Sambucus ebulus* reveals the H antigenic sites, which are not detected by the eel-serum. This lectin has therefore been used in every day groupings in the Montpellier Blood Transfusion Center.

References

1. Janicot, M.: Thése pour le Doctorat d'Etat és Sciences Pharmaceutiques, Montpellier, julliet 1972.
2. Janicot, M., Lalaurie, M., Berlan, J., et Mary, A.: Trav. Soc. Pharm., Montpellier, 38, fasc. 2, 173-182, 1978.
3. Lalaurie, M., Berlan, J., et Janicot, M.: C.R. de Societe de Biologie Tome 175, n° 4, 490, 1981.
4. Soustelle, B.: Thése pour le Doctorat d'Etat és Sciences Pharmaceutiques Montpellier, octobre 1983.
5. Race et Sanger, (traduction Salmon et Mourier). Les groupes sanguins chez l'homme. Masson ed. 1970.

BLOOD GROUP SUBSTANCES FOR DEMONSTRATION OF LECTINS

Hartmut Franz and Jutta Mohr

Staatliches Institut für Immunpräparate und Nährmedien, DDR-1120 Berlin

Dieter Neumann

Institut für Biochemie der Pflanzen der Akademie der Wissenschaften der DDR, Halle

Blood group substances (BGS) are glycoproteins. The A, B and H substances contain the **monosaccharides D-Gal, L-Fuc,** GalNAc and GlcNAc. These substances are therefore reagents for a multitude of lectins. It should be possible to use BGS labelled by common markers such as FITC or colloidal gold for detection of lectins in plant tissues. Preliminary experiments showed a rapid precipitation of BGS A by Helix pomatia lectin and WGA. Thus it is the aim of this paper to demonstrate the detection and localization of wheat germ lectin in situ by means of labelled BGS A.

Lectins, Vol. IV

Material and methods

BGS A was prepared after Bendich et al. (1) from hog gastric mucosa.

Labelling with FITC: 20 mg BGS A were dissolved in 1 ml PBS. Then 1 ml of 0.5 M bicarbonate-carbonate buffer (pH 9.0), containing 2 mg FITC was added and the mixture was stirred for 6 h at room temperature. After that, conjugation was continued overnight at room temperature without stirring. Unbound dye was removed by gel filtration on Sephadex G-50. The first fraction was collected (3 ml). It contained 1.4 mg protein (corresponding ca. 6.5 mg BGS A) and 25 ug dye/ml.

Labelling with colloidal gold: 25 ug BGS A solution (containing 0.25 mg; the amount of BGS A necessary for complete saturation was determined as described by Roth et al. (2)) was dissolved in 10 ml colloidal gold solution of pH 5.6. After shaking for 3 min, 1 ml of 1 % polyethyleneglycol M 20 (PEG) was added followed by centrifugation (60 min, 20.000 rpm). The sediment was suspended in PBS (pH 6.8) containing 0.05 % PEG. After recentrifugation the pellet was resuspended in 1 ml PBS, pH 6.8. The reagent was stable at 4°C for several weeks.

Fluorescence microscopy: Cryosections (10 um) were incubated with BGS A-FITC conjugate for 2 h at room temperature, washed with PBS, mounted in glycerol (pH 7.5, containing 1 % p-phenylenediamine) and viewed with an epifluorescence microscope and appropriate filters for FITC fluorescence.

Electron microscopy: 3 days old germs of Triticum vulgare c.v. Hatri (summer wheat) have been fixed in 4 % formaldehyde/phosphate buffer, pH 7.4 for 2 h. After embedding in araldite ultrathin sections on Ni-grids were incubated with gold labelled BGS A for 2 h. After washing three times using PBS (500 mM NaCl) containing 0.1 % Tween for 10 min and rinsing in aqua dest the sections were stained with 2 % uranyl acetate for 20 min.

Controls: (1) Blocking test: After preincubation with unlabelled BGS A solution (10 mg/ml) and washing using PBS, the sections were incubated with BGS A-gold and BGS A-FITC conjugate, in turn. (2) To exclude nonspecific reactions caused by the marker substances, the preparations were incubated with protein A-gold (electron microscopy) and FITC-labelled IgG (fluorescence microscopy), respectively.

Results

Cryostat sections of 3 day old wheat seedlings, incubated with FITC-labelled BGS A, exhibit the typical FITC fluorescence in round shaped spots in the cytoplasm and sometimes in the vacuole of the coleoptile cells (Fig. 1, 2). The embryo shows only pale fluorescence.

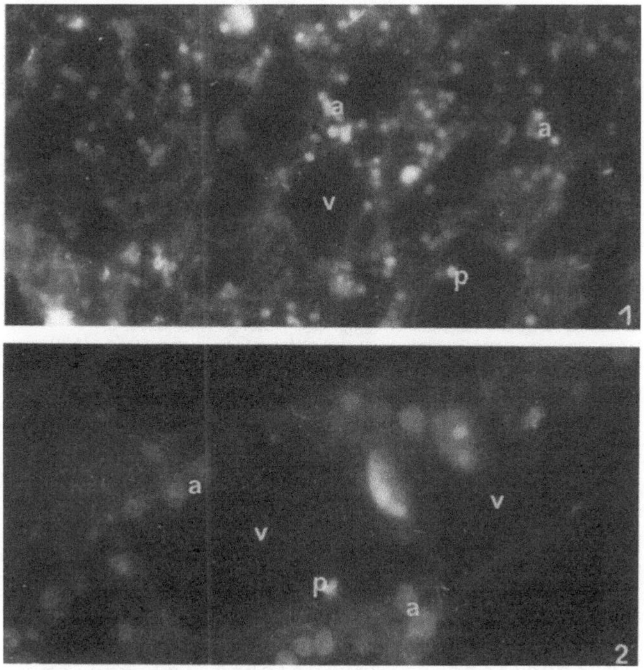

Fig. 1-2. FITC fluorescence in the coleoptile of wheat after incubation with BGS A-FITC (Blood group substance A-FITC. Note the bright fluorescence of amyloplasts (a) and protein lumps (p) in the vacuole (v), x 350 (1), x 750 (2).

Fixation of wheat seedlings with formaldehyde followed by embedding in an epoxy resin for electron microscopy preserves the reactivity of WGA with BGS A and allows the localization of the lectin at the ultrastructural level. In the coleoptile, labelling was observed in the stroma of amyloplasts and in protein lumps in the vacuole (Fig. 3, 4).

In the embryo a different distribution of gold particles is found. Beside slightly labelled proplastids, WGA is localized at the cell wall-protoplast interface and in protein lumps inside the vacuole (Fig. 5, 6). In general, more gold particles could be seen in the coleoptile than in the embryo.

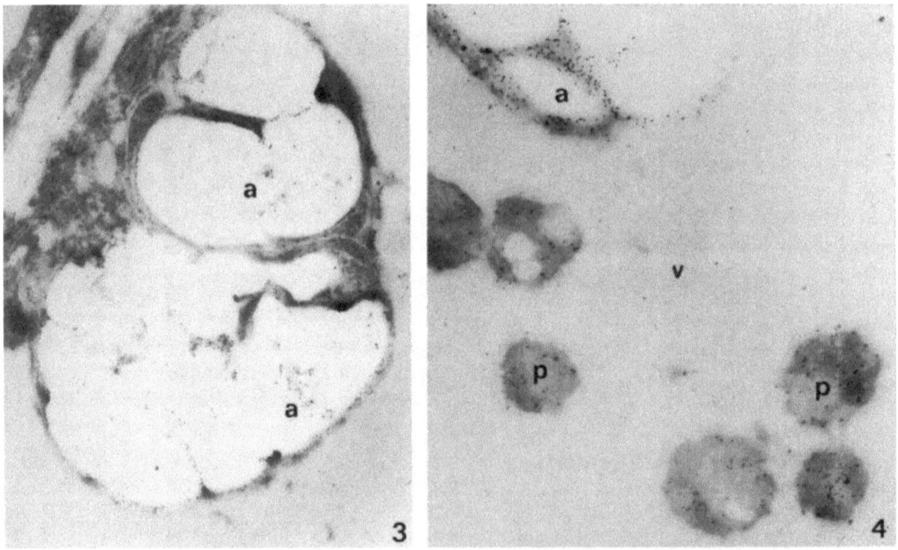

Fig. 3-4. Gold labelling of amyloplasts (a) and protein lumps (p) in the vacuole (v) of the coleoptile after incubation with BGS-gold. The label is concentrated in the stroma of the amyloplasts and the protein lumps. x 20,000 (3), x 16,000 (4).

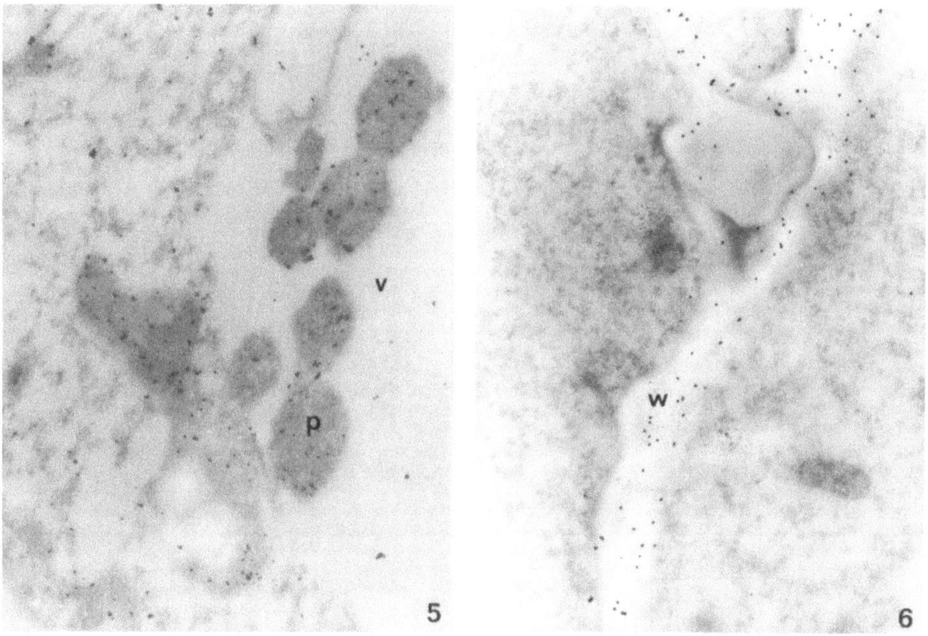

Fig. 5-6. Gold labelling in the embryo after incubation with BGS-gold. Note the labelling of the protein lumps (p) in the vacuole (v) and in the interface between the cell well (w) and the cytoplasme. x 28,000.

Control experiments (Fig. 7, 8) reveal that the labelling observed in the experiments represents the distribution of WGA in germinating wheat.

Discussion

The results reported here demonstrate that WGA is heterogeneously distributed in germinating wheat seeds. This distribution is somewhat different from the distribution of legume lectins so far investigated. Immunohistochemical investigations using anti-lectin antibodies in Phaseolus vulgaris (3), Glycine max (4), Pisum sativum (5) and Cannavalia ensiformis (6) demonstrate an uniform distri-

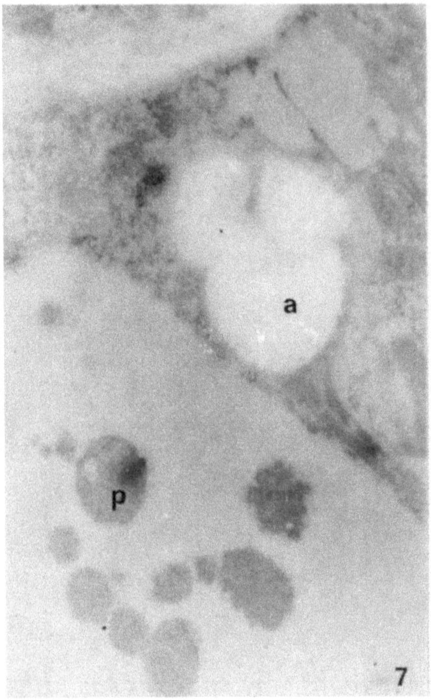

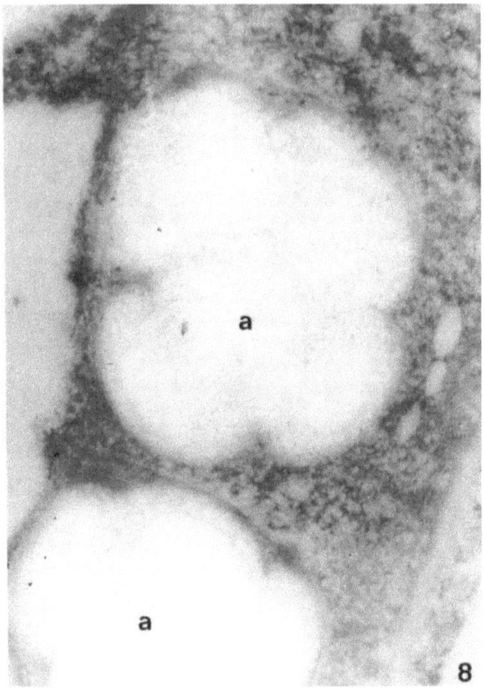

Fig. 7. Incubation with unlabelled BGS A followed by BGS A-gold. There is a very low nonspecific labelling. x 20,000.

Fig. 8. Incubation with protein A-gold. Note the low background labelling in the cytoplasm. x 20,000.

bution of the lectins throughout the protein bodies in the cotyledons of these plants. During synthesis a part of the lectin could be localized in the endoplasmic reticulum and the Golgi apparatus. The vacuolar localization of WGA in our experiments is in accordance with the lectin accumulation in protein bodies which also represents a vacuolar compartment. Recently, Mishkind et al. (7) demonstrated the distribution of WGA by immunocytochemical techniques. They found WGA inside the cell at the interface between cell wall and cytoplasm, as well as in organelles which they called protein bodies.

From the methodological point of view two aspects should be emphasised:

1. Using labelled BGS lectins can by visualized by their ability to bind glycoproteins.
2. Demonstration is possible even after formaldehyde fixation and resin embedding.

References

1. Bendich, A., E.A. Kabat, A.E. Bezer. 1946. In: Kabat, E.A.: Blood group substances, Academic Press Inc., New York 1956, 135-136
2. Roth, J., I.M. Bendayan, L. Orci. 1978. J. Histochem. 26, 1074-1081
3. Matten, J.F., A. Pusztai. 1982. Planta 155, 328-334
4. Horisberger, M., M. Volanthen. 1980. Histochemistry 65, 181-186
5. van Driessche, E., G. Smets, R. Dejaegere, L. Kamarsk. 1981. Planta 153, 287-296
6. Herman, E.M., L.M. Shannon. 1984. Planta 161, 97-104
7. Mishkind, M., N.V. Raikhel, B.A. Pralevitz, K. Keegstra. 1982. J. Cell Biol. 92, 753-764

Part IV

Animal Lectins

C-REACTIVE PROTEIN AS A MODEL OF HOW LECTINS MAY EVOLVE

Irving Kushner

Case Western Reserve University
at Cleveland Metropolitan General Hospital
Cleveland, Ohio 44109 USA

The acute phase response can be defined as the entire array of metabolic and physiologic changes which occur in response to tissue injury or infection (1). Prominent among these changes, in all mammals studied so far, is an alteration in the pattern of synthesis of plasma proteins by the liver. There is decreased synthesis of albumin and several other proteins, and increased synthesis of a large number of plasma proteins, collectively referred to as the *acute phase proteins*. It is important to note that there are great differences between species in the acute phase behavior of various plasma proteins (2). The best studied human acute phase proteins are listed in Table 1. The two most dramatic human acute phase proteins are Serum Amyloid A (SAA) and C-reactive protein (CRP). Their concentrations rise rapidly and markedly following stimulus. In severely ill patients serum concentrations may attain levels 1000 times or more those seen in (some) normal individuals. CRP was the first human protein found to manifest acute phase behavior, and may be regarded as the archetypical acute phase protein (3). Human

Lectins, Vol. IV

Table 1

Best-studied human acute plasma proteins

Group	I	Concentration usually increases about 50%:
		Ceruloplasmin C3,C4
Group	II	Concentration usually increases up to fourfold:
		α_2-acid glycoprotein α_1-antitrypsin α_2-antichymotrypsin Haptoglobin Fibrinogen
Group	III	Concentration may increase up to 1000 fold:
		C-reactive protein (CRP) Serum amyloid A protein (SAA)

CRP is a non-glycosylated protein made up of five identical subunits with an M_r of 21,000, arranged in cyclic symmetry. The name pentraxin has been applied to molecules manifesting this structure (4). While a large number of biological reactivities of CRP have been reported, its precise functional role (or roles) remains unclear.

In considering the possible functions of CRP, it is useful to review the known binding specificities of this protein. These are listed in Table 2. The best defined binding reactivity of CRP is with phosphocholine moieties (5), suggesting to several workers that the major function of CRP may be related its capacity to bind to damaged cell membranes (14-16). However, a number of recent studies have suggested that both human and rabbit CRP have the properties of a lectin, and possess binding affinity for galactosyl residues. Uhlenbruck and his colleagues showed that CRP precipitated with galactans isolated from various snail species (7). Volanakis and Narkates showed weak binding of CRP to agarose, which could be inhibited by galactose

Table 2
CRP binding specificities

Ca^{++}dependent	phosphocholine	(5)
	monophosphates	(5)
	chromatin	(6)
	galactans	(7-11)
Ca^{++}independent	polycations	(12)
	fibronectin	(13)

or derivatives of galactose (8), and found the binding of CRP to artificial phospholipid bilayers could be increased by incorporating galactosyl residues in the bilayers)9). The reaction between CRP and depyruvylated pneumococcal type IV polysaccharide might be due to reactivity with galactosyl residues (10). Finally, we recently have shown *in vivo* binding of rabbit serum CRP to injected kappa carrageenan, a sulfated polygalactan (11). In collaboration with Drs. Uhlenbruck and Solter we also found that human CRP precipitated kappa and iota carrageenans *in vitro*. However, the precise sugar configuration against which CRP is reactive has not been defined.

An understanding of the biological function of CRP in man may perhaps be attained from an analysis of its chatacteristics over the course of evolution. CRP has recently been found by Robey and Liu (17) to be present in the hemolymph of *limulus polyphemus*, the horseshoe crab, where it is a constitive protein, present in high concentration (5 mg/ml). This molecule shows considerable amino acid homology with human CRP, and is immunologically cross-reactive. *Limulus polyphemus*, referred to by some workers as a living fossil, is over a half billion years old. It existed before there were plants on land or fish in the sea. It appears that the limulus CRP molecule is identical to limulin, a lectin discovered by Noguchi over 80 years ago (80). Limulus CPR (or limulin) has distinct binding sites for phosphocholine and for sialic aicd. Some workers now re-

Table 3
Molecular properties of CRP

Source	MW	Subunit	Glycosylated	Reference
Limulus	480,000	24,000	+	(17)
Dogfish	250,000	25,000	+	(20)
Trout	110,000	20,000	?	(21)
Dog	140,000	22,000	+	(22)
Rat	125,000	20,000	+	(23)
Rabbit	110,000	22,000	-	(25)
Man	105,000	21,000	-	(26)

fer to this molecule as limulus lectin (19); it is a constitutive invertebrate recognition protein. What changes have pentraxins undergone in the course of their evolution from this primitive precursor?

A number of characteristics have been conserved over the course of evolution, but considerable diversity has appeared among the descendants of this molecule. Some examples of physicochemical characteristics of CRP's in several different species are shown in Table 3. The subunits are all about the same size, however, the molecular size appears to be relatively constant. The stable characteristics of CRP are shown in Table 4. For practical purposes these have become the definition of CRP; if a molecule did not manifest these stable characteristics we would not recognize it as a CRP. It is of interest that CRP fits the general description of many invertebrate lectins: a molecule manifesting Ca^{++}-dependent binding, made up of subunits with a molecular mass ob about 20,000 Da which are assembled into an active aggregate via noncovalent interactions (27). Variable characteristics of CRP include the presence or absence of glycosylation, the molecular weight of the assembled molecule and the number of subunits making up the basic structure -- there may be 5 or 6. In vertebrates, all of these molecules thus far studied appear to be pentamers.

Table 4

Stable characteristics of CRP

1.	Phosphorylcholine binding (Ca^{++} dependent)
2.	AA sequence
3.	Subunits 20-30,000 D

Of greatest interest is the variation between species in both serum concentration in unstimulated animals and in the degree of increase in concentration following tissue injury. Over the course of evolution, as vertebrates developed increasingly sophisticated defense machanisms, the CRP gene has evolved in a variety of ways. In some vertebrate species, the CRP is a substantial constitutive protein, in others, a trace protein. In some species its concentration rises markedly following tissue injury and infection; in others the increase is minimal. In many species a second pentraxin, SAP, has also evolved. Human SAP does not manifest phosphocholine binding but is clearly a lectin, showing a binding specificity for the pyruvate acetal of galactose (28). In many strains of mouse, SAP behaves as a substantial acute phase protein (29).

A schematic representation of the physiologic behavior manifested by the phosphocholine-binding descendants of the *limulus* CRP gene in some contemporary species is shown in Fig. 1. In this figure the serum concentration of CRP in unstimulated animals is shown as a horizontal line before the vertical arrows, which represent tissue injury. The effect of tissue injury on serum concentration is indicated by the subsequent curve.

The primitive CRP in *limulus polyphemus* hemolymph is a constitutive protein present in relatively high concentration, which apparently does not change in response to tissue injury (17). Dogfish CRP manifests behavior similar to that of *limulus* (20).

In a number of species, including man, rabbit, dog and trout, the molecule which meets the definition of CRP is normally present in only trace amounts, but may display a great increase in concentration following tissue injury or infection. We have become accustomed to regarding this as the normal behavior of CRP, but in fact there is considerable variation. Mouse CRP, is also normally present in serum in trace amounts, but the response to stimulus es extremely modest, rarely exceeding a threefold rise (23). In contrast, rat CRP is normally present in hibh concentration; in response to tissue injury its concentration will merely double at best (24). In the Syrian hamster, Coe has shown that rate of synthesis of the only protein related to CRP, Female Protein, is regulated by sex hormones as well as by inflammatory stimuli (30). While this protein manifests acute phase behavior in the male, in the female its plasma concentration falls following tissue injury!

It is likely that the basic function of CRP in the horeseshoe crab is related to its capacity to bind to foreign chemical configurations, with consequent agglutination, clearance or opsonization, a widely accepted postulated role for invertebrate lectins (31,32). These are effective mechanisms and have apparently worked well for a long time for *limulus*. However, evolution is not only a struggle between species, but also, and ultimately, a struggle between genes for survival. Over the course of evolution, as more sophisticated defense mechanisms evolved, (lymphocytes, polymorphonuclear cells, immunoglobulins and complement) it appears that the CRP gene has successfully adapted and survived. It has changed its size, its carbohydrate content and its behavior. While its function may have changed somewhat, it is unlikely that it has changed substantially. Our view of the function of CRP in man is that it pobably is not very different from what it was in the horseshoe crab. In man, it is an inducible, rather than a constitutive molecule, whose major function is probably still to recognize certain foreign

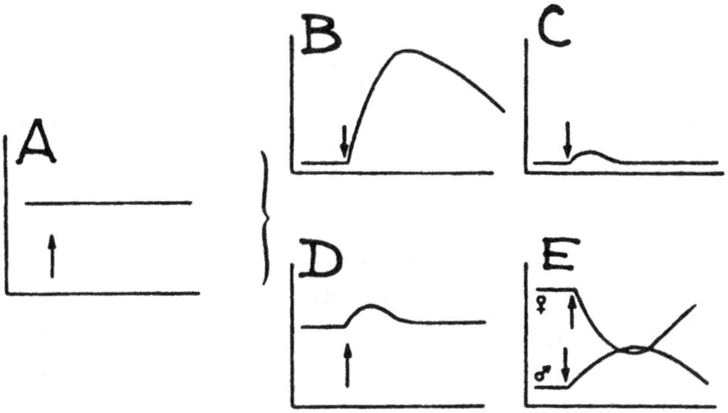

Fig. 1. CRP through evolution. Normal levels in circulation & response to injury. A schematic representation of the physiologic behavior of CRP in some different species. A: Limulus (Horseshoe crab), Dogfish. B: Man, Rabbit, Monkey, Dog, Trout. C: Mouse. D: Rat. E: Syrian Hamster ♀ and ♂

configurations (both endogenous and exogenous) with consequent opsonization, agglutination, and perhaps complement activation.

References

1. Kushner, I.: Ann. N.Y. Acad. Sci. 389, 39-48 (1982).

2. Kushner, I.: In Marker Proteins in Inflammation, Volume 2, P. Arnaud, J. Bienvenu, P. Laurent (eds.), Walter de Gruyter, Berlin, New York, pp 3-14 (1984).

3. McCarty, M.: Ann. N.Y. Acad. Sci. 389, 1-10 (1982).

4. Osmand, A.P., Friedenson, B., Gewurz, H., Painter, R.H., Hofmann, T., Shelton, E.: Proc. Natl. Acad. Sci. U.S.A. 74, 739-743 (1977).

5. Volanakis, J.E., Kaplan, M.H.: Proc. Soc. Exp. Biol. Med. 136, 612. (1972).

6. Robey, F.A., Jones, K.D., Tanaka, T., Liu, T-Y.: J. Biol. Chem. 259, 7311-7316 (1984).

7. Uhlenbruck, G., Solter, J., Janssen, E., Haput, H.: Hoppe-Seyler's Z. Physiol. Chem. Bd. 362, 1167-1169 (1981).

8. Volanakis, J.E., Narkates, A.J.: J. Immunol. 126, 1820-1825 (1981).

9. Narkates, A.J., Volanakis, J.E.: Ann. N.Y. Acad. Sci. 389, 172-182 (1982).

10. Gotschlich, E.C.: Ann. N.Y. Acad. Sci. 389, 163-171 (1982).

11. Preston, J., Kushner, I., Schultz, D., Mahajan, D., Macintyre, S.: Fed. Proc. 42, 421 (1983).

12. Dicamelli, R., Potempa, L.A., Siegel, J., Suyehira, L., Petras, K., Gewurz, H.: J. Immunol. 125, 1933-1938 (1980).

13. Saloneu, E.M., Vartio, T., Hedman, K., Vaheri, A.: J. Biol. Chem. 259, 1496-1501 (1984).

14. Kushner, I., Kaplan, M.H.: J. Exp. Med. 114, 961 (1961).

15. Kaplan, M.H.: Ann. N.Y. Acad. Sci. 389, 419-422 (1982).

16. Volanakis, J.E.: Ann. N.Y. Acad. Sci. 389, 235-250 (1982).

17. Robey, F.A., Liu, T.-Y.: J. Biol. Chem. 256, 969-975 (1981).

18. Noguchi, H.: Zentralbl. Bacterio. Parasitenk. Abt. I Orig. 33, 268 (1903).

19. Vasta, G.R., Marchalonis, J.J., Kohler, H.: J. Exp. Med. 159, 1270-1276 (1984).

20. Robey, F.A., Tanaka, T., Liu, T.-Y.: Biol. Chem. 258, 3889-3894 (1983).

21. Winkelhake, J.L., Chang, R.: Dev. Comp. Immunol. 6, 481-489 (1982).

22. Osmand, A.P., Jones, J.B.: Fed. Proc. 42, 421 (1983).

23. Siboo, R., Kulisek, E.: J. Immunol. Methods 23, 59-67 (1978).

24. DeBeer, F.C., Baltz, M.L., Munn, E.A., Feinstein, A., Taylor, J., Bruton, C., Clamp, J.R., Pepys, M.B.: Immunology 45, 55-70 (1982).

25. Wang, C.M., Nguyen, N.Y., Yonaha, K., Robey, F., Liu, T.-Y.: J. Biol. Chem. 257, 13610-13615 (1982).

26. Oliveira, E.B., Gotschlich, E.C., Liu, T.-Y.: J. Biol. Chem. 254, 489-502 (1979).

27. Marchalonis, J.J., Waxdal, M.J.: In Biomedical Applications of the Horseshoe Crab (Limulidae), Allan R. Liss, Inc., New York, pp 665-675 (1979).

28. Hind, C.R.K., Collins, P.M., Renn, D., Cook, R.B., Caspi, D., Baltz, M.L., Pepys, M.B.: J. Exp. Ped. 159, 1058-1069 (1984).

29. Mortensen, R.F., Beisel, K., Zeleznik, N.J., Phong, T.L.: J. Immunol. 130, 885-889 (1983).

30. Coe, J.E., Margossian, S.S., Slayter, H.S., Sogn, J.A.: J. Exp. Med. 153, 977-991 (1981).

31. Renwrantz, L.: Dev. Comp. Immunol. 7, 603-608 (1983).

32. Vasta, G.R., Marchalonis, J.J.: In Recognition Proteins. Receptors, and Probes: Invertebrates, Alan R. Liss, Inc., New York, pp 177-191 (1984).

MONOCYTE MEMBRANE LECTIN AND IN VITRO PHAGOCYTOSIS OF SENESCENT HUMAN RED BLOOD CELLS

Liliane Gattegno

Department of Cellular Biology, School of Medicine, University of Paris, 93012 Bobigny, France.

Jenny Vaysse, Dominique Bladier

Department of Biochemistry, School of Medicine, University of Paris, 93012 Bobigny, France.

David Aminoff

Department of Internal Medicine, Simpson Memorial Research Institute, University of Michigan, Ann Arbor, MI 48109, USA.

Several mechanisms have been proposed to account for the specific recognition of senescent human red blood cells (RBC) by macrophages with their subsequent sequestration from circulation. New antigenic sites are believed to appear during ageing which react specifically with auto-immune antibodies of the immunoglobulin G class (1-3). Thus opsonized, the senescent erythrocyte can be recognized by the macrophages via the Fc fragment of the immunoglobulin G (1). About 100 IgG molecules have been detected by several authors on the surface membrane of aged RBC (1-4). This auto-immune IgG has been demonstrated to have anti-galactosyl specificity (5,6). Previous work has shown that all the carbohydrate components in the red cell surface decrease significantly and approximately

Lectins, Vol. IV

in the same proportions during in vivo ageing of RBC (7-9) ; desialylation of some surface constituants may also occur (10, 11). The present study examines the ability of galactose-containing glycopeptides isolated from tryptic digest of human RBC (12) to inhibit autologous erythrophagocytosis with human blood monocytes and to compare this activity with neoglycoproteins and galactose-containing monosaccharides and oligosaccharides.

Material and methods

RBC isolation : Human blood samples drawn from healthy donors of the Seine Saint Denis Blood Transfusion Center were collected into heparinized tubes and processed within 30 minutes. Senescent human RBC were separated from young cells taking advantage of differences in the density (13). The top 5 % and the bottom 5 % fractions were enriched respectively with young and old RBC, as shown by the 2.1-fold higher pyruvate kinase activity of the top layer as compared to the bottom layer of the cells.

RBC labeling : young and senescent RBC were washed three times with PBS, then labeled by incubation with an equal volume of $Na_2{}^{51}CrO_4$ in saline (1mCi/ml) at 37°C for 30 minutes. After three washes with PBS, cells were adjusted in Hank's medium to 1.5 10^8 cells/ml.

Monocyte preparation : monocytes were prepared from heparinized peripheral blood ; 3 10^7 mononuclear cells purified by centrifugation over Ficoll-Hypaque (14) were incubated for 1h 30 min at 37°C in Petri dishes in 2 ml of RPMI medium 1640 supplemented with 20 % human AB serum in a 5 % CO_2 humidified atmosphere. Non adherent cells were removed by washing four times with RPMI medium. The adherent monocytes were harvested with the aid of a rubber policeman. In these experiments, 95% of the cells were identified as monocytes either by acridin staining or by phagocytosis of latex particles.

Phagocytosis assay : aliquots of 1 10^5 freshly purified monocytes were allowed to adhere to Falcon culture tubes (12-75mm) in 75 µl serum-free RPMI medium at 37°C for 30 minutes in a 5% CO_2 humidified atmosphere. Autologous ^{51}Cr-labeled young or senescent RBC were added then at a ratio of about 20 RBC per monocyte. In inhibition experiments, the inhibitors were added to monocytes three minutes prior to the RBC. Cells were co-cultured at 37°C for 75 min in a 5 % CO_2 humidified atmosphere. Non-adherent cells were removed by washing and decanting with 2 ml of Hank's medium. This procedure was repeated twice, then adhering RBC were lysed with 2 ml of 140 mM Tris NH_4Cl buffer pH 7.4. Radioactivity of the washed monocytes was counted. The total radioactivity of the labeled RBC which were added in each assay was also counted. The phagocytic activity was expressed as the percentage of total radioactivity that was associated with the monocytes after the lysis step. Each phagocytosis assay was performed in quadriplicate or in quintuplicate. Inhibition studies were made with D-galactose (Carlo Erba), N-acetyl-D-galactosamine (Sigma), D-glucose (Merck), N-acetyl-D-glucosamine (Sigma), L-fucose, D-mannose, lactose, melibiose (Merck) and with synthetic glycoproteins i.e. bovine serum albumin (BSA) (Sigma), derivatized (15) with monosaccharides, 25 moles of D-galactose (β-gal-BSA) or 27 moles of N-acetyl-D-galactosamine (β-galNAc-BSA) per mole of BSA ; and glycopeptides isolated by affinity chromatography on PNA-agarose, from young or senescent RBC and referred as PNA^+ (12).

Microscopic fluorescence analysis : tetra-rhodaminyl-isothiocyanate-β-galactosyl-albumin was a gift from Dr M.A. Deugnier. About 1.6 10^6 plated monocytes were incubated in Petri dishes at 4°C during 45 min with 12-20 µM fluorescent marker in 300µl of RPMI medium, in the absence or presence of galactose at a 0.2M final concentration. In some experiments, galactose was added after the incubation period in order to reverse the specific labeling. The percentage of fluorescent cells was determined with a Leitz microscope.

Results and discussion

Under our experimental conditions (75 min at 37°C) senescent human RBC are phagocytized significantly more by autologous monocytes than are young RBC ($p < 0.001$) (Table I). This observation is in good agreement with those of other authors (16). We have found that longer incubation times, from 4 h to 25 h, lead to higher percentages of phagocytosis of young and senescent RBC by autologous monocytes.

TABLE 1

Percentage of RBC phagocytized by autologous monocytes.

RBC	% of phagocytosis[a] Mean	S.E.M.
Young	1.63	0.21
Senescent	5.71*	0.75

[a]Number of experiments : 22 ; * difference at the $p < 0.001$ level to phagocytosis of young RBC.

In inhibition experiments, much shorter incubation times were used in order to reduce the cellular catabolism of the added carbohydrates which might occur during long incubation periods. Results show (Table 2) that N-acetyl-D-galactosamine, D-galactose, β-D-gal-BSA and β-D-galNAc-BSA specifically inhibit the uptake of senescent RBC by peripheral blood monocytes. These results are in close agreement with those of Bladier et al. using human alveolar macrophages and human RBC (17) and with those of Schlepper-Schafer et al. using rat liver macrophages and rat RBC (18), both describing in their respective model an inhibition of the senescent RBC phagocytosis by β-galactosyl structures. However, this type of inhibition is shown here for the first time in the in vitro model of erythrophagocytosis by human autologous monocytes.

Furthermore, this erythrophagocytosis is inhibited by more

than 50 % in the presence of D-galactose, N-acetyl-D-galactosamine and lactose at millimolar concentrations, by β-bound D-galactosyl and N-acetyl-D-galactosaminyl bovine serum albumin at the micromolar level and by glycopeptides derived from tryptic digests of senescent RBC at the nanomolar level. We have determined that senescent RBC PNA^+ glycoconjugates, corresponding to a 25 nM concentration of galactose or N-acetyl-D-galactosamine (labeled by the galactose oxidase-borotritide method), are inhibitors in the same order of magnitude as 19μM β-D-galNAc-BSA or β-D-gal-BSA and 23 mM D-galNAc. Most significantly, glycopeptides isolated from 3 10^6 young RBC showed no inhibition of erythrophagocytosis (Table 2).

In order to visualize a galactose specific lectin on the human monocyte membrane, plated monocytes were incubated with synthetic tetra-rhodaminyl-isothiocyanate-β-galactosyl-BSA at 4°C (to avoid phagocytosis of the fluorescent marker). The specificity of the labeling was tested either by a parallel incubation with 0.2 M galactose or by a subsequent incubation with 0.2 M galactose. Under these experimental conditions, a mean of 20 % ± 5 % (n = 6) monocytes were able to bind the fluorescent marker specifically.

All these results strongly suggest that a β-galactose specific lectin present on the membrane of human monocytes may be involved in the in vitro phagocytosis of autologous senescent RBC. Whether galactosyl related carbohydrates that might be recognized in vitro by monocytes are located on the immunoglobulin G specific for the senescent antigen ; on the free senescent antigens not yet masked by the IgG ; or on another membrane component remains to be determined.

Acknowledgement. This study was supported by grant HL AM 17881 from the National Institutes of Health.

TABLE 2

Carbohydrate specificity of the in vitro phagocytosis of old RBC by autologous monocytes

Inhibitor	% of inhibition of phagocytosis	Number of experiments
D-Galactose 23 mM	56 ± 6	4
N-Acetyl-D-Galactosamine 23 mM	63 ± 25	4
D-Glucose 23 mM	0	4
N-Acetyl-D-Glucosamine 19 mM	0	5
D-Mannose 19 mM	16 ± 13	4
L-Fucose 20 mM	0	3
Lactose 23 mM	74 ± 2	2
Melibiose 25 mM	27 ± 20	4
β-D-Gal-BSA 19 μM	64 ± 14	4
β-D-GalNAc-BSA 19 μM	74 ± 14	5
PNA^+ tryptic glycoconjugates from 18 10^6 young RBC	55 ± 12	20
PNA^+ tryptic glycoconjugates from 18 10^6 old RBC	74 ± 11*	36
PNA^+ tryptic glycoconjugates from 3 10^6 young RBC	0	4
PNA^+ tryptic glycoconjugates from 3 10^6 old RBC	60 ± 10*	4

The asterisks indicate significant difference at the $p < 0.001$ level to phagocytosis of young RBC.

References

1 - Kay, M.B. : Nature 289, 491-494 (1981).

2 - Lutz, H.U., and Kay, M.B. : Mech. Ag. Dev. 15, 65-75 (1981).

3 - Alderman, E.M., Fudenberg, H.N., and Lovins, R.E. : Blood 58, 341-349 (1981).

4 - Khansari, N., and Fudenberg, H.N. : Cell. Immunol. 78, 114-121 (1983).

5 - Galili, U., Korkesh, A., Kahane, I., Rachmilewitz, E.A. : Blood 61, 1258-1264 (1983).

6 - Aminoff, D. : Erythrocyte Glycoproteins : Recognition and Sequestration of Senescent Erythrocytes. NATO Worshop on the Biology of Aging. Aquafredda de Moratea, Italy, July 8-14, 1984.

7 - Gattegno, L., Bladier, D., and Cornillot, P. : Carbohydr. Res. 51, 555-567 (1976).

8 - Bladier, D., Gattegno, L., Fabia, F. , and Cornillot, P. : Carbohydr. Res. 83, 371-376 (1980).

9 - Baxter, A., and Beeley, J.G. : Biochem. Biophys. Res. Commun. 83, 466-471 (1978).

10 - Gattegno, L., Fabia, F., Perret, G., and Cornillot, P. : Mech. Ag. Dev. 16, 205-219 (1981).

11 - Gattegno, L., Durand, G., Feger, J., Perret, G., Felon, M., and Cornillot, P. : Carbohydr. Res. 117, 255-262 (1983).

12 - Heinrich, C.F., and Aminoff, D. : Carbohydr. Res. 120, 55-66 (1983).

13 - O'Connell, D.J., Caruso, C.J., and Sas, M.D. : Clin. Chem. 11, 771-781 (1965).

14 - Boyum, A. : Scand. J. Clin. Lab. Invest. Suppl. 97, 1-5 (1968).

15 - Westphal, O., and Schmidt, H. : Liebigs Ann. Chem. 575, 84-90 (1951).

16 - Singer, J.A., Walker, W.S., and Morrison, M. : J. Ret. Endoth. Soc. 31, 489-499 (1982).

17 - Bladier, D., Vassy, R., Perret, G., and Gattegno, L. : I.C.R.S. in press.

18 - Schlepper-Schafer, J., Kolb-Bachofen, V., and Kolb, H. : Biochem. Biophys. Res. Commun. 115, 551-559 (1983).

STUDIES ON LECTINS

LXVI. BINDING OF SERUM GLYCOPROTEIN ASIALOOLIGOSACCHARIDES TO THE PORCINE LIVER AND LEUCOCYTE MEMBRANE LECTINS

Karel Bezouška, Hana Karásková, Otto Táborský, Olga Kofroňová*, Josef Voříšek*, Jaroslav Kubrycht+ and Jan Kocourek

Department of Biochemistry, Faculty of Science, Charles University, Albertov 2030, CS-128 40 Praha 2, Czechoslovakia

*Laboratory of Electron Microscopy and +Department of Immunology, Institute of Microbiology, Czechoslovak Academy of Sciences, Vídeňská 1083, CS-142 20 Praha 4, Czechoslovakia

N-Glycosidically bound carbohydrate moieties of the complex type occur as integral part of blood serum and membrane glycoproteins (1). Removal of the terminal sialic acids (desialylation) may serve as an important signal in maintenance of the homeostasis of glycosylated structures (catabolism of serum glycoproteins, aging of erythrocytes, malignant transformation). Desialylated structures are recognized by galactose-specific membrane lectins of hepatocytes (2), lymphocytes (3) or macrophages (4) and catabolized. In our present study we attempted to characterize these recognition systems by following their interactions with well defined serum-glycoprotein-derived asialooligosaccharides (see Fig. 1 on the next page).

Materials and Methods

Materials. Hydrazinium hydroxide, CNBr and monosaccharides were from Merck, Darmstadt, F.R.G. Collagenase was obtained from Sigma, St. Louis, MO, U.S.A., pronase from Serva, Heidelberg, F.R.G., *Clostridium perfringens* sialidase from Boehringer-Mannheim, F.R.G., DEAE- and CM-cellulose (DE32 and CM32) from Whatman, Maidstone, U.K.; Ficoll, Percoll, Sephadex G-25, G-50, G-150, G-200, Sepharose 4B and molecular weight markers - phosphorylase b (94 000), bovine serum albumin (67 000), ovalbumin

Lectins, Vol. IV

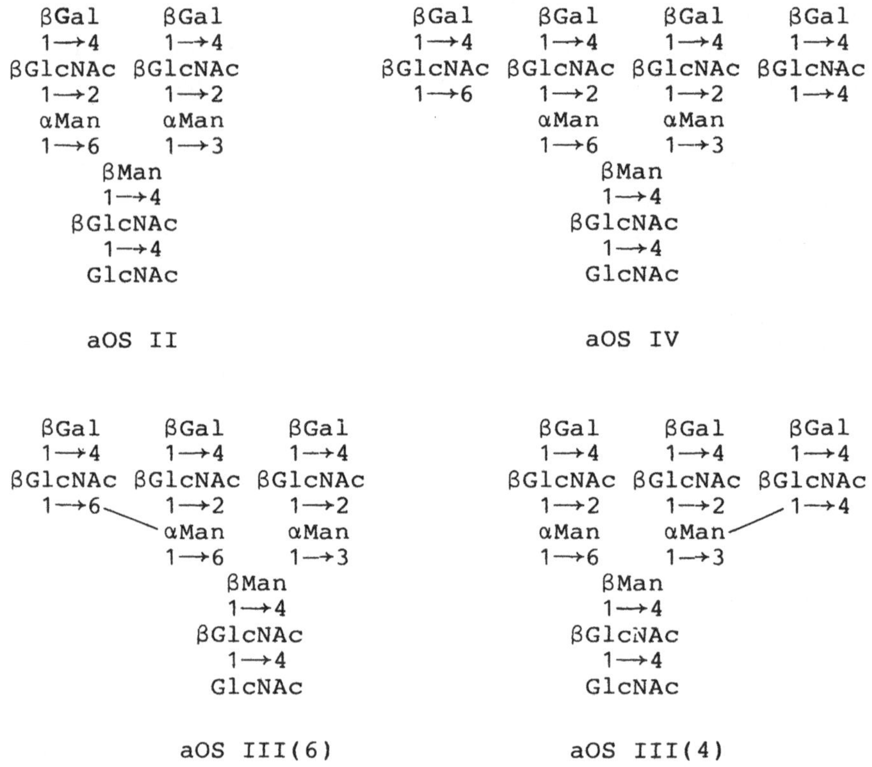

Fig. 1. Four serum-glycoprotein-derived asialooligosaccharides.

(43 000), carbonic anhydrase (30 000), soybean trypsin inhibitor (20 100) and α-lactalbumin (14 400) from Pharmacia, Uppsala, Sweden; Ampholine pH 3.5-9.5 from LKB, Bromma, Sweden, PCS scintillation coctail from Amersham, U.K. Affinity matrices were prepared as described previously (5). Plant lectins were provided by Dr. Karel Filka, Laboratory for Production and Control of Lectin Preparations, Faculty of Science, Charles University, Prague, Czechoslovakia.

Isolation, characterization and labeling of asialoologosaccharides. Glycoproteins were isolated from human serum, fractionated by ammonium sulfate to 40, 60, 70 and 100% of saturation. From the respective fractions α_2-macroglobulin, ceruloplasmin, transferrin and acid α_1-glycoprotein were prepared. The biantennary glycopeptide from transferrin (6), triantennary glycopeptides from α_2-macroglobulin and ceruloplasmin (7) and the tetraantennary glycopeptide from acid α_1-glycoprotein (8) were isolated after two cycles of pronase digestion by gel-filtration (Sephadex G-25, G-50) and ion exchange chromatography (DEAE-cellulose, Dowex 50W) in combination with affinity chromatography on immobilized concanavalin A. After desialylation and final purification by preparative high voltage paper electrophoresis, asialoglycopeptides were converted into asialooligosaccharides and labeled as described (9). The homogeneity of the obtained preparations was checked by thin layer chromatography (10), polyacrylamide electrophoresis in borate buffer (11) and by analysis of the monosaccharide components. The triantennary asialooligosaccharide from α_2-macroglobulin was resolved from that of ceruloplasmin by affinity chromatography on immobilized *Phaseolus coccineus* leucoagglutinin (12).

Preparation of cell suspensions. Rat hepatocytes were prepared by a perfusion with collagenase (13), fractions of pig leucocytes by centrifugation of "buffy-coats" from which adherent cells had been removed through a discontinuous gradient of Percoll (35, 45, 50, 55, 60, 65, 70, 75% of isotonic Percoll). T-lymphocytes were isolated as cells forming rosettes with sheep erythrocytes, B-lymphocytes as cells nonadsorbed on nylon wool (14). Cell suspensions were washed three times with Hanks' buffered saline solution, counted and characterized by trypan-blue exclusion and scanning electron microscopy.

Isolation and characterization of porcine lectins. Porcine liver lectin was isolated from a Triton X-100 extract of the plasma membrane microsomal fraction of pig liver as described (15) by specific elution from the second affinity affinity column with a linear gradient of D-galactose (0.0-0.5 mol/l). Porcine lymphocyte lectins were isolated from a Triton X-100 extract of spleen and lymph node plasma membranes (15,16) by affinity chromatography on asialotransferrin-Sepharose 4B (specific elution with 0.3 M lactose solution). Eluted proteins were then separated on a column of Sephadex G-150 (1x100 cm) into three fractions designated LL I, LL II and LL III in order of their decreasing molecular weight; 10 mg of protein from fraction LL I was incubated in 5 ml of 1.0% (v/v) Triton X-100 solution for 12 h at 4 °C and resolved into nonretained (LL I-1) and retained (LL I-2) fractions by another affinity chromatography on asialotransferrin -Sepharose 4B performed in the presence of 1.0% detergent. Triton X-100 was removed by prolonged dialysis against 1.0% solution of NH_4HCO_3 and the protein was lyophilized. The influence of a nonionic detergent on molecular characteristics of lectins was investigated with 1 mg samples incubated in 1 ml of 0.1%, 0.5% or 1.0% Triton X-100 for 12 h at 4 °C. Distribution of molecular weights was then determined by high resolution analytical gel filtration (TSK-G4000SV column from LKB, Bromma, Sweden was eluted with 0.05 M Tris-HCl pH 7.5 with 0.12 M NaCl and the given concentration of detergent) or electrophoresis in polyacrylamide gels with various pore densities (17).

Analytical methods. Binding assays with cells were done by incubation of 10^6 cells with the individual ^{3}H-asialooligosaccharides (concentration range 10^{-5} to 10^{-10} mol/l). After 2 h at 4 °C, cells were separated by centrifugation, washed and cell-bound radioactivity estimated by liquid scintillation counting. Binding assays with isolated lectins were performed under similar conditions; after 2 h incubation, lectins were precipitated with ammonium sulfate and collected by centrifugation for 10 min at 20 000 ***g***. Results were analyzed according to Scatchard (18).
Alkaline electrophoresis was run in a gradient of polyacrylamide gel (5-20% of acrylamide) in the buffer system of Stewart et al. (19). Gel contained 0.1% (v/v) of Triton X-100 and samples had been preincubated in the indicated concentration of detergent for 12 h at 4 °C before the separation

Table 1

Association constants of the interaction between ^{3}H-asialooligosaccharides and hepatocytes or leucocyte subfractions

Type of cells	K_a at 4 °C ($l \times mol^{-1}$)			
	aOS II	aOS III(6)	aOS III(4)	aOS IV
Hepatocytes	1.0×10^{6}	1.4×10^{7}	1.5×10^{8}	1.4×10^{9}
Leucocytes	1.7×10^{9}	1.3×10^{8}	0.9×10^{7}	2.3×10^{6}
Adherent cells[a]	8.5×10^{4}	5.0×10^{4}	4.7×10^{4}	3.2×10^{4}
Natural killer cells[b]	2.0×10^{7}	1.8×10^{9}	2.1×10^{10}	2.3×10^{7}
Lymphocytes[c]	2.3×10^{9}	1.7×10^{8}	2.2×10^{7}	2.6×10^{6}
T-Lymphocytes[d]	2.0×10^{9}	1.8×10^{8}	2.0×10^{7}	2.4×10^{6}
B-Lymphocytes[d]	2.4×10^{9}	1.5×10^{8}	1.2×10^{7}	2.3×10^{6}
Polymorphonuclear leucocytes[e]	1.1×10^{6}	1.3×10^{7}	1.3×10^{8}	3.3×10^{9}

[a] 45% Percoll fraction isolated from glass adherent cells.

[b] 45% Percoll fraction rich in natural killer activity.

[c] 55% + 60% Percoll fraction.

[d] Isolated from 55% + 60% Percoll fraction as described in the experimental section.

[e] 70% Percoll fraction.

proceeded. The system described by Hashimoto et al. (20) was used for SDS electrophoretic analyses. Isoelectric focusing was performed in a thin layer of polyacrylamide gel (5% of acrylamide, 1.5% Ampholine) for 100 min at a constant power 30 W in an LKB Ultrophor apparatus. Monosaccharide composition was determined by gas-liquid chromatography (21), sialic acids by the thiobarbituric method (22). Amino acid analyses were done on a Durrum D-500 analyzer (U.S.A.). Protein content was determined by a modification of the method of Lowry et al., as described by Hudgin et al. (15).

Results

Binding of ^{3}H-asialooligosaccharides to hepatocytes and leucocytes. Table 1 gives the values of association constants for the

Table 2

Binding capacities of hepatocytes and leucocyte subfractions for ^{3}H-asialooligosaccharides

Type of cells[a]	Number of binding sites per cell (x 10)			
	aOS II	aOS III(6)	aOS III(4)	aOS IV
Hepatocytes	0.44	0.62	0.51	0.43
Leucocytes	0.52	0.64	0.50	0.68
Adherent cells	17.50	16.80	17.50	18.70
Natural killer cells	5.30	5.30	4.90	6.40
T-Lymphocytes	6.40	5.80	5.40	6.60
B-Lymphocytes	0.43	0.48	0.52	0.44
Polymorphonuclear leucocytes	0.05	0.04	0.02	0.04

[a] Cell fractions isolated as given in Table 1.

interactions of individual asialooligosaccharides with rat hepatocytes and subfractions of porcine leucocytes. The data indicate the existence of four distinct recognition systems. Hepatocytes and polymorphonuclear leucocytes bind specifically the more branched asialooligosaccharides, lymphocytes the less branched asialooligosaccharides, natural killer cells prefer triantennary structures and macrophages bind all structures to the same extent. The number of binding sites per cell (the capacity of the recognition system) varies with different cell fractions and is independent on asialooligosaccharide branching (Table 2). All asialooligosaccharides bind probably to the same receptors on the cell surface.

Complementary binding systems on hepatocytes and lymphocytes. Considering the capacity and specificity of the above recognition systems, hepatocytes and lymphocytes have the highest binding capacities for asialooligosaccharides. The values of association constants (Table 1) indicate a possible cooperation of both systems in the catabolism of desialylated structures. Scanning

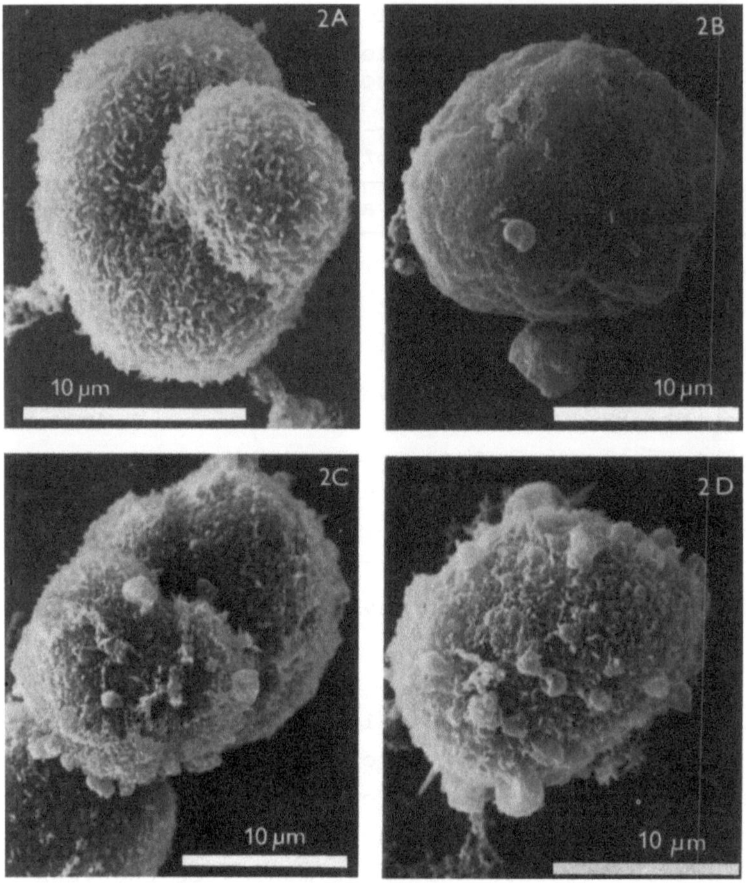

Fig. 2. Binding of latex microspheres coated with aOS II (A), aOS III(6) (B), aOS III(4) (C) and aOS IV to rat hepatocytes. Cells were incubated with the particles for 10 min at 4 °C, centrifuged, fixed with 3% glutaraldehyde, dehydrated with ethanol, dried, coated with gold and examined with scanning electron microscope Tesla.

electron microscopy (Figs. 2 and 3) brought direct evidence for the complementarity of these systems. After incubation of cells with asialooligosaccharide-coated latex microspheres only those derived with more branched asialooligosaccharides bound to the surface of hepatocytes (Fig. 2C, 2D), whereas those derived with less branched structures adhered to the surface of lymphocytes

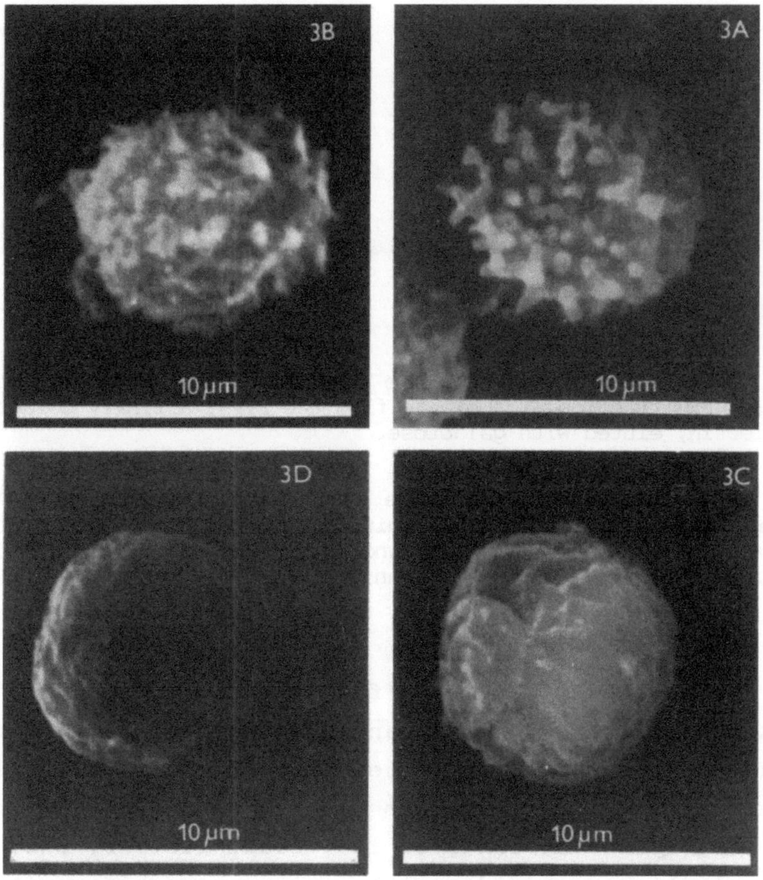

Fig. 3. Binding of latex microspheres coated with aOS II (A), aOS III(6) (B), aOS III(4) (C) and aOS IV to porcine lymphocytes. Cells were treated as described in legend to Fig. 2.

(Fig. 3A, 3B).

Isolation of the pig liver and lymphocyte lectins. For the isolation of the porcine liver lectin, the original procedure (15) was modified by the use of a lower Triton X-100 concentration (0.1%, v/v) for the solubilization of membranes and specific elution with D-galactose. This procedure yielded 2.1 mg of lectin homogeneous on alkaline gradient polyacrylamide gel electrophoresis

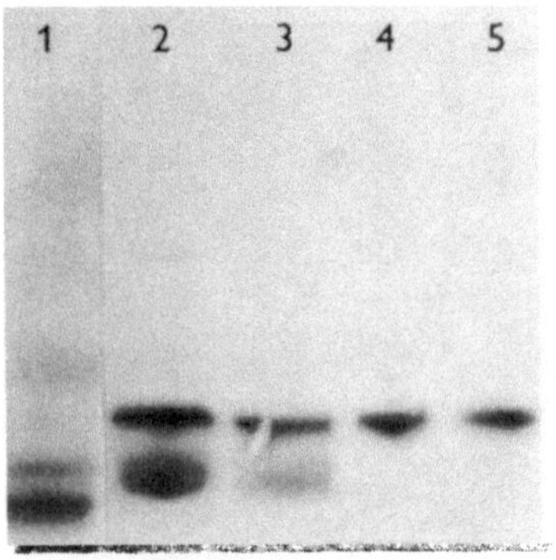

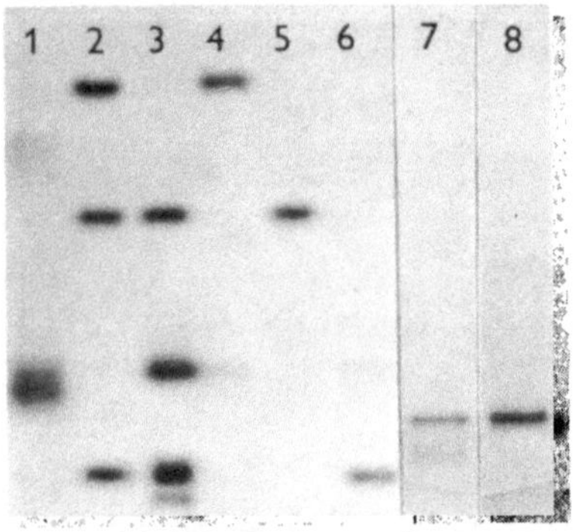

Fig. 4. Purification of porcine liver lectin. (1) Triton X-100 extract, (2) PEG fraction, (3) eluate from the first affinity column, (4,5) purified lectin, eluted with galactose.

Fig. 5. Purification of porcine lymphocyte lectins. (1) Triton X-100 extract, (2,3) eluate from affinity column in 0.1% and 1% Triton X-100, (4,5,6) isolated lectins LL I, LL II and LL III in 0.1% Triton X-100, (7) LL I in 1.0% Triton, (8) purified subunit LL I-2.

(Fig. 4) from 100 g of liver. Affinity chromatography of Triton X-100 extracts from pig spleen and lymph node plasma membranes provided a mixture of three proteins separable by gel filtration on Sephadex G-150 (Fig. 5, lanes 4,5,6). Table 3 gives the values of inhibition constants of individual lectins; only LL I binds asialooligosaccharides with various branchings with different values of association constants. From 500 g spleen or 100 g lymph nodes about 6.6 mg of the lectin LL I was obtained. Lectins were extracted from lymphocyte plasma membranes using the protein : detergent ratio 3:1 (w/w). The change of this ratio to 1:3 brings about dissociation of the original molecule into two fractions (Fig. 5, lane 7). These fractions can be separated on a column of asialotransferrin-Sepharose 4B eluted with 1.0% (v/v) Triton X-100.

Comparision of properties of lectins from pig liver and lymphocytes. The porcine hepatic lectin and the lymphocyte lectin LL I represent two molecular species able to discriminate be-

Table 3

Inhibition constants of lymphocyte lectins LL I, LL II and LL III expressed as concentrations (mol/l) necessary for 50% inhibition of ^{3}H-aOS II binding to the individual lectins

Saccharide	LL I	LL II	LL III
D-Galactose	2×10^{-5}	2×10^{-7}	4×10^{-3}
D-Mannose	$> 10^{-1}$	$> 10^{-1}$	$> 10^{-1}$
L-Fucose	$> 10^{-1}$	$> 10^{-1}$	$> 10^{-1}$
D-Glucose	3×10^{-3}	$> 10^{-1}$	$> 10^{-1}$
Sucrose	5×10^{-2}	$> 10^{-1}$	10^{-1}
Lactose	3×10^{-5}	4×10^{-7}	3×10^{-3}
N-Acetyl-D-glucosamine	$> 10^{-1}$	5×10^{-3}	$> 10^{-1}$
N-Acetyl-D-galactosamine	4×10^{-6}	1×10^{-6}	$> 10^{-1}$
aOS II	4×10^{-10}	9×10^{-7}	3×10^{-3}
aOS III(6)	3×10^{-9}	6×10^{-7}	3×10^{-3}
aOS III(4)	2×10^{-8}	4×10^{-7}	3×10^{-3}
aOS IV	5×10^{-7}	3×10^{-7}	2×10^{-3}

tween asialooligosaccharides with different branching. The hepatic lectin gives a single zone on SDS-electrophoresis with M_r 43 000, whereas lectin LL I yields two zones (Fig. 6). The first component with M_r 43 000, corresponds to LL I-2, a lectin that can bind to asialooligosaccharides even in 1% Triton X-100 solution; the binding of asialooligosaccharides after dissociation is, however, much weaker than with the whole lectin LL I (see Table 6). The second component with M_r 7500 corresponds to LL I-1, not retained in the affinity column. The hepatic lectin and LL I-2 have the same pI (as estimated by isoelectric focusing - Fig. 7) as well as very similar amino acid and sugar compositions (Table 4). The sugar portion of the molecule is represented in both lectins by a single triantennary oligosaccharide. Rabbit antibodies raised against porcine lymphocyte lectin LL I-2 exhibit a strong cross-reactivity with the liver lectin, as proved

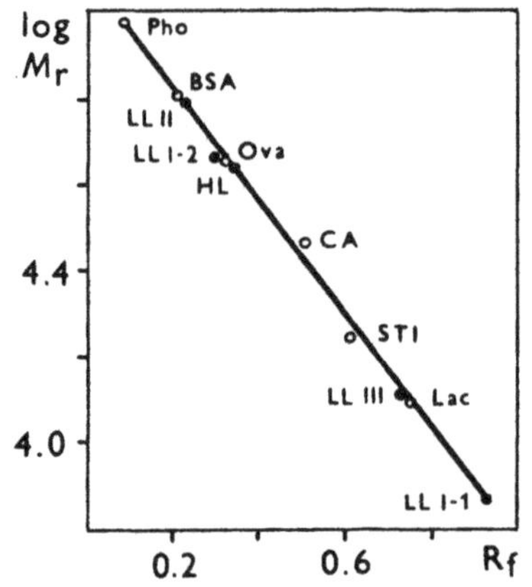

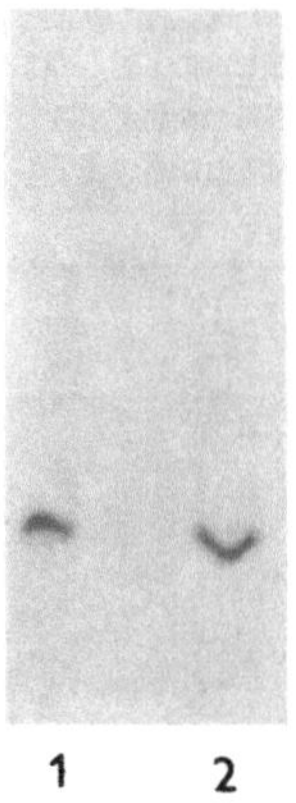

Fig. 6. Molecular weights of subunits of the hepatic lectin and lymphocyte lectin LL I, as estimated by electrophoresis in SDS-medium. List of molecular weight markers is given in Materials and Methods section. Closed circles (●) denote lectins, open circles (o) markers.

Fig. 7. Isoelectric focusing of the porcine hepatic lectin (1) and porcine lymphocyte lectin LL I-2 (2). Two closely related bands were detected with pI around 4.6.

by immunoelectrophoresis and double radial immunodifusion technique (Ouchterlony method).

Subunit structure and H-asialooligosaccharide binding properties of the porcine liver and lymphocyte lectins. Quarternary structures of both lectins are dependent on the concentration of detergent (Triton X-100); higher concentrations cause dissociation into subunits (Table 5). In a low Triton concentration (up to 0.1%), the porcine liver lectin exists as a tetramer of M_r 160 000 (estimated by gel filtration) or 157 000 (electrophoresis) that represents a biologically active functional unit with values of association constants for the interaction lectin -asialoologosaccharide similar to those observed with whole cells Table 6, compare Table 1). Increasing Triton concentration brings about dissociation into subunits with subsequent loss of differentiating capacities towards asialooligosaccharides with different branching. The functionally active porcine lympho-

Table 4

Amino acid and sugar composition of porcine liver lectin and lymphocyte lectin LL I (in moles per one mole of lectin)[a]

	Liver lectin	LL I-2	LL I-1
Asp	11.3	11.4	0.9
Thr	7.7	7.5	2.3
Ser	7.8	7.3	1.1
Glu	12.0	12.1	1.3
Pro	5.1	5.1	-
Gly	9.5	9.4	2.7
Ala	4.4	4.4	1.0
Val	5.4	5.5	1.1
Met	0.8	1.1	-
Ile	1.8	1.8	2.3
Leu	9.0	9.1	4.8
Tyr	2.3	2.2	-
Phe	5.4	5.4	-
Lys	4.2	4.3	-
His	2.7	2.7	-
Arg	4.8	4.8	0.8
Gal	2.95	3.07	6.97
Man	3.07	3.15	3.06
Fuc	-	-	1.03
Sia	2.9	2.7	1.9
GlcNAc	5.11	5.21	11.99
GalNAc	-	-	0.95

[a] Molecular weights of the subunit of liver lectin (43 000), LL I-2 (43 000) and LL I-1 (7500) as estimated by SDS-electrophoresis.

cyte lectin consists of two binding subunits LL I-2 stuck together by two "regulatory" glycopeptides LL I-1. Molecular weight of the whole unit was estimated as 100 000 (gel filtration) or 110 000 (electrophoresis). After dissociation of the lymphocyte lectin LL I into subunits, removal of detergent from the preparation of isolated binding subunits LL I-2 causes reassociation into dimers and tetramers. When isolated by high resolution gel filtration, these reconstituted tetramers exhibit, in spite of their origin, a recognition pattern typical for the liver lectin (Table 6). In a similar way, $(LL\ I)_r$ obtained by the addition of the "regulatory" glycopeptide to the

Table 5

Dissociation of porcine liver and lymphocyte lectins into subunits by nonionic detergent Triton X-100[a]

Lectins	0.1% Triton	0.5% Triton	1% Triton
Hepatic lectin	160 000* 157 000+	80 000, 160 000, 45 000* 85 000, 157 000, 43 000+	45 000* 43 000+
LL I	100 000* 110 000+	100 000, 44 000, 8 000* 110 000, 43 000, 7 500+	44 000, 8 000* 43 000, 7 500+
LL I-2	160 000* 160 000+	80 000, 160 000, 45 000* 83 000, 160 000, 43 000+	45 000* 43 000+

[a] Lectin (1 mg) was incubated in 1.0 ml of Triton X-100 solution of the indicated concentration (v/v) for 12 h at 4 °C. Molecular weight distribution was then estimated by high resolution analytical gel filtration (*), electrophoresis or SDS-electrophoresis (+).

solution of binding subunits of hepatic origin show, after removal of the detergent, binding properties identical to those of the native lymphocyte lectin LL I. Thus, LL I-1 is not an impurity associated with lectin LL I, but represents an essential part of the lectin molecule necessary for its biological properties.

Discussion

In our previous paper (23) we have reported on the existence of two complementary systems involved in the catabolism of desialylated serum glycoproteins localized on the surface of hepatocytes and leucocytes. Resolution of leucocytes into several fractions revealed four recognition systems with different specificities towards desialylated oligosaccharides on lymphocytes, macrophages, natural killer cells and polymorphonuclear leucocytes. These systems can bind not only asialooligosaccharides and asialoglycoproteins, but also defective cells bearing desialyla-

Table 6

Binding of ^3H-asialoologosaccharides to porcine liver and lymphocyte lectins with different subunit arrangements

Subunit[a] arrangement	K_a at 4 °C ($l \times mol^{-1}$)			
	aOS II	aOS III(6)	aOS III(4)	aOS IV
HL_4	0.88×10^6	0.95×10^7	0.75×10^8	0.75×10^9
$(LL\ I\text{-}2)_4$	0.75×10^6	0.80×10^7	0.62×10^8	0.85×10^9
HL_2	1.25×10^6	1.85×10^6	2.50×10^6	0.75×10^6
$(LL\ I\text{-}2)_2$	1.50×10^6	2.10×10^6	3.00×10^6	4.50×10^6
HL	3.00×10^3	5.35×10^3	6.60×10^3	8.10×10^3
LL I-2	2.30×10^3	5.70×10^3	6.80×10^3	7.70×10^3
LL I	0.95×10^9	0.80×10^8	0.78×10^7	1.55×10^6
$(LL\ I)_r$	1.00×10^9	0.87×10^8	0.80×10^7	$1.65.10^6$

[a] HL_4 and $(LL\ I\text{-}2)_4$ are tetramers of the basic binding subunit of hepatic lectin or lymphocyte lectin LL I that occur in 0.1% (v/v) of Triton X-100. HL_2 and $(LL\ I\text{-}2)_2$ are dimeric forms of the binding subunit separated by high resolution gel filtration from the preparations incubated in 0.5% (v/v) solution of Triton X-100. HL and LL I-2 represent dissociated binding subunits as obtained by prolonged incubation of the lectins in 1% (v/v) Triton X-100. $(LL\ I)_r$ is a reconstituted lymphocyte lectin LL I isolated as described in the text.

ted carbohydrate moieties. Immune reactions based on carbohydrate-lectin interactions appeared in early phylogenesis (24) and did not loose their importance even after the development of the specific antibody-antigen recognition system; they are very rapid and cover a broad range of altered structures. From this point of view of special importance appears to be the interaction of natural killer cells with their target cells (25), a reaction representing a significant part of natural antitumor immunity.

A search for the molecular basis of recognition lead us to the investigation of porcine liver and lymphocyte lectins. The binding subunits of both lectins seem to be identical and the specific recognition pattern in lymphocytes is influenced by low-molecular-weight glycopeptides. A more precise physicochemical characterization of these glycopeptides would answer the question how these low-molecular-weight substances can create different binding specificities by different assemblies of the basic lectin subunits.

References

1. Montreuil, J.: Adv. Carbohydr. Chem. Biochem. 37, 187-223 (1980).
2. Ashwell, G., Harford, J.: Annu. Rev. Biochem. 51, 531-534 (1982).
3. Kieda, C.M.T., Bowles, D.J., Ravid, A., Sharon, N.: FEBS Lett. 94, 391-395 (1978).
4. Nagamura, Y., Kolb, H.: FEBS Lett. 115, 59-62 (1980).
5. Cuatrecasas, P., Wilchek, M., Anfinsen, C.B.: Proc. Nat. Acad. Sci. U.S.A. 61, 636-642 (1968).
6. Dorland, L., Haverkamp, J., Schut, B.L., Vliegenthart, J.F.G., Spik, G., Strecker, G., Fournet, B., Montreuil, J.: FEBS Lett. 77, 15-20 (1977).
7. Yamashita, K., Liang, C.J., Funakoshi, S., Kobata, A.: J. Biol. Chem. 256, 1283-1289 (1981).
8. Fournet, B., Montreuil, J.,Strecker, G., Dorland, L., Haverkamp, J., Vliegenthart, J.F.G., Binette, J.P., Schmid, K.: Biochemistry 17, 5206-5214 (1978).
9. Mizuochi, T., Yonemasu, K., Yamashita, K, Kobata, A.: J. Biol. Chem. 253, 7404-7409 (1978).
10. Cahour, A., Debeire, P., Hartmann, L., Montreuil, J.: Biochem. J. 211, 55-63 (1983).
11. Weitzman, S., Scott, V., Keegstra, K.: Anal. Biochem. 97, 438-449 (1979).
12. Cummings, R.D., Kornfeld, S.: J. Biol. Chem. 257, 11230-11234 (1982).
13. Berry, M.N., Friend, D.S.: J. Cell Biol. 43, 506-517 (1969).
14. Wenzel, K.L., Ichiki, A.T.: Clin. Immunol. Immunopathol. 8, 73-78 (1977).
15. Hudgin, R.L., Pricer, W.E., Stockert, R.J., Morell, A.G., Ashwell, G.: J. Biol. Chem. 249, 5536-5543 (1974).
16. Brunette, D.M., Till, J.E.: J. Membr. Biol. 5, 215-224 (1971).
17. Rodbard, D., Chrambach, A.: Electrophoresis and Isoelectric Focusing in Polyacrylamide Gel (Allen, R.C., Maurer, H.R. eds.), p. 28. Walter de Gruyter, New York (1974).
18. Scatchard, G.: Ann. N.Y. Acad. Sci. 51, 660-672 (1949).
19. Steward, F.C., Lyndon, R.F., Barber, J.T.: Am. J. Botany 52, 155-164 (1964).

20. Hashimoto, F., Horigome, T., Kanbayashi, M., Yoshida, K., Sugano, H.: Anal. Biochem. 129, 192-199 (1983).

21. Porter, W.H.: Anal. Biochem. 63, 27-43 (1975).

22. Warren, L.: J. Biol. Chem. 234, 1971-1979 (1959).

23. Bezouška, K., Táborský, O., Kubrycht, J., Pospíšil, M., Kocourek, J.: Biochem. J., in the press (1984).

24. Renwrantz, L.: Dev. Compar. Immunol. 7, 603-608 (1983).

25. Oldham, R.K.: Cancer Metast. Rev. 2, 323-336 (1983).

ENDOGENOUS RECEPTOR FOR CHICK EMBRYONIC β-GALACTOSIDE-BINDING LECTIN

Photochemical cross-linking of chick embryonic β-galactoside-binding lectin to polylactosamino-proteoglycan

Yuko Oda and Ken-ichi Kasai

Department of Biological Chemistry, Faculty of Pharmaceutical Sciences, Teikyo University, Sagamiko, Kanagawa 199-01, Japan

β-Galactoside-binding lectins have been found in various tissues of vertebrates [1,2]. We have isolated and characterized one of these lectins (14 K type) from chick embryonic skin. This lectin may have some fundamental roles in development and differentiation, since the lectin is widely distributed in tissues of the embryo and its content changes markedly during development [3]. To elucidate the function of the lectin, we have attempted to identify the endogenous receptor, i.e., a substance which is present in the skin and contains specific saccharides for the lectin. The approach we have taken was to label the receptor with the lectin by using photoactivatable cross-linking reagents. Amino groups of the lectin were coupled with one end of heterobifunctional cross-linking reagents. The other end carries a photoactivatable aryl azide group. The modified lectin was allowed to react specifically with the dermis (skin) insoluble preparation, then the azide end was activated by UV irradiation to form a covalent bond between the lectin and the nearest neighbor group on the receptor. Successful results were obtained by the use of two reagents; one was methyl-4-azidobenzoimidate [4],

Lectins, Vol. IV

and the other was [^{35}S]N-succinimidyl-3[(2-nitro-4-azidophenyl)-2-aminoethyldithio]propionate, which contains a cleavable disulfide bond and can isotopically label a receptor [5,6]. To characterize the lectin-receptor complex, enzyme treatment was examined. The cross-linked complex was degraded by pronase and by endo-β-galactosidase, which acts on substances containing carbohydrate chains with repeating (Galβ1→4GlcNAcβ1→3) structure, i.e., polylactosaminoglycan [7,8]. This suggests that the endogenous receptor of the lectin is a proteoglycan containing polylactosaminyl structure. Such compounds appears in many cells at an early developmental stage and may have a role in differentiation.

Materials and Methods

1) Coupling of the lectin with methyl-4-azidobenzoimidate : The β-galactoside-binding lectin was purified from 19-day-old chick embryonic dermis as described previously [3]. The lectin was iodinated by the method of Bolton and Hunter [9]. ^{125}I-lectin was coupled with methyl-4-azidobenzoimidate at pH 9.0-9.5 [4].
2) Cross-linking of methyl-4-azidobenzoimidate-lectin to dermis insoluble fraction : The dermis insoluble fraction from which the lectin had been removed was obtained during the process of lectin purification. Methyl-4-azidobenzoimidate-^{125}I-lectin was added to the dermis insoluble fraction and incubated in PBS for 1 hr at 0°C in the dark. Samples were irradiated with a Mineralight UVSL-58 at 366 nm for 10 min at a distance of 5 cm. Some of the cross-linked complexes were treated with 200 µl of endo-β-galactosidase (Escherichia freundii, 0.05 U/ml) (Seikagaku Kogyo, Tokyo) in 0.1 M acetate buffer, pH 6.0. After centrifugation, the pellets were boiled for 5 min in lysis buffer (100 mM Tris-HCl, pH 6.8, 2 % SDS, 20 % glycerol), and subjected to SDS polyacrylamide slab gel electrophoresis. Radioactivity in each gel was detected by autoradiography.
3) Coupling of the lectin with [^{35}S]N-succinimidyl-3[(2-nitro-4-azidophenyl)-2-aminoethyldithio]propionate : The reagent was coupled with the lectin in PBS. The modified lectin was separated from unreacted reagent by gel chromatography. The molar ratio of protein to the reagent was 1 : 0.33.
4) Cross-linking of [^{35}S]N-succinimidyl-3[(2-(nitro-4-azidophenyl)-2-aminoethyldithio]propionate-lectin to dermis insoluble fraction : The procedure for cross-linking was the same as that for methyl-4-azidobenzoimidate-^{125}I-lectin. The cross-linked complexes were treated with following enzymes ; chondroitinase ABC (Proteus vulgaris), 0.25 U/ml; keratanase (Pseudomonas sp.), 0.05U/ml; hyaluronidase (bovine testis), 10 U/ml ; and Pronase P (Streptomyces griseus), 10 µg/ml.

Results

1) Binding and cross-linking of methyl-4-azidobenzoimidate-^{125}I-lectin to the dermis insoluble fraction and enzyme digestion of the complex by endo-β-galactosidase : After irradiation, 15% of added methyl-4-azidobenzoimidate-^{125}I-lectin was bound. Cross-linked products were analyzed by SDS polyacrylamide gel electrophoresis (Fig. 1, lane 3). Two major bands appeared, one corresponding to extremely high-molecular-weight (at the top of the stacking gel), and the other, to lectin monomer (Mr 14,000). Two minor bands appeared, one at the interface of the the gel and the other corresponding to lectin dimer (Mr 28,000). Little binding and no formation of a

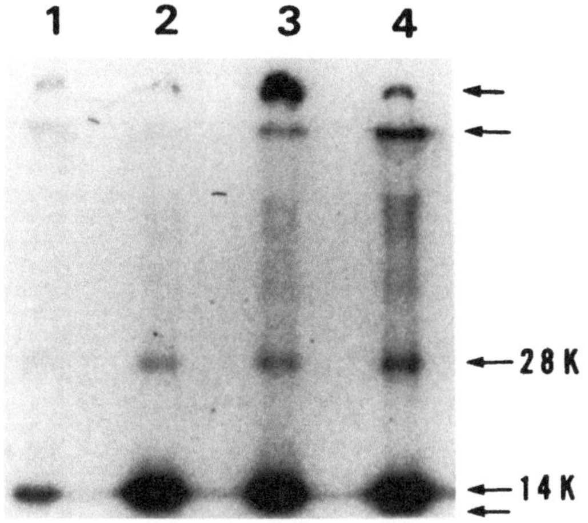

Fig. 1. Cross-linking of methyl-4-benzoimidate-^{125}I-lectin to dermis insoluble fraction and enzyme digestion by endo-β-galactosidase. Cross-linker-^{125}I-lectin was incubated with the dermis insoluble fraction and cross-linked by UV irradiation (lane 3). The reaction was also carried out in the presence to 300 mM lactose (lane 2), or UV irradiation was omitted (lane 1). Cross-linked complex was digested by endo-β-galactosidase (lane 4). Arrows indicate, from top to bottom: the top of the stacking gel, the interface of the gel, Mr 28,000 (corresponding to lectin dimer), Mr 14,000 (lectin monomer), and the dye front.

high-molecular-weight band were seen when the reaction was carried out in the presence of 300 mM lactose (lane 2) or irradiation was omitted (lane 1). Cross-linking was specific for lectin-carbohydrate binding and was dependent on UV irradiation. After endo-β-galactosidase digestion, the cross-linked high-molecular-weight complex at the top of the gel was degraded, forming a band at the interface of the gel and many weak bands in the separating gel. These results indicate that the cross-linked complex may be polylactosaminoglycan.

2) Binding and Cross-linking of [^{35}S]N-succinimidyl-3[(2-nitro-4-azidophenyl)-2-aminoethyldithio]propionate-lectin to dermis insoluble fraction. After irradiation, 60 % of added [^{35}S]N-

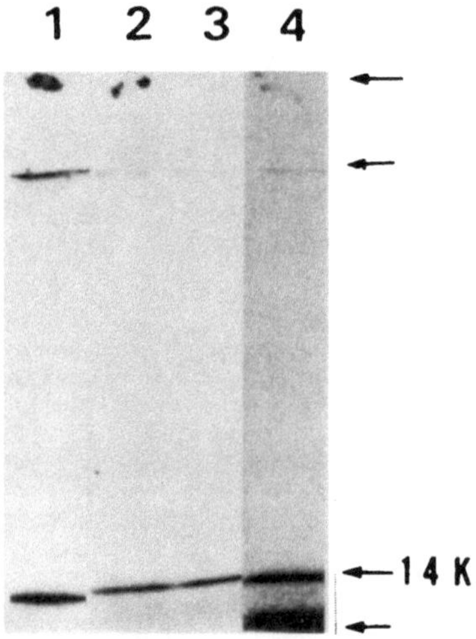

Fig. 2. Cross-linking of (^{35}S)N-succinimidyl-3((2-nitro-4-azidophenyl)-2-aminoethyldithio)-propionate-lectin to dermis diation (lane 1). The reaction was also carried out in the presence of 300 mM lactose (lane 2), or UV irradiation was omitted (lane 3). When the cross-linked complex was reduced with 100 mM dithiothreitol (lane 4), the radioactivity at the top and interface of the stacking gel decreased. Arrow indicate, from top to bottom: the top of the stacking gel, the interface of the gel, Mr 14,000 (corresponding to lectin monomer), and the dye front, respectively.

succinimidyl-3[(2-nitro-4-azidophenyl)-2-aminoethyldithio]-propionate-lectin was bound, while in the presence of 30 mM lactose, only 5.5 % was bound. Little radioactivity bound without irradiation. Cross-linked products were subjected to SDS polyacrylamide gel electrophoresis under non-reducing conditions. UV irradiation resulted in the formation of three bands as shown in Fig. 2 lane 1: two bands were of extremely high-molecular-weight, one at the top of the stacking gel, and the other at the interface, and the third was the lectin monomer (Mr 14,000). Cross-linking was specific because the intensities of the two high-molecular-weight bands were greatly reduced when the reaction was carried out in the presence of lactose (lane 2) or irradiation was omitted (lane 3). When the cross-linked high-molecular-weight complex was reduced with dithiothreitol, most of the radioactivity was found in two bands having high electrophoretic mobility, one at the position of lectin monomer and the other near the dye front (lane 4). The nature of the latter is unclear. It may be proteoglycan that is highly negatively charged, but further detailed study is required. Only a small amount remained at the top of the gel.

3) Susceptibility to enzymatic digestion of [^{35}S]N-succinimidyl-3[2-nitro-4-azidophenyl)-2-aminoethyldithio]propionate-lectin-receptor complex. The experiment using methyl-4-azidobenzoimidate indicated that the endogenous receptor may be polylactosaminoglycan because the receptor-lectin complex was degraded by endo-β-galactosidase. It should be noted that endo-β-galactosidase digests not only polylactosaminoglycan but also keratan sulfate, and the dermis contains chondroitin sulfate, keratan sulfate, and hyaluronic acid, etc. To confirm that the endogenous receptor contains polylactosaminoglycan, we examined its susceptibility to the various enzymes; keratanase (which cleaves only keratan sulfate), chondroitinase ABC (which hydrolyzes chondroitin sulfate A, B, and C), and hyaluronidase and pronase. After enzymatic digestion, a small amount of radioactivity was released only by pronase. Fig. 3 shows the

results of polyacrylamide gel electrophoresis in SDS. Endo- -galactosidase effectively degraded the cross-linked complex which had been located at the top and interface of the stacking gel (lane 3). There was no difference when γ-lactone was added in order to inhibit the trace amount of exo-galactosidase activity present in this enzyme preparation. Protease activity was not detectable in this preparation (lane 4). Keratanase was slightly effective (lane 2). Hyaluronidase and chondroitinase ABC had no effect (lanes 6,7). Pronase cleaved the complex and formed a band near the dye front (lane 5), which seems to be similar to the reduced sample (Fig. 2, lane 4). These results indicate that the lectin was specifically cross-linked to proteoglycan containing polylactosamine structure.

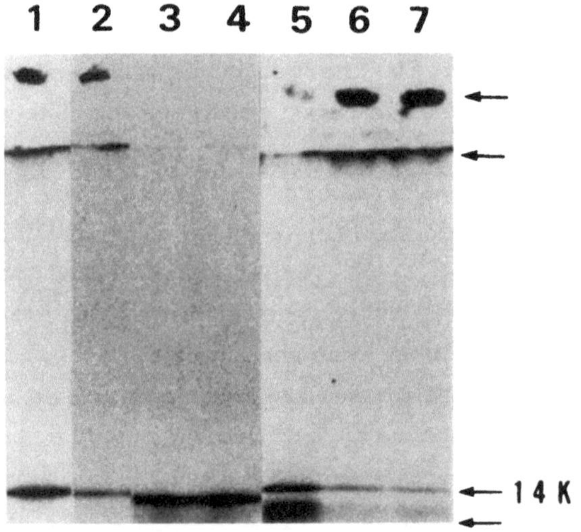

Fig. 3. Susceptibility to enzymes of the cross-linked (^{35}S) N-succinimidyl-3-((2-nitro-4-azidophenyl)-2-aminoethyldithio)-propionate-lectin-receptor complex. Cross-linked complexes were treated with keratanase (lane 2); endo-β-galactosidase in the presence (lane 3) and absence (lane 4) of 10 mM γ-lactone and 20 μg/ml leupeptin; pronase P (lane 5), hyaluronidase (lane 6) and chrondroitinase ABC (lane 7) at 37°C overnight with shaking. Lane 1 show the control (no enzyme treatment). Arrows as in Fig. 2.

Discussion

Beyer et al.[10] have suggested that chick embryonic β-galactoside-binding lectin (CLL-II) binds to intestinal mucin, as hemagglutination activity was inhibited by the mucin. However, no endogenous receptor has yet been unambiguously identified. We choose the dermis insoluble fraction from which the lectin had been removed as a source of "true" endogenous receptor for the following reason: the receptor must be contained in the insoluble fraction because the lectin could not be solubilized by homogenization, but was easily solubilized with excess hapten sugar. The results using methyl-4-azidobenzoimidate and [^{35}S]N-succinimidyl-3[(2-nitro-4-azidophenyl)-2-aminoethyldithio]-propionate showed that the lectin cross-linked to proteoglycan containing polylactosamine structure, suggesting that the lectin binds and interacts with polylactosaminoglycan in chick embryonic skin. Polylactosaminoglycan appears in many cells at early developmental stages, for example, erythrocytes (Ii antigen)[11] and embryonal carcinoma (differential carbohydrate marker) [12,13]. Our finding emphasizes the possible function of the lectin in embryonic development and differentiation.

The use of photochemical cross-linking reagents proved to be very effective. In particular, [^{35}S]N-succinimidyl-3[(2-nitro-4-azidophenyl)-2-aminoethyldithio]propionate allowed us to introduce onto the lectin, in a single operation, a photoactivatable and radioactive group which can be easily transferred to the receptor. This greatly facilitates further analysis of the receptor. The receptor integrated in the insoluble fraction could be detected without the use of a detergent or other solubilizing reagent which may destroy the cellular matrix and the structure of the native receptor. Similar experiments using cultured cell are in progress.

Acknowledgement: We are grateful to Dr. T. Muramatsu for helpful discussions and to Mr. M. Kaneko and Ms. T. Uzawa for technical assistance.

References

1. Barondes,S.H. Science 223, 1259-1264 (1984)

2. Barondes,S.H. Ann. Rev. Biochem. 50, 207-231 (1981)

3. Oda,Y. and Kasai,K. Biochim. Biophys. Acta 761, 237-245 (1983)

4. Ji,T.H. J. Biol. Chem. 252, 1566-1570 (1977)

5. Schwartz, M.A., Prem das, O. and Hynes, R.O. J. Biol. Chem. 257, 2343-2349 (1982)

6. Lahav,J., Schwartz,M.A., and Hynes, R.O. Cell 31, 253-262 (1982)

7. Fukuda,M.N. J. Biol. Chem. 256, 3900-3905 (1981)

8. Fukuda,M.N., Fukuda,M., and Hakomori,S. J. Biol. Chem. 254, 5458-5465 (1979)

9. Bolton,A.E. and Hunter,W.M. Biochem. J.133, 529-539 (1973)

10. Beyer,E.C.,and Barondes,S.H. J. Cell Biol.92, 28-33 (1982)

11. Fukuda,M. Nature 285, 405-407 (1980)

12. Muramatsu,T., Gachelin,G., Damonneville,M. Delarbre,C., and Jacob,F. Cell 18, 183-191 (1979)

13. Muramatsu,H., Muramatsu,T., and Avner,P. Cancer Res. 42, 1749-1752 (1982)

STUDIES ON LECTINS LXIV. FISH OOCYTE LECTINS.

FURTHER PHYSICOCHEMICAL CHARACTERIZATION OF LECTINS FROM OOCYTES OF THE BREAM, *ABRAMIS BRAMA L.*, AND THE PERCH, *PERCA FLUVIATILIS L.* INTERACTION OF THE PERCH LECTIN I WITH COMPONENTS OF THE MATURE EGG JELLY ENVELOPE.

Alexander Krajhanzl, Jan Kocourek

Department of Biochemistry, Faculty of Science, Charles University, Albertov 2030, 128 40 Praha 2.

Petr Pancoska

Department of Chemical Physics, Faculty of Mathematics and Physics, Charles University, 121 16 Praha 2,

Jindrich Nosek

Institute of Molecular Genetics, Czechoslovak Academy of Sciences, 142 20 Praha 4, Czechoslovakia

As revealed by some biochemical and serological studies, eggs of various vertebrate and invertebrate animals represent a particularly rich source of remarkable lectins. Some of them have already been isolated and partially characterized (1-18). In addition to their common source, most of the egg lectins share the affinity to D-galactose or its oligomers and derivatives. At present, the most numerous group is represented by lectins isolated from oocytes of Teleost fishes, families *Clupeidae* (5), *Osmeridae* (15), *Salmonidae* (5-8, 18), *Cyprinidae* (5,6,18), *Cobitidae* (11), *Percidae* (5,6,17,18) and *Sparidae* (9). Fish egg lectins are a relatively homogenous group of proteins for which the pronounced binding affinity for L-rhamnose and human B-erythrocytes are typical, except for the D-glucose and L-fucose binding *Perch* lectins. Some of the fish oocyte lectins also agglutinate *Ehrlich carcinoma* and AH 109A tumor cells (11,15), similarly as the frog egg lectins of the *Genus Rana*.

Lectins, Vol. IV

Out of other biological properties, mitogenic activity, blocking of proteosynthesis in cell-free systems (19), and inhibition of growth of fish bacterial pathogens (8) by some of these lectins have been also described.

In our previous reports the occurence of lectins was shown within cortical vesicles (later granules) of previtellogenic and vitellogenic oocytes in several fish species (20,21). On the basis of these results we have suggested that fish egg lectins, similarly as the *Xenopus laevis* egg lectin (22), take part in the cortical reaction, i.e. in the mechanisms leading to polyspermy blockade.

This paper brings further physicochemical characterization of two fish egg lectins belonging to family *Cyprinidae* (the bream) and *Percidae* (the perch) in addition to some new results of our localization studies supporting our original assumption.

Material and Methods

Isolation of lectins. Fish were obtained from ponds owned by the Technical University of Agriculture, Praha. Fishes were killed by decapitation, the ovaries were collected on ice, delivered to the laboratory and then frozen at -20°C.

All lectins were isolated directly by affinity chromatography of crude ovary extracts in 0.05 M acetate buffer pH 5.5 containing 0.25 M NaCl (buffer A). In case of the bream lectin, the O-α-D-galactosyl polyacrylamide gel was used as described in (6,7). The bream lectin was eluted in a single peak by 10 mM L-rhamnose solution. For affinity chromatography the active fraction, obtained by cold aceton precipitation of crude extract, was alternatively used.

Two lectin preparations from the late vitellogenic perch oocytes were prepared after their adsorption on O-α-D-glucopyranosyl-Separon H 1000 by stepwise elution with 0.1 M D-glucose (fraction A) and 1.0 M D-glucose solution (fraction B). First fraction was prepared by the same procedure (i.e. by elution with 0.1 M D-glucose solution) from early or mid previtellogenic perch oocytes as well. The crude extract from the perch vitellogenic oocytes after adsorption on O-α-D-glucopyranosyl-Separon was used fro affinity chromatography on O-α-L-fucopyranosyl polyacrylamide gel. By this procedure the perch lectin II was obtained by elution with 0.3 M L-fucose solution. The lectin solutions from each chromatography were extensively dialyzed against distilled water and finally lyophilized.

Analytical methods. Methods used for estimation of sedimentation coefficient, amino acid composition, contents of tryptophan, half-cystein (as cysteic acid), methionine (as methionine sulfone), neutral sugar and aminosugar, respectively, and inhibition activity of sugar in erythroagglutination-inhibition tests were performed as described previously (23). From sedimentation coefficient values ($s_{20,w}$) the approximate molecular weight ($^{S}M_r$) were calculated according to equation of Atassi and Gandhi (24). Determination of N-terminal amino acids was done by dansylation technique (25). For discontinuous electrophoresis in the presence of sodium dodecyl sulfate, SDS, the 15% gels of size 12 cm x 8 cm x 0.1 cm and running buffers were prepared according to Laemmli (26). Molecular weight of polypeptide chains, M_r were calculated from migration distances relative to standard proteins. Discontinuous electrophoresis in acidic (pH 4.5) or alkaline (pH 9.3) buffer systems were performed with polyacrylamide gel plates (7-10% polyacrylamide) obtained as given earlier (23). The content of sialic acid was estimated by Svennerholm's method (27), metal content was determined by means of flame atomic absorption spectrophotometer (varian Techtron Model 1200) after hydrolysis of the sample (6 M HCl, 24 H). Erythroagglutination tests with human red blood cells were performed either in test tubes or in wells of Cooke microtiter plates, by the serial dilution procedure (23).

Affinity electrophoresis. Affinity electrophoresis in discontinuous acidic buffer system was performed according to Horejsi and co-workers (28) on cylindric gels containing 0.02, 0.016, 0.011 and 0.0055 M immobilized D-galactose residues, respectively, in addition to various concentration of the free sugar. Association constants were determined graphically from mobilities of the main bands.

Circular dichroism measurements. Circular dichroism (CD) spectra were drawn on CNRS-Roussel Jouan Dichrographe III with thermostated cuvette holder. Cells of 1 mm and 10 mm path length for the range 200-260 nm and 245-340 nm were used, respectively. Solutions were passed through Millipore filters (HAWP 0.45 μ) just before use. The mean residual ellipticity $[\Theta_\lambda]$ (in deg cm^2 dmol^{-1}) was calculated by means of equation

$$[\Theta_\lambda] = 3300 \frac{S \cdot \bar{M}}{l \cdot c} \cdot D_\lambda$$

where M (mean residual weight) is 115, and S, l, c, and D_{λ} stand for sensitivity, optical path length in dm, protein concentration in g per 100 cm^3, and deviation in the experimental curve in dm, respectively. A lectin concentration was determined from absorbance measurement at 280 nm using the value of decadic absorption coefficient, $E_{280}^{1\%}$. Ellipticity values were not corrected for the refractive index. The approximate content of α-helix conformation was calculated from the residual ellipticity at 222 nm relative to 35 000 deg cm^2 $dmol^{-1}$ (29).

Histological procedures. The fate of the perch lectin I in the mature freshly-spawned egg was studied by indirect immunofluorescence method as described previously (20,21). The stripped ripe roe was either fertilized in vitro by perch sperm and allowed to swell in water or it was activated by water without prior fertilization. After swelling of the jelly envelope for 30 min, the ova were frozen in liquid nitrogen, cut in a cryostat and fixed in Boin-d'Hollande solution. Alternatively, mature perch ova, frozen in liquid nitrogen, were sectioned; sections thus prepared were incubated in a drop of water without any fixation procedure. All sections were treated with the antibody as given in our previous reports (20,21).

Results

The beam, *Abramis brama L.*, lectin

Isolation of the lectin. By affinity chromatography on O-α-D-galactosyl polyacrylamide gel, 201.4 mg of the bream lectin was obtained from 1000 g of frozen ovary containing late vitellogenic oocytes. When crude extract was subjected to cold aceton precipitation, the affinity chromatography of the active fraction led to a diminished yield of the lectin, 97 mg from 1000 g of ovary. By ammonium sulfate fractionation (30, 60 and 100% of saturation, respectively) merely inactive fractions were obtained.

The main physicochemical properties. The bream lectin has very similar electrophoretic properties as the roach lectin (30). It is composed of three major polypeptide chains, designated as A, B and C having the following average molecular weight: subunit A 19 000, subunit B 30 000, subunit C 42 000, which were estimated by means of electrophoresis of the nonreduced

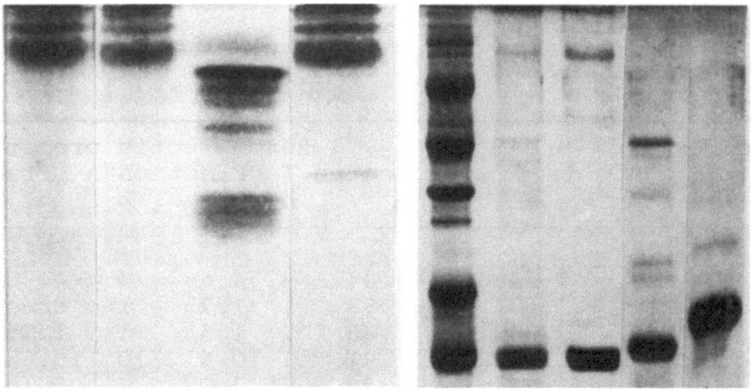

a b c d e f g h i

Fig. 1. Electrophoretic properties of perch lectin preparations a - d: Electrophoresis in alkaline buffer system; e - i: Electrophoresis of reduced samples in the presence of SDS a, f: Lectin I from previtellogenic oocytes; b, g: Fraction A; c, i: Lectin II; d, h: Fraction B; e: Standard proteins: lactalbumin (14 400), soybean trypsin inhibitor (20 100), carbonic anhydrase (30 000), ovalbumin (43 000), bovine serum albumin (67 000) and phosphorylase b (94 000).

bream lectin in the presence of SDS. Electrophoresis of the lectin reduced by 2-mercaptoethanol yielded a rather shifted electrophoretic spectrum composed bands with the following M_r: 20 000, 24 000 and 28 000. In addition, several minor zones with higher mobility could be distinguished. As revealed by gel filtration on a Sephadex G-100 Superfine (column dimensions 1.2 cm x 115 cm, buffer A pH 5.5), the bream lectin subunits exist in solution both in form of monomers and as oligomers with M_r of about 120,00, similarly as the roach lectin polypeptide chains (32).

The bream lectin a glycoprotein containing in case of preparation from vitellogenic oocytes 3.6% of neutral sugar 0.4% of sialic acid, but no amino sugar. It contains metal atoms: 0.14% Ca (i.e. 1 atom per M_r of 29 000), 0.05% Mg (i.e. 1 atom per M_r 46 000) and negligible content of zinc, as well. In agreement with our previous results (7), the amino acid compo-

Table 1.

Amino acid composition of the bream lectin from vitellogenic o-ocytes and of the perch lectin preparations

Amino acid	Bream lectin	Perch lectin preparations					
		Lectin I		Fraction A		Fraction B	
	a	a	b	a	b	a	b
Lysine	5.74	4.54	4	4.68	4	4.20	4
Histidine	3.57	1.59	1	1.75	1	trace	0
Arginine	2.84	1.36	1	1.55	1	1.05	1
Aspartic acid	8.45	11.91	11	10.61	10	10.68	10
Threonine	6.34	6.26	7	7.07	7	9.10	10
Serine	4.42	7.62	9	7.57	9	11.81	15
Glutamic acid	12.13	9.42	8	11.91	10	10.55	9
Proline	8.52	4.65	5	5.03	5-6	3.22	4
Glycine	7.21	9.48	16	8.66	15	9.43	16
Alanine	6.45	9.55	13	8.42	12	9.34	14
Half cystine	9.24	4.91	5	3.61	4	5.91	6
Valine	7.26	3.64	4	4.80	5	3.93	4
Methionine	2.05	2.42	2	1.77	2	4.11	4
Isoleucine	4.05	4.28	5	4.41	5	4.16	5
Leucine	6.33	4.16	5	5.48	6	3.86	4
Tyrosine	2.58	2.16	2	2.41	2	1.75	1
Phenylalanine	2.24	7.15	5	6.41	5	6.90	5
Tryptophan	0.57	4.88	3	3.75	2	n.d.	-

a,b - Expression in molar per cent of the total amino acid residues (a), and in approximately calculated residues of individual amino acids per M_r of 14 000 (b), repectively. For fraction A and B see text; n.d. = not determined.

sition given in Table 1 revealed that the bream lectin is rich in glutamic acid, cysteine and proline residues. Dansylation of the bream lectin yielded three main N-terminal amino acids, glutamic acid, glycine and arginine.

As for erythroagglutination properties, the bream lectin agglutinates human B red blood cells at a minimal concentration 3 µg per ml estimated in test-tubes. Erythrocytes of other human blood groups were not agglutinated even at the lectin concentration of 500 µg/ml. The most potent inhibitor of B-erythrocyte agglutination is L-rhamnose, inhibitory at 50 µm concentration; agglutination was inhibited also by methyl α-D-galactopyranoside already at 80 μM concentration, mellibiose and D-galactose at 1.6 mM, D-galacturonic acid at 1.9 mM at pH 7.0 and by L-arabinose at 3 mM.

Interaction with saccharides. By means of affinity electrophoresis in an acidic buffer, the association constants were graphically obtained. For two main electrophoretic fractions, the following values were found with D-galactose or L-rhamnose:

$$I_K^{Gal} = 45\ M^{-1} \qquad II_K^{Gal} = 35\ M^{-1}$$

$$I_K^{Rha} = 1.4 \times 10^4\ M^{-1} \qquad II_K^{Rha} = 1.0 \times 10^3\ M^{-1}$$

The perch, *Perca fluviatilis L.*, lectins

Isolation of the perch lectin I. By the procedure given above, 4.05 g of the perch lectin I was obtained from 1000 g of the perch ovary containing previtellogenic oocytes. When vitellogenic oocytes have been used as starting material, the first elution step yielded significantly lower amount of the lectin preparation (1.95 g of a fraction A from 1000 g of the ovary). By further elution of the affinity column with 1.0 M D-glucose solution, 1.46 g of a second lectin preparation (fraction B) was isolated from 1000 g of the raw perch ovary.

Physicochemical and serological properties of perch lectins from previtellogenic oocytes. As revealed by electrophoresis of reduced perch lectins (Fig. 1) together with results of sedimentation analysis given in Table 2, both perch lectins are composed of covalently linked, relatively small subunits. Perch lectin I represents a mixture of two oligomers. The major one probably consists of ten subunits with an average M_r of 13 500. Approximately fifteen subunits form a minor oligomer. In a solution of pH 5.5 oligomers possess sedimentation coefficients

Table 2.

Sedimentation analysis of perch lectin preparations

Sample	Medium	s_{20} (20) major	minor	s_{M_r}	a
Lectin I	pH 5.5	7.42	10.38	140,00	210,000
Lectin I	5% 2-mercapto-ethanol, pH 5.5	7.18	1.64	133,000	15,000
Fraction A	pH 5.5	2.88	13.70	30,000	300,000
Fraction B	pH 5.5	7.66	-	145,000	

[a] *- calculated approximately molecular weight from a sedimentation coefficient value (see Materials and Methods).*

Table 3.

Contents of neutral sugar, aminosugar, sialic acid and of some metals in perch lectin preparations (w/w %)

Component	Lectin I	Fraction A	Fraction B
Neutral sugar	1.3	3.1	5.3
Aminosugar	0	n.d.	trace $GlcNH_2$
Sialic acid	0	0.2	0.1
Ca	0.310	0.256	0.164
Mg	0.054	0.052	0.020
Zn	0.011	0.010	0.010

$s_{20,w}$ of 7.42 and 10.38 S. The molecular weight of the perch lectin II basic subinit is somwhat higher (M_r 17 500). The electrophoresis of the nonreduced lectin I in the presence of SDS proved its relatively high native molecular weight, but some minor bands belonging to the subunits and its lower oligomers could be discerned, too. Amino acid composition and contents of neutral sugar, aminosugar, sialic acid and of some metals in the perch lectin I are given in Tables 1 and 3, respectively.

Table 4.

Erythroagglutination properties of perch lectin preparation (in μg/ml)

Preparation	A_1	A_2	B	O
Lectin I	4	0.5	4	0.25
Fraction A	n.a.	n.a.	n.a.	n.a.
Fraction B	4	0.5	4	0.5

n.a. = erythrocytes were not agglutinated even at 0.7 mg/ml concentration

In summary, it can be concluded that the perch lectin I is a Ca-containing glycoprotein rich in glycine, alamine, aspartic acid and serine. The higher content of aromatic amono acid correlates with decadic absorption coefficient $E^{1\%}_{280}$ = 30.2 which is substantially higher than for other fish egg lectins. The polypeptide chain of perch lectin I is terminated by glutamic acid; however, in addition to this, two minor spots belonging to cystein and alanine could be observed on chromatogramms.

As revealed by absorption measurements, the perch lectin I and also II possess, in contradistinction to other fish oocyte lectins, a relatively low solubility in buffer solutions usually used (e.g. in buffer A, pH 5.5 maximally 1.5 mg/ml, 0.1 M phosphate buffer pH 7.0 maximally 2.5 mg/ml). The solubility, which especially decreases after freeze-drying and dialysis against water, can be improved either by dialysis against buffer, by use of acetic acid solutions, or by addition of sucrose to the buffer.

The perch lectin I is partially specific for human A_2 and O erythrocytes, see Table 4. As estimated in wells of Cooke microtiter plates, agglutination of human O-erythrocytes by the perch lectin I is inhibited by D-glucose (even at 0.23 mM concentration), methyl o-D-mannopyranoside (0.45 mM), L-fucose and D-xylose (0.50 mM) and by D-arabinose (0.60 mM).

Chiroptical properties of the perch lectin I from previtellogenic oocytes. CD spectra of the perch lectin I at two different temperatures are illustrated in Fig. 2. In the far ultraviolet region of these spectra the couples of amide chromophore give rise to distinct Cotton effects, the first within the n-II^{-+} transition region with a negative maximum centered at 212 nm, the second one within the region of II-II^{-+} electron transitions as a suggestion of a positive band in the region below 204 nm. The ellipticity values of the main band could be influenced by probable superposition of closely lying bands with opposite sign. The above mentioned superposition lead to rather unusual shape of the n-II^{-+} band. In spite of this, from the position and magnifude of the main dichroic band it can be deduced that in the polypeptide backbone of the perch lectin I predominates probably β-conformation. This assumption is supported by positive ellipticity value in the region below 204 nm. The content of α-helical conformation is approximately 10%. The negative band centered at 235 nm could probably originate from aromatic chromophores interacting with the neighboring peptide (31). The more pronounced near UV region of the perch lectin I CD spectrum appears to be superpositioned from dichroic bands of aromatic amino acids having assymetric enviroments in the lectin polypeptide chain. However, as the CD is present in the region up to 300 nm, contribution of a disulfide chromophore, besides the tryptophan one, appears also likely (32). As in the case of the roach lectin (30), the general features of the perch lectin Å CD spectrum are preserved at an elevated temperature (see Fig. 2). At 55°C the perch lectin I partially precipitates, but the basic character of its CD spectrum remains conserved, too. Appreciable differences in the CD spectrum at 50°C can be observed particularly in the near UV region; this implicates an easy sterical accesibility at least of a part of aromatic chromophores. Small change in aplitude of n-II^{+} band cannot be presumably explained in terms of chan-

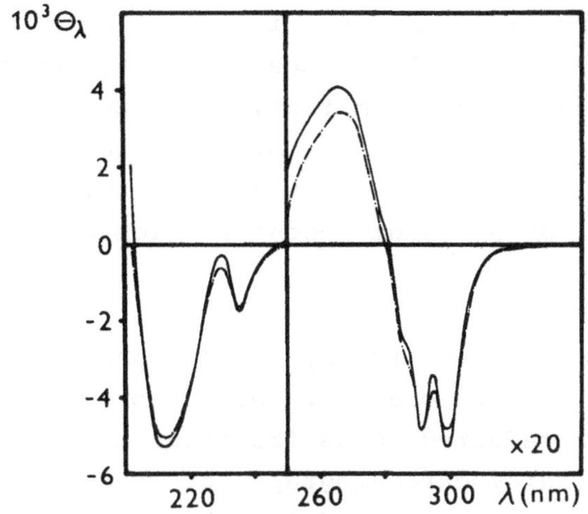

Fig. 2. Circular dichroism spectrum of the perch lectin I from previtellogenic oocytes at pH 6.0 (0.1 M phosphate buffer) and at 5.7°C (———) and 50.6°C (– – – –).

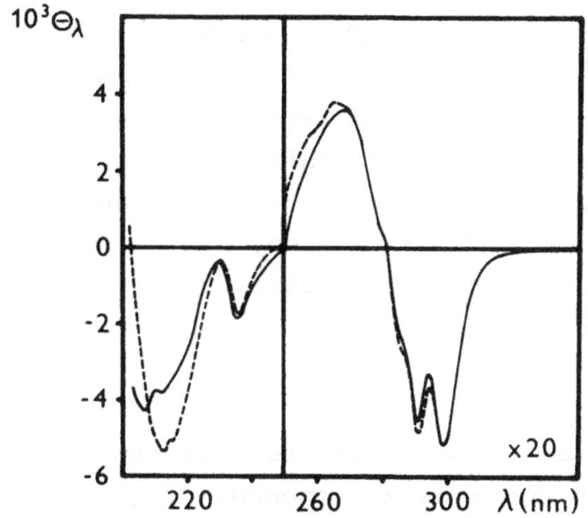

Fig. 3. Perturbation of the CD spectrum of the perch lectin I by the presence of EDTA at 25°C, pH 6.0 (0.1 M phosphate buffer). For comparison the CD curve of the lectin, measured at pH 6.0 and 25°C is given (– – –).

ges of regular conformation ratio. In addition to the relatively thermostable conformation of the lectin I also its pH-stability has been proven in the region of pH 5.0-8.5. On the other hand, the presence of a chelating agent, EDTA, evidently lead to deep conformation changes (Fig. 3). The observed red shift of the n-II $^{+}$ band would be due either to the increasing content of a random-coil conformation, which becomes probably prevailing, or to exciton splitting of degenerate II-II^{+} transitions in an α-helix. In general it appears that Ca^{2+} atoms stabilize the total predominant β-conformation of the lectin I polypeptide backbone.

Basic physicochemical properties of the two lectin components isolated from vitellogenic oocytes. Some properteis of the two components, further referred to as fractions A and B, respively, obtained by the affinity chromatography of crude extract of vitellogenic oocytes, are summarized in Tables 1, 2, 3 and 4. With respect ot the electrophoretic patterns given in Fig. 1 it can be concluded that both fractions are composed mainly of lectin I. However, the fraction A electrophoretically identical with lectin I contains a higher amount of carbohdyrate, see Table 3, but its amino acid composition remains as if the standard deviation of the amino acid determination of about one residue is considered. Despite of this, sedimentation and erythroagglutinating properties of the lectin I and fraction A, both prepared by identical procedures, are quite different, cf. Tables 2 and 4, as well as the solubility of both lectin components (at pH 5.5 only 60 μg of fraction A dissolved in 1 ml of the buffer).

Electrophoresis in alkaline buffer reveals that in addition to lectin I fraction B contains a component of glycoprotein nature (PAS-positive) corresponding to a band with the highest mobility. Under reducing conditions this component corresponds to a M_r of 42 000. Although, in comparison with the perch lectin I, the amino acid composition, the sugar, Ca, and Mg content of fraction B differs, erythroagglutinating and sedimentation properties of these two lectin preparations are lamost identical. The main component of fraction B, lectin I from vitellogenic oocytes, also slightly differs from its counterpart form previ-

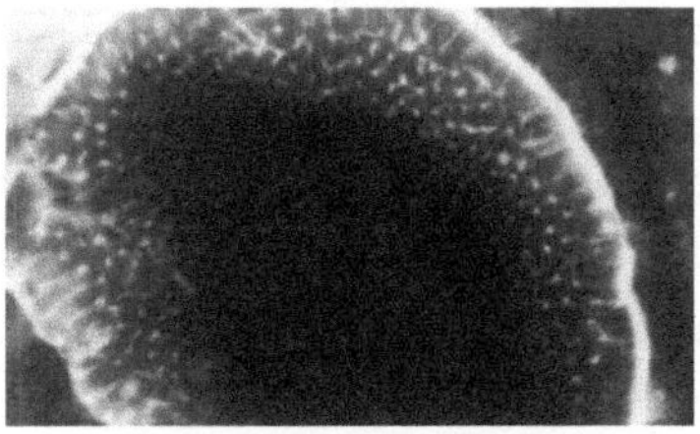

Fig. 4. Immunofluorescence localization of the perch lectin I in mature perch oocytes after in vitro fertilization. The same results were observed in activated and unfixed section of ripe ova, without prior fertilization, after incubation of the section in a drop of water. Cryostat section (10 μm), magnification x 100.

tellogenic oocytes by a somewhat lower mobility in presence of SDS (Fig. 1). Because of affinity chromatography has been used merely for isolation of the lectin preparations, we deduce that the observed 42 kD component has to be present either in form of an alkali nonstable complex with lectin I or to represent a saccharide-binding glycoprotein, a lectin III with and affinity to D-glucose as well. With regard to the unchanged erythroagglutination properties of fraction B, it appears that the complex formation with lectin I would be more probable.

By treatment of vitellogenic oocyte ghosts which remained after first extraction of the perch lectins with sucrose-containing medium (buffer A), additional amounts of lectin preparations could be obtained. Affinity chromatography with stepwise elution (0.1 M and 1.0 M D-glucose solutions) thus yielded two fractions. The first is composed of two components electrophoretically discernable and identical with the proportional mixture of components obtained in the two fractions (i.e. A and B) from the first chomatography of vitellogenic oocyte extract. The second fraction has the same electrophoretic and other physicochemical properties as fraction B.

Localization of the perch lectin I in mature fertilized, activated or unfixed oocytes. Sections from the fertilized eggs and from eggs activated by water, as well as in unfixed sections of ripe ova, cortical granules, forming from cortical vesicles (34) were studied for localization of the lectin. Immunofluorescence staining with an anti-lectin antibody localized the perch lectin I in the outer part of the jelly envelope and inside the pores penetrating the envelope, Fig. 4. The same results had been obtained by direct localization of the perch lectins by means of α-D-mannosyl derivative of FITC-bovine serum albumin (35). This indicates that at least a part of the lectin saccharide binding sites remainst free even after reaction of the lectin with components of the jelly.

Discussion

The aim of this paper was to describe some additional physicochemical properties of lectins from oocytes of the perch and the bream and to present evidence on the changes of properties of the lectins during development of the oocytes with respect to their function in vivo. We folowed up some of the results published in our previous papers (7,20). A more detailed physicochemical studies of lectins from vitellogenic and previtellogenic fish oocytes were prompted first of all by the finding that the erythroagglutinating properties of the extract of the perch female gonads were expressively dependent on the maturity of the oocytes (36).

As for the lectin from the braem vitellogenic oocytes, it has been found that this blood group B specific lectin has a molecular weighy of about 120,000 and is composed of three different polypeptide chains (with predominance of the A-chain) which are probably covalently bound. This characteristics correspond only in rough outlines to properties of the lectin isolated in a very low yield from previtellogenic oocytes (7). By analogy with the lectin from the roach oocytes (30) it appears that also in this case some of the observed differences may, in fact, be due to a somewhat different isolation procedure. In general, however, the serological and carbohydrate specificiteis of the bream lectin for L-rhamnose and for some sterically closely related sugar structures have been confirmed. In contradistinction to the lectin from roach oocytes the lectin from bream oocytes has a relatively high affinity also for O-D-galactose. The complicated subunit composition and heterogeneity of the isolated lectin would indicate that some of the subunits could be degradation products of enzymatic cleavage by proteinases.

In some moredetail the lectin I of the perch oocytes was studied. In one of our first papers (7), the mixture of this lectin (specific for D-glucose) with lectin II (having a higher binding affinity for L-fucose) has been characterized. The lectin I isolated from previtellogenic oocytes is formed by about ten covalently bound subunits with a M_r of 14 000. The conformation of its polypeptide backbone (appearantly with predominance of the β-form) is markedly thermostable and is probably stabilized by Ca^{2+}. The lectin II which is composed of covalently bound subunits with a M_r of 17 500 was not characterized in detail in our present work.

The dependance of some properties of the lectin I on the oocyte maturity was studied in eight preparations obtained by desorption from a column of O-α-D-glucosyl Sepharon H 1000 with 0.1 M solution of D-glucose. As starting material to the individual isolation, ovaria were taken successivelly from fish of approximately the same age in the course of the whole ovarial cycle. It has been shown that on the basis of the physicochemical and erythroagglutinating properties, the preparations thus obtained can be divided into two groups. This division corresponds to the two basic developmental stages of the oocyte, i.e. the previtellogenic and vitellogenic stages. The direct detection of lectin I in envelopes of mature unfixed, activated or fertilized eggs together with the analytical results indicate that lectin I enters during vitellogenesis into a number of interactions with the components of the egg-content and envelope.

With regard to the consistently lower yield of the lectin I from vitellogenic oocytes, the affinity column was checked by elution with 1.0 M D-glucose. Thus two fractions were obtained from which, in fact, the first one was prepared exactly by the same procedure as lectin I from previtellogenic oocytes. In spite of this, the first fraction has in comparison with lectin I from previtellogenic oocytes a number of different properteis, however, its amino acid composition and electrophoretic mobility remain the same or are very similar. Hence we deduced that this fraction contains probably a complex of the lectin I with some carbohydrate substances which occupy a part of the lectin sugar-binding sites. Probably therefore the resulting complex shows a behaviour of a monovalent lectin (does not agglutinate but adsorbs to an affinity carrier) having a substantially lower solubility and different sedementation properties. Up to now it cannot be quite excluded that the lectin I is post-translationally modified by a higher degree of glycosylation in the course of vitellogenesis.

The second fraction shows in comparison with lectin I also some different physicochemical properties including the aminoacid composition; on the other hand, however, the erythroagglutinating and partially also the sedimentation properties remain preserved. The electrophoretic patterns show that the major component of this fraction is again the lectin I, which is presumably modified by attachment of some component of glycoprotein nature having a M_r of about 42 000. This component, however, does not occupy its saccharide-binding sites. Under conditions of discontinuous alkaline electrophoresis (pH 9.3) or in the presence of 2-mercaptoethanol and SDS this complex dissociates. The interactions of lectin I with the components of the ripe oocyte membranes was reliably evidenced by its localization in the unfixed or even fertilized or activated perch oocytes. It is equally possible that the complexes mentioned above are the result of interactions of the partially degraded components of the egg envelopes with lectin I. Proteinases could play a role in the physiologically significant degradation of the egg envelope (37). The presence of two trypsin-like proteinases in the of proenzymes activated by Ca^{2+} have been shown in cortical vesicles of sea urchins (38).

Up to now we do not have any detailed information on the nature of the components interacting with the perch lectin I. According to the previous histochemical studies it could be acid glycosaminoglycans shown to be present in coritcal vesicles of fresh water fish (37). Of particular interest is the finding of Yamamoto (39) who has performed an electrophoretic stüdy on the water soluble proteins of dog salom (*Oncorhynchus keta)* oocytes. The slow moving band which corresponds to carbohydrate protein complex appears to represent the content of cortical vesicles. As well, the cortical vesicles in the eggs of fresh water teleöst fish *Rasbora daniconius* contains polysaccharide protein complexes; their protein part is rich in cystine, whereas the polysaccharid components consists of chondroitin sulfate B and poly-D-galacturonic acid (40). Electron microscopy and autoradiographic studies revealed that in syntehsis of these saccharide components of cortical vesicles the endoplasmatic reticulum and Golgi apparatur are involved. Cortical vesicles are formed in early previtellogenic phase by the fusion of small, electronoptically visible vesicles dispatched from Golgi complex. In the mentioned vesicles both the protein and PAS-positive substances were proven (41). The gradually increasing content of PAS-positive substances forms the most characteristic feature of cortical vesicles during the egg development and maturation, especially in vitellogenic phase (37). As the oocyte grows, the cortical vesicles increase in both size and number and their content becomes less soluble. During the cortical reaction the glycosaminoglycans as well as other components of the cortical granule content become components of the fertilization membrane (37,39).

On the basis of the previous studies as well as our new results we assume that a complex function can be postulated for the fish oocyte lectins:

1. During oocyte development and maturation lectins could play a role of intracellular transport proteins in terms of the hypothesis presented recently by Olden et al. (42). Their sugar binding properties could participate in the dispatching of some glycosaminoglycans, glycoproteins or polysaccharides from the site of their synthesis towards the cortical vesicles.

2. During fertilization leading to exocytosis of cortical vesicles, the cortical reaction, endogenous lectins could take part in the mechanism blocking penetration of supernumerary sperms, preventing polyspermy. In the case of perch oocytes, the basic mechanisms culd be either cross-linking of the oocyte membranes or agglutination of the sperms already entered into the oocyte perivitelline space, with a complex of lectin I and the nonihibitory glycoprotein (fraction B). The agglutination of homologous sperm by the extract of perch oocytes was observed by Kothbauer and Schenkel-Brunner (37,43).

3. In the developing embryo at least some of the lectins could have a protective function. As membrane-bound components with partially free binding sites (18,35) or as free components or the perivitelline space (44) they could bind bacterial pathogens and inhibit their growth, as indicated by some experiments in vitro (8). In pathogen growth inhibition blocking of protein synthesis could become effective, shown hitherto only in the experiments with cell-free systems (19).

References

1. Yokota, M., Sakakibara, H., Kawauchi, H.: Yakugaku Zasshi 95, 50-55 (1975).
2. Sakakibara, F., Takayanagi, G., Kawauchi, H., Watanabe, K., Hakamori, S.: Biochim. Biophys. Acta 444, 386-395 (1976).
3. Sakakibara, F., Watanabe, K., Takayanagi, G., Ise, H., Kawauchi, H.: Tohoku Yakka Daigaku Kenkyu Nempo 23, 125-130 (1976).
4. Sakakibara, F., Takayanagi, G., Ise, H., Kawauchi, H.: Yakugaku Zasshi 97, 855-862 (1977).
5. Krajhanzl, A.: Some lectins from fish roe, RNDr. Thesis, Charles University, Prague 1977 (in Czech)
6. Krajhanzl, A., Horejsi, V., Kocourek, J.: Biochim. Biophys. Acta 532, 209-214 (1978).
7. Krajhanzl, A., Horejsi, V., Kocourek, J.: Biochim. Biophys. Acta 532, 215-224 (1978).
8. Voss, E.W., Fryer, J.L., Banowetz, G.M.: Arch. Biochem. Biophys. 186, 25-34 (1978).
9. Rogers, D.J.: Med. Lab. Sci. 35 239-245 (1978).
10. Sue, H., Takayanagi, G., Koseki, T., Nitta, K., Sakakibara, P., Kawauchi, H.: Yakugaku Zasshi 100, 706-712 (1981).
11. Sakakibara, F., Takayanagi, G., Kawauchi, H.: Yakugaku Zasshi 101, 918-925 (1981).
12. Sasaki, H., Aketa, K.: Exp. Cell Res. 135, 15-19 (1981).
13. Kamiya, H., Shimizu, Y.: Nippon Suisan Gakkaischi 47, 255-259 (1981).
14. Roberson, M.M., Barondes, S.H.: J. Biol. Chem. 257, 7520-7524 (1982).
15. Sakakibara, F., Takayanagi, G., Mukai, C., Sue, H., Terasaki, Y., Nitta, K., Kawauchi, H.: Annu. Rep. Tohoku Coll. Phram. 29, 115-124 (1982).
16. Bretting, H., Stanislawski, E., Jacobs, G., Becker, W.: Biochim. Biophys. Acta 749, 143-152 (1983).
17. Bazil, V., Entlicher, G., Kocourek, J.: Biochim. Biophys. Acta, in preparation.
18. Krajhanzl, A.: Properties of lectins from oocytes of Teleost fishes, their interaction with saccharides and localization, PhD Thesis, Charles University, Prague 1984 (in Czech).
19. Licastro, P., Chiricolo, M., Barbieri, L., Falasca, A., Rossi, C.A., Franceschi, C., Stirpe, F.: Lectins-Biology, Biochemistry and Clinical Biochemistry, Vol. 3, pp. 293-302, Eds. Bøg-Hansen, T.C., Spengler, G.A.; W. de Gruyter, Berlin 1983.

20. Nosek, J., Krajhanzl, A., Kocourek, J.: Lectins-Biology, Biochemistry and Clinical Biochemistry, Vol. 3, pp. 453-459, Eds. Bøg-Hansen, T.C., Spengler, G.A.; W. de Gruyter, Berlin 1983.

21. Nosek, J., Krajhanzl, A., Kocourek, J.: Histochemistry 79, 131-139 (1983).

22. Greve, L.C., Hedrick, J.L.: Gamete Res. 1, 13-18 (1978).

23. Horejsi, V., Kocourek, J.: Biochim. Biophys. Acta 538, 299-315 (1978).

24. Atassi, M.Z., Gandhi, S.K.: Naturwiss. 52, 259-264 (1965).

25. Zanetta, J.P., Vincedon, G., Mandel, P., Gombos, G.: J. Chrom. 51, 441-458 (1970).

26. Laemmli, U.K.: Nature (London) 227, 680-685 (1970).

27. Svennerholm, L.: Biochim. Biophys. Acta 24, 604-611 (1957).

28. Horejsi, V., Ticha, M., Kocourek, J.: Biochim. Biophys. Acta 499, 301-309 (1977).

29. Jirgensons, B.: Biochim. Biophys. Acta 623, 69-77 (1980).

30. Krajhanzl, A., Danisova, A., Pancoska, P., Kocourek, J.: This Proceedings.

31. Woody, R.W.: Biopolymers 17, 1451-1467 (1978).

32. Roche, A.-C., Maurizot, J.-C., Monsigny, M.: FEBS Lett. 91, 233-236 (1978).

33. Jirgensons, B.: Biochim. BCophys. Acta 536, 205-211 (1978).

34. Nosek, J.: Histochem. J. 16, 435-437 (1984).

35. Krajhanzl, A., Nosek, J., Monsigny, M., Kocourek, J.: Histochem. J. 16, 429-431 (1984).

36. Schenkel-Brunner, H., Kothbauer, H.: Z. Immunitätforsch. 154, 155-161 (1978).

37. Guraya, S.S.: Int. Rev. Cytol. 78, 257-360 (1982).

38. Rossignol, D.P., Decker, G.L., Lennarz, W.J.: Cell Fusion: Gene Transfer and Transformation, pp. 5-26, Eds. Beers, R. F., Bassett, E.G., Raven Press, New York 1984.

39. Yamamoto, T.S.: J. Fac. Sci. Hokkaido Univ. Ser. VI. Zool. 20, 245-257 (1976).

40. Krishnamurthy, C.G., Rajasekharasetty, M.R., Gopal Dutt, N. H.: Indian J. Exp. Biol. 10, 29-42 (1972).

41. Droller, M.J., Roth, T.F.: J. Cell Biol. 28, 209-232 (1966).

42. Olden, K., Parent, J.B., White, S.L.: Biochim. Biophys. Acta 393, 31-36 (1982).

43. Kothbauer, H., Schenkel-Brunner, H.: Z. Immunitätsforsch. 154, 87-90 (1974).

44. Ginsburg, A.S.: J. Embryol. Exp. Morphol 9, 173-190 (1961).

STUDIES ON LECTINS. LXV. FISH OOCYTE LECTINS. PHYSICOCHEMICAL STUDY OF A LECTIN ISOLATED FROM OOCYTES OF THE ROACH *RUTILUS RUTILUS L.* AND PARTIAL CHARACTERIZATION OF ITS MAIN POLYPEPTIDE CHAINS

Alexander Krajhanzl, Agata Danisova and Jan Kocourek

Department of Biochemistry, Faculty of Natural Sciences, Charles University, Albertov 2030, CS-128 40 Praha 2,

Petr Pancoska

Department of Chemical Physics, Faculty of Mathematics and Physics, Charles University, CS-121 16 Praha 2, Czechoslovakia

Fish oocyte lectins are a relatively homogeneous class of related proteins with remarkable serological and saccharide specitificites (1). They occur in the egg cortical vesicles (granules) of several fish species (2). In the gonads of the female roach, agglutinin activities for human B and P_1 red blood cells and for some bacteria was found by Köhler and coworkers (3). On the basis of the occurence of the roach lectin in the cortical vesicles (2) we assume that it participates in the cortical reaction; however, a protective function (3) appears also very likely.

The roach lectin has been isolated and partially characterized (4). This report is concerned with further physicochemical characterization of the lectin which has been isolated from the ova of the roach (family *Cyprinidae*) in two basic stages of oocyte maturity. In addition, isolation and partial characterization of the 3 main subunits of the roach lectin are presented.

Lectins, Vol. IV

Materials and Methods

Isolation of the roach lectin and separation of the lectin polypeptide chains. Female roach of about equal size were obtained from ponds owned by the Technical University of Agriculture, Praha. Roach ovaries were stored at -20°C. Roach lectin preparations were isolated by affinity chromatography on O-α-D-galactopyranosyl polyacrylamide gel (1,4). Major polypeptide chains of the roach lectin were separated by gel filtration on a Sephadex G-100 Superfine column (1.2 cm x 115 cm) in the presence of 1% 2-mercaptoethanol, 0.05 M acetate pH 5.5, 0.25 M NaCl (buffer A).

Analytical methods, affinity electrophoresis and CD study: Methods used for physicochemical characterization of the roach lectin preparations were performed as given in our preceding report (1). Some electrophoregramms were evaluated by means of Ultroscan Laser Densitometer equiped with a Recording Intergrator Typ 2220 (LKB).

Results

Isolation of the roach lectins. By affinity chromatography on O-α-D-galactosyl polyacrylamide gel, 479 mg of lectin preparation was obtained from 1000 g of raw frozen ovary containing previtellogenic or early vitellogenic oocytes. The field of the lectin was significantly lower (275 mg from 1000 g of ovary) when mid and late vitellogenic oocytes or spawned ros were used as starting material.

Subunit composition of the lectin. The roach lectin was found to be composed of three major polypeptide chains which could be separated by gel filtration in the presence of 2-mercaptoethanol. From the lectin, irrespective of the stage of oocyte maturity, four fractions were obtained; three of them contained one of the main contains short polypeptide chains. Gel filtration in a nonreducing solution (pH 5.5) gave two fractions differing only in part.

Electrophoresis in the presence of SDS revealed that the main subunit fractions also contained several minor polypeptide chains (designated D,E,F, and C). Theindividual subunits have the following average molecular weights: A 18 900, B 23 900, C 28 400, D 25 400, £ 27 00, F 30 000 and finally G 36 500, presumably a dimer of the A-chain. When reduced samples were submitted to electrophoresis in the presence of SDS, the M_r values were found to be somewhat higher: A 20 600, B 24 300, C 35 700, D and, besides, several minor zones with M_r ranging from 13 300 to 23 200 could be discerned. Fractions obtained by gel filtration of the reduced roach lectin differed also by their acidic electrophoresis patterns. Under conditions of alkaline electrophoresis, the roach lectin precipitated in the running buffer (Tris-HCl, pH 8.3).

Physicochemical properties. The roach lectin represented a mixture of several polypeptide chains (A-G) which in a mably as covalently linked oligomers displaying sedimentation coefficients $s_{20,w}$ of 9.20 and 2.84 S. These values corresponded

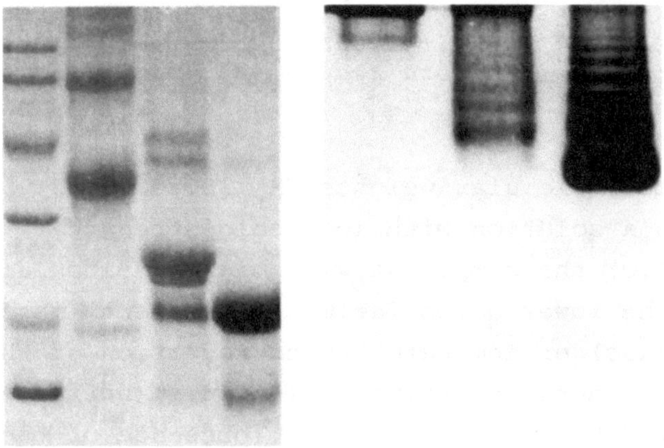

Fig. 1. Electrophoretic properties of gel-filtration fractions Left: Discontinuous electrophoresis of reduced samples in the presence of SDS. Right: Discontinuous electrophoresis in the acidic buffer system.

Left lane: Standard proteins: lactalbumin (14 400), soybean trypsin inhibitor (20 100), carbonic anhydrase (30 000), ovalbumin (43 000), bovine serum albumin (67 000) and phosphorylase b (94 000).

Table 1.

Amino acid composition of the roach lectin and of fractions obtained by gel filtration of the reduced lectin

Amino acid	Rut-P	Rut-V	A	B	C
Lysine	6.36	6.70	7.45	6.77	6.22
Histidine	2.51	trace	3.53	trace	trace
Arginine	3.79	3.66	3.77	3.65	4.30
Aspartic acid	8.85	9.69	9.92	9.80	9.04
Threonine	6.69	6.02	5.50	5.19	8.34
Serine	9.19	8.44	8.31	7.65	9.67
Glutamic acid	9.97	11.24	9.30	14.80	7.15
Proline	5.96	4.20	8.89	8.58	5.48
Glycine	6.35	6.54	6.65	11.95	8.98
Alanine	6.45	5.94	5.99	6.21	7.43
Half cystine	9.35	14.56	8.36	n.d.	9.63
Valine	7.35	6.43	7.70	7.87	7.63
Methionine	1.52	1.75	2.28	n.d.	0.93
Isoleucine	3.38	3.39	2.85	5.08	3.82
Leucine	5.49	4.98	4.96	7.07	4.74
Tyrosine	3.22	3.39	3.49	2.27	3.32
Phenylalanine	2.64	2.46	2.05	3.11	3.32
Tryptophan	0.91	0.60	n.d.	n.d.	n.d.

Expressed in molar per cent of the total amino acid residues. Rut-P, Rut-V = the roach lectin preparations isolated from previtellogenic and late vitellogenic oocytes, respectively. A,B, C = subunit fractions obtained by gel filtration.

to calculated molecular weights ${}^{s}M_r$ 180,000 and 29 000, respectively. In a solution with low ionic strength (1 mM phosphate buffer pH 6.0) the component with higher molecular mass prevailed while the lower molecular mass component (monomers of polypeptide chains) predominated after reduction of the roach lectin by a 5% 2-mercaptoethanol solution in buffer A ($s_{20,w}$ of 2.32 and 8.23 S). The predominant presence of the low molecular weight component in the reducing solution was presumably due to cleavage of the inter-chain disulfide bridges. However, our results indicated that it could be caused by reduction of the intra-chain disulfide linkages in subunits which were probably partially digested by proteolytic enzymes in course of the isolation proceudre. The intra-chain cleavage would lead

Table 2.

Contents of neutral sugar, sialic acid and some metals in roach lectin preparations w/w %.

Component	Rut-P		Rut-V	
	per cent	1 mol/M_r	per cent	1 mol/M_r
Neutral sugar	2.2	8 200	3.8	4 700
Sialic acid	0	-	0.4	58 000
Calcium	0.152	27 000	0.163	25 000
Magnesium	0.026	94 000	0.035	70 000
Zinc	0.076	90 000	0.066	99 000

Rut-P and Rut-V, see Table 1. For calculation of protein molecular weight corresponding to one grammolecula of neutral sugar and sialic acid, D-glucose and N-acetyl neuraminic acid were considered.

to breakdown of the original polypeptide chains into short peptides as observed both in electrophoresis of reduced lectin preparations in the presence of SDS and during gel filtration. In consideration of the changes induced by the ionic strength in the quantitative representation of both components, it could not be excluded that some of the main subunits formed noncovalent aggregates

Amino acid composition, contents of neutral sugar, sialic acid and of some metals in the roach lectin preparations are given in Tables 1 and 2. In summary, it can be concluded that the roach lectin is a Ca-containing glycoprotein rich in glutamic acid, serine and cysteine. Its low content of aromatic amino acids is consistent with the relatively low value of the decadic absorption coefficient: $E_{278}^{1\%} = 5.00$. Out of comparison of the lectin preparation obtained from previtellogenic (designated as Rut-P) and late vitellogenic (Rut-V) oocytes it is obvious that they differ significantly in neutral sugar and sialic acid contents and partially in their amino acid composition.

Much more pronounced are differences in amino acid composition of the subunit fractions obtained by gel filtration of the reduced lectin, but their neutral sugar content is very similar. Out of the fractions obtained by gel filtration, the calcium content was estimated merely for fraction B to be (0.279%), i.e. to a M_r of 14 000 corresponds one Ca atom. The contents of Mg and Zn were negligible in all roach lectin preparations and fractions.

By the dansylation technique three major N-terminal amino acids, lysine, threonine and glutamic acid, were determined for the A, B and C chain. In addition, several minor N-terminal amino acids (glycine, arginine and serine) were also detected, especially in the lectin preparations obtained from late vitellogenic oocytes. Presuably these amino acids terminate the minor polypeptide chains or belong to chains from proteolytic cleavage of the original roach lectin subunits.

Chiroptical properties. CD spectra of the roach lectin prepared from previtellogenic oocytes, were drawn at different temperature (Fig. 2). The far ultraviolet region of the CD spectrum of this lectin exhibits a negative maximum centered at 216 nm which belongs to $n-\Pi^+$ electron transitions. Magnitude and position of this dichroic band suggest that the polypeptide chains of the roach lectin are predominantly in the pleated sheet conformation. Favoured content of β-conformation is strongly supported by 2 positive ellipticity values in the region below 205 nm (band of $\Pi-\Pi^+$ transitions). The amount of helical conformation, as calculated approximately from residual ellipticity at 222 nm, represents about 10%. The near ultraviolet region of the roach lectin CD spectrum corresponds to superposition of dichroic bands of aromatic amino acids, particularly of tyrosine (bands centered at 282 and 287 nm) and phenylalanine (a negative band at 260 nm), whic presumably possess asymmetric environment in the ordered roach lectin polypeptide chains. However, contribution of disulfide chromophor seems

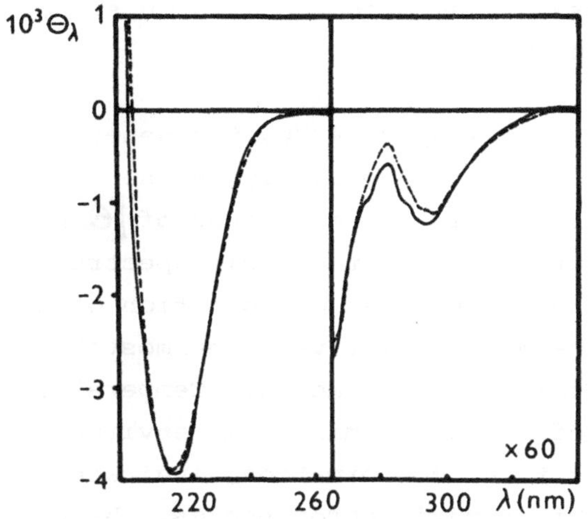

Fig. 2. Circular dichroism spectrum of the roach lectin (Rut-P preparation) at pH 7.0 (0.1 M phosphate buffer) and at 25°C (———) and 47°C (— — —).

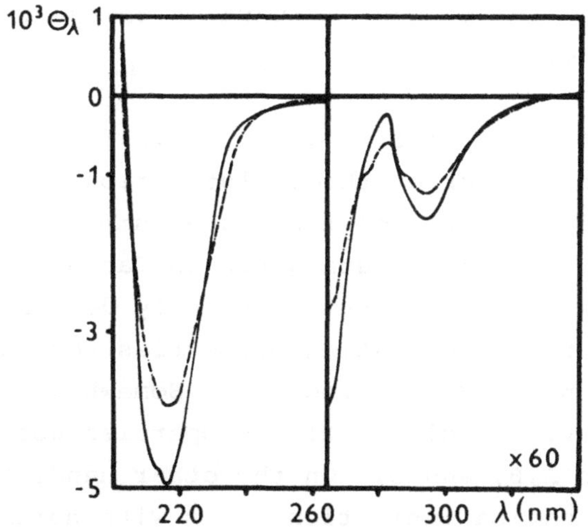

Fig. 3. Circular dichroism spectrum of the mixture of C-, E- and F-chains at 25°C and at pH 7.0 (0.1 M phosphate buffer) (———). For comparison the CD spectrum of the roach lectin, drawn at the same conditions, is given in an equal scale (— — —).

also likely because of the presence of a CD signal at a wave-lenght higher than 300 nm.

At elevated temperatures up to 47°C, the general feature of the spectrum is conserved, especially in the far UV region. Appreciable differences in the magnitude of $[\Theta_\lambda]$ are noticed only in the near ultraviolet part of this spectrum. Hence, it can be concluded that the total conformation of the roach lectin polypeptide chains is relatively thermostable, except for environments of aromatic amino acids. Temperature induced changes in asymmetry of aromatic amino acid environments, together with the conserved total conformation, indicate an easy sterical accesibility of aromatic chromophores. We did not go further into the analysis of the obtained spectra.

In Fig. 3 the CD spectra of the roach lectin and of the mixture of C-, E- and F-chains are compared. In addition to the shoulder in the region under 215 nm and a different magnitude of the ellipticity in the near UV part, both spectra possess very similar features.

Erythroagglutinating properties and interaction with some saccharides. Results of our erythroagglutination study and of erythroagglutination-inhibition tests are summarized in Tables 3 and 4. Comparison of the data given in Table 3 for both lectins preparations (Rut-P and Rut-V) clearly demonstrates that they possess erythroagglutinating properties reproducibly different, except for human B-erythrocytes. Somewhat more pronounced differences in erythroagglutination properties were revealed in subunit fractions A,B, and C. On the other hand, the results of erythroagglutination inhibition tests with human red blood cells revealed approximately the same affinity of both lectin preparations (Rut-P and Rut-V) to the individual carbodhyrates, whereas differences in affinity of fractions A and C were obvious. Enormously high affinity to methyl α-L-rhamnopyranoside is particularly shown by the mixture of subunits C,E and F. In

Table 3.

Erythroagglutination activity of the roach lectin preparations and of subunit fractions A,B, and C from gelfiltration.

Blood group	Rut-P	Rut-V	I C	II B	III A
A_1	-	60	-	-	120
A_2	-	60	-	-	60
B	16	8	4	120	120

The activity was estimated in test-tubes according to Tobiska (1) and is expressed in µg per ml. Erythrocytes were not agglutinated at a lectin concentration of 500 µg/ml.

Table 4.

Lectin-inhibition activity of some monosaccharides.

Sugar	Rut-P	Rut-V	A	B
Me-α-L-Rha	0.06	0.13	0.1	0.0003
Me-β-D-Gal	0.30	0.60	2.5	0.02
Me-α-D-Gal	-	-	-	0.30
L-Ara	0.80	1.50	3.0	0.20
D-Fuc	n.d.	3.0	n.d.	n.d.

The inhibitory concentration is expressed in mM. — = agglutination of B-erythrocytes was not inhibited even at 5 mM sugar concentration; Me-α-L-rhamnopyranoside, Me-β-D-Gal = methyl β-D-galactopyranoside, Me-α-D-Gal = methyl α-D-galactopyranoside, L-Ara = L-arabinose, D-Fuc = D-fucose; n.d. = not determined.

summary, it can be concluded that the roach lectin binds with high affinity to α-L-rhamnopyranose and less intensively to β-D-galactopyranose residues and other C-2 and C-4 sugar analogoues of L-rhamnose. Interaction of the roach lectin (Rut-V preparation) with free L-rhmanose and D-galactose was studied by means of affinity electrophoresis. The dependence of the mobility of individual zones on the free sugar concentration gave two association constants for the subunitsugar complexes. Thus to the complex of the A subunit with L-rhamnose and D-galactose the following association constants correspond ${}^{A}K_a^{Rha}$ = $2.8 \times 10^3 M^{-1}$ and ${}^{A}K_a^{Gal} = 2.6 \times 10^2 M^{-1}$, respectively. Similarly, the association constants of the B subunit were found to be ${}^{B}K_a^{Rha} = 7.5 \times 10^3 M^{-1}$ and ${}^{B}K_a^{Gal} = 1.8 \times 10^2 M^{-1}$. In consideration to the very low mobility of the band corresponding to the mixture of chains C, E and F association constants of these chains with free sugars could not be estimated.

Discussion

This paper is a continuation of one of our early reports on the isolation and characterization of fish oocyte lectins (4). One of its prinical aims was the completion of physicochemical characterization of the lectin of roach oocytes which has many similar features with lectins isolated from oocytes of other fishes of the *Cyprinidae* family (1,4). The results hitherto obtained have shown that this lectin with binding specificity for L-rhamnose is composed of three polypeptide chains with M_r of 19 000, 24 000 and 28 500 predominantly in β-conformation. These polypeptide chains (subunits) display some different physicochemical and erythroagglutinating properties. It appears that the presence of additional minor chains (D-F) and the considerable heterogeneity of the isolated preparation could be a result of some additional modifications of the native lectin subunits by proteinases in course of the isolation procedure. This view would be supported also by the fact that a partial degradation to short peptide chains takes place in a reducing medium. In comparison with the lectin described previously which was isolated from previtellogenic oocytes (4), some minor differences were found in physicochemical properties. These differences are with high probability due to a partial modification of the original isolation procedure. Though it gave a higher yield of lectin, its subunit composition was somewhat different.

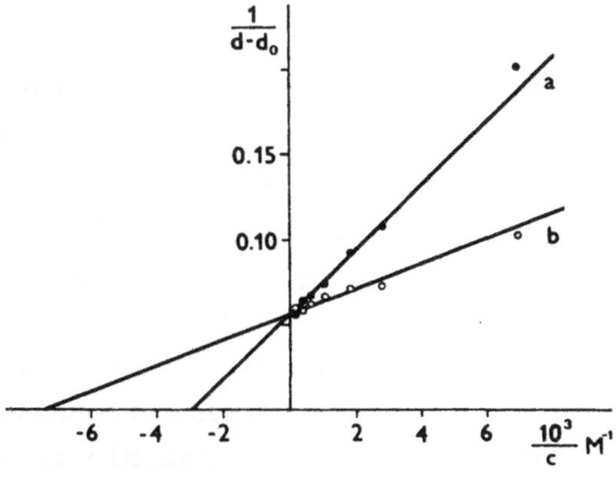

Fig. 4. Estimation of association constants of subunit A-L-rhamnose and subunit B-L-rhamnose complexes by affinity electrophoresis

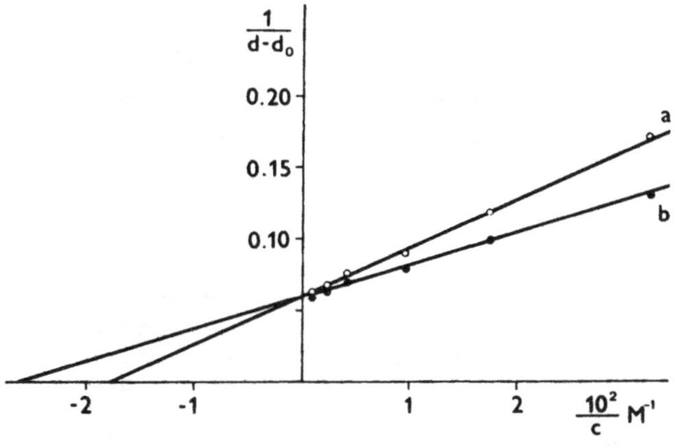

Fig. 5. Estimation of association constants of complexes between subunits A or B and D-galactose by affinity electrophoresis.

The second main goal of the present paper was a comparison of some properties of lectin preparations isolated by the same procedure from oocytes in different stages of maturity. On the basis of the properties found, it was possible to divide the six lectin preparations reliably into two groups, corresponding roughly to the previtellogenic and vitellogenic stages of the oocyte. The differences observed between lectin preparations from previtellogenic and vitellogenic oocytes, epsecially with regard to the yield, neutral sugar and sialic acids contents (and perhaps also to the contents of some amino acids) together with the partially altered erythroagglutinating properties are quite analogous with changes found in the case of the perch lectin I (1). In view of these changes it seems likely that the lectin isolated from vitellogenic oocytes represents either a complex with soluble carbohydrates slightly modulating the conformation of the lectin binding sites or it represents the original polypeptide chains modified post translationally by glycosylation during vitellogenesis. It is possible that free SH groups become available to during the interaction with some components of the oocyte content to stabilize the internal structure of the cortical vesicles (5).

The third aim of this paper was the estimation of the association constants of the lectin-saccharide complexes. The association constants thus obtained correspond roughly to the observed inhibitory activities of the tested saccharides.

Acknowledgement. We are indebted to Prof. Michel Monsigny and Prof. Jean-Claude Maurizot for their kind advices during a stay of A.K. in the Centre de Biophysique Moléculaire, Orléans. This work was in part (CD measurements) supported by a grant from Fondation de la Recherche Moléculaire 1981. Our thants are also due to Mrs. Mitterand and to Ing. P. Smolek for the amino acid and sedimentation analyses.

References

1. Krajhanzl, A., Pancoska, P., Nosek, J., Kocourek, J.: This proceedings (1985).
2. Nosek, J., Krajhanzl, A., Kocourek, J.: Histochemistry 79, 131-139 (1983).
3. Köhler, W., Prokop, O., Kühnemund, O.: Z. Immunitätsforsch. 134, 462-466 (1967).
4. Krajhanzl, A., Horejsi, V., Kocourek, J.: Biochim. Biophys. Acta 532, 215-224 (1978).
5. Guraya, S.S.: Int. Rev. Cytol. 78, 257-360 (1982).

MARINE INVERTEBRATE AGGLUTININS: THE LECTIN FROM *Megapitaria squalida* CLAM

Felipe Ascencio, Jose-Luis Ochoa

Department of Biotechnology
CIB - The Center for Biological Research
Box 128, La Paz 23000, B.C.S., Mexico

We have studied the presence of hemagglutinins in a number of marine invertebrates isolated from along the coastline of the Bay of La Paz in Baja California Sur, Mexico. Athough there is a discontinuity apparent for the presence of lectins in invertebrate with respect to individual animals and with respect to different stages during the life cycle (1-2, and Vasta et al. these proceedings). It has been possible in only one case to assess the presence of lectin activity in every sampling of *M. squalida* clams and for this reason the lectin from this species was chosen for further purification and characterization.

Lectins, Vol. IV

Materials and Methods

Extracts were prepared by cell disruption of desvicerated animals with a "polytron" apparatus (1g of tissue per ml of 0.9 NaCl). The hemagglutinating titers of centrifuged material (3000 x g, 5 min, 4°C) were estimated by the usual two-fold serial dilution procedure using 2% of trypsinized or neurominidized red cells in PBS. Hemagglutination-inhibition test was carried out with simple carbohydrate and glycoprotein solutions (either 0.1 M or 1.0% respectively). With the purpose of estimating the effect of divalent cations on the hemagglutination hemagglutination phenomenon induced by *M. esqualida* clam extracts an aliquot was extensively dialyzed against 0.9% NaCl with 0.1 M EDTA.

For affinity chromatography experiments, the matrix was activated as recommended by (7) and mucin was coupled with a 90% efficiency. A column (5.5 x 3.5 cm) packed with this particular

Table 1

Marine invertebrate organisms studied to detect the presence of hmeagglutinin:

Phylum	Family	Genera	Species
Platyheliminthes	Polycladida	Pseudoceros	canadiensis
			mexicanus
Mollusca	Pelecypoda	Chione	amatusia
		Chione	undatiela
		Megapitaria	aurantica
		Megapitaria	squalida
		Modoiolus	capax
		Ostrea	palmula
		Pinna	rugosa
		Spondylus	calcifer
		Tivela	stultorum
	Gastropoda	Nerita	scabricosta
		Tridochiella	diomedea
		Hypselodores	californiensis
Arthropoda	Brachyura	Aratus	pisonii
	Isopoda	Ligia	fabricius

adsorbent was equilibrated with PBS at a flow rate of 35 ml/h. The active material was eluted by distilled water for being more economical than N-acetyl-D-galactosamine solutions in fractions of 3 ml/tube. For the ion-exchange chromatography a column (10.2 x 2.0 cm) was packed with DEAE-Sephadex A 50 and equilibrated with 0.02 M imidazole pH 7.4 buffer at a flow rate of 20 ml/h. Fractions of 1 ml/tube were collected by applying a reverse flow linear gradient of salt as shown in Fig. 1.

Gel filtration chromatography was carried out with a Sephacryl S 200 column (0.65 x 50 cm) equilibrated with PBS at a flow rate of 3 ml/h and collecting fractions every 0.15 ml. Chromatofocusing on a PBE 94 column (30 x 0.55 cm) was done in the presence of 0.25 M Imidazole pH 7.4 at a flow rate of 35 ml/h and fractions of 2 ml/tube.

The specific activity (hemagglutinating units/mg of protein) was found by determining the protein concentration by the Lowry Test (8). Its carbohydrate content was found by the anthrone method (9).

Results

Table 1 shows the kind of marine invertebrates that were studied. They were collected in several single-day sampling routines and therefore it is not an exhaustive or complete list of all the invertebrate organisms which are found in the Bay of La Paz throughout the year.

Table 2 lists the species which showed hemagglutinating activity towards human blood cells at least in one sampling. The only truly persistent species with repsect to hemagglutinating activity was *M. squalida*.

M. squalida agglutinating activity was further characterized by the hemagglutination inhibition test with simple carbohydrate and commonly used glycoprotein solutions (3). The test showed that *M. squalida* hemagglutinin binds specifically to N-acetyl-galactosamine-containing glycoconjugates as fetuin, mucin and N-acetyl galactosamine in concentrations of 1% or 0.1 M respectively. These concentrations were fully capable of inhibiting units of a crude extract cf. Table 2. Thus the *M. squalida* lectin was found to be specific against trypsin or neuraminidase treated human blood group A and to interact with the "A" determinant.

The purification of *M. squalida* hemagglutinating material was accomplished by two combined methods: affinity chromatography on mucin-Sepharose and ion-exchange chromatography on DEAE-Sephadex A-50. The first adsorbent showed an adsorbent showed an adsorption capacity of 21,000 hemagglutinating units of *M. squalida* crude extract with 83% yield recovery. The specific activity after the purification step on DEAE-Sephadex was 48,500 HA units/mg of protein.

Anthrone analysis showed that the agglutinin is a glycoprotein with about 18% carbohydrate. The individual carbohydrate composition was not determined. By gel filtration a protein band with hemagglutinating activity was detected in the molceular size region of 60.3 KD and by chromatofocusing the agglutinin eluted at a pI of 4.0 pH. Other data appear from Table 3.

Discussion

M. squalida hemagglutinin differs in many respects in comparison to the *Saxidomus giganteus* butter clam lectin reported by Johnson (5) but it resembles those from the Sea Nettle Chrysaora quinquercirrha Nematocyst venom in that it requires the red cells to be treated with trypsin for agglutination (6). A complete characterization of *M. squalida* clam lectin remains to be achieved. The question concerning the use of *M. squalida* clams as a source of lectins is worth mentioning. Table 3 shows the bromatological analysis of deviscerated animals. If we compare these data with the amount of the isolated material obtained after the purification process, we find that 100 μg of pure lectin can be isolated per each gram of the invertebrate organism (on a humid basis). Considering an adult orgnism (5.7 x 7.0 x 3.6 cm) weighing about 10 g without viscera, it is

Table 2

Hemagglutinating activity of marine invertebrate extracts

Species	Human erythrocytes								
	Native			Neuraminidase			Trypsinized		
	A	B	O	A	B	O	A	B	O
P. canadiensis	4	0	8	4	4	255	0	0	256
M. aurantica	4	0	0	16	0	0	256	0	0
M. squalida	0	0	0	2	0	0	32	0	0
T. diomedea	4	0	8	2	2	4	2	2	8
A. pisonii	0	0	0	4	2	4	0	0	0

Table 3

Bromatological analysis of Megapitaria squalida

Humidity	Protein	Fat	Carbohydrate	Ash	Kcal/g
85±0.1	10±0.2	1.3±0.05	1.31±0.01	1.74±0.1	0.688±0.01

The value are given in %. Analysis carried out by micro-Kjeldahl techniques veraged 7 specimens. The caloric value was estimated with an automated adiabatic calorimeter.

thus sufficient to obtain 1 mg of a highly active material as shown here (48,500 HA).

References

1. Yeaton, R.W.: Develop. Comp. Immunol., 5, 391-402 (1981).
2. Michelson, E.H., Dubois, L.: Biol. Bull., 153, 219-227 (1977).
3. Ascencio, F., Ochoa, J.-L.: Calcofi Annual Meeting, Idyll-wild, Calif., Abstract Book (1982).
4. Marchalonis, J.J. and Edelman, G.M.: J. Mol. Biol., 32, 453-465 (1968).
5. Johnson, H.M.: Science, 146, 548-549 (1964).
6. Cobbs, C.S., Gold, P., Calton, J., Burnett, J.W.: Comp. Bio-chem. Physiol., 74C (1), 225-228 (1983).
7. Porath, J., Axen, R. and Ernback, S.: Nature 215, 1491-2 (1967).
8. Lowry, O.H. et al.: Protein Measurement with the Folin Phe-nol Reagent, J. of Biol. Chem., 193, 265-275 (1951).
9. Kabat, E.A., Mayer, M.M.: Immunoquimica Experimental, La Prensa Medica Mexicana, Mexico, 498-500 (1968).

PART V

MICROBIAL LECTINS

SIALIC ACID SPECIFIC LECTINS OF ENTEROTOXIGENIC *ESCHERICHIA COLI*

Torkel Wadström[1], Per-Olof Sjöberg[2] and Mats Lindahl[1]
[1]Department of Veterinary Microbiology, Box 583 and
[2]Institute of Biochemistry, Box 576, Biomedicum, S-751 23 Uppsala

In the last few years several glycoconjugates on mammalian cells have been defined as receptors for bacterial surface proteins (fimbriae). The fimbriae are hairlike homopolymers with carbohydrate binding specificity (1,20). Previously, antigenically different fimbriae of enterotoxigenic *E. coli* such as K99, CFA/I and CFA/II, have been reported to cause haemagglutination of human and bovine erythrocytes. K99, CFA/I and CFA/II exposing bacteria fail to haemagglutinate sialidase treated erythrocytes. These lectin-like fimbriae enable the bacterium to colonize the small intestine of animals (K99) and man (CFA/I and CFA/II) (3,4,15). Other sialic acid specific lectins have been discovered on uropathogenic *E. coli* (9), *Streptococcus sanguis* (14) and *Pseudomonas aeruginosa* (19). Although having sialic acid specificity the lectins on enterotoxigenic *E. coli* strains cause haemagglutination with erythrocytes from different species and have been shown to have minor differences in haemagglutination profile (6).

Lectins, Vol. IV

Materials and Methods

Chemicals. Sialyllactose (NeuAc-Lac) and sialic acid monomers; N-glycolylneuraminic acid (NeuGc) and N-acetylneuraminic acid (NeuAc) were purchased from Sigma Chem. Co., St. Louis, MO, USA. Transferrin was a gift from Kabi, Stockholm, Sweden.

Bacterial strains and growth conditions. Well characterized strains (5,13) were cultured on CFA agar or Minca agar medium as previously described.

Haemagglutination test. Haemagglutination was performed on glass depression slides (5) or in a Payton 300A aggregometer (Payton Assoc. Inc., Buffalo, NY, USA) (10). Monosaccharides and glycoconjugates were mixed with suspensions of human erythrocytes ($5 \cdot 10^6$ cells ml^{-1}) in (1:1) 0.01 M PBS (pH 6.8) and 0.3 M sucrose containing 0.1 M α-methylmannoside, in a cell aggregometer at 800 rpm and 20°C. Bacterial cells ($4 \cdot 10^7$ fimbriated cells ml^{-1}) were added and the increase in transmitted light (ΔT min^{-1}) at 609 nm was measured. An increase in transmittance is a measure of the decreasing number of particles (erythrocytes or clumps of erythrocytes) scattering light, as the haemagglutination reaction is completed. The increase in transmittance is constant for a few minutes after a lag-phase of 1-2 min. The increase in transmittance per minute, during the constant phase, in the presence of differing concentrations of potential inhibitors, was plotted against the log concentration of inhibitor, giving linear plots for each inhibitor (Fig. 1). The concentration of sialic acid giving half the value for increase in transmittance per minute of the reaction without inhibitor was regarded as the concentration giving 50 % inhibition.

Results

Haemagglutination profile. The haemagglutination profile of the different lectins is shown in Table 1. Interestingly the K99, CFA/I and CFA/II-exposing organisms show quite different haemagglutination profiles. Moreover two antigenically different CFA/II-exposing strains only agglutinate preferentially bovine erythrocytes of specific individual animals (6).

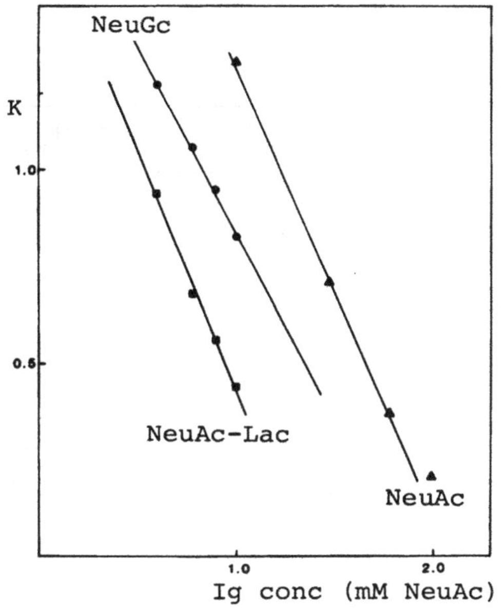

Fig. 1. Haemagglutination profile of strain C91f in the presence of sialyllactose (NeuAc-Lac), N-glycolylneuraminic acid (Neu Gc) and N-acetylneuraminic acid (NeuAc) at different concentrations.

Table 1a

Haemagglutination profile of enterotoxigenic E. coli strains.

Fimbriae	Strain	Erythrocytes			
		Bovine	Human	Sheep	Horse
CFA/II	(C91f)	+++	-	-	-
CFA/II	(B303d)	+++	-	-	-
CFA/I	(1628-15)	+	+++	-	-
K99	(B117)	++	+++	+	+++

The following signs, i.e. +, ++ and +++, indicate low, intermediate and strong haemagglutination reactions, respectively, where negative sign (-) indicates absence of haemagglutination reaction.

Table 1b

Haemagglutination reaction of enterotoxigenic E. coli strains with erythrocytes from different cows.

Cow No.	Fimbriae, strain	
	CFA/II (C91f)	CFA/II (B303d)
93	+	-
94	++	+++
95	+	-
96	++	+
97	+	-
98	-	-
99	-	+++

The following signs, i.e. +, ++ and +++, indicate low, intermediate and strong haemagglutination reactions, respectively, where negative sign (-) indicates absence of haemagglutination reaction.

Table 2.

Concentration of sialic acid (mM), giving 50% haemagglutination inhibition[a].

Fimbriae	Strain	NeuAc	NeuGc	NeuLac	Transferrin
CFA/II	(C91f)	32.9	17.3	6.2	1.5×10^{-2}
CFA/II	(B303d)	22.7	18.0	2.6	7.1×10^{-2}
CFA/I	(1628-15)	38.2	26.3	12.9	18.0×10^{-2}
K99	(B117)	11.8	6.8	3.5	ND[b]

[a] *see Materials and Methods.*
[b] *ND = not determined.*

Inhibition with glycoconjugates. N-acetyl galactosamine, N-acetyl glucosamine and fucose were not inhibitory at 0.1 M concentration.

Transferrin was the most potent inhibitor for all strains, with a concentration giving 50 % inhibition, a hundred times lower than the other inhibitors.

NeuGc was only a slightly stronger inhibitor than NeuAc for strain B303d while this difference was greater for all other strains (Table 2).

Discussion

Some plant lectins such as wheat germ agglutinin (WGA) recognize terminal sialic acid residues (12,18) and bind to cell membrane gangliosides and certain sialic acid-rich glycoproteins such as fetuin and bovine submaxillary mucin but usually not to sialidase-treated (asialo) glycoconjugates. However a lectin from the horse shoe crab, called limulin, is the only extensively studied lectin up to now which shows a high specificity for sialic acid glycoconjugates. It binds mainly to N-glycolyl neuraminic acid containing sialoglycoconjugates such as certain gangliosides. Furthermore one hydroxyl group (group 4) should be free to react with limulin and NeuGc4Ac-GM_3 does not react with limulin (11). Moreover, free NeuAc and NeuGc inhibit haemagglutination only at high concentrations (200 mM). Preliminary studies show that another sialic acid specific lectin (from blue mussels), edulin (8), is also inhibited with sialic acid monomers (20 mM).

Furthermore the N-acetyl groups are necessary for recognition by limulin but N-acetyl hexosamines were not inhibitors in contrast to a limulin-like lectin isolated from Japanese horse show crab.

We have previously shown that sialic acid-rich glycoconjugates such as GM_2 gangliosides and bovine submaxillary mucin are potent inhibitors of K99 and CFA/I lectins (5) as in the case for limulin.

Since treatment of erythrocytes with sialidase from C. perfringens reduces haemagglutination with K99, CFA/I and two different CFA/II strains (5,6) it is obvious that they all belong to a group of sialic acid specific microbial lectins. Most interestingly, certain uropathogenic E. coli strains cause sialidase sensitive haemagglutination and possess a NeuAc-α-2-3 Gal specific lectin (17).

Titration of different sialoglycoconjugates in the aggregometer haemagglutination inhibition assay, which is used to define effective inhibitor concentration of oligosaccharide which reduces haemagglutination by 50%, shows that NeuAc-Lac is a potent inhibitor for K99, CFA/I and CFA/II lectins (Table 2). Similar aggregometer assay studies on D-mannose specific lectins of E. coli, Klebsiella pneumoniae and Salmonella typhi murium (7) show inhibition with oligomannosides at similar millimolar concentrations. Moreover, these studies also showed that trisaccharides were up to 30 times more inhibitory than α-methyl-D-mannopyranoside. The minimal sugar recognizing sequence of uropathogenic E. coli with p-fimbriae lectin is Gal-α-1-4 Gal as a terminal or internal part of the saccharide chain (9) and recently the Shigella cytolytic toxin was shown to recognize the same sequence but prefers it in terminal position (2).

Further studies on the sugar recognition requirements for E. coli sialic acid specific lectins are required but it is quite obvious already today that these lectins may become important lectins with a number of applications in such as: a) purification of sialoglycoconjugates from various body fluids, b) probes for identifying specific cell membrane glycoconjugates and c) histopathology studies. Studies on maturation of villi and crypt cells in the gastrointestinal tract and work on identification of unique epithelial cells such as the M cell were recently performed with conventional commercially available plant lectins (13).

Other studies on the gut mucus barrier has used such lectins to identify different mucus composition in different parts of the intestinal tract. It is tempting to speculate that purified lectins from different mucosa-associated pathogens such as enterotoxigenic E. coli and S-fimbriated uropathogenic E. coli will become more specific tools for similar studies in the future. The possibilities to produce and purify these lectins on a large scale for such purposes make them attractive as generally available tools for such studies. Differences in terminal sialic acid compounds on transformed cells suggest that these lectins may become important tools also in diagnostics of certain tumors.

Acknowledgements. This study was supported by a grant from the Swedish Medical Research Council (16X-04723). The authors are indebted to Professor Jerker Porath and Dr. Bo Ersson for stimulating discussions and support.

References

1. Brinton, C.C., Trans. N.Y. Acad. Sci., 27, 1003 (1965).
2. Brown, J.E., Karlsson, K.-A., Lindberg, A., Strömberg, N. and Thurin, J., In: Chester, M.A., Heinegård, D., Lundblad, A. and Svensson, S., Eds., Proc. 7th Int. Symp. Glycoconjugates, Lund, 1983, p. 678.
3. Evans, D.G., Evans, D.J., Jr., Tjoa, W.S. and Du Pont, H. L., Infect. Immun., 19, 727-763 (1978).

4. Evans, D.G. and Evans, D.J., Jr., Infect. Immun., 21, 638-647 (1978).

5. Faris, A., Lindahl, M. and Wadström, T., FEMS Microbiol. Lett., 7, 265-269 (1980).

6. Faris, A., Sellei, J., Lindahl, M. and Wadström, T., Zbl. Bakt. Hyg. I. Abt. Orig. A 253, 175-182 (1982).

7. Firon, N., Ofek, I. and Sharon, N., Carb. Res., 120, 235-249 (1983).

8. Hardy, S.W., Fletcher, T.C. and Gerrie, L.M., Biochem. Soc. Trans., 4, 473-475 (1976).

9. Leffler, H. and Svanborg Edén, C., FEMS Microbiol. Lett., 8, 127-134 (1980).

10. Lindahl, M. and Wadström, T., Vet. Microbiol., 9, 249-257 (1984).

11. Maget-Dana, R., Roche, A.-C. and Monsigny, M., In: Progress in Clinical and Biological Research, Vol. 29 (Cohen, E., Ed.), Alan R. Liss, Inc., N.Y., 1979, pp. 567-578.

12. Monsigny, M., Roche, A.-C., Sene, C., Maget-Dana, R. and Delmotte, F., Eur. J. Biochem., 104, 147-153 (1980).

13. Morris, J.A., Thorns, C.J. and Sojka, W.J., J. Gen. Microbiol., 118, 107-113 (1980).

14. Murray, P.A., Levine, M.J., Tabak, L.A. and Reddy, M.S., Biochem. Biophys. Res. Commun., 106, 390-396 (1982).

15. Ørskov, I., Ørskov, F., Smith, H.W. and Sojka, W.J., Pathol. Microbiol. Scand. Sect. B, 83, 31-26 (1975).

16. Owen, R.L. and Bhalla, D.K., Am. J. Anat., 168, 199-212 (1983).

17. Parkkinen, J., Finne, J., Achtman, M., Väisänen, V. and Korhonen, T.K., Biochem. Biophys. Res. Commun., 111, 456-461 (1983).

18. Peters, B.P., Ebisu, S., Goldstein, I.J. and Flasher, M., Biochem., 18, 5505-5511 (1979).

19. Rampal, R. and Pyle, M., Infect. Immun., 41, 339-344 (1983).

20. Wadström, T. and Trust, T.T.J., Med. Microbiol., Vol. 4, 287-334 (1984).

K99 - A SIALIC ACID SPECIFIC HAEMAGGLUTININ OF ENTEROTOXIGENIC E.COLI BINDS TO GLYCOPHORIN FROM HUMAN ERYTHROCYTES.

Mats Lindahl[1], Reinhard Brossmer[2] and Torkel Wadström[1].

1) Department of Veterinary Microbiology, Biomedicum, Box 583, S-751 23 Uppsala, Sweden.

2) Institut für Biochemie II, Universität Heidelberg, Im Neuenheimer Feld 328, 6900 Heidelberg 1, Fed.Rep.Germany.

The discovery of mannose specific fimbriae in the mid 1950s (5) opened a new research field in bacterial pathogenesis. Fimbriae are hair-like protein polymers which have carbohydrate binding properties, thus they are lectins.

A considerable number of new fimbriae with different specificities have been found. In contrast to plant lectins, sialic acid specificity seems to be important for many fimbriae. Sialic acid specificity has been found for three *E.coli* fimbriae, (15,16,21) fibrillae of *Streptococcus sanguis*, (19) fimbriae of *Neisseria gonorrhoeae* (3) and *Pseudomonas aeruginosa* (23). These bacteria colonize and cause infection of mucosal surfaces in man and animals. Their fimbrial surface lectins enables the colonization by specific interaction with terminal sialic acid residues, which are common in the mucins which cover the mucosal surfaces (1,8). It is not clearly understood what is the relative importance of fimbrial binding specificity in the colonization and infection, where one bacterium cause infection only in a certain tissue in a certain animal. The carbohydrate specificity may perhaps only be a part of the explanation. The O-acetylation of sialic acids may play an important role (24),

Lectins, Vol. IV

as well as the linkage of terminal sialic acid to its oligosaccharide (21), in the bacterial discrimination of tissues.

K99 is a lectin found on enterotoxigenic E.coli. Bacteria producing this lectin has been found to cause diarrhoea in young pigs, calves and lambs (20). K99 enables colonization of the small bowel mucosa. The lectin is a homopolymer of a subunit with a molecular weight of 18500 daltons and an isoelectric point of 9.5 (11). Bacteria exposing this lectin agglutinates erythrocytes from horses, pigs and humans (17).

The haemagglutination by K99 possessing bacterial cells has been shown to be inhibited by the GM_2 ganglioside, and sialidase treatment of erythrocytes was shown to partially inhibit haemagglutination (7). Furthermore, haemagglutination was found to be inhibited by sialoglycoproteins and N-acetylneuraminic acid (NeuAc) monomers (16). In order to better understand the protein-carbohydrate interaction and reveal the receptor structure for this lectin, we began this study using different sialic acid derivatives to inhibit haemagglutination.

Material and Methods

Bacterial cells. The bacterium, an enterotoxigenic E.coli strain (B117), kindly donated by Dr. J.A.Morris, Weybridge, Surrey, England, was cultivated on a minimal-saltmedium (Minca) at 37°C, over night, for optimal expression of the K99 lectin (13). This strain only produces K99 fimbriae besides type 1, (i.e.mannose-specific lectin).

Erythrocytes. Erythrocytes were collected in a heparinized syringe and washed 5 times in 10 volumes of phosphate buffer, pH 7,4, according to Sörensen, and finally suspended in 0,01M sodiumphosphate buffered saline, pH 6,8, to a concentration of $5x10^7$ cells/ml.

Fab fragments. Anti-N Fab fragments were prepared by digestion of rabbit IgG directed against human N-antigen with mercuripapain (22).

Haemagglutination inhibition test. Inhibitors were mixed with suspensions of human erythrocytes ($5x10^6$ cells/ml) in (1:1) 0,01M PBS, pH 6,8, and 0,3M sucrose containing 0,1M α-methyl-mannoside, in a cell aggregometer (Payton 300A, Payton Assoc. Inc., Buffalo, NY, USA) at 400 rpm and 20°C. Bacterial cells ($4x10^7$ fimbriated cells/ml) were added and the slope, i.e. the increase of transmitted light per minute at 609 nm, was measured. An increase in transmittance is a measure of decreasing number of particles (erythrocytes or clumps of erythrocytes) scattering light, as haemagglutination reaction is completed. The increase in transmittance is constant for a few minutes after a lagphase of 1-2 minutes. The slope during the constant phase, obatined after addition of inhibitors at different concentrations, was plotted against the log concentration of inhibitor (slope/logc), giving linear plots for each inhibitor. The concentration inhibitor reducing the increase in transmittance per minute with 50 % was defined as the concentration giving 50 % inhibition (2).

Chemicals. Human glycophorin was kindly donated by Dr.A.W. Bernheimer, New York University, School of Medicin, New York, USA. Rabbit anti-human N-antigen serum was purchased from DADE AG, Düdingen, Switzerland. The sialic acid derivatives and the sialooligosaccharides were prepared at the University of Heidelberg. All other chemicals were of analytical grade.

Results

Purified human glycophorin was found to be a potent inhibitor, in the aggregometer assay (see M&M) giving 50 % inhibition at a concentration at $1,4x10^{-4}$ mM. In another experiment Fab fragments directed against the N-antigen were used to block the binding sides on the erythrocytes. The Fab fragments gave inhibition at $1,6x10^{-3}$ mM.

NeuGc was found to be a stronger inhibitor than NeuAc (see Table 1). An increased inhibition was found when NeuAc was 4-O-acetylated or 2-α-benzylated. 7-O, 9-O acetylated and deacetylated (5 noracetyl) NeuAc were found to have a decreased inhibiting potency, since all of these derivatives showed no inhibition at all, at the concentration used. The 2β-methyl-NeuAc and the methylester were also less inhibitory. Structures of the sialic acid derivatives used are given in Table 2. Tri- and tetrasaccarides were found to be stronger inhibitors than NeuAc monomers.

Table 1

Haemagglutination inhibition with NeuAc-derivatives

Inhibitor	Concentration giving 50 % inhibition (mM)
NeuAc	11,8
NeuGc	6,8
4-acetyl-NeuAc	4,0
9-acetyl-NeuAc	>3,0[a]
4,7-acetyl-NeuAc	>2,5[a]
2- β-methyl-NeuAc	13,5
2- α-benzyl-NeuAc	0,4
2- α-benzyl-5-noracetyl-NeuAc	>6,0[a]
2- α-benzyl-NeuAcmethylester	6,9
2- α-benzyl-9-acetyl-NeuAc	3,2
NeuAcα2-3gal β1-4glc	3,5
NeuAcα2-6gal β1-4glc	3,1
NeuAcα2-8NeuAcα2-3gal β1-4glc	0,8

[a]No inhibition at this concentration, haemagglutination reaction appeared as the control.

Discussion

The finding that glycophorin and Fab fragments directed against this glycoprotein on the erythrocyte inhibit haemagglutination with the lectin strongly suggests glycophorin as the principal receptor for K99 on erythrocytes. Similar oligosaccharides has been found on glycophorin of human, porcine and equine origin (12,14,9) reflecting the haemagglutination pattern of K99 (17). However, the haemagglutination of equine erythrocytes is much stronger than haemagglutination of human erythrocytes which would be explained by the stronger affinity of K99 to N-glycolylneuraminic acid, since this sialic acid is much more abundant on equine than on human erythrocytes (9).

The affinity of K99 is enhanced when the sialic acid is linked to di- or trisaccharides, or when a benzyl group is 2α-linked to NeuAc. This enhancement is due to the strengthening of the protein-carbohydrate interaction confered by the additional groups. The influence of the penultimate sugar is due to the additional hydroxylgroups and in the case of α2-8-disialyl lactose additional carboxyl- and carbonylgroups, which will strengthen the binding. The benzyl group implies a hydrophobic patch near the binding site, it is possible that charge-transfer interactions can occur between the benzyl group and a phenylalanine or tyrosine in the protein (11).

The additional hydroxyl group on N-glycolylneuraminic acid enhances affinity, compared to NeuAc, probably by enabling formation of a new H-bond. The O-acetylation at C_4 also enhances affinity by confering a polar-charge interaction with the carbonyl group. The importance of the carboxyl group of NeuAc is shown by the decreased affinity for the methylester. This can be explained by the fact that K99 is a basic protein and is likely to expose a positively charged amino acid which can form an electrostatic bond with the carboxylgroup of sialic acid. Moreover, the lowered affinity of 2 β -methyl-NeuAc also emphasizes the influence of the carboxyl-group since this derivative has its carboxylgroup in another position than NeuAc. Thus the α2-linked NeuAc is a more favourable receptor for the lectin. The N-acetyl group of C_5 was shown to be of great importance since 2α-benzyl-5-noracetyl NeuAc did not give any inhibition at all. This implies the importance of the carbonylgroup at this position.

The pronounced decrease in affinity for NeuAc with O-acetyl substituents at C_7 and C_9 can be due either to steric hindrance by the acetyl groups or to the loss of the hydroxyl groups. The fact that the lectin has a higher affinity for neuraminic acid with a N-glycolyl group may be part of the explanation why K99 possessing enterotoxigenic E.coli cause diarrhoea in pigs, calves and lambs but not in humans, since it has been found that primates has a lower NeuGc/NeuAc ratio than other animals (24).

The increased affinity for sialic acid with an O-acetyl group at C_4, and decreased affinity when O-acetylation occurs at C_7 or C_9, might also play a part in the choice of host for this organism since different species (e.g. cow and horse) has been found to have differences in their O-acetylation of NeuAc (4). The prevalence of K99 producing E.coli to very young animals may be due to lower degree of O-acetylation which has been shown in fetal mucins (6).

In conclusion, this bacterial surface lectin cause haemagglutination by binding to glycophorin. It has stronger affinity for sialic acids with O-acetyl group at C_4 and N-glycolyl group at C_5 respectively, while O-acetylation at C_7 or C_9 decreases affinity markedly. This specificity may play a role in the choice of host and tissue, in colonization and infection. Furthermore, the penultimate sugar can increase the activity especially if it is a NeuAc residue.

Table 2

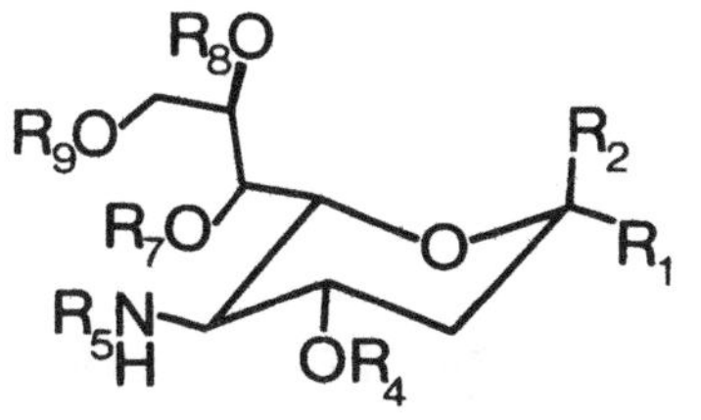

NeuAc	R_1=COOH, R_2=OH, R_4=H, R_5=$COCH_3$, R_7=R_8=R_9 =H
NeuGc	R_1=COOH, R_2=OH, R_4=H, R_5=$COCH_2OH$, R_7=R_8= R_9=H
4-acetyl-NeuAc	R_1=COOH, R_2=OH, R_4=R_5=$COCH_3$, R_7=R_8=R_9=H
9-acetyl-NeuAc	R_1=COOH, R_2=OH, R_4=H, R_5=$COCH_3$, R_7=R_8=H, R_9=$COCH_3$
4,7-acetyl-NeuAc	R_1=COOH, R_2=OH, R_4=R_5=R_7=$COCH_3$, R_8=R_9=H
2-β-methyl-NeuAc	R_1=OCH_3, R_2=COOH, R_4=H, R_5=$COCH_3$, R_7=R_8= R_9=H
2-α-benzyl-NeuAc	R_1=COOH, R_2=$OCH_2C_6H_5$, R_4=H, R_5=$COCH_3$, R_7 =R_8=R_9=H
2-α-benzyl-5-amino-NeuAc	R=$OCHC_6H_5$, R_2=COOH, R_4=R_5=R_7=R_8=R_9=H
2-α-benzyl-NeuAc-methylester	R_1=$OCH_2C_6H_5$, R_2=$COOCH_3$, R_4=H, R_5=$COCH_3$, R_7=R_8=R_9=H
2-α-benzyl-9-acetyl-NeuAc	R_1=$OCH_2C_6H_5$, R_2=$COOCH_3$, R_5=$COOCH_3$, R_7= R_8=H, R_9=$COCH_3$
NeuAcα2-3galβ1-4glc	R_1=-3gal 1-4glc, R_2=COOH, R_4=H, R_5=$COCH_3$, R_7=R_8=R_9=H
NeuAcα2-6galβ1-4glc	R_1=-6gal 1-4glc, R=COOH, R_4=H, R_5=$COCH_3$, R_7=R_8=R_9=H

Acknowledgements. The skilful technical assistance of Ms L. Eriksson is gratefully acknowledged. This work was supported by a research grant from Swedish Medical Research Council (MFR 16X-04723).

References

1. Allen,A. Brit.Med.Bull. 34, 28-33, (1978).

2. Born,G.V.R. and Cross,M.J., J.Physiol., 168, 178-195, (1963).

3. Buchanan,T.M., Pearce,W.A. and Chen,K.C.S. In: Immunbiology of Neisseria gonorrhoeae (Brooks,G.F., Gotschlich,E.C., Holmes,K.K., Sawyer,W.D. and Young,F.E., eds.) ASM, Washington D.C., 242-249, (1978).

4. Buscher,H.-P., Casals-Stenzel,J. and Schauer,R. Eur.J. Biochem., 50, 71-82 (1974).

5. Collier,W.A. and de Miranda,J.C., Antonie van Leeuwenhoek, J.Microbiol.Serol., 21, 135-140, (1955).

6. Culling,C.F.A., Reid,P.E., Dunn,W.L. and Clay,M.G. J.Clin. Pathol., 30, 1963-1067, (1977).

7. Faris,A., Lindahl,M. and Wadström,T. FEMS Microbiol.Lett., 7, 265-269, (1980).

8. Forstner,J.F., Jabbal,I., Qures,R., Kells,D.I.C. and Forstner,G.G., Biochem.J., 181, 725-732, (1979).

9. Fukuda,K.,Tomita,M. and Hamada,A., J.Biochem., 87, 687-693, (1980).

10. Gaastra,W. and de Graaf,F.K., Microbiol.Rev., 46, 129-161, (1982).

11. de Graaf,F.K., Klemm,P. and Gaastra,W., Infect.Immun., 33, 877-883,(1980)

12. Gahmberg,C.G. and Andersson,L.C., Eur.J.Biochem., 1221, 581-586, (1982)

13. Guinée,P.A.M., Jansen,W.H. and Agterberg,C.M., Infect. Immun., 13, 1369-1377, (1976).

14.Kawashima,I., Fukuda,K., Tomita,M. and Hamada,A., Biochem. J., 91,865-872, (1982).

15.Lindahl,M., Faris,A. and Wadström,T., Lancet, 8292, 280, (1982).

16.Lindahl,M., and Wadström,T., IRCS Med.Sci., 11, 790, (1983).

17.Lindahl,M., and Wadström,T., In: Proceedings of the Fourth International Symposium in Neonatal Diarrhea, VIDO,(Acres, S.D. ed.) Stuart Bundle Press, (1983).

18.Moon,H.W., Nagy,B., Isaacson,R.E. and Ørskov,I., Infect. Immun., 15, 614-620, (1977).

19.Murray,P.A., Levine,M.J., Tabak,L.A. and Reddy,M.S., BBRC, 106, 390-396, (1982).

20.Ørskov,I., Ørskov,F., Smith,H.W. and Sojka,W,J., Pathol. Microbiol.Scand.Sect.B,83, 31-36, (1975).

21.Parkkinen,J., Finne,J., Achtman,M., Väisänen,V. and Korhonen,T.K., BBRC, 456-461, (1983).

22.Porter,R.R., Biochem.J., 73, 119-127, (1959).

23.Rampal,R. and Pyle,M., Infec.Immun., 41, 339-344, (1983).

24.Schauer,R., Adv.Carbohydrate Chem.Biochem., 40, 131-234, (1982).

Part VI

Plant Lectins

CACTI LECTINS

Edgar Zenteno and Jose-Luis Ochoa

Department Biotechnology
CIB-The Center for Biological Research
Box 128, La Paz, BCS, Mexico

The presence of hemagglutinating substances (lectins) in practically all kingdoms, but not all organisms, is apparently a fact (1-4). In spite of this, only a handful of specimens serves as a commercial supply of this kind of valuable biological reagents thus giving rise to high qutation prices.

As the nature of its source varies, so does the method employed in lectin isolation. For instance seeds require grinding and sometimes defatting, other animal or plant tissues can be simply extracted with a blender or a homogenizer, then the substance of interest is fractionated by ammonium sulfate precipitation and further purified by gel filtration, ion-exchange and/or affinity chromatography techniques.

The reason why the most commonly employed material as a source of lectins are edible seeds probably lies behind earlier reports showing that lectins were responsible of toxic and other deleterious physiological effects in foodstuffs (2-3). However, very recently the economical interest set upon lectins moved many research groups to explore alternate sources to substitute

Lectins, Vol. IV

the expensive agglutinins available in the market. For developing countries this task is particularly important as food and health care are still priorities in goverment programs. At the Center for Biological Research in the Peninsula of California in Mexico, we decided to investigate the regional flora and fauna, terrestrial and marine, to determine the presence or activity of lectin-like substances in their components (5,6). As a result, we introduce here from the rich variety of organisms found in the California Pensinsula, a group of specimens belonging to the cactacea family which have showed to possess hemagglutinating activity (6).

The cactacea is a family of succulent plants (7,8) characterized by their ability to store water in their tissues. In the botanical sense they are calssified as Xerofites, that is plants which have adapted themselves to live in dry areas. These plants are relatively new on earth (about 20,000 years old) and because of their recent evolution the cacti are an exceedingly varied group well suited to the different environments where they are found.

Most cacti are stem succulent and therefore possess a well developed stem containing plenty of soft tissue (7,8). The surface of the stem has usually a waxy outer surface or epidermis which tends to reduce water loss, as a result they can swell but shrink to a limited degree in times of drought. Most cacti are clothed in spines to protect them against the sun, to condense moisture, and to hold easily detachable joints that when catched by the fur of animals can be transported for miles before being rubbed off then to grow into a new plant. The fruits and berries containing the seeds are of a great beauty and their colour, smell and fleshy texture when ripe, attract animals and birds that in turn help to scatter the seeds. The cacty, with few exceptions come from new world plants and can be found from the south of Canada to the tip of south America, in West Indies and the Caribbean Islands as well as in the Galapagues

Islands in the Pacific (7,8). They grow the same near the shore or in the high mountainous areas 4,000 m above the sea level. With such a variation of habitats and climatic conditions some species are considered only endemic, but there are others that have an qually wide range from hot, dry desert regions to the colder areas a thousands m up where they have to endure snow and freezing conditions in the winter.

Our interest in studying the cacti as an alternate source of lectins motivated the basis of a research program oriented to find convenient means for its rational explotation. For a start we selected a few of the most abundat cacti in the neighbourhoods to La Paz, BCS, Mexico and developed a method for extraction of agglutinins from their seeds and/or stems.

Materials and Methods

The seeds and cacti selected for this study were collected either near La Paz city or San Carlos harbour. The ground seeds were extracted in saline solution (0.9% NaCl) 1:10 w/v with a blendor, 5 min at 4°C, and rapidly centrifuged at 6000 rpm in a clinical centrifuge to obtain a clear supernatant. Then solid ammonium sulphate to a total concentration of 40-80% saturation was added and the suspension centrifuged again to obtain a pellet which was dialysed against PBS (0.02 M potassium phosphate, 0.9% NaCl, buffer pH 7.4) the protein concentration was determined colorimetrically with the folin reagent and the hemagglutination assay performed as usual with the double- serial dilution procedure.

Chromatographic Procedures: Ion exchange chromatography was run in columns of sulphopropyl-Sephadex (Pharmacia Fine Chemical) equilibrated with 0.05 M citrate buffer. The hemagglutinating material was applied until the activity started to appear in the eluate. Then the column was washed with 2-3 bed volumen before applying a stepewise gradient of Na_2SO_4. Hydrophobic chromatography was run in columns packed with phenyl-Sepharose (Pharmacia Fine Chemicals, Uppsala, Sweden) equilibrated with 0.5 M Na_2SO_4. A linear decreasing gradient after washing the column with 2-3 bed volumes of the original buffer was then applied and the fractions analyzed for both UV absorption and hemagglutinating activity. Affinity chromatography in stroma columns was performed as described elsewhere (11) using PBS washed human red blood cells type "o". The total hemagglutinating units adsorption capacity was found by saturation of the column and then determining the total amounts of hemagglutinating units eluted by distilled water and 0.2 M glycine-HCl pH 2.5 buffer.

Results and Discussion

Table I shows the hemagglutination activity of several extracts obtained from either seeds or stems of various cacti. It is apparent that a serological species specificity exists since the subceptibility of the different human blood groups or erythrocyte species varies. In most cases hemagglutination provoked by cacti extracts with red cells treated with trypsin is enchanced (Table 2).

Regarding chemical specificity, the cacti agglutinins show a rather complex behaviour: either they are not inhibited at all, or they can be blocked by galactose, glucose or fucose. To illustrate this we have summarized our observations in Table 3.

In other experiments we assayed the immunosuppression properties of the cacti extracts, which we found of particular inte-

Table 1

Hemagglutination capacity of cacti extract against native cells

Species	Human blood cells			Animal blood cells		
	A	B	O	Donkey	Sheep	Rabbit
Pachydereus pringleii	232	232	476	116	476	116
Macharocereus eruca	71	625	151	1	303	9
Cholla opuntia	0	0	0	0	188	0
Echinocereus engelmanii	952	476	454	111	666	2
Lophocereus shottii	1600	400	400	100	1600	10
Stenocereus thurberi	0	0	0	0	0	0
Macharocereus gummosus	Lysis	Lysis	Lysis	Lysis	Lysis	Lysis

Minimum protein extract amount required to induce agglutination of a 2% cell suspension.

Table 2

Hemagglutination capacity of cacti extracts against trypsinized cells.

Species	Human blood group			Animal blood cells		
	A	B	O	Donkey	Sheep	Rabbit
Pachycereus pringlei	58	116	58	116	232	116
Macharocereus eruca	9	9	9	9	75	6
Cholla opuntia	0	0	0	0	0	0
Echinocereus engelmanii	118	230	230	20	472	7
Lophocereus shottii	50	10	10	45	800	10
Stenocereus thurberi	0	0	0	0	0	0
Macharocereus gummnosus	Lysis	Lysis	Lysis	Lysis	Lysis	Lysis

Details as for Table 1.

Table 3

Inhibition of hemagglutinating capacity of cacti extracts by 0.1 M sugars.

Carbohydrate	*P. prinlei*	*M. eruca*	*L. shottii*	*S. thurberi*
None	64	64	64	64
D-ribose	16	64	64	64
D-galactose	16	8	64	64
N-acetyl-D-galactosamine	32	16	64	64
D-glucose	64	16	64	32
D-glucosamine	32	32	64	8
D-mannose	64	32	64	4
N-Acetyl-D-mannosamine	32	64	64	16
L-fucose	64	8	64	4

rest (listed in Table 4). We have further purified *M. eruca* agglutinin only from the stem because we have failed to determine the seed production season. The stems are very spinous and, as mentioned earlier, they are covered by a thick layer of wax. After removing the wax and spined the stems first looking yellowish change rapidly to a dark-brown colour indicating a process of oxidation. This is a common characteristic of polyphenol or tannin-rich plants. To avoid the negative effects that such oxidation could bring upon lectin activity the cacti juice is obtained with a food processor and fractionated with ammonium sulphate (60% saturation) very quickly. Ammonium sulphate precipitation has a double advantage as a first step in a purification of this kind: first, helps to concentrate and stabilize proteinous material; second, eliminates most of polyphenol and other oxidizing substances that could provoke interference in the following biological and chemical analysis.

After an extensive dialysis against 0.05 M citrate buffer, pH 5 the crude extract is poured into a column packed with SP-Sephadex (30 x 2.4 cm) equilibrated with the same buffer. The hemagglutinating material adsorbed under such conditions is eluted by 0.5 M Na_2SO_4 and directly applied to a Phenyl-Sepharose column (10. The separation of several hemagglutinating fractions of *M. eruca* extract by hydrophobic interaction chromatography on Phennyl-Sepharose is shown in Table 5.

From Table 5 it appears that the fractions have emerged from the column in order of its carbohydrate content; the most hydrophobic being those with the highest amount of sugar. This behaviour correlates well with a hypothesis put forward by other authors who claim that sugars induce the exposure of hydrophobic sites in glycoproteins (12-13). Fractions C and F were selected for further purification not only because they are the mayor fractions but also because they elute quite separated from one another and we could expect meaningful differences in their biological properties. Indeed, when applied separately

Table 4

Immunosuppression capacity of cacti extracts in sensitized mice

	Protein concentration mg/0.1 ml
Pachycereus pringleii	0.54
Macharocereus eruca	0.10
Cholla opuntia	0.5
Echinocereus engelmanii	0.17
Lophocereus shottii	0.5
Stenocereus thurberi	0.3
Macharocereus gummosus	0.05

The protein concentration is the amount required to reduce mice humoral immunoresponse against donkey erythrocytes to 50% (9).

Table 5

Fractionation of hemagglutinating activity of M. eruca extracts on phenyl-Sepharose.

Fraction	Chemical analysis Protein (%)	Sugar (%)	Hemagglutinating activity Yield (%)
A	84	16	-
B	82	18	-
C	82	18	27
D	80	20	12
E	76	24	12
F	74	26	19

to a strom-column for purification by affinity chromatography (11), fraction F is mainly eluted by distilled water while fraction C required a lowering of the pH. Fractions obtained from stroma columns exhibit some differences in protein and carbohydrate proportion which could be regarded as microheterogeneity since they are immunological indistinguishable. Other biological properties such as hemagglutination, mitogenicity, toxicity and immunosuppression support further this consideration. First, the mitogenic activity of *M. eruca* purified fractions is very weak for fraction C, and practivally none for fraction F. Second, more than 88% of the cells are still viable after incubation with 10-100 μg of either lectin fraction during a 3 day-culturing period. Third, the immunosuppression effect provoked when samples of 5 or 3 μg/ml are injected to sensitized mice against human erythrocyges is about 75-87.5%.

In sum, we have showed that the hemagglutinating activity of cacti extracts may be attributed to lectins. The span of lectin specificity among the cacti appears complex. Unfortunately the most potent hemagglutinating extract is obtained from a cactus *(M. eruca)* which is almost extinguished. Therefore we are not recommending it as a potential supply of lectins althoug, as is being suggested here, it may possess some intersting biological properties.

Hevertheless it should be concluded that most cacti are a potential source of lectin material. The presence of agglutinating substances in the seeds as well as in the juicy stem calls for a biological role of still unknown nature. Maybe they do act as cross-linking substances that help to set an intricate net of glycoproteins and polysaccharides which in turn is responsible for the water storage capacity.

Acknowledgements. We thank Dr. Lorentz Larsson and Pharmacia Fine Chemicals for the valuable chemicals they have kindly provided us to escape our national crisis. Thanks also due to Ms. Linda Cesna and Reni Storgaard for help in the preparation of this manuscript.

References

1. Gold, E.R. and Balding, P.: Receptor-specific proteins plant and animal lectins, (AM Elsevier, New York) 1-440 (1975).

2. Goldstein, L.J. and Hayes C.E., the lectins: Carbohydrate binding proteins, Adv. carbohyd. Chem. Biochem., 35, 127-340, (1978).

3. Lis, H., and Sharon, N.: Lectins in higher plants, the biochemistry of plants (P.K. Stumpf and E.E. Conn) Acad. press. NY. 371-488 (1981).

4. Cohen, E. (ed.). Biomedical perspectives of agglutinating of invertebrate and plant origin Ann. N.Y. Acad. Sci. 234, 1-412 (1974).

5. Zenteno, E. and Cordoba, F. In lectins, Biology Biochemistry, Clinical Biochemistry. Vol. 2 (T.C. Bøg-Hansen, ed.) Walter de Gruyter, Berlin, 721-728 (1982).

6. Herrera, E.M., Montano, C., Ochoa, J.L. and Cordoba, F.: Lectins, Biology Biochemistry, Clinical Biochemistry Vol.2 (T.C. Bøg-Hansen, ed.). Walter de Gruyter, Berlin, 693-700 (1982).

7. Lamb, E. and Lamb, B.. Cacti and succulents, MacMillan Publ. Co. Inc. New York, 6-217 (1970).

8. Lamb, E. and Lamb, B. Colorful cacti of the American deserts MacMillan Publ. Co. Inc., New York, 7-235 (1974).

9. Calderon, R.A. and Cordoba, F.: Eur. J. Immunol., 6, 522-525 (1976).

10. Ochoa, J.L.: Biochimie, 60, 1-15 (1978).

11. Ochoa, J.L. In: Immobilized cells and organelles (C.B. Mattiasson, Ed.). CRC Press Inc. Uniscience Titles, Vol. 2, pp. 125-131 (1983).

12. Mizrahi, A., O'Malley, J.A., Carter W.A., Takatsuki, A., Tamura, G. and Sulkowski, E.: J. Biol. Chem 253, 7612-7615 (1978).

13. Sachdev, G.P., Zodrow, J.M. and Carubelli, R.: Biochim. Biophys. Acta, 580, 85-90 (1979).

INHIBITION STUDIES ON THE SPECIFICITY OF THE *VICIA GRAMINEA* LECTIN

Marie José Prigent

INSERM U. 180, Faculté de Médecine Lariboisière-Saint-Louis
45, rue des Saints Pères, 75006 PARIS, France.

Vincent Verez-Bencomo, Pierre Sinaÿ

Laboratoire de Biochimie Structurale, ERA 739, UER de Sciences Fondamentales et Appliquées, 45040 ORLEANS, France.

Jean Pierre Cartron

INSERM U. 76, Centre National de Transfusion Sanguine,
6, rue Alexandre Cabanel, 75015 PARIS, France.

Glycophorin A, the best characterized red cell membrane glycoprotein, is a single polypeptide chain of 31,00 Kd containing 60 % by weight of carbohydrates (1). The entire amino acid sequence of glycophorin A is known (2) and it has been clearly established that the blood group MN determinants present on this glycoprotein are carried by the five N-terminal amino acid with three identical sialotetrasaccharides NeuAcα2-3Galβ1-3(NeuAcα2-6)GalNAc α-linked by O-glycosidic bond to ^{2}Ser, ^{3}Thr and ^{4}Thr residues (3,4). Glycophorin A differ by amino acid substitutions at positions 1 and 5, Ser-$\overset{T}{\text{Ser}}$-$\overset{T}{\text{Thr}}$-$\overset{T}{\text{Thr}}$-Gly in M individuals (A^M) and Leu-$\overset{T}{\text{Ser}}$-$\overset{T}{\text{Thr}}$-$\overset{T}{\text{Thr}}$-Glu in N individuals (A^N) (5,6). This is the only structural difference between M and N antigens. *Vicia graminea* agglutinin (VGA), which is specific for the N blood group interacted with this cell surface glycoprotein, before and after neuraminidase treatment of erythrocytes (7,8). The lectin recognized a complex determinant which involved both a carbohydrate and a peptide moiety

Lectins, Vol. IV

(9,10). In order to clarify further the characteristics of the epitope for this lectin, we have determined by agglutination inhibition tests the relative inhibitory potency of several synthetic glycopeptides containing the T-hapten structure β-D-Gal(1-3)-α-D-GalNAc analogs of the N-terminal end of glycophorin A^M and A^N (11,12) after neuraminidase treatment. One aim of this study was also to evaluate the effect on the lectin of clusters of carbohydrate chains present on the peptide backbone, and also the influence of the N-terminal amino acid residue.

Material and Methods

Red cells : MM and NN erythrocytes from blood group O donors all types as ss, were from the Centre National de Transfusion Sanguine, Paris, France.
Lectins : *Vicia graminea* agglutinin (VGA) was purified as previously described (13) from seeds cultivated under artificial conditions (Phytotron du CNRS, Gif sur Yvette, France) and used at 1 mg/ml in 100 mM Tris-HCl buffer pH 7.5 containing 50 mM NaCl.
Neuraminidase treatment of erythrocytes : The release of sialic acids from intact red cells was carried out by incubating 0.8 unit neuraminidase from *Vibrio cholerae* (Behring, FRG) and 1 ml of washed packed red cells suspended in 150 mM NaCl containing 10 mM $CaCl_2$.
Synthetic oligopeptides : Sixteen glycopeptides containing the T-hapten structure were chemically synthetized as described elsewhere (12,13). The pentapeptides and glycopeptides containing N-acetyl-D-galactosamine residues were a generous gift from Dr. Pavia (Avignon, France) and were also obtained by chemical synthesis (14,15).
Inhibition of hemagglutination : Inhibition tests were carried out as previously described with the synthetic peptides and glycopeptides diluted in PBS (16). When necessary the pH was re-adjusted to 7.2 to avoid hemolysis. The lectin concentration was four times the concentration necessary for hemagglutination. The results are expressed as the lowest concentration of inhibitor (in mM) for complete inhibition of 4 agglutinating doses.

Results and Discussion

The lectin from *Vicia graminea* exhibits a blood group N specificity towards native human erythrocytes but agglutinates strongly both N and M red cells treated with trypsin or neuraminidase. However, inhibition studies have shown that the lectin still recognizes better N than M desialylated glycopeptides derived from the human red cell membrane (for review see 17). The erythrocyte membrane receptors for VGA have been recently characterized as glycophorin A and B on native N red cells and glycophorin B only on native M red cells (7,8). The specificity of VGA is rather complex since it involves both a sugar sequence β-D-Gal(1-3)-α-D-GalNAc (the T-hapten) and a peptide moiety (9, 10, 17). So far no simple sugar alone (even the T-hapten disaccharide) has been found to inhibit the lectin. Several synthetic glycopeptides, corresponding to the five N-terminal amino acids of glycophorin A from N and M individuals were used in inhibition of agglutination. The results are presented in Table 1. The inhibition tests were carried out with native and neuraminidase treated N and M red cells. Glycopeptides 1, 2, 3, 4, and 5 which carry one or several T-hapten disaccharides linked to serine or threonine are not inhibitory supporting the view that the oligosaccharide chain(s) alone are not a sufficient requirement for binding. Similarly, the pentapeptides Leu-Ser-Thr-Thr-Glu and Ser-Ser-Thr-Thr-Gly are not inhibitory at 0.0023 mM concentration (not shown) suggesting that the peptide sequence itself, either derived from N or M typical amino-acid sequences, do not represent an adequate determinant for binding.

The best inhibitors of VGA are represented by glycopeptides carrying the T-hapten structure and a terminal leucine residue (structures 7, 8, 12) whereas those carrying a terminal serine residue are inactive (structures 9, 10, 11, 13). These results indicate clearly that the determinant specific for VGA includes both carbohydrates and an hydrophobic amino acid. Obviously, the presence of N-acetylneuraminic-acid is not necessary for the N specificity.

These observations correlate well with previous findings showing that Edman degradation of N and M erythrocyte glycopeptides abolish the VGA activity (17,18). Further examination of Table 1 shows that on the basis of the galactose concentration, glycopeptide 6 is respectively 3-, 5- or 30-times less inhibitory than glycopeptide 7, 8 or 12 (as seen using native N red cells). The results demonstrate that the highest VGA activity is obtained with glycopeptides carrying a leucine at the N-terminal end together with a cluster of T-hapten disaccharides at position 2, 3 and 4. The spatial arrangement of the three carbohydrate chains may constitute the receptor for VGA. In addition, since glycopeptide 12 had the highest activity, it is possible that the Glu residue at position 5 favors the best overall structure adapted to VGA.

Our results bring strong experimental support to the proposal (18,19) that the receptor site for VGA includes a glycopeptide with a specific conformation which is determined by a heavy cluster of T-hapten disaccharides in close vicinity to a hydrophobic area of an amino acid. Interestingly, a similar determinant (cluster of T-hapten and a valine residue) is present in the region between residue 10 and 16 of glycophorin A (1,2) and is uncovered after removal of sialic acids. Such region of the molecule might represent a potential binding site for VGA to neuraminidase-treated N and M erythrocytes, although the valine is N-protected. Synthesis of glycopeptides corresponding to the 10-17 segment of glycophorin A may clarify this point. The results presented in this paper have demonstrated the usefulness of synthetic glycopeptides in elucidation of the structural requirement of VGA lectin since most of the structures used have not been obtained yet by any chemical or enzymatic fragmentation of any biological material. Further investigation using these substances, together with other synthesized clustered glycopeptides, might provide more information on the combining site of this lectin.

TABLE 1

Inhibitory effect of synthetic glycopeptides on agglutinating activity of vicia graminea lectin

N°	INHIBITOR	Minimum concentration of inhibitor (mM) for complete inhibition			
		NN	MM	NN(n)	MM(n)
		$x10^3$	$x10^3$	$x10^3$	$x10^3$
1	T Ser	>10		>10	
2	T Thr	> 9.7		> 9.7	
3	T T Ser-Thr	> 0.8		> 0.8	
4	T T Thr-Thr	> 0.8		> 0.8	
5	T T T Ser-Thr-Thr	> 0.5		> 0.5	
6	T Leu-Ser	1.6		1.6	
7	T T Leu-Ser-Thr	0.3		0,62	
8	T T T Leu-Ser-Thr-Thr	0.17		0.34	
9	T Ser-Ser	> 1.25	> 1.25	> 1.25	> 1.25
10	T T Ser-Ser-Thr	> 1.25	> 1.25	> 1.25	> 2.25
11	T T T Ser-Ser-Thr-Thr	> 1.1	> 0.5	> 1.1	> 1.1
12	T T T Leu-Ser-Thr-Thr-Glu(A^N)	0.02	0.08	0.1	
13	Ser-Ser-Thr-Thr-GLy(A^M)		> 0.5	> 0.5	

T = β-D-Gal(1-3)-α-D-GalNAc
(n) = red cells treated with neuraminidase.
> = indicates that inhibition was not obtained at the concentration mentioned

Acknowledgments

We thank the Ministère des Relations Extérieures (France) for a fellowship to one of us (VB), and INSERM for financial support (C.R.L. n°821027).

References

1. Tomita, M., V.T. Marchesi. 1975. Proc. Natl. Acad. Sci. USA 72, 2964-2968.
2. Tomita, M., H. Furthmayr, V.T. Marchesi. 1978. Biochem. 17, 4756-4770.
3. Thomas, D.B., R.J. Winzler. 1969. J. Biol. Chem. 244, 5943-5946.
4. Prohaska, R., T.A.W. Koerner, I.M. Armitage, H. Furthmayr. 1981. J. Biol. Chem. 256, 5781-5791.
5. Dahr W., G. Uhlenbruck, E. Janssen, and R. Schmalisch. 1977. Hum. Genet. 5, 335-343.
6. Wasniowska K., Z. Drzenieck, and E. Lisowska. 1977. Biochem. Biophys. Res. Commun. 76, 385-390.
7. Duk, M., E. Lisowska, M. Kordowicz, and K. Wasniowska. 1982. Eur. J. Biochem. 123, 103-112.
8. Prigent M.J., D. Blanchard, and J.P. Cartron. 1983. Arch. Biochem. Biophys. 222, 231-234.
9. Lisowska E., and M. Duk. 1975. Eur. J. Biochem. 54, 469-474.
10. Dahr W., G. Uhlenbruck, and G.W.G. Bird. 1975. Vox Sang. 28, 389-391.
11. Verez-Bencomo V., and P. Sinaÿ. 1983. Carbohydr. Res. 116, 9-11.
12. Verez-Bencomo V., and P. Sinaÿ. 1984. Glycoconj. J. 1, 5-8.
13. Prigent M.J., C. Lombart and A. Asseraf. 1982. FEBS Lett. 150, 85-88.
14. Ferrari B., and A.A. Pavia. 1983. Int. J. Peptide Protein Res. 22, 549-559.
15. Ferrari B., and A.A. Pavia. 1984. Int. J. Peptide Protein Res. (in press).
16. Cartron J.P., B. Ferrari, M. Huet, and A.A. Pavia. 1984. Exp. Clin. Immunogenet. 1, 112-116.
17. Lisowska E. 1977. In "Handbook Series in Clinical Laboratory Science" (Greenwalt J. and Steane E.A., eds) vol. 1, section D; pp 281-299. CRC Press, Cleveland, Ohio.
18. Dahr W., and G. Uhlenbruck. 1978. Hoppe Seyler's Z. Physiol chem. 359, 835-844.
19. Lisowska E., and K. Wasniowska. 1978. Eur. J. Biochem. 88, 247-252.

ULTRASTRUCTURAL LOCALIZATION OF PEA LECTIN IN THE EMBRYO AND COTYLEDONS DURING DEVELOPMENT BY THE COLLOIDAL-GOLD METHOD

Gerda Smets[1], Edilbert Van Driessche[2,3] and Sonia Beeckmans[2]

[1]Laboratorium voor Pathologie, Vrije Universiteit Brussel, Laarbeeklaan 103, B-1090 Brussels, Belgium

[2]Laboratorium voor Chemie der Proteïnen, and [3]Laboratorium voor Plantenfysiologie, Vrije Universiteit Brussel, Instituut voor Moleculaire Biologie, Paardenstraat 65, B-1640 Sint-Genesius-Rode, Belgium.

Despite the abundant use of lectins as "Tools" in different areas of research, and the fact that several lectins have been thoroughly characterized, relatively little unequivocal information is available about the physiological role of plant lectins (for reviews, see Refs. 1-15). Since any proposed function must be consistent with the physical location of the lectin, both in space and time, histological localization studies of these molecules (16-20) will undoubtedly lead to a better understanding of their physiology. In our previous study on the localization of pea seed lectin in mature seeds, the unlabeled peroxidase-antiperoxidase procedure was used (16). An important imperfection of this method is the poor resolution and the poor preservation of ultrastructural detail after the immunocytochemical localization on ultra-thin sections. Furthermore, the background staining is inevitably high. Previously we noticed that the peroxidase-antiperoxidase procedure was unsuitable for the detection of lectin which might be associated with the cell walls (16). In an attempt to overcome the problems just mentioned, colloidal gold was used as a cytochemical marker. Since gold granules are particulate, they allow a fine localization

Lectins, Vol. IV

and a precise identification of the marked structures (21, 22). Colloidal gold has also been successfully used as a marker for the localization of ConA (17), PHA (18) and SBA (19).
In this paper we describe the localization of pea lectin in the embryo of mature and immature seeds, and in the cotyledons of immature seeds at different stages of the maturation of the seeds.

Materials and Methods

Antibodies against highly purified pea lectin were prepared as described previously (16).

Plant material for immunocytochemical localization was prepared as follows. Mature (dry) seeds were imbibed for 24 hours in glassdistilled water at room temperature. The embryo was removed and integrally fixed for 12 hours in 0.2 M cacodylate buffer pH 7.2, containing 2.5% glutaraldehyde. After the fixation, the material was washed in bidistilled water, dehydrated in an ethanol gradient and embedded in Low Viscosity Spurr Embedding Standard medium (firm) (23). Ultra-thin sections (1000 Å thick) were obtained with a diamond knife and mounted on uncoated golden grids (150 Mesh).
Immature seeds were removed from the pods and dissected in embryo and cotyledons. The time between the harvest of the pods and the fixation was less than 1/2 hour. Embryos were integrally fixed, while cotyledons were cut into small cubes. After fixation, the material was further treated as described above. Duplicate samples were always post-fixed with 1% OsO_4 for 3 hours, washed with bidistilled water, and dehydrated and embedded as described above.

For the immunocytochemical staining, the grids were covered for 10 minutes at room temperature with a drop (100 μl) of normal goat serum (Nordic), diluted in phosphate buffered saline (PBS) pH 7.2 to obtain a 1% (v/v) solution, washed with PBS and then covered with 100 μl affinity purified anti-pea lectin antibodies (33 μg/ml) for 12 hours at room temperature. The grids were again washed with PBS and incubated for 1 hour on a drop of goat-antirabbit immunoglobulin-G coupled to 20 nm colloidal gold (Janssen Pharmaceutica, Belgium), diluted 1/100 in PBS. Finally, the grids were rinced with PBS and then with bidistilled water. Contrasting was performed with 0.5% aqueous uranyl acetate (30 minutes, twice) and then with 0.26% lead citrate (30 minutes) under a constant nitrogen flow. The immunocytochemical staining was performed on one side of the grids, while the contrasting was done on the backside.
In control experiments, the anti-pea lectin antibodies were

either omitted or replaced by the same concentration of antibodies isolated from pre-immune rabbit sera by affinity chromatography on Sepharose-protein A (16).
The results were examined on a transmission electron microscope Zeiss EM 109.

Results

Sections were made from the embryo of mature and immature seeds at different levels of the root tip, at the region just above the root tip, at different places along the shoot, at the shoot tip and leaf primordia. Lectin localization was only possible when the tissues had been fixed with glutaraldehyde : post-

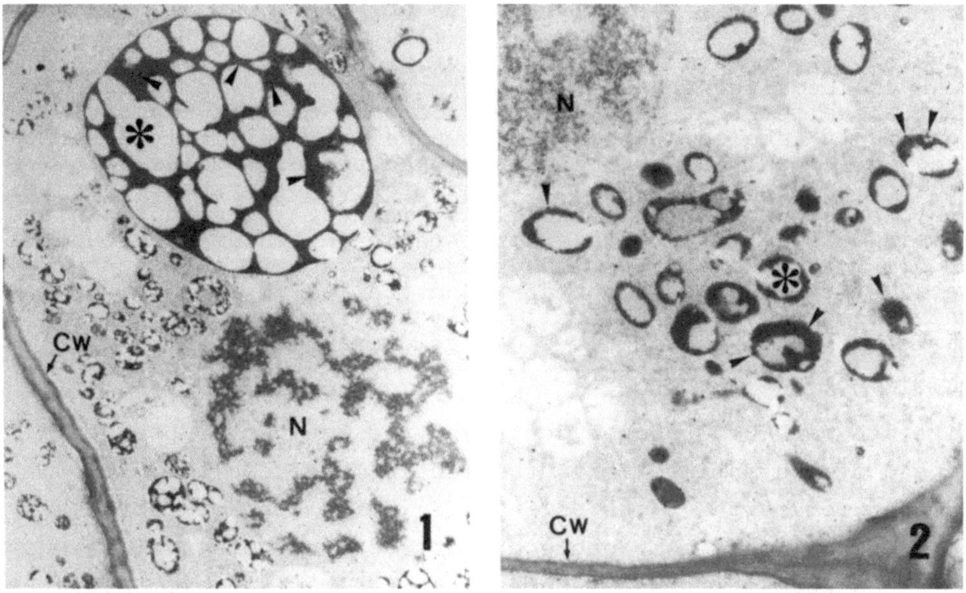

Figures 1,2: Immunocytochemical localization of lectin in embryos from mature (24h imbibed) pea seeds. Sections were made through the root tip. Some cells (Fig. 1) show fusing protein bodies, thus forming huge organelles filled with lectin. Note that the cell wall and all the cell organelles are devoid of gold granules. Magnifications: x3600 (Fig.1); x5300 (Fig.2). CW: cell wall; N: nucleus; *: protein body; ►: gold granule.

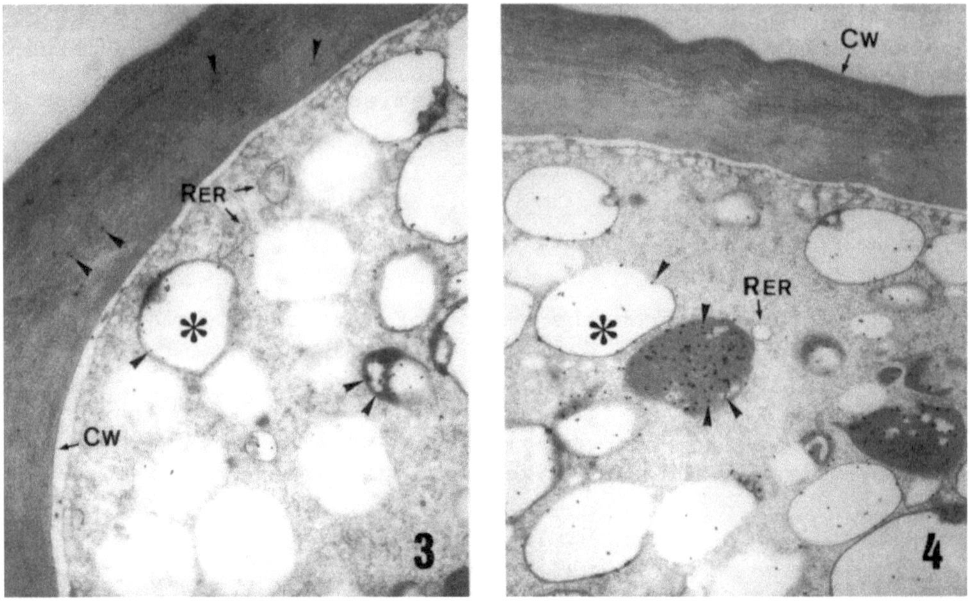

Figures 3,4: Localization of lectin in embryos from mature (24h imbibed) pea seeds. The section shown in Fig. 3 was made just above the root tip. The rhizodermis of this region is positively labeled. Label was found either uniformly distributed in, or only at the borders of the protein bodies. The section depicted in Fig. 4 was made through the shoot. In this zone, the rhizodermis is much less stained. Note that, in the resting tissue, the rough endoplasmatic reticulum was found to be completely negative. Analogous results were obtained with sections through the shoot apex and the leaf primordia. Magnifications: x14500. CW: cell wall; RER: rough endoplasmatic reticulum; ✱: protein body; ►: gold granule.

fixation with OsO_4 either destroyed the antigen or prevented access of the antibodies to the lectin molecules (24, 25). However, omission of the OsO_4-postfixation step was not disadvantageous since a comparable ultrastructural conservation was obtained when the tissues were exclusively fixed with glutaraldehyde.

In the root tip of the embryo from mature seeds (Figs. 1 and 2) labeling was restricted to the protein bodies, while the cell walls and all other organelles were devoid of label. Fusing protein bodies were also observed. In a small region just

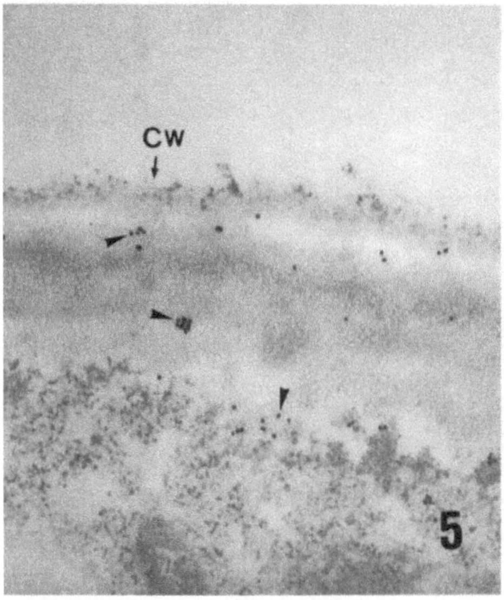

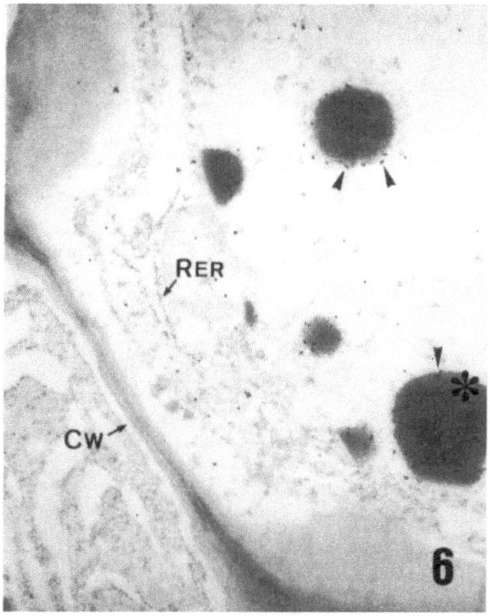

Figures 5,6: Localization of lectin in embryos from immature pea seeds at different stages of development. The walls of the cells of the embryos from young, immature seeds, which do not yet contain protein bodies, were found positive (Fig. 5). At later stages of development, when the embryo cells contain already protein bodies (Fig. 6), label is found in the latter organelles, whereas the cell walls are negative. Magnifications: x24000 (Fig.5); x19000 (Fig.6). CW: cell wall; RER: rough endoplasmatic reticulum; ✱: protein body; ►: gold granule.

above the root tip, the same intracellular labeling pattern was found; in addition the outer cell walls of the rhizodermis were intensively labeled (Fig. 3). The cells of the shoot, shoot apex and leaf primordia contained much less protein bodies, but gold granules were present in all of them (Fig. 4). Although the cell walls were mostly negative, some scarcely distributed gold granules were consistently observed in the cell walls of the epidermis of the embryo.

In the embryo and cotyledons of immature seeds at later stages of development, essentially the same labeling was observed as in mature seeds. However, in the very young embryo, at a stage

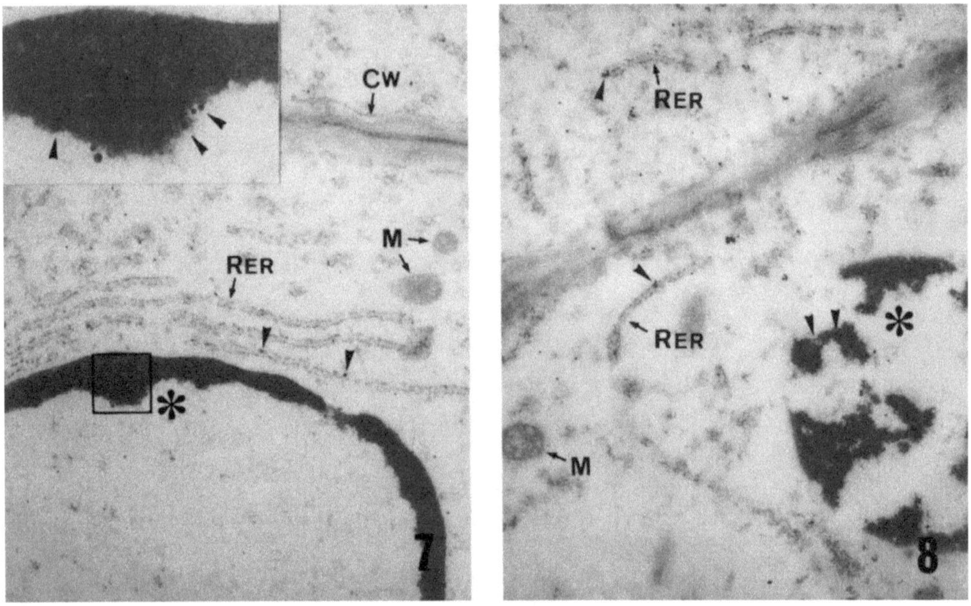

Figures 7,8: Localization of lectin in the cotyledons from immature pea seeds. The tissue had the same age as the tissue shown in Fig. 6. The cells of these cotyledons contain already many protein bodies. The rough endoplasmatic reticulum is positively labeled. All other cell organelles are devoid of gold granules. Magnifications: x14500. CW: cell wall; RER: rough endoplasmatic reticulum; M: mitochondrium; *: protein body; ►: gold granule.

of development where the reserve protein synthesis had not started yet, the cells showed very little intracellular detail. In these tissues, gold granules were only found in the cell walls (Fig. 5). At later stages of development, when protein bodies were present in the cells of both embryo (Fig. 6) and cotyledons (Figs. 7 and 8), the cell walls were negative and the majority of the labeling was detected in the protein bodies. At the latter stage of development, the rough endoplasmatic reticulum was also positively labeled, especially in the parenchyma cells of the cotyledons.

Discussion

The results presented in this paper confirm our previous findings that in the mature seed, the lectin is localized in the protein bodies (16). By using the colloidal-gold method we could now also demonstrate lectin in the outer cell walls of a restricted zone of the rhizodermis situated just above the root tip of the embryo of the mature seed. Since it has been reported that during water uptake of dry soybean seeds, lectin is released in the environment (26-28), the possibility can not be excluded that the cell wall associated lectin represents a pool of molecules which are destined to be exudated. However, if so, it is not obvious why lectin in association with the outer cell walls should be restricted to a well-defined zone of the root epidermis. Nevertheless, our results clearly indicate that the immunogold method can be successfully used on ultrathin sections as a cytochemical marker for the detection of lectin which is associated with the cell wall.
The presence of lectin in the cell walls of the embryo and cotyledons in very young seeds at a stage when protein bodies are not observed, and, consequently, the lectin is not accumulating in reserve organelles, is of special interest. It might indicate that the lectin can fulfil different development-dependent functions. The lectin which accumulates in the protein bodies during the later stages of seed maturation might, together with the albumins and globulins, function as reserve proteins. Indeed, a parallellism was found not only in the synthesis of lectins and reserve proteins during seed maturation, but also in their disappearance during germination (29, 30). On the other hand, lectin which is synthesized in the very early stages of seed development and which is found in the cell walls might represent receptors for glycosylated molecules or enzymes participating in cell wall metabolism. Finally, lectin found, after imbibition of the dry seeds, in the outer cell walls of the rhizodermis of the embryo, might act as a receptor to attach symbiotic bacteria to the roots during the early stages of

germination (31-38).
The labeling of the rough endoplasmatic reticulum, especially seen in parenchyma cells of the cotyledons in immature seeds, is in agreement with our current knowledge of the synthesis and transport of lectins and storage proteins to the protein bodies (39-45).
In order to get a better understanding of the physiological function of leguminous lectins, we feel that the following major questions should be answered in the future : (a) Does the cell wall associated material which we could localize in the very young embryos really behave as a lectin, or is it an immunologically cross-reacting cell wall component which disappears or becomes masked during seed maturation ? (b) Are lectins (or cross-reacting material) bound in the cell walls by a receptor-dependent mechanism or by less specific interactions with other cell wall components ? (c) By which mechanisms are lectins (or cross-reacting material) channeled to their final intracellular position ? (d) Which mechanisms govern the development-dependent expression of the lectin-genes ? It should be mentioned in this respect that the presence of very low quantities of lectin (or cross-reacting material) has been demonstrated during the whole life cycle of Pisum, in contrast to the huge quantities of lectin which is synthesized during seed maturation (Hosselet et al., these Proceedings).

References

1. Lis, H., Sharon, N.: Ann. Rev. Biochem. 42, 541-574 (1973)
2. Lis, H., Sharon, N.: The Antigens (M. Sela, ed.), Vol. IV, pp. 429-529 (1977) (Academic Press, N.Y.)
3. Sharon, N., Lis, H.: Science 177, 949-959 (1972)
4. Liener, I.E.: Ann. Rev. Plant Physiol. 27, 291-319 (1976)
5. Brown, J.C., Hunt, R.C.: Int. Rev. Cytol. 52, 277-349 (1978)
6. Coulet, M.: Ann. Immunol. (Paris) 1, 3-16 (1979)
7. Goldstein, I., Hayes, C.E.: Adv. Carbohydrate Chem. Biochem. 35, 128-270 (1978)

8. Uhlenbruck, G.G.: Funkt. Biol. Med. 2, 40-48 (1983)

9. Lotan, R., Nicolson, G.: Biochim. Biophys. Acta 100, 119-130 (1963)

10. Pistole, T.G.: Ann. Rev. Microbiol. 35, 85-112 (1981)

11. Albersheim, P., Anderson-Prouty, A.J.: Ann. Rev. Plant Physiol. 26, 31-52 (1975)

12. Pusztai, A., Cray, R.R.D., Grant, G., Stewart, J.C.: Seed Proteins (J. Dussant, J. Mossé, J. Vaughan, eds.) pp. 53-78 (1983) (Academic Press)

13. Schmidt, E.L.: Ann. Rev. Microbiol. 33, 355-376 (1979)

14. Barondes, S.H.: Ann. Rev. Biochem. 50, 207-231 (1981)

15. Uhlenbruck, G.G.: Naturwissenschaften 68, 606-612 (1981)

16. Van Driessche, E., Smets, G., Dejaegere, R., Kanarek, L.: Planta 153, 287-296 (1981)

17. Herman, E.M., Shannon, L.M.: Planta 161, 97-104 (1984)

18. Manen, J.F., Pusztai, A.: Planta 155, 328-334 (1982)

19. Horisberger, M., Vonlanthen, M.: Histochemistry 65, 181-186 (1980)

20. Mishkind, N., Raikhel, N.V., Palevitz, B.A., Keegstra, K.: J. Cell Biol. 92, 753-764 (1980)

21. Horisberger, M.: Biol. Cellulaire 36, 253-258 (1979)

22. Horisberger, M.: Scanning Electron Microscopy (O. Joharied, ed.) Vol. II, pp. 9-31 (1981) (SEM Inc. AMF O'Hare; Chicago Il. 60666, U.S.A.)

23. Spurr, A.R.: J. Ultrastruct. Res. 26, 31-48 (1969)

24. Baskin, D.B., Erlandsen, S.L., Parsons, J.A.: J. Histochem. Cytochem. 27, 1290-1292 (1979)

25. Erlandsen, S.L., Parsons, J.A., Rodning, C.B.: J. Histochem. Cytochem. 27, 1286-1289 (1979)

26. Fountain, D.W., Foard, D.E., Replogle, W.D., Yang, W.K.: Science 197, 1185-1187 (1977)

27. Hwang, D.L., Yang, W.K., Foard, D.E., Lin, K.T.D.: Plant Physiol. 61, 30-34 (1978)

28. Causse, H., Rougé, P.: Lectins: Biology, Biochemistry, Clinical Biochemistry (T.C. Bøg-Hansen, G.A. Spengler, eds.) Vol. 3, pp. 559-572 (1983) (W. deGruyter)

29. Rougé, P., Plantefol, M.L.: C.R. Acad. Sci. Paris 282, 621-623 (1976)

30. Rougé, P., Plantefol, M.L.: C.R. Acad. Sci. Paris 278D, 449-452 (1974)

31. Hamblin, J., Kent, S.P.: Nature (Lond.) 245, 28-29 (1973)

32. Bohlool, B.B., Schmidt, E.L.: Science 185, 269-271 (1974)

33. Wolpert, J.S., Albertsheim, P.: Biochem. Biophys. Res. Comm. 70, 729-737 (1976)

34. Dazzo, F.B., Yanke, W.E., Bull, W.J.: Biochim. Biophys. Acta 539, 276-286 (1978)

35. Kijne, J.W., Van der Schaal, I.A.M., De Vries, G.E.: Plant Sci. Letters 18, 65-74 (1980)

36. Kato, G., Mayarama, Y., Nakamura, M.: Plant Cell. Physiol. 22, 759-772 (1981)

37. Kijne, J.W., Van der Schaal, I.A.M., Diaz, C.L., Van Iren, F.: Lectins: Biology, Biochemistry, Clinical Biochemistry (T.C. Bøg-Hansen, G.A. Spengler, eds.) Vol. 3, pp. 521-529 (1983) (W. deGruyter)

38. Van der Schaal, I.A.M.: Ph. D. Thesis, State University Leiden, the Netherlands (1983)

39. Chrispeels, M.J.: Phil. Trans. R. Soc. London B304, 309-322 (1984)

40. Bollini, R., Chrispeels, M.J.: Planta 146, 487-501 (1979)

41. Higgins, T.J.V., Spencer, D.: Plant Physiol. 65, 205-211 (1981)

42. Hurkman, W.J., Beevers, L.: Plant Physiol. 69, 1414-1417 (1982)

43. Bollini, R., Van der Wilden, W., Chrispeels, M.J.: Plant Physiol. 55, 82-92 (1982)

44. Chrispeels, M.J., Higgins, T.J.V., Craig, S., Spencer, D.: J. Cell Biol. 93, 5-14 (1982)

45. Chrispeels, M.J.: Planta 158, 140-151 (1983)

BIOLOGICAL ACTIVITIES OF MISTLETOE LECTIN I (ML I) AND ITS A AND B CHAINS (A REVIEW)

Hartmut Franz
Staatliches Institut für Immunpräparate und Nährmedien, DDR-1120 Berlin

At the 5th lectin meeting we presented a method for preparation of the A and B chain of mistletoe lectin I (ML I) without denaturating agents (1). Using this method we have been able to investigate biological activities of both chains. In this paper the hitherto existing results are summarized (see Table 1).

Structural studies

Electronmicroscopic visualization of ML I reveal that ML I consists of A and B chains.
The properties of these chains resemble thos of the A and B chains of ricin and abrin. They are predominantly bound by disulfide bridges, forming the monomer AB. AB is in equilibrium with the dimer $(AB)_2$. Recently both monomer and dimer have been visualized electronmicroscopically (2). After negative staining with uranyl acetate, monomers are visible as nearly triangular or rounded particles with dimensions of 80-90 Å. The dimers show a rodlike profile and seem to be composed of two triangular parts (dimensions 175 x 80 x 60 Å) which are dislocaled against each other. For details see (2). Following treatment with sodium

Lectins, Vol. IV

dodecylsulfate (SDS), particles were found to be either smaller (< 50 Å) or larger (> 200 Å) than the rodlike particles (dimer). This result has been interpreted in the following way:
The dimers of ML I are split by SDS not only into the monomer but in some cases also into the constituent A and B chains, because not all A and B chains of the monomer of ML I are connected by disulfide bounds (2). Beside the major band (ML I) in SDS PAGE, also two faint bands at mol. wt. 29 000 and 34 000 have been proved. Similar results have been published by Olsnes et al. (3).

Cytotoxic activity

Like ricin and abrin, ML I is cytotoxic preferentially to some types of cancer cells (for summary see (4)). Corresponding to other toxic lectins, ML I can bind to glycoconjugates on the cell surface via the B chain. This fixation allows the invasion of the A chain into the cell. On the ribosomal level, the A chain inhibits the protein synthesis. The result of the interaction (60 min, 20°C) of ML I and its chains on Ehrlich Mice Ascites tumor cells (EMA) is shown in Fig. 1. A concentration of 0.5 µg/ml ML I cause cell destruction and deformation of the cell organelles (5). The nuclei remain unchanged (Fig. 1a). It is difficult to imagine that such serious alterations are caused only by the mechanism described above. Immediate membrane alterations seem to be likely, too. The incubation of EMA with the A chain alone at a concentration of 1 - 10 µg/ml does not give any alterations (Fig. 1b). The same result we got using the B chain (Fig. 1c). Higher concentrations (10 µg/ml and more) cause enlargement of mitochondra. Thos it seems that it is the combined A and B chains that exhibit the cytotoxic capacity.

Macrophage activation, lymphokine and mitogenicity.

The influence of ML I and its A and B chains on human leu-

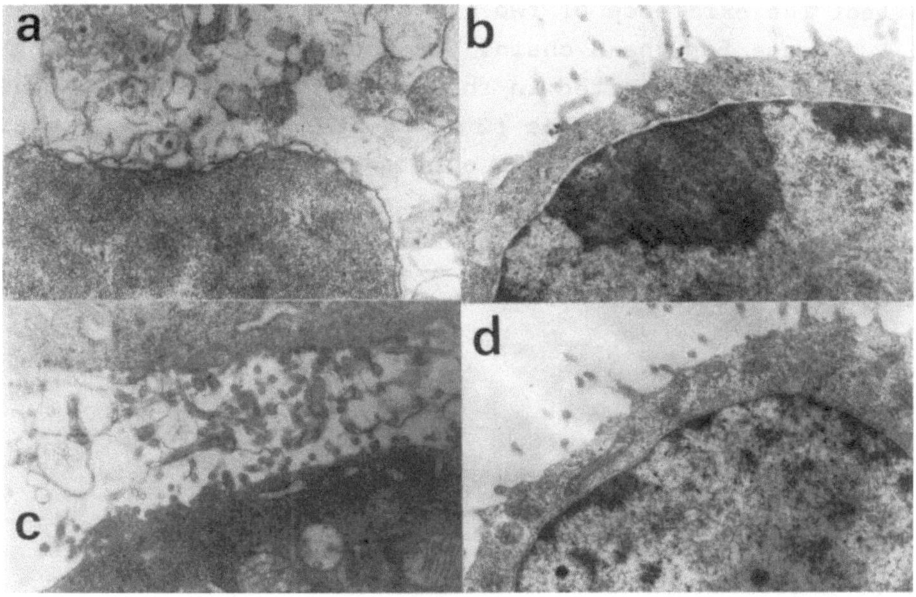

Fig. 1
Electronmicroscopical visualization of interaction of ML I or its A chains with EMA (magnification 20 000 x)

a) ML I (0.5 µg/ml destroys cell membrane and cytoplasma
b) No alteration after A chain treatment (1-10 ug/ml)
c) Enlargement after the B chain (1-10 ug/ml) interaction
d) Control (untreated EMA)

cocytes and peritoneal macrophages (guinea pig) were investigated together with Metzner et al. (6). These cells are agglutinated by ML I (2 µg/ml). Concentrations up to 1 µg per 10^7 macrophages are not cytotoxic (trypan blue and ethidium bromide test). Agglutination of macrophages was also observed with the B chain (10 µg/ml). In dependence on concentration a decreas of surface charge of macrophages was obtained as could be demonstrated using the macrophage electrophoretic mobility test (MEM test). The diagram (Fig. 2) shows two peaks. This could possibily reflect the existence of two types of receptors, one for ML I and one for the B chain. The spontaneous migration of macrophages is inhibited in the order: ML I⟩⟩ B chain⟩ A chain. Pretreatment of leucocytes (2 - 4 x 10^6/ml) with the B chain (0.5 - 5 µg/ml) increase the spontaneous phagocytoses up to 10 %. Perhaps the mechanism of action of the B chain is similar to that of lectin type macrophage activating lymphokines (MAF).

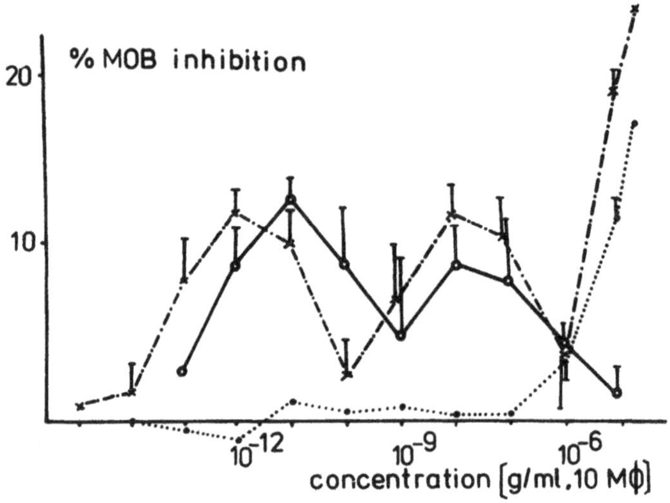

Fig. 2
Influence of ML I and its isolated A and B chains on the electrophoretic mobility (MOB) of macrophages of guinea pig (M Φ). For experimental details see (6)

ML I	————————	(n = 5)
A chain	················	(n = 4)
B chain	x—·—·x—·—·x	(n = 5)

The A chain does not influence the phagocytotic activity. Recent experiments (7) demonstrate that ML I and its B chain diminish the incorporation of tritiated thymidine into human mononuclear cells (after 72 h, from 10 μg to 1 ng/10^6 cells (ML I) and 10 μg to 0.1 μg/10^6 cells (B chain). This means that ML I is 100 times more active than the B chain. Moreover, production of a MSF(macrophage stimulating factor)-like factor was evident in response to the B chain at a concentration of 1 μg/10^6 cells. In contrast to the B chain, the A chain stimulates lymphocytes at concentrations between 0.1 μg and 1 μg/10^6 cells. The stimulation index (3.5) was however remarkably lower than that of PHA (23, 2). The same concentration of A chain produced a lymphokine (MSF). The interaction of this lymphokine with indicator cells (macrophage) could be inhibited by L-fucose and by monoclonal antibodies directed against MSF-like factors. In general, these results show that the B chain as well as the A chain of ML I are especially cytoactive substance. The B chain activates macrophages, acts as a lymphokine and can liberate lymphokine like factors whereas the A chain is weekly mitogenic and releases a lymphokine. In the foot pad swelling (injection of the reactent into the foot pad of mice) only the intact lectin ML I is active. Both types of chains are inactive (8).

Cytotoxic conjugates containing the A chain.

The alteration of ribosomal leading to an inhibition of protein synthesis by the A chain makes this glycoprotein a candidate for the preparation of immunotoxine. We have begun such experiments using monoclonal antibodies to leukaemia cells and these studies are still in progress. Moreover we coupled the A chain of ML I to histamine by means of 1 ethyl-3(3-dimethylaminopropyl)-carbodiimide-hydrochloride (EDC, Serva) in order to get a cytotoxic reagent against cells bearing histamine receptors (9). This conjugate is toxic to lymphocytes. After incubation for 60 min (trypan blue exclusion test) 27 % of spleen cells, 60 % of T and 15 % of B lymphocytes are killed (Fig. 3).

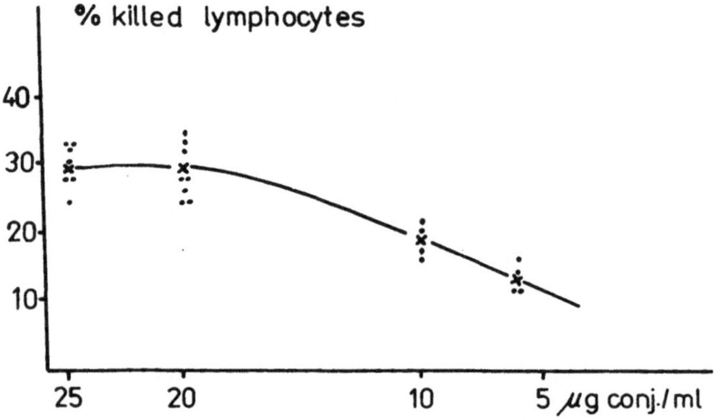

Fig. 3
Killing of spleen cells by means of A chain-histamine conjugate (trypan blue exclusion test). The conjugate is cytotoxic for histamine receptor bearing cells.

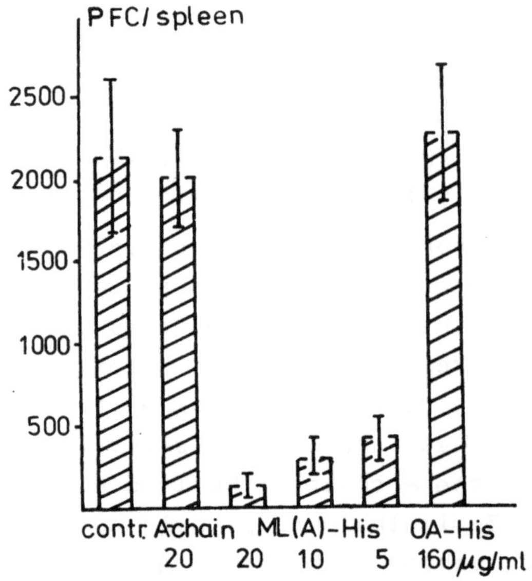

Fig. 4
Immune response to sheep red cells in irradiated mice after transfer of lymphocytes pretreated with different substances. Pretreatment with A chain-histamine conjugate (ML(A)-His) inhibits the formation of plaque forming cells (PFC). Pretreatment with an ovalbumin-histamine conjugate or the A chain does not show any effect

Table 1
Comparison of biological activities of mistletoe lectin I and its A and B chains. Especially remarkable are the macrophage activation and the lymphokine liberation by the B chain and the mitogenic capacity of the A chain.

	Mistletoe lectin I (ML I)	A chain	B chain
Characteristic	conjugate of lectin and enzyme (?)	enzyme (?)	lectin
mol. wt.	120 000	29 000	34 000
binding of D-Gal.-containing glycoconjugate	+	-	-
cell agglutination	+	-	+
inhibition of protein synthesis			
- ribosomal level	+	+	-
- cellular	+	-	-
immediate activation of macrophages (MEM test) increase of leucocyte phagocytosis	-	-	+
mitogenicity	-	+	-
lymphokine liberation	no tested	+	+
foot pad swelling	+	-	-
potential constituent of immunotoxine and other affinotoxine	-	+	-

experiments demonstrate:
1. Conjugate of the A chain of ML I and histamin are cytotoxic to cells bearing histamine receptors.
2. The described conjugates act as immunosuppressiva, possibly because immunocompetent cells bear a high degree of histamine receptors. We classify the cytotoxic The A chain alone is completely nontoxic. In a second experiment we treated spleen cells with the conjugate and transferred these cells together with sheep erythrocytes into irradiated syngenic recipients (Fig. 4). Neither the A chain nor histamine conjugated to ovalbumin showed any effect. The histamine A chain conjugates produce a significant suppression of antibody production. Antihistaminica block the H2 receptor (e.g. cimetidin) and an ovalbumin-histamine conjugate inhibits the toxic effect of the A chain-histamine conjugate. In summary, our first A chain-histamine-conjugate as an "affinotoxin".
It cannot be excluded that conjugates consisting of an antigen and the A chain produce an antigen specific immunosuppression.

Is there a relation between the protein complex of Vester and mistletoe lectin?
Vester isolated a complex of numerous proteins (VP) from mistletoe, which he described as highly cancerotoxic (10). This statement is controversal. Recent investigations (4) suggest that there exists a relationship between "VP" and mistletoe lectine for two reasons
1. The molecular weights of the components of this complex agree over a wide range with those of mistletoe lectine and their chains (Table 2).
2. VP agglutinates human erythrocytes. The agglutination is inhibited by D-galactose.

Perhaps the investigations of Vester have been the first experiments using chains of mistletoe lectin.

Table 2
Comparison of molecular weights of the Vester proteins and of mistletoe lectins and their chains shows an extensive correspondence. Therefore it is obvious to consider the Vester proteins as mistletoe lectins or their chains. The proteins 9 and 10 are probably products of enzymatic cleavage of mistletoe lectins.

Components according to Vester (10)		Components according to Franz et al. (11)	
number	mol. wt.	mol. wt.	possible origin
1	125 000	120 000	ML I (dimer)
2	70 000	63 000	ML I (monomer)
3	63 000	60 000	ML II
4	50 000	50 000	ML III
5	48 000		
6	33 000	34 000	B chain of ML I
7	30 000	30 000	chain of ML II
8	24 000	25 000	chain of ML III
9	18 000	) probably fragments	
10	14 000	) of chains	

Conclusion

The results existing till now demonstrate a multitude of biological activities of ML I and its chains. It cannot be excluded that the products of enzymatic cleavage of both A and B chains are biologically active, too. The next few years can be expected to generate many significant and interesting results.

References

1. Franz, H., Ziska, P., Kindt, A.: Lectins, Vol. III
Editors T.C. Bøg-Hansen, G.A. Spengler
Walter de Gruyter, Berlin - New York 645-651 (1983)

2. Lutsch, G., Noll, F., Ziska, P., Kindt, A., Franz, H.: FEBS Letters 170, 335-338 (1984)

3. Olsnes, S., Stirpe, F., Sandvig, K., Pihl, A.: J. Biol. Chem. 257, 13263-13270 (1982)

4. Franz, H.: Pharmazie (in press)

5. Bergmann, P., Ziska, P., Franz, H.: 11. Congr. "Electronmicroscopy" 16.-18.1.1984 Dresden GDR, (in press)

6. Metzner, G., Franz, H., Kindt, A., Fahlbusch, B., Süss, J.: Immunbiology (in press)

7. Metzner, G., Franz, H., Kindt, A., Schumann, J., Fahlbusch, B.: in preparation

8. Walzel, H., Mix, E., Jenssen, H.-L., Ziska, P., Franz, H.: Acta histochem. 71, 41-42 (1982)

9. Eckert, R., Pfüller, U., Kindt, A., Reichelt, E., Franz, H.: in preparation

10. Vester, F.: Krebsgeschehen 9, 106-114 (1977)

DETERMINATION OF LECTIN CONTENTS IN COMMERCIAL MISTLETOE PREPARATIONS FOR CANCER THERAPY USING THE ELISA TECHNIQUE

Peter Ziska and Hartmut Franz
Staatliches Institut für Immunpräparate und Nährmedien
DDR-1120 Berlin

Extracts of mistletoe (Viscum album L.) from various sources have been used in the treatment of human cancer since 1921 (1,2), but the effect on malignant tumours is still contestable (3,4). On the other hand it is well known that the mistletoe contains compounds with toxic properties against tumour cells. Samuelsson isolated and estimated the amino acid sequence of toxic polypeptides, called Viscotoxins (5,6). Vester described the isolation of a protein-preparation containing 10 compounds with differing toxicity (7). We reported the isolation of three lectins (ML I, ML II, ML III) from mistletoe extracts (8,9). ML I is a toxic lectin very similar to ricin (10). On the other hand ML II and ML III posses only about 10 per cent of the toxicity of ML I, as demonstrated by inhibition of protein synthesis in tumour cells (11). Since extracts of mistletoe, marketed under the names of Iscador[R], Helixor[R], and Isorel[R] are widely used in the treatment of cancer, it is of interest to estimate the lectin content in these preparations. For this purpose we have developed an ELISA.

Lectins, Vol. IV

Material and methods

Preparation of anti-ML I. 20 µg of the purified ML I suspended in 1 ml 0.15 M NaCl and 1 ml Freund's Complete Adjuvant was injected subcutaneously into a rabbit at multiple sites on the back. Subsequent injections were given in Freund's Incomplete Adjuvant at 8-day intervals over 3 weeks. The two last injections were given without any adjuvant. One week after the last injection, the rabbit was bled. The blood was allowed to clot overnight at 4 °C and, after centrifugation at 1000 x g for 10 min, the serum was removed and the immunglobulin-fraction was enriched by chromatography using DEAE cellulose (12). The immunglobulin-fraction was conjugated with horseradish peroxidase as described by Wilson and Nakane (13,14).

Enzyme-linked-immuno-assay. Polystyrene plates (VEB Berlin-Chemie, GDR) were used for the assay. PBS (pH 7.2) containing 0.5 M NaCl, 0.4 g $Na_2HPO_4 \cdot 2 H_2O$ and 0.2 g NaN_3 was used as binding buffer. All other reagents were diluted in 0.15 M NaCl with 0.1 M D-galactose and 0.1 % Tween 20 added. (NaCl-Tween)

Anti-ML I antibody was diluted to a final concentration of 250 ng protein/ml PBS. Plates were coated with antibody by overnight incubation at 4 °C with 200 µl per well. The antibody was removed, the wells were washed four times with NaCl-Tween each of 3 min and 200 µl ML I in NaCl-Tween or the probe at an appropriate dilution was added per well. Plates were incubated at 37 °C for 2 h and washed four times with NaCl-Tween. 200 µl peroxidase conjugate at a suitable dilution (1:1000) in NaCl-Tween (containing 1 % v/v foetal calf serum) was added to each well and incubation continued for 3 h at 37 °C. The plates were washed again. Orthophenylenediamine 4 mg and 15 µl H_2O_2 (30 %) were combined in 10 ml fresh 0.1 M citrate buffer, pH 4.5. The enzyme reaction was allowed to proceed for 20 - 30 min at room temperature before stopping the reaction with 50 ul of 2.5 N H_2SO_4. The colour was read quantitatively on a Titertak Multiscan spectrophotometer at 492 nm. For each determination, a standard curve was generated by plotting the optical density against the concentration of ML I (2.5 - 20 ng/ml; fourfold determinations; Fig. 1).

Results

The ELISA described above does not allow differentiation between the three lectins ML I, ML II, and ML III, because ML II and ML III cross-reacted with the anti-ML I antibody. ML II and ML III cross-reacted to 12.5 % and 25 %, respectively with the ML I immunoassay (Table 1). However the total quantity of ML II and ML III in mistletoe extracts amounts only to about 15 per cent of the total quantity of isolated ML I.

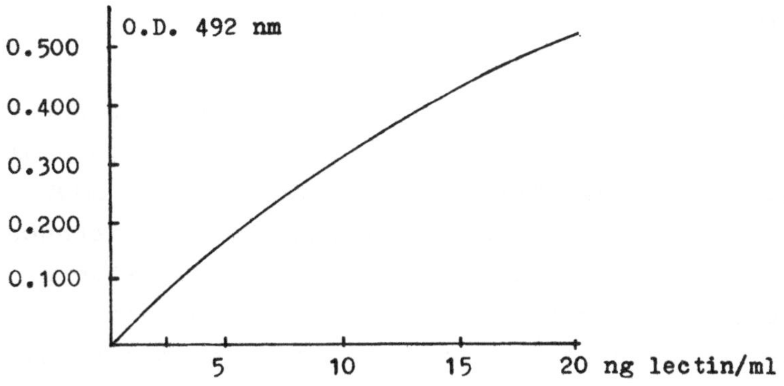

Fig. 1. ELISA titration curve of ML I against anti-ML I. Mean absorbance of fourfold determinations of ML I concentrations. For details see materials and methods.

Table 1

Cross-reaction of ML II and ML III with anti-ML I antibody using the ELISA technique.

lectin	concentration	O.D. 492 nm	corresponds to	%
ML I	20 ng/ml	0.430	-	
ML II	20 ng/ml	0.125	2.5 ng ML I/ml	12,5%
ML III	20 ng/ml	0.200	5.0 ng ML I/ml	25 %

The cross-reactivity of purified ML II and ML III was expressed relatively to ML I.

We have determined the contents of mistletoe lectins in 3 specimens of Isorel[R] (Novipharm GmbH, Pörtschach, Austria), 12 lots of Helixor[R] (Helixor-Heilmittel GmbH, Rosenfeld,FRG) 4 lots of Iscador[R] (Weleda GmbH, Arlesheim, Switzerland),and 6 specimens of Iscador[R] (Research Lab. Hiscia, Arlesheim, Switzerland). The testet preparations are extracts or juice of mistletoe. Details are not known. Sources of the material are mistletoe grown on trees as different as the fir (A, abietis), the apple (M, mali), the pine (P, pini), and the oak (Qu. quercus).

The results of our investigations concerning the total content of lectins, mainly ML I, are summarized in Table 2. The lectin content varied between 0 and about 6000 ng/ml. Looking for explanations concerning these differences, we determined the content of lectins in a mistletoe grown on one tree, an apple-tree, in relation to season (Table 3). We observed a seasonal varation, and the content of lectin in mistletoe harvested in the "cold" winter 1982 was greater than in the "warm" winter 1983. But these differences could not explain the different lectin contents in the commercial preparations.

Table 2

Measured lectin content of commercial mistletoe preparations

Sign of the product	Lot no.	ng lectin/ml
Iscador (Weleda)		
M Fr F 5 %	1 2287	920
M 5 %	2 2229	145
M F 5 %	2F 2354	630
Qu 5 %	2 2389	400
Iscador (Hiscia)		
M-F	St 3 2-2227	7
M-N	St 3 2-2231	0
M-T	St 3 2-2231	0
Qu c. Hg-T	St 3 2-2163	3
Qu c. Hg-N	St 3 2-2163	6
Qu c. Hg-F	St 3 2-2237	5
Isorel	specimens	
M St 24	"	12
M St 36	"	16
M St 60	"	750
Helixor		
A 50 mg	CH.B. 820931	4400
A 100 mg	CH.B. 830471	2600
M 50 mg	CH.B. 820825	3775
M 50 mg	CH.B. 821033	4050
M 100 mg	CH.B. 830469	1500
P 100 mg	CH.B. 830260	2000
P 100 mg	CH.B. 830473	6400
M 20 mg	CH.B. 346	230
M 30 mg	CH.B. 363	675
M 50 mg	CH.B. 331	1040
M 50 mg	CH.B. 348	1100
P 50 mg	CH.B. 353	650

Abbreviations for mistletoe grown on different trees:
A (abietis) = fir-tree; M (mali) = apple-tree; P (pini) = pine; Qu (quercus) = oak

Table 3

Time study of the lectin content of one mistletoe grown on an apple-tree.

Month and year of harvest		µg lectin/g material
November	1982	600
January	1983	1000
March	1983	400
April	1983	450
September	1983	340
October	1983	340
December	1983	480

1 g of dried mistletoe was extracted with 10 ml PBS for 16 h at 4 °C. Proper dilutions of the filtrates were used for lectin determination by the ELISA technique.

Table 4

Lectin content of mistletoe grown on different kinds of trees, harvested on one day

kind of tree	µg lectin/g material
crataegus oxyacantha (kind of hawthorn)	475
robinia pseudoaccacia (acacia)	775
acer saccharinum (maple)	500
quercus rubra (kind of oak)	1250
populus (poplar)	1800
tilia (lime-tree)	2000

1 g of dried mistletoe was extracted with 10 ml PBS for 16 h at 4 °C. Appropriate dilutions of the filtrates were used for lectin determination by the ELISA technique.

Looking for further reasons with regard to differences of the lectin content, we investigated some mistletoe material grown on different kinds of trees (acer, crataegus, populus, quercus, robinia,and tilia). The plant material was harvest on one day (february 1984) in the Botanical Garden in Dresden (GDR). We found a four-fold variation. The results are summarized in Table 4.

Discussion

The observed lectin amounts differ considerably. Thus some reasons for the different content of lectin in the commercial mistletoe preparations could be:

- the month in wich the mistletoe harvest took place
- the kind of tree on which the mistletoe has grown
- the concentration of crude material in the extracts
- the manufacturing process.

There is no proof that the lectins are effective components of the mistletoe preparations used for cancer therapy. Moreover, the exact composition of these preparations is still unknown, and the various components of the plant extracts may alter from one lot to another. Maybe the determination of lectins in mistletoe extracts can be a useful criterion in quality control. In any case, the toxic mistletoe lectin content should be taken into consideration in cancer therapy.

References

1. Leroi, R.: Helv. chir. Acta 44, 403-414 (1977).
2. Salzer, G., Müller, H.: Prax. Pneumol 32, 721-729 (1978).
3. Berger, M., Schmähl, D.: J. Cancer Res. Clin. Oncol. 105, 262-265 (1983).

4. Hülsen, H., Mechelke, F.: Arzneim.-Forsch./Drug Res. 32, 1126-1127 (1982).

5. Samuelsson, G.: Svensk. farm. tidskr. 65, 481-494 (1961).

6. Samuelsson, G., Iayavardene, A.L.: Acta Pharm. Sviecia 11, 175-184 (1974).

7. Vester, F.: Krebsgeschehen 5, 106-114 (1977).

8. Ziska, P., Franz, H., Kindt, A.: Experientia 34, 123-124 (1978).

9. Franz, H., Ziska, P., Kindt, A.: Biochem. J. 195, 481-484 (1981).

10. Stirpe, F., Legg, R.F., Onyon, L.J., Ziska, P., Franz,H: Biochem. J. 190, 843-845 (1980).

11. Stirpe, F., Ziska, P.: 1982, unpublished observations.

12. Nowotny, A.: Basic Exercises in Immunochemistry, Springer, Berlin, Heidelberg. New York 1979.

13. Nakane, P.K., Kawaoi, A.: J. Histochem. Cytochem. 22, 1084-1091 (1974).

14. Wilson, M.B., Nakane, P.K. 1978. In: Immunofluorescence and Related Staining Techniques (W. Knapp., K. Holuber and G. Wick, eds.) Pub. Elsevier, Amsterdam. pp. 215-224.

CARBOHYDRATE BINDING PROTEINS FROM *DATURA STRAMONIUM* SEEDS

Willem F. Broekaert, Willy J. Peumans
KULeuven, Laboratorium voor Plantenbiochemie,
Leuven, Belgium

Anthony K. Allen
Department of Biochemistry, Charing Cross Hospital Medical School, London W.6.

Solanaceae lectins are an homogeneous class of related proteins occurring in diverse tissues of a number of species (1). Within the genus Solanum, lectins are prominent in vegetative tissues (leaves, stems, tubers) but are present in minute amounts in the seeds. Jimson weed (Datura stramonium) lectin on the contrary, is only found in seeds where it occurs in reasonable amounts. Although there are undoubtely considerable differences between Solanaceae seed and non-seed lectins, they have a fairly similar molecular structure. Basically, Solanaceae lectins, which are glycoproteins with a high carbohydrate content (30-50 %), are built up of polypeptide chains with two very distinct domains. One of these domains comprises a glycine/cystine rich sequence (which is believed to contain the carbohydrate binding sites) whereas the second consists of a serine/hydroxyproline rich polypeptide (which is extensively glycosylated) (2,3).

Although the Datura stramonium lectin ("DSL") has been studied in detail with respect to some of its biochemical and physicochemical properties, its molecular structure is still poorly

Lectins, Vol. IV

understood. Several authors isolated lectin preparations (using different affinity chromatography procedures) that yielded two bands (Mr 45000 and 40000) upon SDS-polyacrylamide gel electrophoresis (5,6,7,8). Some of them reported the presence of contaminating lower molecular weight proteins in their preparations (3,4,5), but these 'impurities' received little attention. At present it is generally accepted that DSL is a dimeric protein composed of two non-identical subunits (Mr 45000 and 40000) (5,7,8). In this report, a new separation technique is described by which DSL can be separated into three fractions, representing all possible combinations two and two of both subunits. In addition, we present evidence that Datura seeds contain non-haemagglutinating carbohydrate binding proteins, which copurify with DSL.

Materials and Methods

Analysis methods. Agglutination assays were done with trypsin-treated human blood-group A erythrocyes as previously described (9). Protein was determined by the method of Lowry (10), using bovine serum albumin as a standard.

SDS-polyacrylamide gel electrophoresis. Lectin samples were analysed by SDS-PAGE by using a discontinuous system (11) on 12.5- 25% (w/v)-acrylamide gradient gels. The sample buffer contained 1%-β-mercaptoethanol.

Analysis of amino acids and sugars. Samples were hydrolysed in 3-M toluene-4-sulphonic acid for 24 h at 110 °C under vacuo. Amino acids were analysed on a Locarte Mini amino acid analyser. Sugars were analysed on a Perkin-Elmer F.33 gas chromatograph after methanolysis and trimethylsilylation. The two sets of analysis were related by means of internal standards. (For more details , see (12)).

Deglycosylation of glycoproteins. Dry samples were deglycosylated in trifluoromethanesuphonic acid/anisol (2/1,v/v) by the method of Edge et al.(13).

Preparation of oligomers of GlcNAc. A crude mixture of oligomers of GlcNAc was prepared by controlled hydrolysis of chitin from crab shells, according to Rupley (14).

Isolation of *Datura stramonium* lectin.
Method 1 (large scale purification): *Datura* seeds were ground in a coffee mill and extracted overnight in 0.1-M acetic acid. The suspension was centrifuged (20 min.,20 000g) and the supernatant brought at pH 4 with 4-N NaOH. Then, the extract was centrifuged again (30 min.,20 000g) and the supernatant applied to a column of chitin (Sigma Chemical Co.). Unbound proteins were eluted by extensive washing with 1-M NaCl and the lectin was desorbed with 0.5-M acetic acid. The lectin was concentrated by $(NH_4)_2SO_4$-precipitation (65% saturation) and redissolved in a minimal volume of phosphate buffered saline (PBS). After dialysis against the same buffer, the lectin fraction was further purified by gel filtration on Sephadex G-100. The lectin tends to aggregate when applied as a concentrated solution (up to 20 mg/ml). It elutes in the excluded fraction. This fraction was collected and stored at -20 °C.
Method 2 (small scale purification):*Datura* seeds were homogenized (in a mortar) in 50-mM sodium acetate buffer(pH 4) containing 0.1-M NaCl. The homogenate was centrifuged (10 min., 10 000g) and the supernatant applied to a column of chitin. After extensive washing with 1-M NaCl, the lectin was desorbed with a crude mixture of oligomers of GlcNAc. This fraction was dialysed against PBS and stored at -20 °C.
Method 3 (small scale purification):An extract, prepared as in method 2, was applied to a column of fetuin-agarose (Sigma Chemical Co.). The column was thoroughly washed with PBS and the lectin was desorbed with 0.2-M acetic acid. This fraction was dialysed against PBS and stored at -20 °C.

Results

We isolated DSL by three different procedures (using either chitin or fetuin-agarose as an affinity matrix, see Materials and Methods) and analysed the resulting fractions by gradient polyacrylamide gel electrophoresis. As shown in Fig.1A, the preparations yielded a complex pattern in which four groups of polypeptides (A,B,C and D) could be discerned. The A,B,C and D groups migrated with an average apparent Mr of 58,49,36 and 31 kd respectively, and could further be subdivised into two or three distinct bands (A_1, A_2, A_3, B_1, B_2, C_1, C_2, C_3, D_1, D_2) (Fig.1). All these polypeptides exhibited specificity for oligomers of GlcNAc, since they could specifically be desorbed from a chitin column with a crude mixture of oligomers of GlcNAc.

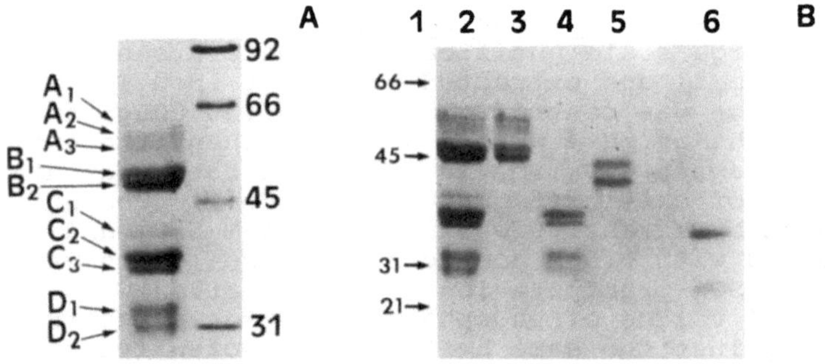

Fig.1:(A) SDS-PAGE pattern of affinity-purified fraction, with designation of the different bands. (B) SDS-PAGE patterns of affinity-purified fraction(lane 2), A/B fraction(lane 3), C/D fraction(lane 4), A/B fraction after deglycosylation(lane 5), C/D fraction after deglycosylation(lane 6). The Mr of the Mr-reference proteins(lane 1) is indicated in kd.

The elution behaviour of our preparation upon ion exchange chromatography on sulphopropyl-Sephadex, is shown in Fig.2, together with the SDS-PAGE analyses of different fractions. The separation yielded four peaks. The first one contained all A and B bands, whereas the second peak mainly consisted of C_1 and C_2 bands. In the third peak, C_2, C_3 and D_1 **bands** were prominent and the fourth peak (which was the smallest) contained merely D_2 bands.

We used trifluoromethanesulphonic acid to deglycosylate material from the first peak (the A/B fraction) and the third peak (the C/D fraction). As shown in Fig.1B, the deglycosylated A/B fraction yields only two distinct bands upon SDS-PAGE of Mr 46 and 41 kd respectively. The SDS-PAGE pattern of the C/D fraction was also markedly reduced in complexity, as only two distinct bands of Mr 33 and 28 kd respectively could be discerned after deglycosylation. We concluded that our affinity-purified preparation consisted of four different types of polypeptide backbones that are each glycosylated.

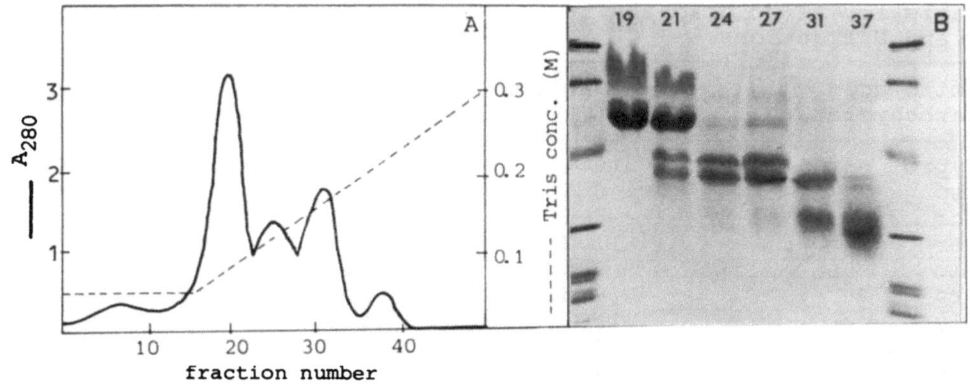

Fig.2: (A) Elution pattern of affinity-purified fraction upon ion exchange chromatography on sulphopropyl-Sephadex. The column (2.7 x 40 cm) was eluted with a linear gradient (1000 ml) from 0.1 to 0.5-M Tris-acetate, pH 5. (B) SDS-PAGE patterns of different fractions of the eluate. Numbers on top of the lanes correspond to the fraction numbers of Fig.2A.

Material of the first peak (the A/B fraction), the second peak (the C fraction) and the fourth peak (the D fraction) of Fig.2A were further purified on sulphopropyl-Sephadex (under the same conditions as specified in Fig.2A) and subsequently subjected to amino acid and carbohydrate analyses. The composition of the individual fractions is given in Table 1. The figures reveal that the amino acid composition of the A/B fraction is very similar to that of the C fraction (when compared on a molar basis). The A/B fraction, however, obviously has a higher arabinose content. In contrast, the D fraction has a totally different amino acid composition; it lacks hydroxyproline, has a fairly low cysteine content and is especially rich in serine and glycine.

Furthermore, we observed that the C and D fractions exhibited a specific agglutination activity, which was two and three orders of magnitude, respectively, lower than that of the A/B fraction.

Table 1: Amino acid and carbohydrate composition of the A/B, C and D fractions.

Amino acid Carbohydrate	A/B fraction	C fraction	D fraction
Hypro	10.1	7.1	0.0
Asp	6.6	6.3	7.2
Thr	5.6	6.1	4.1
Ser	11.2	10.1	22.2
Glu	9.6	9.6	15.1
Pro	7.8	9.4	4.0
Gly	13.7	15.3	23.0
Ala	4.8	4.6	5.5
Cys	14.3	11.0	3.8
Val	2.0	1.8	1.9
Met	0.6	0.5	0.3
Ile	0.9	0.8	1.1
Leu	2.5	2.5	2.1
Tyr	3.1	3.1	2.0
Phe	1.3	1.3	1.0
His	1.1	1.3	1.3
Lys	1.4	1.5	5.7
Arg	3.3	3.2	0.5
Trp	N.D.	N.D.	N.D.
Ara	39.5	11.9	25.3
Gal	2.7	2.0	1.0

Results are expressed as mol % of the total amino acid composition. The amount of carbohydrate is expressed as mol % , relative to total amino acid content. N.D. means 'not determined'.

Recently we developed a new and efficient method for further separation of the A/B fraction, which is based on hydrophobic interaction chromatography on phenyl-Sephadex. As shown in Fig.3A the A/B fraction eluted from the column in three peaks. SDS-PAGE analyses revealed that peak 1, 2 and 3 contained bands A, A+B and B respectively (Fig.3B). The specific agglutination activity of the three individual fractions was determined but little, if any, difference could be detected. The minimal concentration required for agglutination was about 100 ng/ml.

We used also phenyl-Sephadex to purify our C and D fractions from contaminating A and B bands . C and D fractions were applied to a column of phenyl-Sephadex (equilibrated with water)

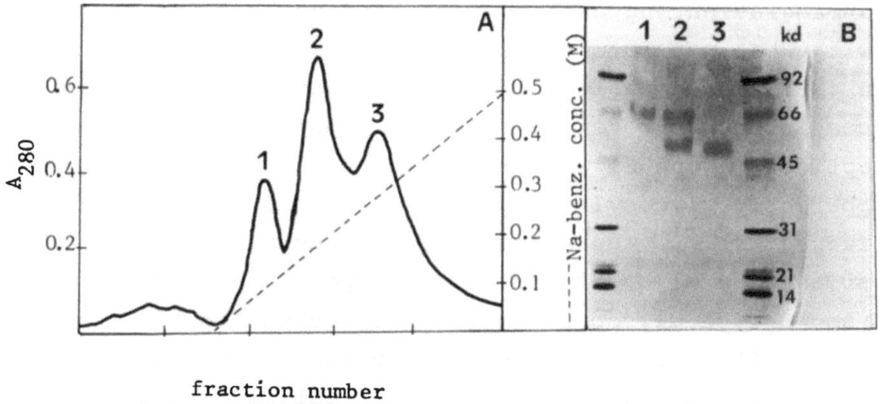

Fig.3: (A) Elution pattern of the A/B fraction upon hydrophobic interaction chromatography on phenyl-Sephadex. The column (2.7 x 20 cm) was equilibrated with water and eluted with a linear gradient (500 ml) from 0.0 to 0.5-M sodium benzoate. Protein was monitored by measurement of A_{280} after extensive dialysis against water. Sodium benzoate was monitored with a refractometer. (B) SDS-PAGE patterns of the peak fractions. Numbers in top of the lanes correspond to the peak numbers.

and the column was washed with water. Both C and D polypeptides passed through the column whereas the A and B bands were completely retained (results not shown). After this purification step, the C and D fractions were found to be completely devoid of agglutination activity. The purified fractions, however, retained their affinity for fetuin, since more than 70% of the total protein content was recovered after purification on a fetuin-agarose column.

Discussion

Datura stramonium seeds contain different types of carbohydrate binding proteins, which share in common their specificity towards oligomers of GlcNAc. A clear distinction can be drawn between haemagglutinating and non-haemagglutinating carbohydrate binding proteins .

The haemagglutinating carbohydrate binding proteins are synonyme with what was formerly designated as *Datura stramonium* lectin (DSL). A comparison of SDS-PAGE patterns of native and deglycosylated DSL revealed that DSL consists of two types of polypeptides (A and B, with Mr 46 and 41 kd respectively) that are apparently glycosylated in three (A) and two (B) different ways respectively. The A and B type glycopolypeptides migrate with an average Mr of 58 and 49 kd respectively. Obviously, the A and B type glycopolypeptides correspond to the two types of DSL-subunits (Mr 45 and 40 kd) reported in previous papers (5,7,8). Discrepancies between our Mr estimations and those reported in the literature (5,7,8), and differences in obtained resolution,are probably due to differences in applied methodology of gel electrophoresis.
In contrast to (8), we found that our DSL preparation was not homogeneous. Indeed, hydrophobic interaction chromatography allowed us to separate DSL in three fractions, probably representing AA, AB and BB type dimers. This,however, provides no unequivocal proof that these three types of dimers occur also *in vivo*. The main importance of the achieved separation is that purified A and B type glycopolypeptides are now available for further biochemical characterisation. Moreover, the fact that the A and B type glycopolypeptides can associate in all posible combinations, already indicates that they might have a similar structure.

We have also drawn our attention to the smaller proteins, which copurifie with DSL, but can not be considered as true haemagglutinins. At present,we do not know why the C and D type proteins fail to agglutinate erythrocytes. A possible explanation is that they are monovalent,or that they do not recognize any receptor on the erythrocyte membrane.
From an amino acid point of view, the C polypeptide resembles closely the A/B polypeptides. It might be possible therefore, that these proteins are proteolytic degradation products of DSL-subunits. The D polypeptide, on the contrary, has a quite

different amino acid composition. Its unusually high serine content, together with the fact that its carbohydrate moiety consists mainly of arabinose, suggest that there might occur serine-arabinosyl linkages in this protein, which have not been described as yet.

Acknowledgements. This work was supported in parts by grants from the National Fund for Scientific Research (Belgium), of which W.Peumans is Research Associate and W.Broekaert Research Assistant

References

1. Kilpatrick, D.C., Jeffree, C.E., Lockhart, C.M., Yeoman, FEBS Lett. 113, 129-133 (1980).
2. Allen, A.K., Desai, N.N., Neuberger, A., Creeth, J.M.: Biochem. J. 131, 665-674 (1978).
3. Desai, N.N., Allen, A.K., Neuberger, A.: Biochem. J. 197, 345-353 (1981).
4. Horejsi, V., Kocourek, J.: Biochim. Biophys. Acta 532, 98-104 (1978).
5. Kilpatrick, D.C., Yeoman, M.M.: Biochem. J. 184, 1151-1153 (1978).
6. Kilpatrick, D.C., Yeoman, M.M., Gould, A.R.: Biochem. J. 184, 215-219 (1979).
7. Crowley, J.F., Goldstein, I.J.: FEBS Lett. 130, 149-152 (1981).
8. Crowley, J.F., Goldstein, I.J., Arnarp, J., Lönngren, J.: Arch. Biochem. Biophys. 231, 524-533 (1984).
9. Broekaert, W.F., Nsimba-Lubaki, M., Peeters, B., Peumans, W.J.: Biochem. J. 221, 163-169 (1984).
10. Lowry, O.H., Rosebrough, N.J., Farr, A.L., Randall, R.J.: J. Biol. Chem. 193, 265-275 (1951).
11. Laemmli, U.K.: Nature (London) 227, 680-685 (1971).
12. Ashford, D., Desai, N.N., Allen, A.K., Neuberger, A., O'Neill, M.A., Selvendran, R.R.: Biochem. J. 201, 641-645 (1982).
13. Edge, A.S.B., Faltinek, C.R., Hof, L., Reichert, L.E., Weber, P.: Anal. Biochem. 118, 131-137 (1981).
14. Rupley, J.A.: Biochim. Biophys. Acta 83, 245-255 (1964).

STUDIES ON LECTINS

LXIII. ISOLATION AND CHARACTERIZATION OF A LECTIN FROM THE STINKHORN MUSHROOM (*PHALLUS IMPUDICUS* L. EX PERS.)

Gustav Entlicher, Katarína Jesenská, Ladislava Jarošová-Dejlová, Michal Jarník and Jan Kocourek

Department of Biochemistry, Faculty of Science, Charles University, Albertov 2030, CS-128 40 Praha 2, Czechoslovakia

The presence of a nonspecific lectin in the stinkhorn mushroom was reported by Mustier and Coulet already in 1965 (1). They found the lectin in the "eggs" of the mushroom, especially in the spore-producing layer - the gleba. As found by paper electrophoresis the lectin migrated in the region of β- and γ-globulins (1). However, up to the present the lectin has neither been isolated nor characterized. In the present paper we describe the isolation and characterization of this lectin. We found that its erythroagglutinating activity was inhibited by fetuin. This property was utilized in the main purification step described here.

Materials and Methods

Fruit-bodies of the stinkhorn (*Phallus impudicus* L. ex Pers.) in young closed state of development ("eggs") were collected in recreation woods in the close proximity to Prague (Czechoslovakia) in the years 1979 - 1983. Immediately after collection the "eggs" were peeled, i.e. the outer and inner membrane were removed together with the gel and discarded. Residues were stored frozen at -20°C.

Lectins, Vol. IV

Extraction and fractionation with ammonium sulfate. In a typical run, 300 g of frozen, peeled "eggs" were homogenized in a mixer with 900 ml of 0.9% NaCl. The homogenate was centrifuged at 3 400 *g* for 20 min, the supernatant was poured off and the sedimented residues were extracted with saline once more under the same conditions. Proteins contained in the pooled supernatants were fractionated by salting out with solid $(NH_4)_2SO_4$. The fraction corresponding to 20-70% saturation was collected at 3 400 *g* for 20 min, dissolved and dialyzed against distilled water. Proteins precipitating during dialysis were removed by centrifugation and the soluble proteins were freeze-dried.

Affinity chromatography. Sepharose-fetuin was prepared from CNBr-activated Sepharose 4B (Pharmacia, Uppsala) and fetuin (SEVAC, Prague) as prescribed by the producer of CNBr-activated Sepharose. About 9 mg of fetuin was bound to 1 ml of Sepharose. 200 mg of the lyophilized active protein fraction (20-70% saturation) was dissolved in 14 ml of saline. A small amount of precipitate was removed by centrifugation and the clear solution was applied to a column of Sepharose-fetuin (17 x 200 mm, 40 ml), washed with 0.25 M sodium tetraborate and equilibrated with saline. Inactive proteins were eluted with saline at a flow rate of 40 ml/h and 10 ml fractions were collected at room temperature. After the absorbance of the eluate had decreased to about 0.2, the lectin was eluted with 0.25 M sodium tetraborate at 37°C at a flow rate of 72 ml/h and 6 ml fractions were collected. Fractions containing the lectin were pooled, dialyzed against 0.01 M pyridine-formiate buffer pH 4.9 (3 x 3 l) and the solution was freeze-dried.

Titration assays. The erythroagglutinating activity as well as the erythroagglutination inhibition activity were assayed as described by Kubánek et al. (2). To study the effect of metal ions on erythroagglutinating activity 0.15 M solutions of salts were used. These solutions were used for the dissolution of the lectin as well as for dilution. Red blood cells were added in saline. The effect of the following salts was studied: $MgCl_2$, $ZnCl_2$, $MnCl_2$, $CaCl_2$, $CuCl_2$, $NiCl_2$, $CoCl_2$. In a similar way the effect of EDTA on erythroagglutinating activity was examined. The EDTA solution was titrated with 0.1 M NaOH to pH 7.0 before use.

Glycopeptide inhibitors. Mixtures of glycopeptide inhibitors from bovine erythrocytes and pig spermatozoa were prepared by the procedure described earlier (3).

Electrophoreses. Discontinuous polyacrylamide gel electrophoresis was performed in an apparatus designed by Davis (4) at pH 8.9 (5) or at pH 4.3 (6). Discontinuous electrophoresis on polyacrylamide gel in the presence of sodium dodecyl sulfate (SDS) was performed according to Laemmli (7). Samples, together with standards, were electrophoresed according to their presumed M_r in 7.5 - 15.0% polyacrylamide gels. For discontinuous affinity electrophoresis of the lectin subunit in polyacrylamide gel in alkaline medium the method of Hořejší et al. (8) was used. Fetuin bound to Sepharose 4B was used as affinity ligand with 9 and 3 mg of fetuin per 1 ml of Sepharose. A layer of homogeneously packed beads of Sepharose-fetuin formed 1/8 of the total volume of affinity gel rods.

Protein and amino acid analyses. Protein in eluates was estimated by the optical absorption at 280 nm. For amino acid analyses on an amino acid analyzer, the lectin (1 mg) was hydrolyzed with 0.5 ml of distilled 6 M HCl in a sealed tube under nitrogen at 110 °C for 24, 48 and 72 h, respectively. Dry weight, tryptophan (9, 10) and amide nitrogen (11) were estimated independently. The content of 1/2-cystine and methionine was determined after oxidation to cysteic acid and methionine sulfone, respectively, with performic acid according to the method of Schram (12). The 1/2-cystine content was also determined after S-sulfonation by the method described by Liu and Inglis (13) and after S-carboxymethylation according to Liu et al. (14) and Hirs (15). For determination of S-carboxymethyl cystein, data presented by Spackman et al. (16) and Cole et al. (17) were used. The N-terminal amino acid was determined according to the method of Zanetta et al. (18).

Carbohydrate analysis. The total neutral sugar content was determined by the phenol-sulfuric acid method of Dubois et al. (19) with D-galactose as standard. For amino sugar analysis the lectin (2 mg) was hydrolyzed with 0.5 ml of 4 M HCl in a sealed tube under nitrogen at 110 °C for 10 h. Amino sugars were determined on an amino acid analyzer (20).

Metal content. Cu, Ca, Mg, Mn, Zn, Co, Ni and Fe were determined by atomic absorption spectroscopy. A Varian-Techtron atomic absorption spectrophotometer, model 1 200, equipped with hollow cathode lamps was used.

Determination of isoelectric point. The isoelectric point of the lectin was determined by electrophoresis on cellulose acetate strips (Oxoid, Great Britain) in various 0.1 M buffers. Electrophoresis was carried out on strips 3 x 16 cm at 3 mA per strip for 1 h. The effect of the electroendoosmosis on the migration of the lectin was estimated by the use of blue dextran and used for corrections of migration distances.

Determination of free -SH groups. The number of free -SH groups in the lectin molecule was estimated by the method of

Habeeb (21).

S-Sulfonation and fractionation by gel filtration on Sephadex G-100. Disulfide bridges in the lectin molecule were split and -SH groups were blocked by S-sulfonation. The method of Cole (22) was modified. 27 mg of the lectin was dissolved in 800 µl of 0.9% NaCl and 68 µl of 0.1 M $CuSO_4$. 135 mg of crystalline Na_2SO_3 was added and the mixture was kept for 4 h at room temperature. The mixture was then applied to a Sephadex G-100 column (12 x 780 mm, 87 ml) equilibrated with a buffer which was 0.01 M in formic acid and 0.75% in pyridine (pH 6.3). Elution was carried out with the same buffer at a flow rate of 4.6 ml/h and 2.4 ml fractions were collected. Fractions corresponding to individual protein peaks were combined and freeze-dried.

Reassociation of the lectin subunit. The instability of the S-sulfo derivative of a protein in the presence of 2-mercaptoethanol (22, 23) was utilized for reassociation of the S-sulfo derivative of the lectin subunit. 1 µl of 2-mercaptoethanol was added to 1 mg of the sulfonated subunit of the lectin dissolved in 100 µl of 0.01 M pyridine-formiate buffer pH 4.9. The solution was mixed, incubated for 1 h at room temperature and freeze-dried. The obtained preparation was dissolved in water (for electrophoreses) or saline (for determination of erythroagglutinating activity) to a concentration of 0.1% and oxygen was bubbled through the solution for 1 min. The aqueous solution was then supplemented with components of sample solutions (5, 7) for electrophoreses.

Results

The lectin of stinkhorn was isolated from peeled "eggs" of this mushroom by a sequence of steps involving extraction with 0.9% NaCl, fractionation by ammonium sulfate precipitation (20-70% saturation) and affinity chromatography on Sepharose-fetuin. Table 1 summarizes data pertaining to the purification of the lectin.

During the affinity chromatography on Sepharose-fetuin, the lectin has been only retarded on the column, not tightly bound. When eluted with saline, it appears in a very broad peak after the elution of inactive material. A change of the elution

Table 1

Purification of the lectin from stinkhorn

	weight	erythroagglutinating activity (titer)[a]
peeled "eggs"	300 g	-
combined extracts	1 240 ml	32
precipitate after dialysis	0.3 g	128
protein fraction 20-70% saturation	1.3 g	512
affinity purified lectin	78 mg	2 560

[a] Erythroagglutinating activity (titer) of the extract or of a 1% solution of the freeze-dried preparations.

medium from saline to 0.25 M sodium tetraborate together with an increase in the temperature to 37 °C (to prevent crystallization of the salt) results in elution of the lectin in a discrete, relatively sharp peak. Unmodified Sepharose 4B does not retard the lectin.

The lectin preparation obtained by affinity chromatography is homogeneous on discontinuous polyacrylamide gel electrophoresis at alkaline as well as at acidic pH (Fig. 1). The presence of isolectins could not be detected.

The approximate isoelectric point of the lectin is 6.0 in acetate, maleate and phosphate buffer as shown by the electrophoretic mobility of the lectin on cellulose acetate strips (Fig. 2). At pH values above the isoelectric point, the lectin shows an anomalous migration behaviour.

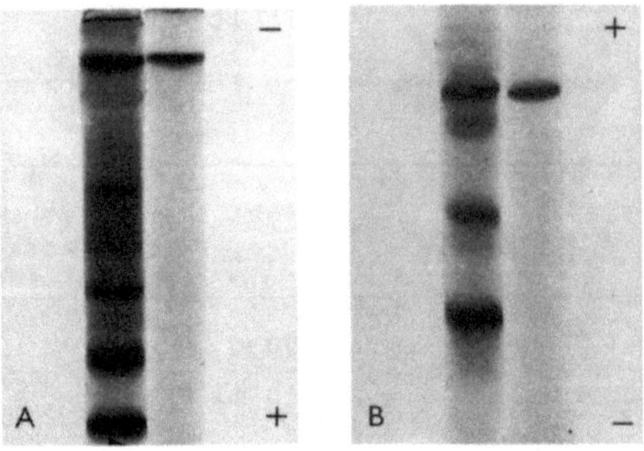

Fig. 1. Discontinuous polyacrylamide gel electrophoresis at (A) pH 8.9 and (B) pH 4.3. Gels on the left: protein fraction (20-70% saturation). Gels on the right: affinity purified lectin.

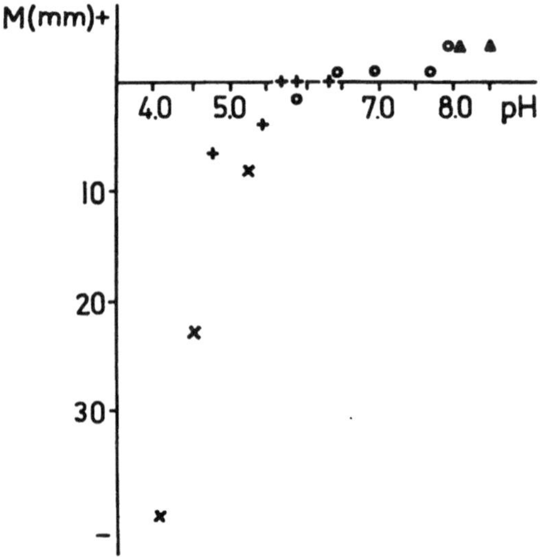

Fig. 2. Isoelectric point of the stinkhorn lectin. M (mm) - corrected migration distance in mm. × - acetate buffer, + - maleate buffer, o - phosphate buffer, Δ - carbonate buffer.

The lectin agglutinates human erythrocytes of all types of the AB0 system nonspecifically at a minimum concentration of 1.8 µg/ml. The presence of some metal cations increases its erythroagglutinating activity: Co^{2+} (64 times), Ca^{2+} (8 times), Zn^{2+} (2 times) and Mg^{2+} (2 times); the presence of some other cations (Mn^{2+}, Cu^{2+}) as well as of EDTA decreases the erythroagglutinating activity. In spite of this fact, none of the activating metal cations or any other of four additional cations examined could be found in the lectin in a significant amount.

The lectin is stable in solution at 60 °C and 70 °C; at 80 °C, however, its erythroagglutinating activity decreased rapidly.

None of the examined 25 common monosaccharides, oligosaccharides (including amino sugars), polysaccharides and glycoproteins inhibited the erythroagglutinating activity of the lectin. However, the activity was inhibited by fetuin and a glycopeptide mixture from bovine erythrocytes and pig spermatozoa.

Amino acid and carbohydrate composition of the lectin is summarized in Table 2. The results of the analysis show that the stinkhorn lectin is a glycoprotein containing 13.9% of neutral sugar and 0.9% glucosamine; relatively high is its content of aspartic acid, threonine, proline, glycine, alanine and valine. Isoleucine was found as the only N-terminal amino acid by the dansylation technique.

The lectin has a M_r of 75 000 as determined by discontinuous SDS-electrophoresis in the absence of 2-mercaptoethanol. A very similar value of M_r (74 000) was found by thin-layer gel filtration. By discontinuous SDS-electrophoresis in the presence of 2-mercaptoethanol a subunit M_r of 21 000 was found. The native lectin molecule is a tetramer. Mono-, di-, tri- and tetramers can be found on electrophoresis when the reduction is incomplete. The tetrameric structure is stabilized by six -S-S- bridges (see Table 2) - no free -SH groups are present. As the dissociation into subunits takes place in the presence

Table 2

Composition of the stinkhorn lectin

	g/100 g of the lectin[a]	Residues per mole of the lectin subunit (21 000)	
		actual	nearest integer
Lys	2.58	3.71	4
His	1.37	1.85	2
Arg	3.55	4.28	4
Asp	10.27	16.20	16
Thr	7.66	13.50	14
Ser	5.25	10.49	10
Glu	5.32	7.59	8
Pro	6.61	12.06	12
Gly	5.25	14.69	15
Ala	5.26	12.40	12
1/2-Cys[b]	1.73	2.99	3
Val	6.74	12.08	12
Met	0	0	0
Ile	3.93	7.29	7
Leu	5.32	9.87	10
Tyr	3.71	4.30	4
Phe	4.62	5.87	6
Trp	5.43	5.65	6
GlcNAc	0.88	0.84	1
GalNAc	0	0	0
Neutral sugar	13.92	16.24	16
Total	99.40	161.90	162
Amides (NH_2)	1.37	17.98	18

[a]Values corrected for a moisture content of 9.5%.

[b]An average of determinations by different methods (see Materials and Methods).

of 2-mercaptoethanol under nondenaturing conditions, noncovalent bonds participate relatively insignificantly in the stabilization of the tetrameric structure.

By S-sulfonation of the native lectin under nondenaturing conditions and gel filtration, a S-sulfonated subunit was prepared (Fig. 3).

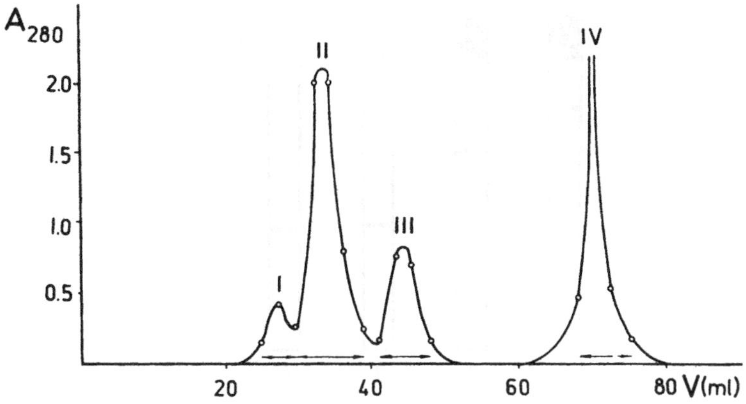

Fig. 3. Gel filtration of S-sulfonated stinkhorn lectin on Sephadex G-100. Conditions of fractionation are described under Materials and Methods. Arrows indicate the pooled fractions.

Fractions I, II, III and IV correspond to the residual tetramer, the mixture of trimer and dimer, the pure monomer and the low molecular weight, nonprotein material, respectively. The S-sulfonated subunit, soluble in aqueous media, has a very low erythroagglutinating activity (titer 10) but retains the ability (the degree of affinity not estimated) to interact with fetuin (Fig. 4).

S-sulfo groups introduced into the molecule of the subunit can be removed easily by reaction with thiols. Thus the -SH groups can be regenerated. A reassociation will take place and leads to a mixture of all oligomers up to the tetramer, however, formation of the dimer is prefered. Erythroagglutinating activity (titer) of the reassociated preparation increases to 40 compared with titer 10 found for the S-sulfonated subunit.

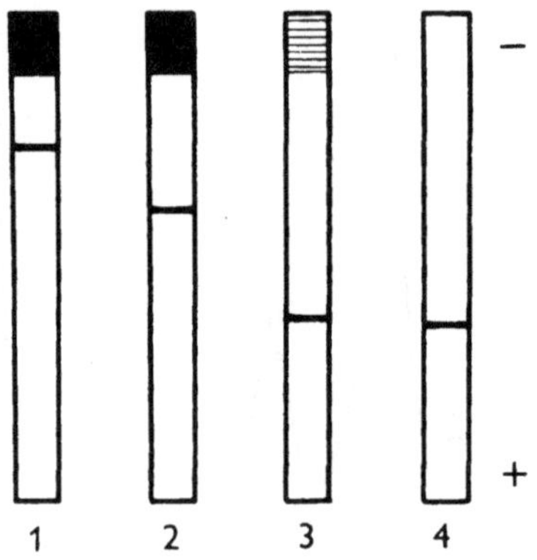

Fig. 4. Discontinuous affinity electrophoresis on polyacrylamide gel at pH 8.9 of S-sulfonated subunit of the stinkhorn lectin. Concentration of fetuin bound to Sepharose 4B: 9 mg/ml (1), 3 mg/ml (2), unmodified Sepharose 4B (3), gel without affinity ligand (4). A layer of homogeneously packed beads of Sepharose-fetuin formed 1/8 of the total volume of affinity gel rods.

Discussion

The finding that the lectin is present only in the inner parts of the stinkhorn "eggs" greatly facilitated its isolation. The gel present in the outer parts which would complicate the extraction and fractionation procedures could be removed. The lectin is present also in upper parts of the fruit-bodies in the opened state of development. However, in contrast to the "eggs", collection of this material is impeded by a very unplesant odor.

As far as the sterical arrangement of the stinkhorn lectin molecule is concerned, it is clear that the lectin molecule is a tetramer. It contains twelve 1/2-cystine residues, i.e. three 1/2-cystine residues per subunit. If two of the three 1/2-cystine residues of one subunit would form an intrachain bridge, then only one 1/2-cystine residue would remain available to establish an interchain bond. Only dimers could thus be formed - which is in disagreement with our finding of the tetrameric structure of the lectin. Theoretically two possibilities of the arrangement of the tetrameric structure remain: "square" with doubled -S-S- bridges between the two pairs of subunits and tetrahedron with isologous bonding (Fig. 5, A and B).

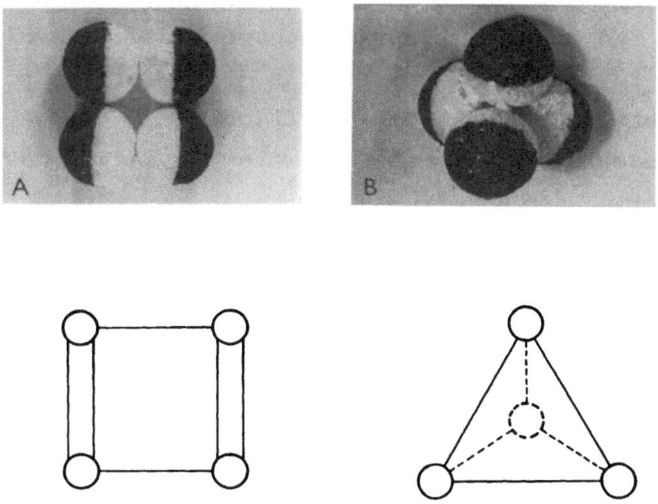

Fig. 5. Possible sterical arrangements of the native stinkhorn lectin molecule.

As a preferential formation of a dimer was never observed during the partial dissociation, we suppose that the subunits are bound in a homogeneous manner and that the tetramer have the tetrahedron arrangement (Fig. 5 B).

Some physicochemical and biological properties of the stinkhorn lectin are similar to another mushroom lectin, the *Agaricus bisporus* lectin. The *Agaricus bisporus* lectin is also a tetramer with subunits of M_r 16 000; its native molecule has a comparable M_r (64 000); it is also nonspecific in erythroagglutination and no simple saccharide was found which would inhibit this activity (24, 25). In contrast to the stinkhorn lectin, the *Agaricus bisporus* lectin contains 4 free -SH groups and no -S-S- bridges. Hence, the subunits are bound by noncovalent bonds. The exceptionality of the stinkhorn lectin lies in the fact that its subunits are bound mainly by -S-S- bonds. This finding opened the possibility to obtain the subunit under nondenaturing conditions and therefore in a relatively native state. There is some possibility, that the S-sulfonated subunit of the stinkhorn lectin behaves as a monovalent lectin in binding to saccharide structures. However, a definite proof of this property is very difficult due to the fact that no simple saccharide interacting with the lectin has been found so far.

References

1. Mustier, J., Coulet, M.: C.R. Soc. Biol. (Paris) 159, 1566-1571 (1965).
2. Kubánek, J., Entlicher, G., Kocourek, J.: Acta biol. med. germ. 41, 771-780 (1982).
3. Entlicher, G., Kocourek, J.: Methods Carbohydr. Chem. 7, 191-194 (1976).
4. Davis, B.J.: Ann N.Y. Acad. Sci. 121, 404-427 (1964).
5. Steward, F.C., Lyndon, R.F., Barber, J.T.: Am. J. Botany

52, 155-164 (1965).

6. Reisfeld, R.A., Levis, N.J., Williams, D.E.: Nature (London) 195, 281-283 (1962).

7. Laemmli, U.K.: Nature (London) 227, 680-685 (1970).

8. Hořejší, V., Tichá, M., Holý, A., Tichý, P.: Anal. Biochem. 125, 358-369 (1982).

9. Spies, J.R., Chambers, D.C.: Anal. Chem. 20, 30-39 (1948).

10. Spies, J.R., Chambers, D.C.: Anal. Chem. 21, 1249-1266 (1949).

11. Hirs, C.W.H., Stein, W.H., Moore, S.: J. Biol. Chem. 211, 941-950 (1954).

12. Schram, E., Moore, S., Bigwood, E.J.: Biochem. J. 57, 33--37 (1954).

13. Liu, T-Y., Inglis, A.S.: Methods Enzymol. 25, 55-60 (1972).

14. Liu, W., Trzeciak, H., Schüssler, H., Meienhofer, J.: Biochemistry 10, 2849-2855 (1971).

15. Hirs, C.W.H.: Methods Enzymol. 11, 199-203 (1967).

16. Spackman, D.H., Stein, W.H., Moore, S.: Anal. Chem. 30, 1190-1206 (1958).

17. Cole, R.D., Stein, W.H., Moore, S.: J. Biol. Chem. 233, 1359-1363 (1958).

18. Zanetta, J.P., Vincendon, G., Mandel, P., Gombos, G.: J. Chromatogr. 51, 441-458 (1970).

19. Dubois, M., Gilles, K.A., Hamilton, J.K., Rebers, P.A., Smith, F.: Anal. Chem. 28, 350-356 (1956).

20. Havlíková, M., Smolek, P., Entlicher, G., Kocourek, J.: J. Chromatogr. 154, 336-339 (1978).

21. Habeeb, A.F.S.A.: Methods Enzymol. 25, 457-464 (1972).

22. Cole, R.D.: Methods Enzymol. 11, 206-208 (1967).

23. Dixon, G.H., Wardlaw, A.C.: Nature (London) 188, 721-724 (1960).

24. Sage, H.J., Vasques, J.J.: J. Biol. Chem. 242, 120-125 (1967).

25. Sage, H.J., Connett, S.L.: J. Biol. Chem. 244, 4713-4719 (1969).

STUDIES ON LECTINS. LXII. A NONSPECIFIC ERYTHROAGGLUTINATING LECTIN AND A BLOOD-GROUP A SPECIFIC LECTIN IN THE MUSHROOM *AGROCYBE AEGERITA* (BRIG.) SING.

Marie Tichá, Vlasta Dudová and Jan Kocourek

Department of Biochemistry, Faculty of Science, Charles University, Albertov 2030, CS-128 40 Praha 2, Czechoslovakia

Jindřich Volc

Department of Experimental Mycology, Institute of Microbiology, Czechoslovak Academy of Sciences, Vídeňská 1083, CS-142 20 Praha 4, Czechoslovakia

Erythroagglutinating activities of fruit-bodies extracts of various species of higher fungi have been known for a long time (1); the presence of erythroagglutinating substances - lectins - was detected in about 230 species of higher fungi (2). On the other hand, there exist only a few reports on the occurence of erythroagglutinating substances in mycelial cultures of higher fungi (3,4). No attempts have been made to compare properties of lectins present in fruit bodies and mycelial cultures. The occurence of an erythroagglutinin in extracts of fruit bodies of *Agrocybe aegerita* has been reported only by Merland and Coulet (5). In the present communication we described the isolation and characterization of a blood-group nonspecific lectin from fruit bodies of *Agrocybe aegerita* and results of the investigation of lectin formation in mycelium during cultivation under different conditions.

Lectins, Vol. IV

Materials and Methods

Fruit bodies of Agrocybe aegerita have been obtained from the cultivation farm Novy Jicin - Bocheta, Czechoslovakia. The mycelial culture of *Agrocybe aegerita* (Brig.) Sing., strain III, was obtained from the Institute of Microbiology, Czechoslovak Academy of Sciences, Praha.

Culture conditions. Submerged culture: mycelium was cultuvated in a medium containing 1.5% corn-steep, 3% D-glucose and 0.15% $MgSO_4$, with shaking for 15 to 36 days at 30°C. Surface culture: mycelium was grown in Roux flasks on the same medium at laboratory temperature for 31 to 42 days to the stage of full development of fruit bodies.

Ammonium sulfate fractions. From fruit bodies: frozen fruit bodies (200-500 g) were homogenized in 0.15 M NaCl (400-1000 ml); insoluble material was removed by centrifugation. The clear supernatant was saturated to 80% with ammonium sulfate. The precipitate, dissolved in water, was dialyzed extensively against deionized water and then freeze-dried. From mycelial culture mycelia were harvested by filtration and extensively washed with water. The washed mycelia were homogenized in 0.1 M phosphate buffer pH 6.5 with an Ultra-Turrow homogenizer. Preparation of the ammonium sulfate fraction was the same as described for fruit bodies.

Affinity chromatography. The crude protein fraction corresponding to about 200 g of frozen fruit bodies or 100 mg of ammonium sulfate fraction of surface mycelial culture was dissolved in phosphate buffered saline and applied to a column (1.8 x 20 cm) of O-β-lactosyl polyacrylamide (6). The column was washed with phosphate buffered saline at a rate of 16 ml/h until the presence of the anti-A-specific lectin was detected in eluate. The nonspecific lectin was then released by a 0.2 M lactose solution. Fractions containing the nonspecific and the anti-A-specific lectins, respectively, were combined, dialyzed and freeze-dried.

Analytical methods. Discontinuous polyacrylamide gel electrophoresis was performed in alkaline (7) or acidic buffer systems (8). Polyacrylamide gel electrophoresis in the presence of sodium dodecyl sulfate (SDS) was carried out as described by Weber and Osborn (9). Affinity electrophoresis on polyacrylamide gels was performed by the method of Horejsi et al. (10) with lactosyl ligands immobilized in the form of soluble O-β-lactosyl polyacrylamide copolymer (11). Analytical electrofocusing in polyacrylamide gel was carried in the pH range 3.5-9.5 on Ampholine (LKB) polyacrylamide gel plates according to the manufacturer's instructions. Molecular weight of the lectin was established by means of gel filtration on Sephadex G-100 in phosphate buffered saline. Total neutral sugar content was determined by the phenol/H_2SO_4 method (12). Protein in chromatographic fractions was estimated by measuring absorbance at 280 nm or by the method of Lowry et al. (13). The erythtroagglutinating activity and inhibitory activity of carbohydrates were assayed by the test-tube serial dilution method, as described previously (14).

Polysaccharide fractions from fruit bodies and mycelial culture. Hot-water-soluble and dilute-alkali-soluble-fractions were isolated ad described by Kusunose et al. (15).

Results

In the extracts of fruit bodies of *Agrocybe aegerita*, the presence of two lectins differing in their physico-chemical properties, erythroagglutinating activities and saccharide-binding specificities was detected. The more abundant lectin (termed "nonspecific") agglutinated human erythrocytes of all types of the ABO system as well as rabbit and chicken erythrocytes. The other lectin present in extracts of fruit bodies (anti-A-specific-lectin) agglutinates rabbit erythrocytes and, after trypsinization, only human A_1 and A_2 erythrocytes.

Isolation and characterization of the nonspecific lectin. Affinity chromatography on O-β-lactosyl polyacrylamide copolymer (6) was used for the isolation of the nonspecific lectin from fruit bodies. The lectin was fully adsorbed to the affinity carrier and was eluted by lactose solution in a sharp peak.

The affinity-purified lectin yielded a single sharp band on polyacrylamide gel electrophoresis in alkaline medium (Fig. 1, sample 5). The lectin was not stained by the Schiff reagent (16). Isoelectric focusing on Ampholine polyacrylamide gel plates (pH 3.5-9.5) showed a single zone with pI of 4.0. Electrophoresis in the presence of SDS of both reduced and nonreduced samples yielded a single zone corresponding to an approximate molecular weight of 22 000 (Fig. 2). Molecular weight as determined by gel-filtration on Sephadex G-100 in phosphate buffered saline was approximately 44 000. Affinity electrophoresis on polyacrylamide gel containing immobilized lactosyl residues is shown in Fig. 1, sample 6; dissociation constant of the complex lectin-immobilized lactose was determined to be 3.9×10^{-4}M.

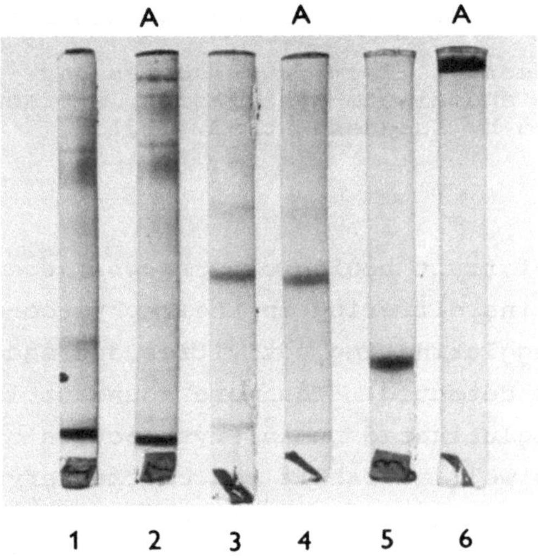

Fig. 1. Affinity electrophoresis in polyacrylamide gel in alkaline buffer.
1,2. - Ammonium sulfate fraction from fruit bodies.
3,4. - Ammonium sulfate fraction from surface mycelial culture (S-1).
5,6. - Nonspecific lectin isolated from fruit bodies.
1,3,5.-Control gels.
2,4,6.-Affinity gels (A) containing immobilized β-lactosyl ligands.

The affinity-purified lectin agglutinates human erythrocytes of all types of the ABO system at the minimum concentration of 15 μg/ml. Its erythroagglutinating activity is inhibited most effectively by lactose, soluble blood-group A substance, fetuin and the glycoprotein from edible birds´nest (Table 1). The lectin is thermostable; its erythroagglutinating activity does not change even after incubation at 80°C for 15 min.

Properties of the anti-(blood-group A)-specific lectin. The second lectin present in extracts of fruit bodies was only slightly retarded on O-β-lactosyl polyacrylamide copolymer and was eluted on the trailing edge of a peak of inactive proteins in fractions with almost zero absorbance at 280 nm.

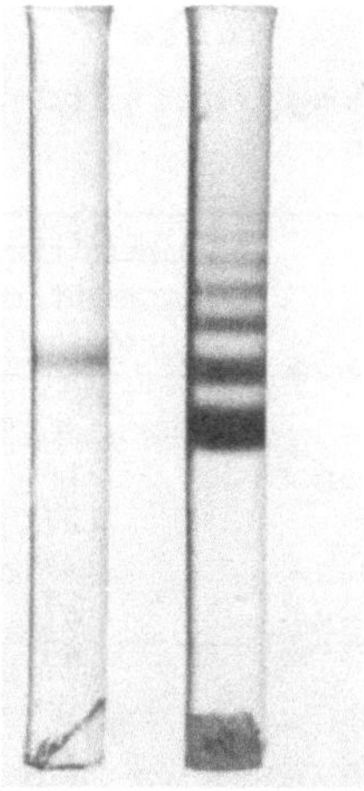

1 2

Fig. 2. Electrophoresis of the nonspecific lectin isolated from fruit bodies (polyacrylamide gel containing SDS).
1. - Nonspecific lectin isolated from fruit bodies (the same with reduced or nonreduced sample).
2. - Lysozyme and its covalent oligomers.

The lectin thus obtained yielded a very faint diffuse band on polyacrylamide gel electrophoresis in the acidic buffer system when gels were stained were stained with Coomassie Blue R after fixation with formaldehyde or trichloroacetic acid. Electrophoresis in the presence of SDS of both reduced and nonreduced samples showed a single zone corresponding to molecular weight of 17 000.

The lectin agglutinates rabbit erythrocytes and after trypsinization human A_1 and A_2 erythrocytes. Of the tested saccharides, glycoproteins and polysaccharides, only soluble blood-group A substance and hog blood group substance A+H inhibit the erythroagglutinating activity (Table 1).

Lectins in mycelial cultures. For investigation of lectin formation in the mycelium during cultivation, submerged and surface mycelia were cultivated for different periods of time. Re-

Table 1

Inhibition of the erythroagglutinating activity of lectins from fruit bodies and mycelium

Inhibitor	Minimum concentration[a] for inhibition of nonspecific lectin[b] mM (mg/ml)	anti-A-spec. lectin[c] mM (mg/ml)
Lactose	0.9	-
p-Nitrophenyl β-D-galactopyranoside	16	-
Methyl α-D-galactopyranoside	27	-
Methyl β-D-galactopyranoside	27	-
D-Galactose	27	-
N-Acetyl-D-galactosamine	45	-
N-Acetyl-D-glucosamine	-	-
D-Glucose	-	-
Methyl α-D-glucopyranoside	-	-
D-Mannose	-	-
L-Arabinose	-	-
L-Fucose	-	-
L-Rhamnose	-	-
Melibiose	-	-
D-Glucuronic acid	-	-
Soluble blood-group A substance	25×10^{-4}	3×10^{-4}
Hog blood-group substance A+H	75×10^{-4}	25×10^{-4}
EBN glycoprotein	0.6	-
Desialylated EBN glycoprotein	1.25	-
Fetuin	1.25	-
Acid α_1-glycoprotein	2.5	-
Ovomucoid	2.5	-
Desialylated fetuin	5	-
Mannan	-	-
Glycogen	-	-
Polysaccharide from mycelial culture	-	-
Polysaccharide from fruit bodies	-	-

a Inhibitory activity is expressed as the minimum concentration of sugar (mM) or glycoprotein (mg/ml) inhibiting erythroagglutination by 0.06% lectin solution (nonspecific lectin) or 0.002% lectin solution (anti-A-specific lectin).

b Lectin isolated from fruit bodies; type B erythrocytes were used.

c Lectin obtained from surface mycelial culture or from fruit bodies after removal of nonspecific lectin; type A_1 trypsinized erythrocytes were used.

- No inhibition observed; the highest concentration used was 55 mM or 10 mg per ml.

EBN - Glycoprotein from edible birds'nest.

Table 2

Erythroagglutinating activity of ammonium sulfate fractions obtained from mycelial cultures of different periods of time

Mycelial culture	Human erythrocytes	
	A_1 Tryp	^{B}Tryp
Submerged 15 d	0	0
Submerged 22 d	8	0
Submerged 29 d	8	0
Submerged 36 d	8	0
Surface S-1	64	0
Surface S-2	128	8
Fruit bodies	32 768	16 348

Agglutinating activity is expressed as titer of 1% lectin solution.
S-1 Stage of primordia formation
S-2 Fully developed fruit bodies.

sults of erythroagglutination assays of ammonium sulfate fractions obtained from the mycelial cultures are summarized in Table 2. The presence of the nonspecific lectin was detected only in the surface mycelial culture in the stage of fully developed fruit bodies (S-2); the lectin was not found in any of the submerged mycelial cultures grown for 15-36 days. On the other hand, the presence of the anti-A-specific lectin was detected both in the submerged (after the 20th day of cultivation) and in the surface mycelial cultures. The content of the lectin was considerably higher in the surface cultures than in the submerged ones.

Analogously with the anti-A-specific lectin from fruit bodies, the erythroagglutinating activity of the anti-A-specific lectin in the surface mycelial culture (S-1) was inhibited only by soluble blood-group A substance.

The chromatographic behavior of the mycelial anti-A-specific lectin on the O-β-lactosyl polyacrylamide copolymer was the same as in the case of the anti-A-specific lectin from fruit bodies; the lectin was retarded on the affinity carrier and a part of it could be separated from nonactive proteins. The electrophoretic properties of the mycelial lectin were also very similar to that of the anti-A-specific lectin from fruit bodies; only electrophoresis on polyacrylamide gel in the presence of SDS revealed a single zone, molecular weight of which corresponded to 17 000. Affinity electrophoresis on polyacrylamide gel containing immobilized β-lactosyl ligands of ammonium sulfate fractions of mycelial culture (Fig. 1, samples 3,4) as well as partially purified anti-A-specific lectin, did not reveal any retarded affinity zone. The presence of the nonspecific lectin in mycelial culture at the stage of fully developed fruit bodies was proven by affinity electrophoresis, affinity chromatography on O-β-lactosyl polyacrylamide copolymer as well as by gel chromatography on Sephadex G-100.

Discussion

In extracts of fruit bodies of *Agrocybe aegerita* we have detected the presence of two lectins greatly differing in their properties. The nonspecific lectin agglutinates human erythrocytes irrespective of blood type, rabbit and chicken erythrocytes; its activity was considerably higher against trypsin treated cells. Lactose was found to be the most powerful inhibitor. Other derivatives of D-galactose have been found to possess lower inhibitory activity. With regard to this property, the lectin resembles most to the lactose-type lectin from fruit bodies of *Laccaria amethystina* (17). Such difference between inhibition by lactose and D-galactose and its derivatives was not observed in the case of *Russula nigricans* (unpublished), *Pleurotus ostreatus* (18), *Clitocybe nebularis* and *Fomes fomentarius* (14). In contradistinction to the lactose-type lectin from *Laccaria amethystina* (17), the nonspecific lectin from *Agrocybe aegerita* is a dimer composed of two subunits bound non-covalently.

Extracts of *Agrocybe aegerita* fruit bodies contain a hemolysin which, contrary to the lectin, is thermolabile. The hemolytic activity of extracts is destroyed by an incubation at 65°C for 20 min. The lectin seems to be very thermostable. This property might be a common feature of lectins isolated from fruit bodies of higher fungi; thermostability has been described in the case of lectins from *Agaricus campestris* (19), *Pleurotus ostreatus* and *Pleurotus spodoleucus* (18) and the lectin from *Phallus impudicus* (Entlicher et al., these proceedings).

The second lectin present in extracts of fruit bodies of *Agrocybe aegerita* agglutinates trypsinized A_1 and A_2 human erythrocytes and rabbit erythrocytes; its activity was not inhibited by any of the tested saccharides and glycoproteins. Only soluble blood-group A substance was found to possess inhibitory activity. Similar properties of this kind have been described for the lectin from fruit bodies of *Flammulina velutipes* (20); it agglutinates rabbit erythrocytes and, according to our experiments, also trypsinized human A_1 and A_2 erythrocytes; Tsuda did not find any inhibitor of the erythroagglutinating activity (20).

As already mentioned, little attention has been paid to the presence of lectins in mycelial cultures of higher fungi; Pardoe et al. (4) detected erythroagglutinating activity in mycelial cultures of *Fomes fomentarius*. Banerjee et al. (3) tested submerged mycelial cultures of 7 species of higher fungi; extracts agglutinated in most cases only sheep erythrocytes. However, nothing has been reported about the presence of lectins in fruit bodies of these species. Our experiments have proven that the nonspecific lectin is formed in surface mycelial cultures only at the stage of fully developed fruit bodies. Submerged mycelial cultures and surface mycelial cultures up to the stage of primordia formation, contain only the anti-A-specific lectin. We have not succeeded to prepare a polysaccharide fraction from either the fruit bodies or the mycelial cultures which would interact with the isolated lectins.

References

1. Gold, E.R., Balding, P.: Receptor-Specific Proteins, Plant and Animal Lectins, Excerpta Medica, Amsterdam, American Elsevier, New York 1975.

2. Hořejší, V.: unpublished review.

3. Banerjee, P.C., Ghosh, A.K., Sengupta, S.: Appl. Environment. Microbiol. 44, 1009-1011 (1982).

4. Pardoe, G.I., Uhlenbruck, G., Anstee, D.J., Reifenberg, V.: Z. Immunitätsforsch. Allergie Klin. Immunol. 139, 468-485 (1969).

5. Merland, A.-M., Coulet, M.: C.R. Soc. Biol. 155, 2348-2352 (1961).

6. Hořejší, V., Kocourek, J.: Biochim. Biophys. Acta 297, 346-351 (1973).

7. Davies, B.J.: Ann. N.Y. Acad. Sci. 121, 404-427 (1964).

8. Reisfeld, R.A., Lewis, V.J., Williams, D.E.: Nature (London) 195, 281-283 (1962).

9. Weber, K., Osborn, M.: J. Biol. Chem. 244, 4406-4412 (1969).

10. Hořejší, V., Tichá, M., Kocourek, J.: Biochim. Biophys. Acta 499, 301-308 (1977).

11. Hořejší, V., Smolek, P., Kocourek, J.: Biochim. Biophys. Acta 538, 293-298 (1978).

12. Dubois, M., Gilles, A.K., Hamilton, J.K., Rebers, P.A., Smith, F.: Anal. Chem. 28, 350-356 (1956).

13. Lowry, O.H., Rosebrough, N.J., Farr, A.L., Randall, R.J.: J. Biol. Chem. 193, 265-275 (1951).

14. Hořejší, V., Kocourek, J.: Biochim. Biophys. Acta 538, 299-315 (1978).

15. Kusunose, H., Matsumara, H., Sawamura, H.: Agric. Biol. Chem. 44, 2743-2751 (1980).

16. Zacharius, R.M., Zell, T.E., Morrison, J.H., Woodlock, J.J.: Anal. Biochem. 30, 148-152 (1969).

17. Guillot, J., Genaud, L., Gueugnot, J., Damez, M.: Biochemistry 22, 5365-5369 (1983).

18. Kogure, T.: Vox Sang. 29, 221-227 (1975).

19. Sage, H.J., Connet, L.: J. Biol. Chem. 244, 4713-4719 (1969).

20. Tsuda, M.: J. Biochem. (Tokyo) 86, 1463-1468 (1979).

LECTIN INVOLVED IN FAST MOBILIZATION OF SEED RESERVES? RYE GERM AGGLUTININ (RGA) ACTIVATE THE ENDOGENOUS ACID PHOSPHATASE

Mirosława Ferens and Bronisława Morawiecka
Institute of Biochemistry, University of Wrocław,
50-137 Wrocław, Poland

The question what is the physiological role of lectins, although fascinating, is not answered so far. The findings concerning the localization of these proteins together with storage proteins and numerous hydrolytic enzymes in legume protein bodies (1) and the previous data concerning the interaction of lectins with some acid phosphatases (2-5) could give a new approach to the problem. We have found the increase of rye germ acid phosphatase (AcPase) activity in the presence of several plant lectins and the changes of its K_m and V_{max} values in the presence of conA (6).
On the other hand it is well known that acid phosphatase is deeply engaged in the mobilization of phosphorus reserves during early stages of germination (7,8). As the protein bodies are thought to be the autophagic organelles (9) and the process of reserve mobilization is in fact a controlled autolysis of their content, the new question arises: is the third component of protein bodies - a lectin - inwolved in process of germination? It was therefore interesting to check if there is an effect of activation analogous to that previously found (6) in the presence of AcPase and lectin isolated from the same tissue and what is the role of the carbohydrate part of enzyme in the interaction. The agglutinin was iso-

Lectins, Vol. IV

lated from rye germs in two laboratories so far (10,11) but data reported in these two papers differ concerning the molecular mass, amino-acid composition and sugar content of the lectin. We have developed a procedure for common isolation of both proteins: AcPase and RGA. We have also examined some physicochemical properties of the lectin and its effect on AcPase activity.

Material and Methods

Analytical. Protein was determined according to Lowry et al. (12). The acid phosphatase activity was determined as it was described earlier (6) Hemagglutination tests were carried out on microtiter plates with 2-fold serial dilution of samples in PBS. Trypsinated and glutarylated rabbit erythrocytes were used in 1% suspension. The final sample volume was 40 µl; agglutination was controlled visually after 30 min.

Purification procedure. Step 1. Proteins were extracted from the germ flour by shaking for 2 h at room temperature with 10 vol of 100 mM acetate buffer pH 5.1 (buffer A). The extract was centrifuged at 4000xg, 30 min, 5°C. The supernatant was precipitated with solid $(NH_4)_2SO_4$ to 60% of saturation. The pellet after centrifugation was suspended in buffer A and centrifuged again. Step 2. The supernatant from step 1 was precipitated with cold ethanol to 60% concentration and stored overnight. The pellet after centrifugation was suspended in 100 mM acetate buffer pH 5.8 containing Mg^{2+}, Mn^{2+}, Ca^{2+} in 1 mM concentration (buffer B), centrifuged and dialysed against the same buffer. Step 3. The supernatant from step 2 was applied to a conA-Sepharose 4B column equilibrated with the same buffer. The agglutinin does not bind to conA. AcPase binds to the column and was eluted with 5% α-metyl--D-glucoside solution. After dialysis and freeze-drying AcPase was further purified on Bio-gel P-150 (6). Step 4. The material not absorbed on conA-Sepharose was dialysed against PBS applied to affinity chromatography on N-acetyl-D-glucosamine-Spheron. The bound lectin was eluted from the column with 100 mM CH_3COOH, dialysed to water and freeze-dried. Electrophoresis in polyacrylamide gels was performed at pH 4.5 in 15% gel (13) and with SDS in 9% gel (14) Gels were stained for proteins with Coomasie Brilliant Blue R-250 and for glycoproteins (15). Phosphorylase b (94 000), BSA (67 000), ovoalbumin (43 000), carbonic anhydrase (30 000), soybean trypsin inhibitor (20 100) and α-lactoalbumin were used as molecular mass standards. Amino acid composition was determined after 24 h hydrolysis. Cysteine was determined as cysteic acid after hydrolysis in performic acid. Tryptophan was determined spectrophotometrically.

Table 1

Purification of Rye Germ Agglutinin

Purification step	Total protein mg	Agglutination activity µg of protein	Specific activity per mg of protein	Total activity $\times 10^{-6}$	Yield %
Crude extract	4300	0.8	1250	5.4	100
precipitation	2600	0.6	1670	4.3	81
Ethanol-precipitation	560	0.12	8330	4.6	87
ConA-Sepharose not-binding fraction	360	0.08	12500	4.5	83
GlcNAc-Spheron	20	0.005	250000	5.0	92

Average data of 4 preparations from 100 g of germ flour are presented.

Agglutination activity is defined as the smallest amount of protein which agglutinates rabbit erythrocytes (see Methods).

Results

In our purification procedure, we have obtained about 20 mg of purified protein from 100 g germ flour (Table 1). Agglutination occurs at lectin concentrations as low as 0.005 µg/40 µl. The hemagglutination was tested with the following sugars: mannose, glucose, galactose, arabinose, N-acetyl-galactosamine and N-acetyl-glucosamine, but inhibited by addition of N-acetyl-glucosamine only. The purified lectin is homogenous in SDS-PAGE and in ultracentrifugation. Its molecular mass in SDS-PAGE is 18000. The sedimentation coefficient is s_{20} = 3,8 S and the molecular mass calculated from ultracentrifugation data is about 40000. Therefore we conclude that the protein has a dimeric structure. RGA does not stain with basic fuchsin in polyacrylamide gels, thus it is not a glycoprotein.

Table 2
Amino acid composition of RGA and WGA

Amino acid	RGA	RGA (10)	RGA (11)	WGA form I (16)
Lys	7	8	20	6.8
His	2	2	10	0
Arg	4	3	12	4.0
Asp	16	16	43	14.9
Thr	5	4	13	5.0
Ser	11	15	27	13.7
Glu	16	18	43	16.4
Pro	7	12	20	5.2
Gly	39	40	99	41.0
Ala	8	8	27	10.1
1/2 Cys	26	38	50	37.7
Val	3	1	6	0.9
Met	2	2	5	2.0
Ile	3	2	6	2.1
Leu	4	4	10	4.0
Tyr	6	8	21	8.0
Phe	2	1	7	3.3
Try	2	1		3.2
Total	163	182	428	176
Calculated mol.mass.	17536	8273	-	-

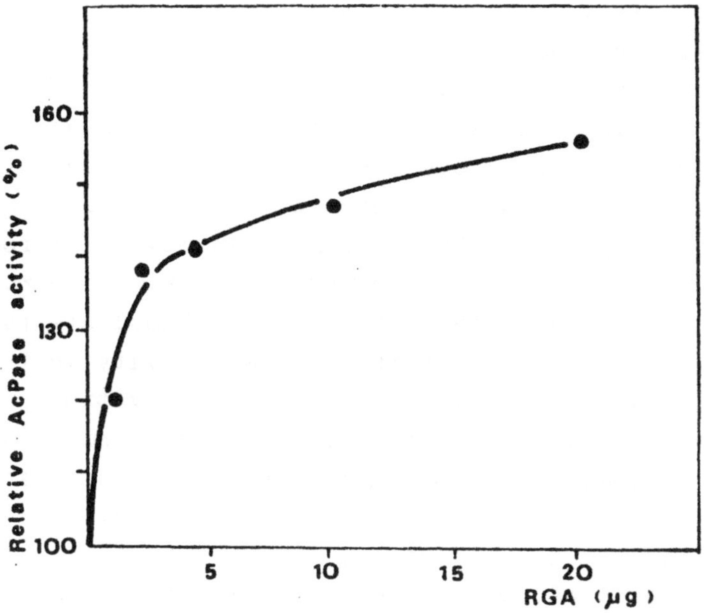

Fig.1. Changes in rye germ AcPase activity caused by RGA. 2,5 µg of AcPase was preincubated with different amounts of RGA for 10 min at 37°C before substrate was added

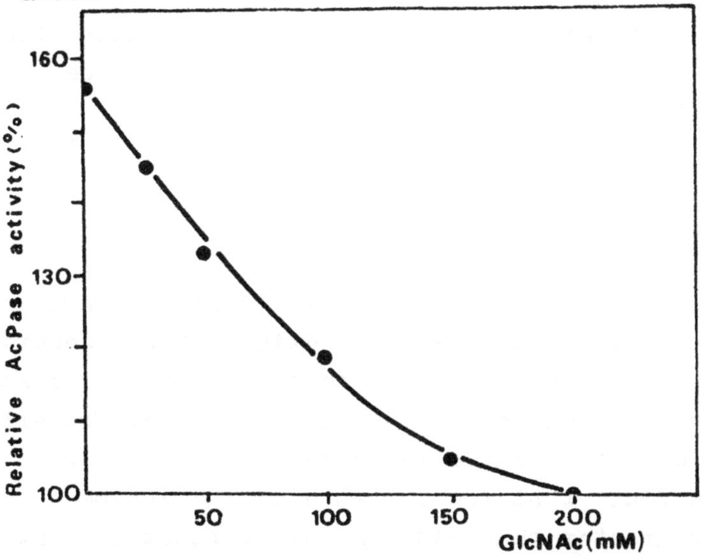

Fig.2. Effect of GlcNAc on the AcPase RGA interaction. Samples containing 2.5 µg of AcPase, 20 µg of RGA and different amounts of GlcNAc was preincubated 10 min at 37°C before substrate was added

In Table 2 we have shown the amino acid composition of RGA. The molecular mass of the lectin monomer calculated from the amino-acid composition is 17500.
We have examined the effect of the RGA on the acid phosphatase from the same tissue. We found a 60% increase of the enzyme activity in the presence of 10-fold excess (w/w) of lectin (Fig.1). When the enzyme/lectin molecular ratio was 1:1 a 40% increase of the AcPase activity was found.
In order to examine the role of the carbohydrate part of the enzyme we have determined the AcPase activity in the presence of RGA and GlcNAc (Fig.2). The enzyme activation decreased when the sugar concentration was increased and it was completely blocked at 200 mM sugar concentration. These data show that the carbohydrate moiety of the enzyme is engaged in the process of interaction.
On the other hand, we have examined the effect of AcPase on the hemagglutinating activity of RGA. 0.25 μg of the lectin was preincubated with increasing concentrations of AcPase (from 0.25 μg to 25 μg) and was analyzed in the hemagglutination test. With a high AcPase concentration, the agglutination became weaker and it disappeared at the AcPase concentration of 15 μM.

Discussion

The lectin which we have purified from rye germs is almost identical with the WGA-like lectin that was described by Peumans et al. (10) with respect to molecular mass, amino acid composition and lack of carbohydrate. Properties of the lectin do not resemble those described by Kubánek et al.(11).
We have found that the increase of the enzyme activity with RGA was greater than with any other plant lectin. The activation was found when the enzyme lectin ratio was 1:1, while in the case of exogenous lectins the high excess of added

lectin was necessary for it. Two-fold higher sugar concentration was necessary for blocking the AcPase RGA interaction than in the case of conA and other exogenous lectins. It can be suggested that the interaction between these two endogenous proteins is more specific and much stronger than with exogenous lectins.

The RGA/AcPase interaction reported here could hardly be regarded only as another example of artificial glycoconjugate binding to lectins, but it could be of physiological importance. As both proteins could be localized in protein bodies, like it was reported in legume seeds (1) it is reasonable to expect that lectin/enzyme complexes exist in vivo. During the first days of germination the increase of AcPase activity occurs, and it is only partially due to de novo synthesis (17). On the other hand it was reported, that the agglutinating activity in rice seeds decreased during first three days of germination (18). Probably it could be caused by its binding to some endogenous glycoproteins. It was found recently in our laboratory, that some grass seed RNases could be also activated in the presence of lectins (Morawiecka and Wiśniowska, unpublished).

The problem requiring further study is whether such interaction is common for glycoenzymes engaged in germination and whether it could be a system for fast mobilization of seed reserves.

Acknowledgments

This work was supported by the grant no. R.1.9. of the Ministry of Science, High Education and Technics

References

1. Chrispeels,M.J.: Phil.Trans.R.Soc.Lond. B304, 309-322, 1984
2. Lorenc-Kubis,I., Bøg-Hansen,T.C.: Lectins, Biology, Biochemistry, Clinical Biochemistry, vol.1,W.de Gruyter Berlin, p.157-167, 1981
3. Lorenc-Kubis,I., Morawiecka,B., Wieczorek,E., Wiśniowska J., Wierzba,E., Ferens,M., Bøg-Hansen,T.C. in: Lectins, Biology, Biochemistry, Clinical Biochemistry, vol.1, W. de Gruyter, Berlin, p.169-178, 1981
4. Lorenc-Kubis,I., Morawiecka,B., Bøg-Hansen,T.C.: Lectins, Biology, Biochemistry, Clinical Biochemistry, vol.2, W. de Gruyter, Berlin, p.509-515, 1982
5. Morawiecka,B., Arabska,E., Vth International Lectin Meeting Abstracts, 126-127, Bern, Switzerland, 1982
6. Ferens,M., Morawiecka,B.: Phytochemistry - in preparation
7. Yamagata,H., Tanaka,K., Kasai,Z.: Plantand Cell Physiol. 21, 1449-1460, 1980
8. Murray,D.R., Ann.Bot. 45, 273-281, 1980
9. Van der Wilden,W., Herman,E., Chrispeels,M.J.: Proc. natn.Acad.Sci. USA 77, 428-432, 1980
10. Peumans,W.J., Stinissen,H.M., Carlier,A.R.: Biochem.J. 203, 239-243, 1982
11. Kubánek,J., Entlicher,G., Kocourek,J.: Acta biol.med. germ. 41, 771-780, 1982
12. Lowry,O.H., Rosenbrough,N.I., Farr,A.L., Randall,R.J.: J.Biol.Chem. 193, 265-275, 1951
13. Reisfeld,R.A., Lewis,V.J., Williams,D.E.: Nature (London) 195, 281-283, 1962
14. Weber,K., Osborn,M.: J.Biol.Chem. 244, 4406-4412, 1969
15. Zacharius,R.M., Zell,T.L., Morrison,J.H., Woodlock,J.J.: Anal.Biochem. 30, 148-152, 1969
16. Rice,R.H., Etzler,M.E.: Biochemistry 14, 4093-4099,1975
17. Bailley,K.M., Phillips,I.D.J., Pitt,D.: J.Exp.Bot. 27, 324-336, 1976
18. Peumans,W.J., Stinissen,H.M., in: Chemical Taxonomy, Molecular Biology and Function of Plant Lectins, pp.99-116, A.R.Liss,N.Y., 1983

LECTIN AS A STORAGE PROTEIN IN ELDER (*SAMBUCUS NIGRA*) BARK

Willy Peumans, Makuta Nsimba-Lubaki and Willem Broekaert
Katholieke Universiteit Leuven, Laboratorium voor Plantenbiochemie, B-3030 Leuven (Heverlee), Belgium

Although in the past phytohemagglutinins have been studied mostly in dry seeds, and especially in seeds of legumes (1), evidence is accumulating now that they occur also in various types of vegetative tissues of a number of plant species. Well-known examples of typical non-seed lectins are the tuber, fruit and leaf lectins from Solanaceae species (2), phloem lectins from Cucurbitaceae species (3), pokeweed root lectins (4), *Bryonia dioica* root stock agglutinin (5), *Griffonia simplicifolia* leaf lectins (6) and black locust bark lectin (7). Recently we observed that elder bark contains considerable amounts of a lactose/N-acetylgalactosamine-specific lectin (8). This lectin, called *Sambucus nigra* agglutinin (SNA), can be purified in a single step by affinity chromatography on fetuin-agarose (8). SNA has been characterized into some detail with respect to its biochemical, physico-chemical and biological properties. Basically, it is a tetrameric protein composed of 2 A- and 2 B-subunits of Mr 34,500 and 37,500 respectively. Amino acid analysis revealed that the lectin is rich in asparagine/aspartic acid, glutamine/ glutamic acid, leucine and valine (8). It contains also the sulfur amino acid cysteine and methionine. In addition, SNA contains about 5.6 % covalently bound carbohydrate including 1.5 % glucosamine.

In the present report we present evidence for a remarkable circa-annual rhythm of lectin accumulation and disappearance

Lectins, Vol. IV

in elder bark. An unspecific storage role of the bark lectin is presented on the basis of the seasonal variation of the lectin content.

Materials and Methods

Materials. Bark (and other tissues too) were taken from locally growing elder trees at different times throughout the year. The age of stems and branches was determined on the basis of the branching pattern or by counting the annual rings (especially for ages above 5 years, the latter method had to be used).

Extraction. For quantitative determinations of the lectin content, small strips (0.3 cm x 1 cm) were cut from stems or branches with a scalpel blade. After removal of the outer suberinized layer the tissue was homogenized in 10 volumes of phosphate-buffered saline (v/w) and the homogenate centrifuged at 10 000 x *g* for 4 min. The resulting supernatant was used in agglutination assays.

Quantitative agglutination assays. The lectin content of bark extracts was estimated by determining the agglutination titer with trypsin-treated human erythrocytes of blood-group A.

Results

Distribution of SNA over different tissues. We found that SNA represents up to 5 % of the total (soluble) protein in the bark which corresponds to a lectin content of about 2 mg/g fresh weight. Thus, elder bark is a particularly rich source of lectin. To find out whether SNA is a bark-specific lectin, its presence and relative concentration was determined in other parts of the trees. As shown in Table 1, the lectin does not occur exclusively in bark (of stems and roots) but is found also in seeds and all other vegetative parts of the tree. However, the lectin concentration in the latter tissues is much lower than the maximal concentrations found in bark during winter or autumn.

Table 1. Distribution of SNA in different tissues of elder trees at different times of the year

Tissue	Titer of extracts		
	Early summer	Late summer	Late autumn
stem - bark	160	2,500	10,000
stem - wood	10	20	160
stem - marrow	5	10	10
root - bark	320	2,500	7,500
root - wood	40	80	10
root - tips	1	1	1
leaves (young)	320	-	-
leaves (expanded)	80	80	10
leaves - petioles	20	20	2.5
flowers - peduncles	80	-	-
flowers - petals	320	-	-
flowers - calices	80	-	-
flowers - styles	240	-	-
flowers - stamens	240	-	-
flowers - pollen grains	0	-	-
fruit juice	-	160	-
seeds	-	80	-

Titers are expressed as the highest dilution at which the extracts still agglutinated a 1%-suspension of trypsin-treated human group-A red blood cells

Table 1 indicates that the lectin content of bark tissues from both stems and roots is much higher at the end of the growing season than in the summer. Samples were taken with regular intervals (of about 14 days) throughout the year and the lectin content quantitatively determined. The results of

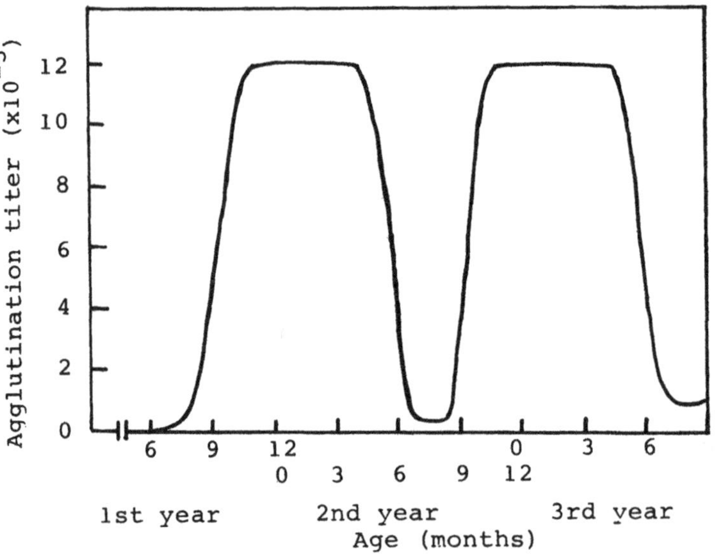

Fig.1. Agglutination titer of extracts from elder bark as a function of the age of the shoots over a 2.5 year period. Titers are expressed as the highest dilution at which the extracts still agglutinated a 1%-suspension of trypsin-treated human group-A red blood cells

our study are summarized in Figs. 1 and 2, which clearly demonstrate that the lectin content of elder bark changes periodically. Indeed, as shown in Fig.1, the lectin content of the bark of a new shoot (which starts growing around the beginning of May) remains at a low level during the summer but increases dramatically during the autumn Then it remains at the same high level during the resting season until it starts to decrease during late spring and early summer (of the second growing season). From then on, the cycle apparently starts again. Indeed, during the second year also the lectin content remains low during summer and rapidly increases again at the end of the growing season to about the same level as the first year. After this high level has been maintained during the resting season, resumption of growth during spring of the third year is accompanied by a second decrease of lectin con-

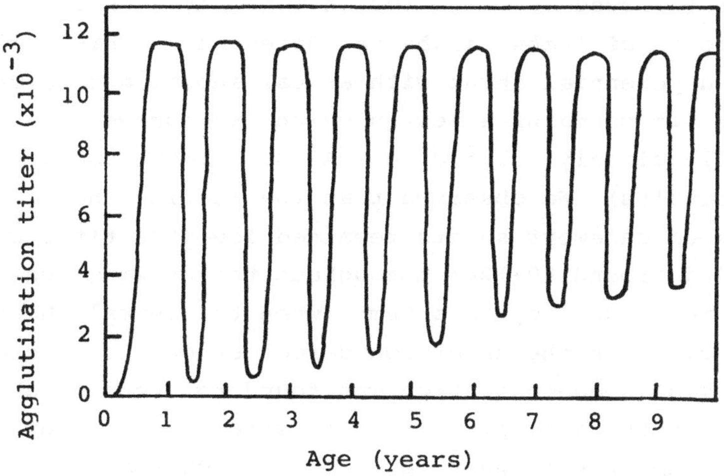

Fig.2. Circa-annual rhythm of lectin accumulation and disappearance in elder bark. Bark samples were taken with intervals of about 1 month over a one-year period from branches of different age (which were all growing on a single tree) and their lectin content determined. The interpretation of the Fig. is as follows : the first peak shows the lectin content of a new shoot from June 1983 to Juli 1984 whereas the ninth peak shows that of a branch sprouted in 1975 over the same period.

tent of the bark. The observed cycle of lectin accumulation at the end of the growing season and disappearance during the period of rapid growth of elder trees seems to be repeated year after year. Indeed, analyses of bark samples from branches of different age up to 9 years (Fig.2) demonstrated a circa-annual rhythm of lectin accumulation and disappearance. In addition, Fig.2 indicates that the lectin content of bark tissue of stems of increasing age was virtually the same during the winter maximum, whereas during the summer minimum, the lectin content apparently increased as a function of age.

A similar circa-annual rhythm was observed in bark tissue of elder roots. However, because of practical reasons, the root system could not be studied in much detail.

Finally, we followed also the lectin concentration in bark of stems and roots of *Sambucus ebulus* (Danewort) . This *Sambucus* species is a perennial shrub with annual shoots and perennial root stocks. It contains a lectin which is biochemically and serologically closely related to SNA (M. Nsimba-Lubaki; unpublished results). We observed that the lectin concentration in the bark of danewort shoots remained low (the titer of extracts varied around 80-120) throughout the growing season and did not increase during the autumn (when the annual shoots eventually die). In the underground root stocks, on the contrary, the lectin concentration was found to increase about 20-fold during the autumn. Titres of extracts from root bark collected in the winter went up to 10,000, whereas during the summer, titres did not exceed 500.

At present, the disappearance of SNA during spring and early summer is not understood at the molecular level. It seems likely, however, that the lectin is, together with most of other proteins which accumulate during autumn, degraded *in situ*. Studies are underway to follow the accumulation and disappearance of SNA in elder bark at the cellular and subcellular level.

Discussion

Bark tissue of stems and roots of elder trees contain considerable amounts of lectin, especially during the resting season. This lectin resembles in several aspects the bark lectin from black locust (*Robinia pseudoacacia*), which was the first lectin to be isolated from bark tissue (7). Indeed, both lectins are tetrameric glycoproteins composed of two different types of subunits, and have a fairly similar amino acid composition, except for the sulfur amino acids (7,8). In addition, both lectins occur in high concentrations as they represent up to

5% of the total protein content of the respective bark extracts (7,8). The presence of such high concentrations of lectin in living bark of some trees leaves intriguing questions about their possible physiological role(s). We believe that SNA fulfils an unspecific storage role for the following reasons. First, from an amino acid point of view, SNA (and black locust bark lectin) strongly resemble typical (seed) storage proteins, especially those of the globulin type (9,10). The latter proteins are characterized by high contents of asparagine/aspartic acid and glutamine/glutamic acid, amino acids which are believed to represent an important storage form of nitrogen in seeds (9,10). Second, although SNA is not exclusively located in the bark, its concentration is by far highest in this tissue, which can be considered as a typical vegetative storage tissue. In this context it should be mentioned that along with SNA, large amounts of a globulin-like protein are accumulated in the bark (Peumans; unpublished results). Third, SNA represents an important fraction of the total protein content, which argues against a specific function of this protein. Fourth, the circa-annual rhythm of lectin accumulation and disappearance during the autumn and early summer respectively is exactly what has to be expected for a protein with an (amino acid) storage role. Finally, an additional argument is provided by the observation that in *Sambucus ebulus* the lectin (which is closely related to SNA) does not accumulate in the bark of the annual shoots, but only in that of the (perennial) roots.

These arguments do not prove but indicate that SNA fulfils a storage function. It is only tempting to speculate on the basis of biochemical and physiological data that this protein is somehow involved in the accumulation and mobilization of storage material. This, however, does not exclude that SNA may have another role or fulfils diverse and multiple functions.

Acknowledgements. This work was supported in part by grants from the National Fund for Scientific Research (Belgium), of which W. Peumans is Research Associate and W. Broekaert Research Assistant.

References

1. Goldstein, I.J., Hayes, C.E.: Adv. Carbohydr. Chem. Biochem. 35, 127-140 (1978).
2. Allen, A.K., Desai, N.N., Neuberger, A., Creeth, J.M.: Biochem. J. 171, 665-674 (1978).
3. Sabnis, D.D., Hart, J. W.: Planta 142, 97-101 (1978).
4. Waxdal, M.J.: Biochemistry 13, 3671-3676 (1974).
5. Peumans, W.J., Nsimba-Lubaki, M., Carlier, A.R., Van Driessche, E.: Planta 160, 222-228 (1984).
6. Lamb, J.E., Shibata, S., Goldstein, I.J.: Plant Physiol. 71, 879-887 (1983).
7. Horejsi, V., Haskovec, C., Kocourek, J.: Biochim. Biophys. Acta 532, 98-104 (1978).
8. Broekaert, W.F., Nsimba-Lubaki, M., Peeters, B., Peumans, W.J.: Biochem. J. 221, 163-169 (1984).
9. Bailey, C.J., Boulter, D.: Eur. J. Biochem. 17, 460-466 (1979).
10. Sawai, H., Morita, Y.: Agric. Biol. Chem. 32, 76-80 (1970).

THE REMOVAL OF *Amaranthus leucocarpus* LECTIN AFFECTS THE NUTRITIONAL VALUE OF THE SEED MEAL.

A.M. Calderón de la Barca, E. Zenteno, and J.-L. Ochoa

Department of Biotechnology
CIB-The Center for Biological Research, Box 128, La Paz, B.C.S., 23000, Mexico.

M. Valencia
CIAD, Box 1735, Hermosillo, Son., Mexico.

R. Carbajal
INER, Tlalpan, Mexico, D.F.

The toxicity of lectins seems to be related to secondary effects provoked after their binding to the cell surface (1-4). That is, the agglutinating activity present in tomatoes, carrots, green peas, grapes, cherries, corn flakes, rice krispies, garlic, mushrooms etc. (5) may induce some deleterious gastrointestinal symptoms. Very recently we have been studying the lectin present in *Amaranthus leucocarpus* (6, 7) which is an edible vegetal species considered a potential source of non-animal protein (8). This lectin is a glycoprotein with 10% in weight of carbohydrate and a molecular size of about 70 KDa and pI 4.9-5.2 (6). The biological properties of this lectin and the nutritional value of the seeds meal as shown in this report, are closely related with each other.

Materials and Methods

Seeds of *Amaranthus leucocarpus* var. "alegria" were acquired in the local marked of Tulyehualco, Mexico and finely ground (Mill Arhur H. Thomas Co., USA) witha 40 mesh. One part of flour was extracted with 10 parts of 0.9% NaCl during 2 h at 4°C under constant stirring. After adjusting the pH to 4 with 4M acetic acid, the mixture was allowed to stand overnight at 4°C. A clear supernatant was decanted and the sediment centrifuged and washed three times with 0.2 M acetate buffer pH 4 (3000 x g, 15 min, 4°C). All the supernatants were pooled and concentrated by ultrafiltration with an Amicon Apparatus equiped with a PM10 Diaflo membrane (FTS Systems, Inc., Stone Ridge, N.Y., USA).

Hemagglutination activity in the supernatant was found by the two-fold serial dilution procedure with a 2% erythrocyte suspension of human blood group 0 in PBS. Specificity activity is defined ad defined as the number of hemagglutinating units per mg of protein.

Trypsin-inhibitor capacity was estimated by determining the rate of hydrolysis of N-benzoil-DL-arginine-P-nitroanilide (BAPA) (Sigma, St. Louis, MO, USA) at 405 nm in the presence of 20 µg of trypsin during5 min. The specific inhibitor activity is defined as the number of inhibitor units per mg of protein in the crude extract. One inhibitor unit is every 0.1 M BAPA not hydrolyzed under such conditions.

Stability and toxicity. The temperature and pH stability of the hemagglutinating active substance was determined by heating and pH adjustment (with 0.2 M HCl or NaOH) at different intervals. The toxicity was found by studying its effects on mice after intraperitoneal injection at several doses. Protein quality of the flours was estimated by means of the protein efficiency ratio (PER) according to standard procedures (7).

Results and discussion.

First we studied the lectin stability under a number of experimental conditions listed in Table 1, then its susceptibility to protease digestion by trypsin. It is apparent on one hand that the lectin activity is thermolabile and that boiling after 3 min irreversibly destroys its hemagglutinating capacity. On the other hand, at low and high pH it loses its capacity to bind to cell surfaces but this effect is reversible and the hemagglutinating capacity can be completely restored as long as the treatment is not too prolonged. Thus it could be expected that if amaranth seeds were eaten raw, the lectin would be able to

Table 1

The stability of Amaranth leucocarpus lectin.

Treatment	HA activity
Control	2^5
Heating, 65°C 15 min	2^5
65°C, 30 min	2^3
100°C, 1 min	2^4
100°C, 3 min	0
pH 4 - 8*	2^5

HA activity = titre of sample with 0 erythrocytes.

** = the HA activity was less at both extremes.*

Table 2

Toxocity of A. leucocarpus lectin injected intraperitoneally into mice

Sample	Specific activity	Dose mg/kg	Mortality %
Purified lectin	2048	40	0
Phaseolotoxin A*	333	50	13
		60	83
		100	100
Phaseolotoxin B*	166.6	125	42.8
	250	150	40

**From Ref. 10.*

Table 3

Effect of the extraction of A. leucocarpus lectin on seed meal protein quality.

Sample	Food intake (g)	Weight gain (g)	PER
Extracted seed meal	349.42	103.07	2.15
Whole seed meal	306.34	82.09	1.95

PER = protein efficiency ratio measured as in (7)

Table 4

Protein efficiency rates for some common foodstuffs.

Cassein	2.5
Soya protein	1.7
Defatted milk	3.1
Defatted soya	2.1
Egg	3.3
Weat flour	0.6

withstand the pH conditions in the digestive tract. However, as the hemagglutinating material present in *A. leucocarpus* is completely destroyed by cooking or extreme pH treatments, the flour can be recommended as a protein supplement (8).

It is not noteworthy that both the flour (7) and the crude extract of *A. neucocarpus* seeds possess very low trypsin-inhibitor activity (as low as 3 trypsin inhibitor units). Moreover, the agglutinating material favours lymphocyte proliferation (6) and is apparently not toxic as suggested by the low cell death rate observed in cell culture experiments with different amounts of lectin added. Furthermore, laboratory animals injected aliquots at different concentration of *A. leucocarpus* lectin indicate that this lectin is apparently innocuous (Table 2).

These results are in contrast with those of another variety of Amaranth (*Amaranthus caudatus)* which was found lethal to cells in culture (9) and with another well konwn toxic lectin isolated from *Phaseolus vulgaris* (10).

The effects of the removal of lectin activity of the seed meal employed in a diet formula given to laboratory animals (7) are shown in Table 3. The gain of weight observed in the animals is slightly different being highest for the extracted meal, thän for the whole seed meal. Extracted Amaranth seed meal possesses an elevated PER value which is regarded as the most suitable index of the quality of the protein contained in any particular foodstuff.

It is of interest to note that Amaranth seed meal scores well in protein quality above the average of some common foodstuffs (Table 4), therefore it may be indeed recommended as a food complement.

References

1. Stillmark, H.: Der Giftige eisweisskroper ricin. Seine wirkung auf das blut (Inaug. Diss. Dorpat), (1888).

2. Olsnes, S. and Pihl, A.: Abrin, Ricin and their associated agglutinins. In: Receptors and recognition (P. Cuatrecasas Ed.) Chapman and Nall, London, Series B, Vol. I, 131-173, (1976).

3. Turner, R.H. and Liener, I.R.: J. Agric. Food. Chem. 23, 484-487, (1975).

4. Higuchi, M., Suga, M., Iwai, K. Agric. Biol.: Chem. 47, 1879-1886, (1983).

5. Nachbar, M.S., and Oppenheim, J.D.: Am. J. Clin. Nutr. 33, 2338-2345, (1980).

6. Zenteno, E. and Ochoa, J.-L.: Can. J. of Biochemistry (submitted), (1984).

7. Calderon de la Barca, A.M., Valencia, M., and Ochoa, J.-L.: J. Food Sci. (in Press), (1984).

8. Sanchez-Marroquin, A.: Potencialidad industrial del Amaranto Publicacion CESTEM, Mexico, D.F., Mexico, (198o).

9. Badui-Dergal, S.: Quimmica de los alimentos, Ed. Alhambra Mexicana, S.A., Mexico, (1981).

10. Pardoe, G.I., Bird, G.W.C., Uhlenbruck, G., Sprenger, I., and Heggen, M.: Immunotaetsforsch. Allerg. Klin. Immunol. 140, 374-394, (1970).

11. Jaffe, W.G., and Gaede, K.: Nature, 183, 1329-1330, (1959).

MITOGENIC, IMMUNOSUPPRESSIVE AND PHAGOCYTIC ACTIVITY OF MACHAEROCEREUS ERUCA AND AMARANTHUS LEUCOCARPUS LECTINS

Edgar Zenteno and Jose Luis Ochoa

Centro de Investigaciones Biológicas de Baja California
Apdo. Post. 128 - La Paz, Baja California, México

Cecilia Parra, Luis Felipe Montaño, Ignacio Rayon

Instituto Nacional de Enfermedades Respiratorias, Departamento de Inmunología, Unidad de Investigación
Calzada de Tlalpan 4502, 14080 México D.F., México

Guadalupe Maldonado, Blanca Ruíz, Roger Carvajal

Departamento de Bioquímica, Facultad de Medicina, U.N.A.M.
Ciudad Universitaria - 04510 México D.F., México

Several models have been proposed to explain the immunosuppression induced by lectins based on their effect upon the immune response cells. It has been shown that Con A induces the proliferation of T suppressor cells capable to regulate, specifically and non-specifically, the antibody production of B cells (1); that pokeweed mitogen (PWM) inhibits the spontaneous citotoxic capacity of monocytes by inducing a circulating suppressor cell (2); that peanut and soybean agglutinins activate blastogenesis in direct relation to the presence or absence of prostaglandin E_2 (3); and that phytohaemagglutinin (PHA) increases the level of the prostaglandin precursor arachidonic acid (4).

The purpose of this study is to describe the effect of two lectins on the cells of the immune-response system. One of

Lectins, Vol. IV

these lectins is extracted from a Mexican cactus (*machaerocereus eruca*), and the other from the seeds of Amaranto (*amaranthus leucocarpus, var. alegria*). We report on their agglutinating capacity, mitogenic activity, immunosuppression properties and their effect on the phagocytic cells.

Material and methods

Lectins. Seeds from *phaseolus coccineus* (alubia) and *phaseolus limmensis* (lima bean) were obtained from the Graduate School of Agriculture, Chapingo, México; those of *lens culinaris* (common lentil) and *amaranthus leucocarpus* were purchased in a México city local market. Lectins from the above seeds were purified by affinity chromatography using physically entrapped stroma derived from type A human erythrocytes (5). *machaerocereus eruca* lectin was extracted by hydrophobic chromatography interaction on Phenyl Sepharose (Sigma, St.Louis Missouri) (These proceedings). The molecular weight was determined by SDS-PAGE on 10% gels, according to Laemmli (6) using the Bio-Rad molecular weight kits (Bio-Rad Lab., Richmond, CA). The sugar specificity was determined in the lectin-induced haemagglutination by testing the inhibiting capacity of 100 ul of 0.1 M solutions of the following sugars: glucose and mannose (*lens culinaris*, 45 Kd), galactose and fucose (*machaerocereus eruca*, 80 Kd), N-acetyl-D-galactosamine (*amaranthus leucocarpus*, 70 Kd; *phaseolus limmensis*, 114 Kd). *phaseolus coccineus*, 112 Kd, was not inhibited by any of this sugars. The protein concentration was determined by the method of Lowry (7). The capacity of the different lectins to induce haemagglutination is reported as specific activity :

$$\frac{\text{agglutinating titre per milliliter}}{\text{lectin protein concentration in mg/ml}}$$

Mice. Two-day old and six-week old female Balb-c and CD-1 animals were obtained from the breeding colony.

Mitogenic assay. Thymocytes and spleen cells were obtained by disgregation of the tissues in RPMI-1640 medium, and resuspended to a concentration of $4x10^6$ cells/ml. A suspension of bone marrow cells was obtained by flushing marrow plugs from the femurs with RPMI-1640 medium. The cells were washed and resuspended to $4x10^6$ cells/ml. Cell viability was determined by trypan blue dye exclusion. A total of 0.1 ml of each cell suspension was dispensed into multiple tissue culture plate wells, and 0.1 ml of culture medium containing 1-100 µg of each lectin was also added. Cultures were incubated for 72 h and pulsed for the last 18 h with 1 uCi of Methyl-^{3}H-Thymidine. Cultures were harvested and counted in a liquid scintillation system. All experiments were carried out in triplicate.

Immunosuppression assay. Six-week old Balb-c mice were injected intraperitoneally with 300 ug of lectin. Three days later the animals were immunized with fresh sheep red blood cells (SRBC). Five days later the animals were bled and their serum (heat inactivated) tested for the presence of antibodies to SRBC by a direct haemagglutination assay. Results are expressed as % of immunosuppression (I.S.)

$$\% \text{ I.S.} = 100 - \frac{\text{agglutinating titre} \times 100}{\text{serum control titre}}$$

Phagocytic activity. Six-week old CD-1 mice were killed by cervical dislocation and 5 ml of Hank's solution containing 1 U/ml of heparin was injected intraperitoneally and collected The cells were washed after lysing erythrocytes and resuspended to 2×10^6 cells/ml in RPMI-1640 medium. One ml of the cell suspension was added to flat bottom tubes each containing a coverslip for macrophage attachment, and incubated for 3 h. Non-adherent cells were washed out. Viability of this macrophage-enriched population surpassed 95%. Each tube was incubated with 20 µg of the lectin for 3 h. At the end the cells were washed to eliminate lectin excess and exposed to fresh CD-1 red blood cells or SRBC at a macrophage:erythrocyte ratio of 1:10 in RPMI-1640 medium and cultured for 1 h. Non-phagocytized erythrocytes were eliminated by extensive washing, and the adherent cells fixed for 12 h with absolute methanol. After fixation the cells were stained by the Giemsa procedure for microscopic examination. Results are expressed as phagocytic index, P.I.= No. of phagocytized erythrocytes per 100 macrophages.

Results

Haemagglutinating activity. All the lectins haemagglutinated human red blood cells (Table I). *phaseolus limmensis* agglutinated type A human erythrocytes but had no effect on B and O red cells. The four remaining lectins agglutinated human A, B and O erythrocytes as well as mouse and chicken red blood cells.

Mitogenic activity. The optimal mitogenic dose for each lectin was determined after testing several protein concentrations (1-100 ug/ml). Table II shows that at the optimal dose *lens culinaris* and *phaseolus coccineus* lectins are T cell mitogens, whereas *phaseolus limmensis* lectin did not show any

Table 1

Lectin haemagglutinating activity shown as specific activity.

Lectin	with erythrocytes from human (type A)	mouse	chicken
amaranthus leucocarpus	1863	1863	440
lens culinaris	476	476	238
machaerocereus eruca	7441	3720	930
phaseolus coccineus	6368	1542	771
phaseolus limmensis	476	0	0

mitogenic activity in adult or newborn animal cells. *machaerocereus eruca* lectin proved to be a poor T cell mitogen, whereas *amaranthus leucocarpus* lectin showed excellent mitogenic activity for bone marrow cells and in addition, was capable of stimulating total spleen cells.

Table II

Lectin mitogenic activity on Balb-c mouse spleen, thymocytes and bone marrow cells.

Lectin	conc. (ug/ml)	adult mice spleen cells	newborn mice bone marrow	thymocytes
none	-	2210±210	1215±125	1432±159
a. leucocarpus	15	19439±716	28956±2816	7554±380
l. culinaris	56	42115±2307	1240±275	48680±3070
m. eruca	25	3223±699	670±240	10325±276
ph. coccineus	50	38610±860	N.D.	17400±311
ph. limmensis	35	1014±278	300±76	N.D.

Counts are expressed as mean ± standard error mean; the control consisted of 4 x 10^5 cells cultivated under the same conditions as those of lectin-stimulated cells.

Table IV

Phagocytic activity of lectin-incubated mouse macrophages for homologous and heterologous.

Lectin	phagocytic index homologous red cells	SRBC
none	73.0	69.0
amaranthus leucocarpus	37.0	72.0
lens culinaris	73.0	87.0
machaerocereus eruca	0.0	83.0
phaseolus coccineus	75.0	87.0
phaseolus limmensis	73.0	70.0

Previous data regarding the dose-response activity of phaseolus limmensis lectin showed that 20 µg induced optimal phagocytosis. The strain was CD-1. The control was CD-1 peritoneal macrophages not exposed to lectin.

Discussion

Both new lectins we describe, *amaranthus leucocarpus* and *machaerocereus eruca*, showed a good haemagglutinating activity similar to that of most lectins (5). The sugar recognized by the *amaranthus leucocarpus* lectin on the erythrocyte surface was N-acetyl-D-galactosamine, whereas *machaerocereus eruca* lectin recognized galactose-fucose. It was interesting to observe that the *amaranthus leucocarpus* lectin was capable of agglutinating type A human erythrocytes. This blood group contains, in the terminal galactose of the H-chain, a N-acetyl-D-galactosamine residue (8). If we compare the agglutinating activity of the *amaranthus leucocarpus* lectin to that of the *phaseolus limmensis* lectin which also recognizes N-acetyl-D-galactosamine, we can conclude that the latter lectin, probably recognizes the sugar only in the terminal position whereas

Immunosuppressive activity. All the lectins except *phaseolus limmensis* inhibited the primary immune response of mice to SRBC when administered two days before immunization with SRBC (Table III).

Phagocytic activity. Table IV shows that *amaranthus leucocarpus* and *machaerocereus eruca* lectins inhibited the phagocytic activity of peritoneal macrophages to homologous erythrocytes; the former induced a 50% diminution, whereas the latter blocked entirely the ingestion capacity of macrophages. The three remaining lectins did not show any blocking effect. It was interesting to observe that when heterologous erythrocytes (SRBC) were used, none of the lectins were capable of blocking the phagocytic activity of the peritoneal macrophages. Although the table shows a slight increase in phagocytic activity for *lens culinaris*, *machaerocereus eruca* and *phaseolus coccineus* lectins, the differences were not statistically significant.

Table III

Lectin immunosuppressive activity in SRBC immunized mice.

Lectin	serum antibody titre	% of I.S.
none	128	0.0
amaranthus leucocarpus	8	93.5
lens culinaris	16	87.5
machaerocereus eruca	8	93.5
phaseolus coccineus	4	96.0
phaseolus limmensis	128	0.0

The intraperitoneal injection of 300 µg of each lectin was the optimal concentration for intuction of the immunosuppressive state in the animals. The control consisted of untreated Balb c mice.

amaranthus leucocarpus lectin and the remaining lectins recognize terminal and non-terminal sugars since they were capable of agglutinating A, B and O human erythrocytes as well as mouse and chicken red blood cells.

A great number of lectins, particularly plant lectins, have the capacity to induce the blastogenic transformation of several species lymphocytes (9). Some, such as the *phaseolus limmensis* lectin act on human though not on mouse lymphocytes (10) whereas others, such as *lens culinaris* and *phaseolus coccineus* lectins, act on cells of both species (10) as our results have confirmed.

The *amaranthus leucocarpus* lectin induced the blastoid transformation of spleen cells. Previous experiments had showed that the *amaranthus leucocarpus* lectin did not stimulate human peripheral blood mononuclear cells (Parra C., unpublished results), an observation which led us to believe that the lectin acts only on immature cells. We therefore tested newborn bone marrow-derived cells and thymocytes, the results of which showed that this lectin stimulates preferentially resting B cells in a way similar to that of LPS (11,12). The addition of mouse accessory cells to macrophage-depleted spleen cell cultures reduced the blastogenic activity of the lectin (data not shown). The satisfactory mitogenic activity shown by the lectin on total spleen cell cultures derives from the presence of resting B cells in the spleen, but the reduction in blastogenic activity observed in total spleen cell cultures, as compared to bone marrow-derived cell cultures, is explained by the higher concentration of accessory cells (5%) (8).

The *machaerocereus eruca* lectin proved to be a poor splenocyte mitogen, but it did induce a good blastogenic response on newborn mice thymocytes, which is to say that it acted upon immature T cells. The accessory cells seem not to interfere with the mitogenic activity of this lectin; accessory cells added to macrophage-depleted spleen cell cultures in the presence of the lectin did not modify the response. It is possible that the sugar recognized by the lectin on immature T cells is lost during the maturing process, such as occurs with other membrane antigens, such as the T-10 antigen on human maturating T cells (13). *machaerocereus eruca* lectin does not stimulate the blastogenic response of human peripheral blood mononuclear cells (data not shown).

Some lectins such as Con A and PWM are known to suppress the normal immune response by either activating or blocking different immune-system cell populations (1,14). Both *amaranthus leucocarpus* and *machaerocereus eruca* lectins blocked the production of antibodies for the T cell dependent antigen SRBC. This antigen requires the presence of accessory and T helper cells for optimal antibody response (15). The mechanism of action of the *machaerocereus eruca* lectin can be the stimulation of T suppressor cells, similar to that of Con A (1). Mouse immature T_s cells have a $Lyt^{2+,3+}$ phenotype which is retained even after maturation (16). Our results have shown that this lectin acts on immature T cells, although we do not know if it acts on $Lyt^{2+,3+}$ cells or on the precursor T cells which are $Lyt^{1+,2+,3+}$ (16). The latter seems improbable because, should it be the case, we would expect more cooperation and therefore a better antibody response.

The immunosuppression induced by the *amaranthus leucocarpus* lectin a mitogen for immature B cells, could be as has been shown for other B cell mitogens, the consequence of the exhaustion of B reactive cells which cannot be activated when the antigen is injected two days later (17). On the other hand, a direct action of the lectin on the accessory cells by either blocking the antigen presentation or by inhibiting the synthesis and secretion of soluble factors needed to mount a good primary immune response could not be excluded. Therefore it became necessary to study the role of this suppressive lectins on the phagocytic activity of the macrophages. Previous reports have shown that some lectins inhibit or stimulate the phagocytic activity of macrophages depending on whether they recognize galactose derivatives or galactose. Our results showed that, the *machaerocereus eruca* lectin favoured the phagocytic activity of heterologous red blood cells, while the *amaranthus leucocarpus* lectin did not inhibit phagocytosis, thus reinforcing the possibility that this lectin recognizes sugars that are not related to the stimulation or inhibition of macrophage phagocytic activity for heterologous red cells.

Highly interesting results were obtained when homologous erythrocytes were used. We found that the phagocytic activity of macrophages incubated with the *machaerocereus eruca* lectin was completely abolished, whereas macrophages stimulated with the *amaranthus leucocarpus* lectin showed only a 50% reduction in their phagocytic activity. The most likely explanation for this phenomena is that these two lectins are bound to the macrophage by means of a galactose bridge which induces the redistribution of macrophage surface receptors and thereby probably enhances the phagocytic activity of foreign particles by the macrophages. In addition, these lectins appear to have no effect on the putative "self" recognizing receptor present on the macrophage membrane, thus explaining why homologous red cells were not phagocytized and why heterologous erythrocytes ingestion was slightly enhanced.

In summary our data shows that *amaranthus leucocarpus* and *machaerocereus eruca* lectins both possess haemagglutinating properties, are mitogenic to different immune cell populations, are capable of producing immunosuppression for a T dependent antigen and finally, modify the phagocytic capacity of macrophages.

References

1. Haynes, B.F., Fauci, A.S.: J.Immunol. 120, 700-708 (1978).
2. Muchmore, A.V., Decker, J.M., Blease, R.M.: J.Immunol. 122, 61-64 (1979).
3. Novodrodsky, A., Rubin, A.L., Stenzel, K.H.: J.Immunol. 122, 1-7 (1979).
4. Parker, C.W., Stenson, W.F., Huber, M.G., Kelly, J.P.: J. Immunol. 122, 1572-1577 (1979).
5. Ochoa, J.L., Kristiansen, T.: FEBS lett. 90, 145-148 (1980)
6. Laemmli, U.K.: Nature (London) 227, 680-685 (1970).
7. Lowry, H.O., Rosenbrough, N.J., Farr, A.L., Randall, R.J.: J.Biol.Chem. 193, 265-275 (1951).
8. Stites, P.D., Fudenberg, H.H., Stobo, J.D., Wells, J.V.: Basic & Clinical Immunology, Third edition, Lange Medical Publications, New York 1982
9. Ling, N.R., Kay, J.E.: Lymphocyte Stimulation, Second edition, North-Holland Publishing Company, Amsterdam · Oxford 1975
10. Lis, H., Sharon, N.: Lectins in Higher Plants. The Biochemistry of Plants, P.K. Strumpf & E.E. Conn eds., Academic Press, New York 1981
11. Gronowicz, E., Coutinho, A.: Eur.J.Immunol. 4, 771-775 (1974).
12. Phillips, R.A., Melchers, F.: J.Immunol. 122, 1473-1475 (1979).
13. Reinherz, E.L., Schlossman, S.F.: N.E.J.M. 303, 370-373 (1980).
14. Moretta, L., Mingari, M.C., Moretta, A., Cooper, M.D.: J. Immunol. 122, 984-990 (1979).
15. Murhead, D.Y., Cudkowicz, G.: J.Immunol. 120, 579-585 (1978).
16. Janeway, C.A.: Transplant.Proc. 10, 355-359 (1978).
17. Diamanststein, T., Keppler, W., Blitstein-Willinger, E.S., Ben-Efraim, S.: Immunology 30, 401-407 (1976).

FURTHER STUDIES ON SPERM-REACTIVE LECTINS IN INDIAN PLANTS

Rajindar S. Sandhu, Sushil K.Copra, Jatinder S. Arora, Ranbir S. Reen
Department of Biology, Guru Nanak Dev University, Amritsar (143005), India

Lectins obtained from plant seeds have been extensively studied as agglutinins of erythrocytes and other animal cell types (1-4). A number of lectins are known to bind differentially to embryonic malignant and differentiated cells besides discriminating between sub-populations of lymphocytes (5,6). Similarly considerable attention has been paid to the study of lectins as polyclonal mitogens (7). However, phytolectins have remained largely unexplored as potent macromolecular cell surface probes for post-mitotic specialized cells like the spermatozoa. In our earlier report we have described 10 sperm-reactive agglutinins out of 50 new hemagglutinins detected in a survey of seeds of Indian plants (8). In further continuation of this work 25 additional phytolectins have been discovered which agglutinate one or more types of sperms and they have been characterized with respect to their biological properties and sugar specificities.

Lectins, Vol. IV

Materials and Methods

Collection of nature and dry seeds of wild plants was made from the field during the two main ripening seasons, that is, April to June and October to December in the years 1979-1982 from the following places in India: Jammu sub-mountainous region, Shivalik range, Parts of Kumaon and Garhwal Himalayas, Darjeeling and Shillong in North-eastern Himalayas, Western ghats, Balley forest in Mysore and Silent valley in Kerala. Species of *Grotalaria* were procued from the Regional Research Laboratory (CSIR), Jammu. Seeds were stored at room temperature. Their extracts were prepared in 0.05 M Tris buffered saline (TBS), pH 7.2 and dialysed as described earlier (8).

Spermatozoa suspension: Freshly ejaculated human semen was allowed to liquefy for 30 minutes at 37°C and thereafter diluted 20 times in Tris buffered saline. The sperms were separated by centrifugation at 400 xg for 10 minutes. Samples of bull semen were obtained from a local artificial insemination centre. Rabbit, guinea pig and rat sperms taken directly from cauda epididymis by puncturing were immediately suspended in TBS and centrifuged at 400 xg for 5 minutes. In all the cases, the sperms were washed thrice and a final suspension containing $1x10^7$ sperms per ml were reconsituted in TBS. Whenever needed one ml of standard suspension of spermatozoa set at pH 5.0 in normal saline was mixed with 0.5 units of neuraminidase (Type V, Sigma) and the mixture was incubated for 20 minutes at 37°C. The desialized cells thus obtained were washed thrice in TBS before using them in agglutination assays.

Agglutination: Sperm agglutination assays were carried out in microtiter plates as described earlier (8).

The active extracts were tested for their heat stabilities at elevated temperatures and their titre as well as protein concentrations were determined. Twenty four different mono-, di- and tri-saccharides were tested each at final concentrations ranging between 1.25 mM to 25 mM. Besides, native and desialized fetuin was tested at concentrations of 25 to 1000 μg per ml.

Results

The seed extracts of 181 plant species belonging to 114 genera and 46 families were tested for sperm-agglutination against human, bull, rabbit, guinea pig and rat spermatozoa. As many as 18 plant species representing 8 genera of 3 different families revealed lectin activity against one or more spermatozoa

Table 1

Distribution of Lectins in Major Plant Taxa

Taxon	Number positive/Number tested			
	Family	Genera	Species	Per cent positive
<u>Dicots</u>				
Fabaceae	1/1	5/19	15+7*/54	33.3
Other Legumes	1/2	1/15	1/38	2.6
Non-Leguminous	0/38	0/70	0/77	0
<u>Monocots</u>	1/3	2/6	2/7	28.6
<u>Gymnosperms</u>	0/2	0/4	0/5	0
Total	3/46	9/114	25/181	13.8

* Effective on neuraminidase treated spermatozoa only.

types. Besides, 7 *Crotalaria* species were found to agglutinate the neuraminidase treated (desialized) sperms. Data on the overall prevalence of phytolectins in major plant taxa are summarized in Table I.

Except for three all the sperm-agglutinins were found in the family Fabaceae. These included 8 species of the genus *Crotalaria*. But of the 37 species tested from the other two legume families one *Caesalpinia sepiaria* was found to possess lectin activity. The other two non-fabaceous lectins belonged to the monocot family Araceae. The list of sperm-agglutinins encountered in the present study along with data on their biological specificities are given in Table II. It is apparent that most of the sperm-agglutinins reacted with 2-4 types of sper-

Table II: Biological and Sugar Specificities of Sperm-Reactive Phytolectins.

Sr. No.	Source of Lectin	Stable at °C	Agglutination of Sperms					Hapten inhibitor
			Human	Bull	Rabbit	Guinea pig	Rat	
	Family Araceae							
1.	*Arisaema consanguineum*	70	+	+	+*	+	+	Asialofet
2.	*Amorphophallus bulbifer*	65	+	+	+	+	+	Asialofet
	Family Fabaceae							
3.	*Crotalaria wightiana*	70	+	+	+	+	+	GalNAc
4.	*C. walkeri*	75	+	+	+	+	+	GalNAc
5.	*Butea bojapatra*	85	+	+**	+	+	-	GalNAc
6.	*B. frondosa*	85	+	+	+	+	-	Lac
7.	*B. monosperma*	85	+	+	+	+**	-	GalNAc
8.	*Crotalaria medicaginea*	75	+	+	+	+*	-	GalNAc
9.	*Glycine hispida*	80	+	+	+	+**	-	GalNAc
10.	*Crotalaria anagyroides*	75	+	+**	+	-	-	GalNAc
11.	*Erythrina indica*	65	+	-(+)	-	+	-	GalNAc
12.	*E. umberosa*	65	+	+	-	+**	-	GalNAc
13.	*Crotalaria mucronata*	80	+	-(+)	-	-	-	GalNAc

14.	*C. steriata*	70	-(+)	+**	-	-	-	GalNAc
15.	*C. verrucosa*	70	-(+)	-(+)	-	+	-	GalNAc
16.	*C. brevifolia*	70	-(+)	-(+)	-	+	-	GalNAc
17.	*Dumasia villosa*	65	-	-(+)	+	-	-	GalNAc
18.	*Crotalaria candicans*	70	-(+)	-(+)	-	-	-	Gal
19.	*C. laburnifolia*	75	-(+)	-(+)	-	-	-	GalNAc
20.	*C. usaramoensis*	70	-(+)	-(+)	-	-	-	GalNAc
21.	*C. brownei*	65	-(+)	-	-	-	-	GalNAc
22.	*C. barbata*	70	-(+)	-	-	-	-	GalNAc
23.	*C. sericea*	65	-(+)	-	-	-	-	GalNAc
24.	*C. incana*	65	-	-(+)	-	-	-	GalNAc
	Family Caesalpinioideae							
25.	*Caesalpinia sepiaria*	60	-	-(+)	-(+)	+	-	GalNAc

+ and - signs denote positive and negative agglutination reactions respectively. Agglutination pattern mixed type unless marked with esterisk (*) which indicated head-to-head agglutination; it may change to mixed type following neuraminidase treatment (**).

+ sign within parenthesis indicated agglutination of desialized (Neuraminidase treated) spermatozoa.

GalNAc = N-acetyl-D-galactosamine; Gal = Galactose; Lac = Lactose; Asialofet = Asialofetuin matozoa while 6 phytolectins were found to be monospecific. Seed extracts of *Arisaema consanquineum*, *Amorphophallus bulbifer* and two species of *Crotalaria* reacted with all the 5 types of spermatozoa. Similarly *Crotalaria medicaginea*, *Glycine hispida* and three species of *Butea* agglutinated spermatozoa of human, bull, rabbit and guinea pig but not the rat. Two other species of *Erythrina* were found to be reactive with human and guinea pig spermatozoa only.

Amongst the monospecific lectins, *Caesalpinia sepiaria*, *Crotalaria brevifolia* and *C. verrucosa* were active on guinea pig sperma, *C. mucronata* on human sperms, *C. steriata* on bull spermatozoa and *Dumasia villosa* was found to be reactive with rabbit spermatozoa only. Besides, 7 additional species all belonging to the genus *Crotalaria* revealed lectins active on desialized sperms of one or more types, which were otherwise found to be inactive against the normal cells. (*C. candicans*, *C. laburnifolia* and *C. usaramoensis* agglutinating human and bull spermatozoa; *C. sericea*, *C. browni* and *C. barbata* agglutinating human sperms; and *C. incana* agglutinating bull sperms).

Usually, a mixed type of agglutination involving both the head and tail regions of the spermatozoa was observed (Fig. 1). However, 9 lectins induced head-to-head agglutination (Fig. 2) of guinea pig, bull and rabbit spermatozoa (3 species of *Crotalaria*, 2 of *Butea* and one each of *Caesalpinia*, *Erythrina*, *Glycine* and *Arisaema*). No tail-to-tail type of agglutination was observed with any of the phytolectins tested in this study. The desialized spermatozoa invariably responded with the mixed type of agglutination (Fig. 3).

Sugar inhibition tests revealed that except for two monocot lectins, *Arisaema consanguineum* and *Amorphophallus bulbifer*, all the other were inhibited by galactose or its derivatives; the most effective being N-acetylgalactosamine abolishing sperm-agglutination at concentration ranging between 1.25 mM

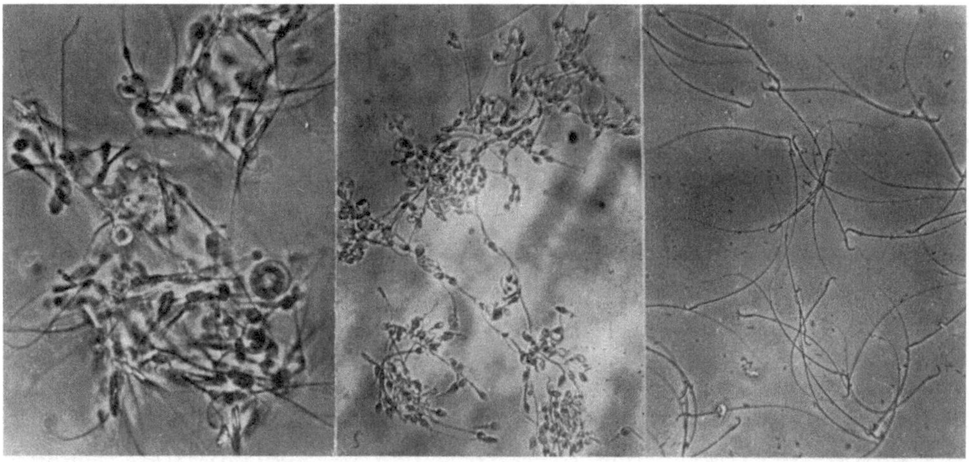

Fig. 1. Mixed-type agglutination of human spermatozoa with Amorphophallus bulbifer, of rabbit spermatozoa with Crotalaria wightiana, and of rat spermatozoa with Crotalaria walkeri.

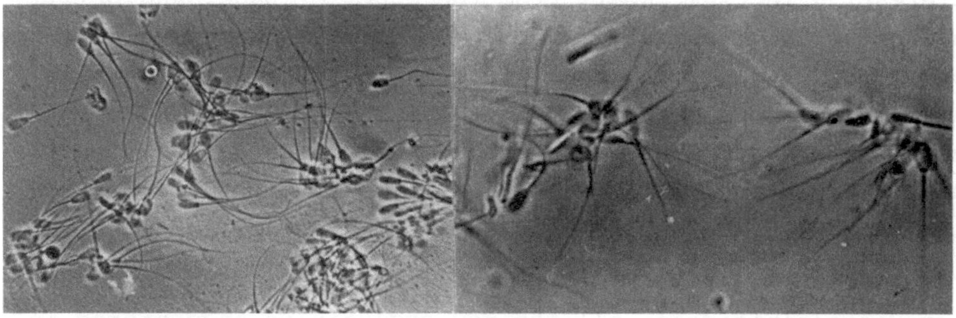

Fig. 2. Head-to-head agglutination of bull spermatozoa with Butea bojapatra and of rabbit spermatozoa by Arisaema consanguinem.

to 10 mM. Asialofetuin inhibited not only the two monocot lectins mentioned above but all the rest, while native fetuin inhibited 9 of the lectins.

In order to check thermal stability the extracts were incubated from 5 to 30 minutes at elevated temperatures viz. 55°C,

60°C, 65°C, 70°C, 75°C, 80°C, 85°C and 90°C. The highest temperature of 85°C was tolerated by 3 species of *Butea*. Two lectins, namely, *Crotalaria mucronata* and *Glycine hispida* tolerated incubation at 80°C. Four species of *Crotalaria* were stable at 75°C and 7 other species of this genus as well as *Arisaema consanguineum* withstood 70°C. Seven lectins were found to be stable at 65°C and not at higher temperatures. The *Caesalpinia sepiaria* lectin was affected adversely at 65°C but not at 60°C.

Discussion

The present study has revealed 25 new sources of sperm-agglutinins which react with one or more types of spermatozoa. These include 6 species of *Crotaliaria* which exclusively agglutinate neuraminidase treated spermcells. In an earlier report from this laboratory 10 sperm-reactive lectins have been described (8). Our studies demonstrate that a systematic screening in Indian plants holds considerable promise for the discovery or new and interesting lectins. Apart from the paper of Badel and Brilliantine (9) in which out of 50 legume species tested 23 have been reported to agglutinate spermatozoa, reports on lectin-sperm interaction are sporadic Con A, RCA, SBA (10,11,12) WGA, *Dolichos biflorus* and *Ulex europeus* (13) have been employed to localize lectin acceptors on the plasma membrane of sperma (see also repo ts by Malmi and Naaby-Hansen, these proceedings).

It is noreworthy that except for two monocot lectins and one from Caesalpinioideae, all the new agglutinins encountered in the present study belong to the family Fabaceae. The observed preponderance of fabaceous lectins is in accord with their generally higher prevelence in these legumes (1). The two monocot lectins, namely from *Arisaema consanguineum* and *Amorphophallus bulbifer*, represented the family Araceae. So far WGA is the only monocot lectin known to agglutinate sperms (14). However, unlike the latter the two new monocot lectins detected in this study are inhibited by fetuin and not by N-acetylglucosamine. Since desialized fetuin with its exposed sub-terminal galactose

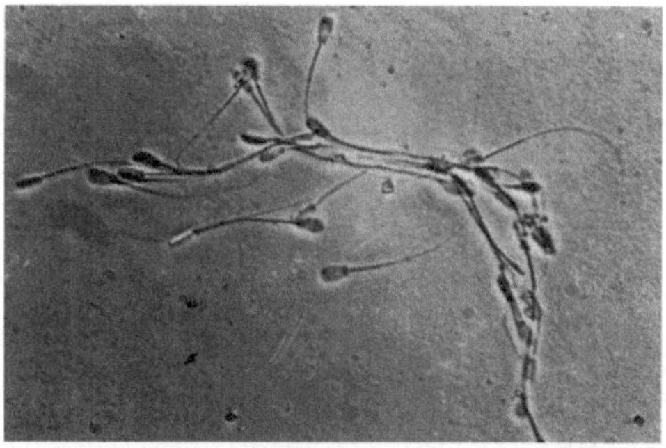

Fig. 3. Miexd-type agglutination of neuraminidase-treated bull spermatozoa with Butea bojapatra (compare Fig. 2).

moieties is much more efficaceous as inhibitor, these two lectins can be placed in the category of galactose binding lectins although due perhaps to the presence of extended binding sites for which a single galactose moiety cannot act as the hapten inhibitor (15). It is not yet clear why galactose binding lectins are more abundant than those which show affinity for mannose or its derivatives, in plants.

The spermatozoa employed in the study clearly show differential response to various lectins, the most sensitive being those of guinea pig and human. These were amenable to agglutination by 14 and 13 phytolectins respectively, while the bull and rabbit sperms responded to 11 lectins each. Among rodents rat and guinea pig presented a contrasting picture in regard to agglutinability of their erythrocytes and sperms. The erythrocytes of rat are more susceptible to lectin induced agglutination than those guinea pig (16), while the behavious of the spermatozoa is just the reverse. Differences in sperm morphology may be one important factor contributing to the agglutinability. The larger dome-shaped head of guinea pig sperm as compared to the thin sickle shaped sperm of rat might be responsible for the better presentation of lectin acceptor sites. It is noteworthy that as many as 5 lectins agglutinated the guinea pig sperms giving an exclusive head-to-head pattern. Similarly from among 6 monospecific phytolectins, 3 exclusively agglutinated the guinea pig sperms, while one each was specific for human, bull and rabbit spermatozoa. No lectin was found to agglutinate rat sperms alone. The comparison of biological action spectra of various lectins and the patterns of agglutination of normal and desialized sperms suggest that the lectin acceptors may either be distributed over the entire plasma membrane of sperm or in other cases restricted to head regions only. For example, as many as 9 lectins induced the head-to-head pattern of agglutination with guinea pig, bull and rabbit spermatozoa. Interstingly, in most of these cases it changed to the mixed type agglutination involving both the head and tail regions when desialized sperms were tested.

Lectins generally possess better heat stability than the bulk of the seed proteins. In this respect 3 species of *Butea* namely *Butea bojapatra*, *B. frondosa* and *B. monosperma* warrant special mention as they were found to withstand incubation at as high a temperature as 85°C.

Acknowledgements. The study was in part financed by Department of Science and Technology, Govt. of India, New Delhi. One of us (JSA) is a senior research fellow of CSIR.

References

1. Gold, E.R., Balding, P.: Receptor-specific proteins. Plant and Animal Lectins, American Elsevier Publishing Company Inc., New York pp 151-236 (1975).
2. Loor, F.: Exp. Cell Res. 82, 415-425 (1973).
3. Oikawa, T., Yanagimachi, R. and Nicolson, G.L.: Nature 241, 256-259 (1973).
4. Majerus, P.W., Brodie, G.N.: J. Biol. Chem. 247, 4253-4257 (1972).
5. Rapin, A.M.C., Burger, M.M.: Adv. Cancer Res. 20, 1-91 (1974).
6. Reisner, Y.: In: Bøg-Hansen, T.C. (ed.) Lectins Biology, Biochemistry, Clinical Biochemistry. W. de Gruyter, Berlin Vol. 3 pp 677-697 (1983).
7. Sharon, N.: In: Pepys, M.S., Marilyn, L.B. (eds.) Advances in Immunology. Academic Press, N.Y. Vol. 34, pp 213-298 (1983).
8. Sandhu, R.S., Reen, R.S., Singh, S., Singh, J., Chopra, S.K.: In: Bøg-Hansen, T.C. (ed.) Lectins-Biology, Biochemistry, Clinical Biochemistry. W. de Gruyter, Berlin, Vol. 2, pp 679-692 (1982).
9. Badel, P., Brilliantine, L.: Proc. Soc. Exp. Biol. Med. 130, 621-623 (1969).
10. Edelman, G.M., Millette, C.F.: Proc. Natl. Acad. Sci. USA. 68, 2436-2440 (1971).
11. Talbot, P. and Chacon, R.: J. Cell Biol. 83, 208a (1979).
12. Kashiwabara, T., Tanaka, R., Matsumoto, T.: Nature 207, 831-832 (1965).
13. Nicolson, G.L., Yanagimachi, R.: Science 177, 276-279 (1972).
14. Schwarz, M.A. and Koehler, J.K.: Biol. Reprod. 21, 1295-1307 (1979).
15. Kabat, E.A.: J. Supramol. Struct. 8, 79-88 (1978).
16. Reen, R.S.: Ph. D. thesis, Guru Nanak Dev University, Amritsar, India (1982).

NEW LYMPHOCYTE-STIMULATING LECTINS FROM INDIAN PLANTS

R.S. Sandhu, S.S. Kamboj, R.S. Reen
Department of Biology, Guru Nanak Dev University, Amritsar, India

R.J. Tatake and D. Subrahmanyam
Ciba-Geigy Research Centre, P.O. Box 9002, Bombay, India

R. Somasundaram and S.G. Gangal
Immunology Division, Cancer Research Institute, Parel, Bombay-12, India

The binding of lectins to lymphocytes is of considerable biological significance in as much as such ligands may act as pomitogens (1) besides being discriminatory for cell surface antigens crucial to the identity of subpopulations of immunocompetent cells (2). *Phaseolus vulgaris* (PHA) is routinely used as a mitogen in leukocyte culture media for karyotype studies (3) and in the determination of the immune status in various disorders like Hodgkin's disease (4). Nearly 20 other phytolectins are known to transform lymphocytes (5). In our survey of seeds of Indian cultivated and wild plants (6) many lectins were found also to agglutinate lymphocytes of man and some animal species (7). The present investigation was undertaken to explore the mitogenic poteintial of 22 lymphoagglutinins including 6 which agglutinate only neuraminidase treated lymphocytes. At least 14 of these lectins have turned out to be phytomitogens of varying potency for human lymphocytes as revealed by ^{3}H-TdR uptake.

Lectins, Vol. IV

Materials and Methods

Crude extracts were prepared by homogenizing soaked seeds (1:5 w/v) in 0.05 M Tris-buffered saline pH 7.2 at 4°C. Supernatants were incubated at elevated temperatures (60°C-75°C) for 5-10 minutes to denature extraneous proteins. The extracts were then subjected to differential ammonium sulphate precipitation. Lectin titres were determined by both hemagglutination and lymphocyte agglutination assays. Human lymphocytes were separated by Bøyum method (8) on a Histopaque discontinuous gradient. They were cultured in RPMI-1640 medium supplemented with foetal calf serum, pencicillin and streptomycin. PHA and Con A (Sigma) served as positive controls which were added at their optimum concentrations i.e. 0.5 and 5.0 μg/ml respectively. Each culture received 0.5 μCi of ^{3}H-TdR (BARC, Bombay) after 48 h of growth and harvested at 72 h with an automatic cell harvestor on glass fibre filters. The radioactivity was recorded as DPM in a scintillation counter. The mitogenic index (M.I.) represented the ratio of radioactivity incorporated in the presence and in the absence of lectin. Eahc lectin was tested on 3-5 different subjects and the highest M.I. value obtained with the best responder has been taken for comparison.

Results

The data on the mitogenic index (M.I.) of 22 lectins are summarized in Table 1. Apparently, 14 lectins induced significant stimulation of lymphocytes. Excepting 2 species of *Datura* (Solanaceae) and one of *Arisaema* (Araceae) the mitogenic lectins belong to the family Fabaceae. The values of mitogenic index varied from 10.9 for *Datura metel* while the positive controls containing PHA or Con A gave M.I. values of 136.1 and 18.3 repsectively. At least 8 of these lectins, namely, *Crotalaria candicans*, *Phaseolus coccineus* (Syn. *P. multiflours)*, *Erythrina subrosa*, *Canavalia obtusifolia*, *Arisaema consanquineum*, *Lablab niger*, *Datura stramonium* and *Crotalaria walkeri* were found to be highly mitogenic. The stimulation of lymphocyte proliferation with these lectins was in the range of 47.17 - 116.67 per cent of that obtained with PHA. With another 4 lectins viz., *Butea boiapatra*, *B. frondosa*, *B. monosperm* and *Erythrina indica*, the M.I. values varied between 20.0 and 40.9 and were higher than that of Con A used as the second positive control. The M.I. of *Crotalaria verrucosa* and *Datura metel*

Table 1

Mitogenicity of lymphoagglutinins as revealed by invitro incorporation of ^{3}H-TdR by human lymphocytes

Lymphoagglutinins	Lectin preparation Protein µg/0.2 ml			Mitogenic index		Hapten inhibitor
	MLAC	Range tested	Optimum	Absolute	Relative	
Phaseolus vulgaris (PHA)	-	-	0.1	136.1	100.0	Fetuin
Canavalin ensiformis (Con A)	-	-	1.0	18.3	13.44	Man
Crotalaria candicans (*)	250.0	10-250	100.0	158.8	116.67	GalNAc
Phaseolus coccineus	6.8	3-250	10.0	138.6	101.83	Fetuin
Erythrina subrosa	120.9	30-300	150.0	101.5	74.57	GalNAc
Canavalia obtusifolia	8.7	10-250	250.0	94.8	69.65	Man
Arisaema consanguineum (*)	80.6	10-250	10.0	96.5	70.90	Fetuin
Lablab niger	120.7	3-300	30.0	68.0	49.96	U
Datura stramonium	5.5	10-100	10.0	66.4	48.78	GlcNAc
Crotalaria walkeri (*)	84.0	10-250	100.0	64.2	47.17	GalNAc
Butea frondosa	10.8	10-250	50.0	40.9	30.05	GalNAc
B. boiapatra	33.3	10-250	30.0	27.0	19.83	GalNAc
Erythrina indica	193.5	150-300	300.0	20.0	14.69	GalNAc
Butea monosperma	40.3	10-250	50.0	20.0	14.69	GalNAc
Crotalaria verrucosa (*)	10.5	10-100	50.0	12.3	9.03	GalNAc
Datura metel	3.4	10-250	250.0	10.9	8.00	GlcNAc
Erythrina umberosa	157.3	3-300	300.0	1.6	1.17	GalNAc
E. arborescens	150.0	10-250	10.0	1.4	1.02	GalNAc
Glycine hispida (*)	1.7	3-300	3.0	1.2	0.88	GalNAc
Parkia biglandulosa (*)	137.5	3-300	30.0	1.1	0.80	U
Dolichos lablab	213.3	10-250	10.0	1.1	0.80	Man
Abrus precatorius	8.9	3-300	30.0	0.9	0.66	GalNAc
Bauhinia tomentosa	4.4	10-250	10.0	0.8	0.58	GalNAc
Abrus fruticulosus	8.5	3-300	300.0	0.7	0.51	GalNAc

(*) React with desialized human lymphocytes; MLAC, Minimal lymphocyte agglutinating concentration; U, Undetermined

was relatively low. The remaining 8 lectins failed to stimulated the proliferation of lymphocytes. Apparently, of the 6 lectins which agglutinated the desialized lymphocytes 4 turned out to be mitogenic for normal lymphocytes.

The sugar specificities as determined by agglutination inhibition assays revealed that a majority of the new phytomitogens responded to N-acetylgalactosamine or galactose. As expected species of *Canavalia* and *Datura* were found to be specific for mannose and N-acetylglucosamine respectively while *Lablab niger* did not respond to any of the test sugars. Desialized fetuin inhibited *Phaseolus coccineus* and *Arisaema consanquineum*. Except for *Dolichos lablab* (specific for mannose) and *Parkia biglandulosa* (undertermined sugar specificity) all lectins lacking mitogenicity were inhibited by N-acetylgalactosamine in the lymphocyte agglutination.

Discussion

The present study has revealed the mitogenic nature of as many as 14 lymphoagglutinins tested from Indian plants. Although the mitogenic potential of the different lectins varied considerably, the association of this property with lymphoagglutination has been found to be more than 60 per cent. A comparison of mitogenic indicies of various lectins (Fig. I) suggests that 8 of these may be regarded as strong mitogens and another 4 as fairly effective phytomitogens. The former include *Crotalaria candicans*, *Phaseolus coccineus*, *Erythrina subrosa*, *Canavalia obtusifolia*, *Arisaema consanguineum*, *Lablab niger*, *Datura stramonium* and *Crotalaria walkeri*, while the latter category includes *Butea frondosa*, *B. bojapatra*, *B. monosperma* and *Erythrina indica*. The mitogenicity of another two lectins, that is *Crotalaria verrucosa* and *Datura metel* is lower than that of Con A which severed as the second positive contro. Interestingly, the lectins which agglutinate only the desialized but not the untreated cells may also be strongly mitogenic as is the case with *Crotalaria candicans*. Apparently, a

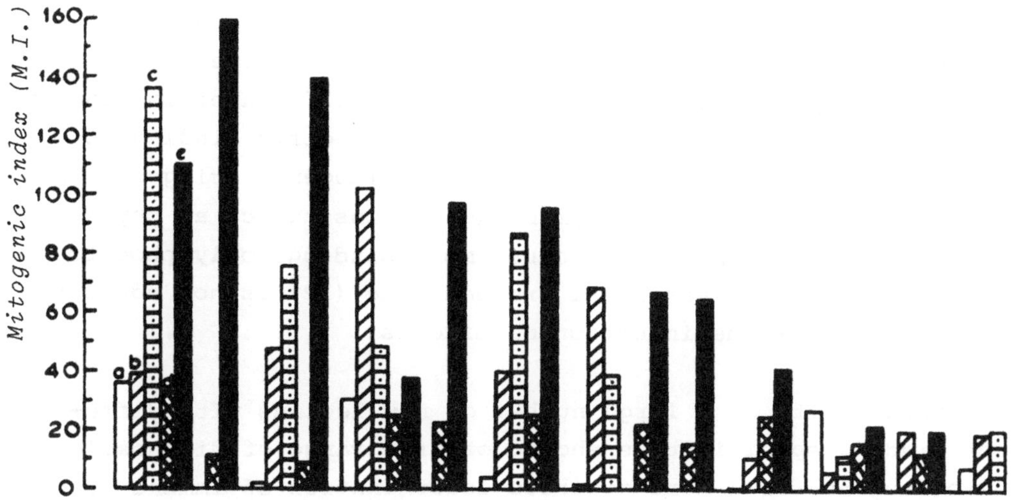

Fig. 1. The proliferative response of human peripheral blood lymphocytes from healthy individuals a,b,c,d and e, to 12 different lymphoagglutinins partially purified from Indian plants. The bars depict the highest mitogenic index (^{3}H-TdR incorporation) achieved in each individual at the following lectin concentration given as µg proteiner per 0.2 ml of culture medium: 1. PHA (0.1, 2); 2. Crotalaria candicans (250, 100); 3. Phaseolus coccineus (30, 15, 3, 10); 4. Erythrina subrosa (300, 150, 100); 5. Arisaema consanguineum (50, 10); 6. Canavalin obtusifolia (15, 3, 30, 50, 250); 7. Lablab niger (30); 8. Datura stramonium (10); 9. Crotalaria walkeri (100, 10); 10. Butea frondosa (3, 150, 50); 11. Butea bojapatra (30, 150, 30, 50, 100); 12. Butea monosperma (50); 13. Erythrina indica (300, 100, 250,); The M.I. for Canavalia ensiformis, Con A (1.0) was about 20 (not shown).

limited binding of these lectins may be insufficient to agglutinate but adequate enough to impart the mitogenic trigger to resting lymphocytes. Most of the phytomitogens are of Fabaceous origin with diverse sugar binding properties. This may be suggestive of the plural nature of receptors mediating the lectin-induced mitogenic trigger. Of the two species of *Datura* studied, the mitogenic nature of *D. stramonium* is already known (9). Notably, the lectin from *Arisaema consanguineum* represents a new monocot phytomitogen. So far among the monocots, the rice bran lectin has been shown to possess mitogenic property (10) while WGA acts as an antimitogen (11). Apparently for each lectin there was an optimum concentration giving the highest M.I. value beyond which the phenomenon of self-inhibition was observed. This may be attributed to non-specific binding of excess lectin which interfered with the mitogenic trigger. The failure of the remaining 8 lymphoagglutinins to act as phytomitogens is not surprising because more binding to lymphocyte surface although an essential prerequisite (12) is not adequate in itself for the induction of mitosis.

The availability of a large number of new lectins with different binding specificities should offer a range of diagnostic phytomitogens for a more accurate determination of immune status in various immunological malfunctions, such as, Hodgkin's disease (4), systemic lupus erythematosus (13) and chronic lymphotic leukemia (14). The use of a battery of mitogens should be preferred to reliance on PHA alone in order to provide for the known marked variations in immune response of different individuals even to any one lectin preparation.

Acknowledgements. The authors acknowledge the financial support from University Grants Commission and Department of Science and Technology, New Delhi and the grant of a DAE Senior Research Fellowship (SSK).

References

1. Andersson, J., Melchers, F.: In: Poste, G., Nicolson, G.L. (eds.). Dynamic Aspects of Cell Surface Organization, North Holland, Amsterdam, N.Y. Vol. 3, pp. 604-606, 1977.

2. London, J., Berrih, S., Bach, J.F.: J. Immun. 121, 438-443 (1978).

3. Wedner, H.J., Parker, C.W.: Prog. Allergy 20, 195-197 (1975).

4. Holm, G., Mellstedet, H., Bjorkholm, H., Johansson, E., Killander, D., Sundbald, R., Soderberg, G.: Cancer 37, 751 (1974).

5. Lis, H., Sharon, N.: In: Marcus, A. (ed.) Biochemistry of Plants, Academic Press, N.Y., Vol. 6, pp. 371-448, 1981.

6. Sandhu, R.S., Reen, R.S., Singh, S., Singh, J., Chopra, S. K.: In: Bøg-Hansen, T.C. (ed.). Lectins - Biology, Biochemistry, pp. 679-692, 1982.

7. Sandhu, R.S., Kamboj, S.S., Reen, R.S., Arora, J.S.: Vox, Sang., 43, 345-350 (1982).

8. Bøyum, A.: Scand. J. Clin. Lab. Invest. 21 Suppl. 97, 77-79 (1968).

9. Kilpatrick, D.C.: Plant Sci. Lett. 15, 279-284 (1979).

10. Tsuda, M.: J. Biochem. 86, 1451-1461 (1979).

11. Greene, W.C., Parker, C.M., Parker, C.W.: J. Biol. Chem. 251, 4017 (1976).

12. Novogrodsky, A., Katchalski, E.: Proc. Nat. Acad. Sci. USA. 70, 2515-2518 (1973).

13. Nischinarita, S., Takizawa, I., Hirai, H., Morita, K., Amaki, I.: J. Rehumatol. 8, 19-30 (1981).

14. De Vries, J.E., Vyth, F.A., Meadelsohn, J.: Clin. Exp. Immunol. 43, 302-310 (1981).

IS POTATO LECTIN A PRECURSOR OF CELL WALL EXTENSIN ?

Edilbert Van Driessche[1,2], Sonia Beeckmans[1], Robert Dejaegere[2] and Louis Kanarek[1]

[1]Laboratorium voor Chemie der Proteïnen, and [2]Laboratorium voor Plantenfysiologie; Vrije Universiteit Brussel; Instituut voor Moleculaire Biologie; Paardenstraat, 65; B-1640 Sint-Genesius-Rode; Belgium.

Lectins have been isolated and partially characterized from several species of the *Solanaceae* such as *Datura stramonium* (1-8), *Lycopersicon esculentum* (9-13) and *Solanum tuberosum* (14-32). From these, potato lectin (*Solanum tuberosum* agglutinin, STA) has been most thoroughly investigated. This lectin is a very highly glycosylated glycoprotein containing at least 50% of carbohydrate (14). It was shown by Ashford *et al.* (27) that the arabinofuranosides are mainly present as unbranched chains, which are linked to the polypeptide through hydroxyproline, while nine or ten serine residues are substituted with single α-galactopyranoside residues. Upon deglycosylation, the protein retains both its sugar-binding capacity and its haemagglutinating activity, indicating that the carbohydrate moiety of the protein molecule is not essential in order to function as a lectin (28). The high content of hydroxyproline, proline and glycine in potato lectin (14, 17) suggests that its secondary structure may be largely determined by these amino acids; they are known to induce an ordered structure leading to the formation of a left-handed open helix of the poly(L-proline)-II type (33-35). In this structure there are 3 residues per turn and the rise per residue along the axis is 3.12 Å instead of 1.5 Å in the more classical α-helix.

Lectins, Vol. IV

Since the poly(L-proline)-II conformation displays a very characteristic circular dichroism spectrum (33-36), this technique was used to study the structure of potato lectin. Furthermore, the effects of denaturating agents such as $CaCl_2$, LiCl and guanidine-HCl on the conformation of both the native and the reduced or carboxymethylated potato lectin were investigated. From the comparison of the structural features of potato lectin with those of other hydroxyproline-rich glycoproteins it may be suggested that potato lectin represents a salt-extractable precursor of the cell wall glycoprotein extensin.

Materials and Methods

Preparation of the lectin. Potato lectin was purified by affinity chromatography on chitin and gelfiltration on Sephadex G-150 as was described previously (31). Thiourea, 5 mM, was used as antioxidant during the extraction of the potato tubers. The amino acid analysis of the lectin was performed on a Durrum 500-D single-column amino acid analyzer and was found to be in general agreement with that published by Allen *et al.* (14, 17), except that our preparation contained no ornithine or methionine; phenylalanine was completely absent. The purity of our protein preparations was checked by SDS-polyacrylamide gel electrophoresis as described previously (31).

Structural modification of the lectin. Potato lectin was reduced by incubating the protein for 10 h. at room temperature in 0.2 M Tris-HCl, pH 8.5, containing 0.7 M β-mercaptoethanol, either in the presence or in the absence of 5 M guanidine-HCl. Renaturation of the reduced potato lectin was accomplished by dialysis, first against 0.1 M sodium acetate pH 3.5 at 4°C, and afterwards against 10 mM potassium phosphate pH 7.3, initially at 4°C and then at 25°C.
The reduced potato lectin was carboxymethylated as follows.
(a) When guanidine-HCl was used as denaturant during the reduction, the pH of the protein solution was lowered to 3.5 by the addition of 6 N HCl, and the solution was dialyzed extensively against 5 M guanidine-HCl pH 3.5. To 4 ml of lectin solution (0.5 mg/ml) was added 1 ml of 5 M guanidine-HCl containing 0.3 M iodoacetic acid and 0.5 M Tris, pH 8.1. The carboxymethylation was allowed to proceed for 30 min. at 25°C. The reaction was stopped by lowering the pH to 3.5 with 6 N HCl and the solution was dialyzed against 10 mM potassium phosphate pH 7.3, initially at 4°C and afterwards at room temperature.
(b) When the reduction was performed in the absence of guanidine-HCl, the pH was lowered to 3.5 with 6 N HCl and the solution

was dialyzed against 10 mM acetic acid, pH 3.3. The carboxymethylation was performed as described above, omitting the guanidine-HCl.

Analytical methods. Circular dichroism spectra were recorded on a Cary 61 spectropolarimeter. Measurements were made at 25°C. The mean amino acid residue ellipticity, $[\theta]$, in degrees $cm^2 dmole^{-1}$, was calculated as :

$$[\theta] = \frac{MRW}{100} \times \frac{\theta}{c.d}$$

where c is the protein concentration in g/ml and d is the optical path length in dm. The mean residual weight (MRW) was 110. Protein concentrations were calculated from the UV-absorption of the sample at 280 nm, using an extinction coefficient of :

$$E^{1mg/ml}_{280\ nm} = 1.5.$$

The carbohydrate moiety of the lectin was not considered a chromophore, since neither arabinose nor galactose absorb light in the region of the spectrum where the measurements were performed (37).
UV-absorption spectra were recorded on a Shimadzu 210-UV (Bausch and Lomb) double-beam spectrophotometer.
Haemagglutinating activity was determined with rabbit erythrocytes as described before (31).

Results

Structure of native lectin. The circular dichroism spectrum of native potato lectin is shown in Fig. 1 and its major features are compared with those of poly(4-hydroxy-L-proline) and poly(L-proline) in Table 1. The similarities of the spectrum of potato lectin with those of the synthetic polymers indicate that a good part of the potato lectin molecule seems to have an ordered structure leading to the formation of a left-handed open helix of the poly(L-proline)-II type. Our circular dichroism spectrum of native potato lectin is qualitatively similar to that reported recently by Matsumoto *et al.* (30).
The circular dichroism spectrum in the near UV-region, where aromatic amino acid residues and the S-S bridges contribute when present in an asymmetric environment, is shown in Fig. 3.

Denaturation of potato lectin in concentrated salt solutions as monitored by circular dichroism. When potato lectin is incub-

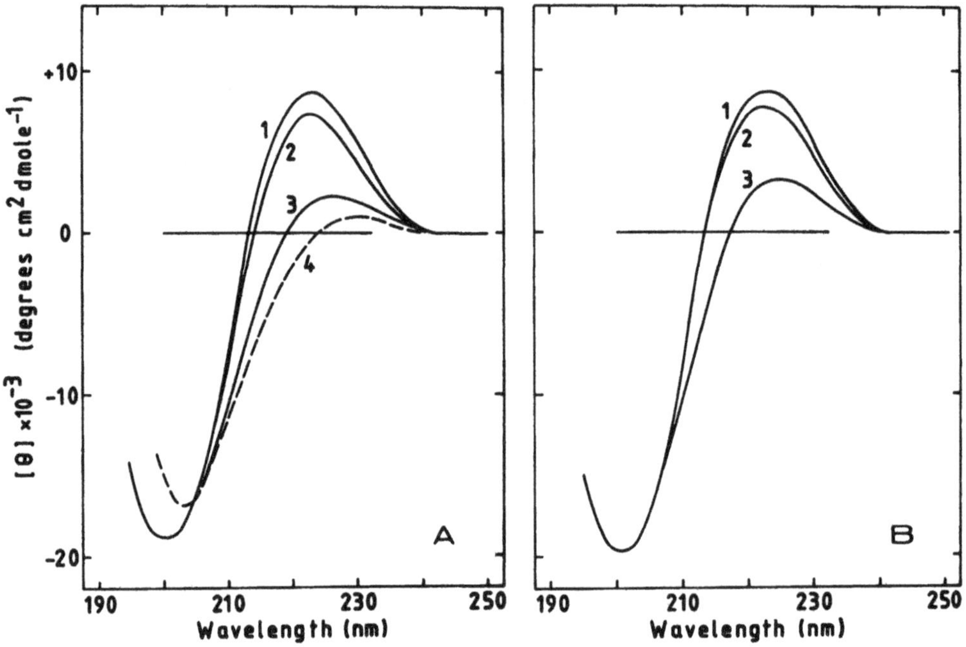

Figure 1. Structural analysis by means of circular dichroism in the far UV-region of potato lectin in high salt solutions : the effect of denaturation by calcium chloride and lithium chloride, and the renaturation from calcium chloride, lithium chloride and guanidine-HCl is shown.
Protein concentrations in this study ranged from 0.05 to 0.4 mg/ml.

A. Spectrum of native potato lectin in water (1), and spectrum of the lectin in respectively 2 M (2), 4 M (3) and 6 M (4) calcium chloride. When the latter solution was dialyzed against water, the spectrum of native lectin (1) was obtained. The same spectrum (1) was also recorded after dialysis of potato lectin which was incubated in 5 M guanidine-HCl.

B. Spectrum of native potato lectin in water (1), and spectrum of the lectin in respectively 4 M (2) and 8 M (3) lithium chloride. When the latter solution was dialyzed against water, the spectrum of native lectin (1) was obtained.

ated in increasing concentrations of either $CaCl_2$ or LiCl, changes in the circular dichroism spectrum are observed (Fig.1), which are comparable with those induced by $CaCl_2$ in the spectra

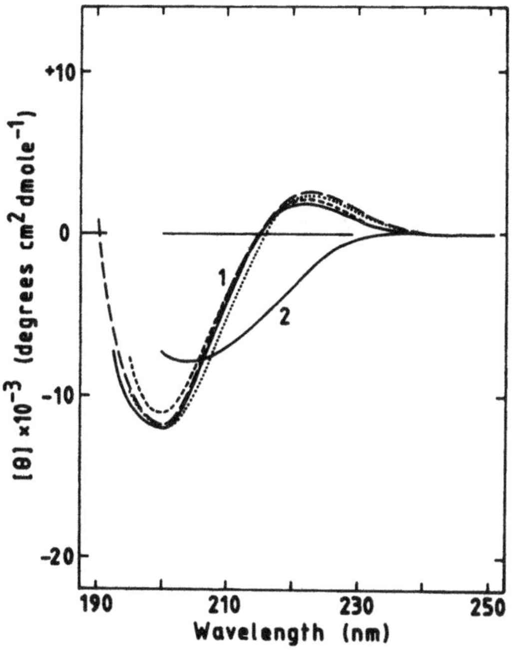

Figure 2. The effect of reduction and carbomethylation on the circular dichroism spectrum of potato lectin in the far UV-region.

(1) Spectrum of carboxymethylated potato lectin in 10 mM potassium phosphate pH 7.3 (——). The lectin was reduced by β-mercaptoethanol in the presence of guanidine-HCl, followed by carboxymethylation with iodoacetic acid according to the procedure described in Materials and Methods.

Spectrum of reduced potato lectin in 0.1 M sodium acetate pH 3.2 (--). The lectin was reduced by β-mercaptoethanol in the presence of guanidine-HCl and dialyzed against the Na-acetate buffer; in these conditions no S-S bridges can be formed.

Spectrum of reduced and renaturated potato lectin in 10 mM potassium phosphate pH 7.3 (···). The lectin was reduced in the absence of guanidine-HCl and renaturated by dialysis against the phosphate buffer.

Spectrum of carboxymethylated potato lectin in water after renaturation from calcium chloride (— —). The carboxymethylated lectin was first incubated in 6 M calcium chloride, and renaturated by dialysis against water.

(2) Spectrum of carboxymethylated potato lectin in the presence of 6 M calcium chloride.

of poly(L-proline) and poly(4-hydroxy-L-proline) (33, 34). According to these authors, the most obvious change accompanying the disruption of the poly(L-proline)-II type structure is the loss of the positive band in the highly concentrated $CaCl_2$ solutions. In 6 M $CaCl_2$, where poly(L-proline)-II adopts an unordered structure which is characterized by the complete absence of the positive band, the spectrum of potato lectin still displays a positive contribution (Fig.1A and Table 2). The

Table 1.
Circular dichroism of potato lectin as compared to poly-(4-hydroxy-L-proline) and poly(L-proline).

Feature[a]	*Unit*	Potato lectin	Poly(4-hydroxy-L-proline)[b]	Poly(L-proline)[b]
$[\theta_\lambda]_{max}$	*degrees* $cm^2 dmole^{-1}$	8500	7260	1650
λ_{max}	*nm*	222.5	225	228
λ_{cross}	*nm*	213.5	219	224
$[\theta_\lambda]_{min}$	*degrees* $cm^2 dmole^{-1}$	-19600	-46200	-42900
λ_{min}	*nm*	200	205	206

[a] Subscripts max, min and cross refer to the observed maximum, minimum and crossover, respectively.

[b] Values are according to Mattice and Mandelkern (33). The mean residue ellipticities are calculated from the relationship $[\theta] = 3300\ \Delta\epsilon$ where $\Delta\epsilon$ is expressed in $cm^2 mmole^{-1}$.

disorganization of the poly(L-proline)-II conformation by $CaCl_2$ is accompanied by a shift of the positive maximum to longer wavelengths, as is true for potato lectin. A comparable disorganization of the molecule is accomplished by LiCl, although at higher concentrations (Fig.1B and Table 2). When the denat-

Table 2.
Circular dichroism of potato lectin in water, $CaCl_2$ and LiCl solutions.

Potato lectin in :		λ_{max} *(nm)*	$[\theta]_{max}$ *(degrees* $cm^2 dmole^{-1}$*)*
H_2O		222.5	8500
$CaCl_2$	2 M	222.5	7480
	4 M	225	2160
	6 M	230	910
LiCl	4 M	222.5	7940
	8 M	225	3230

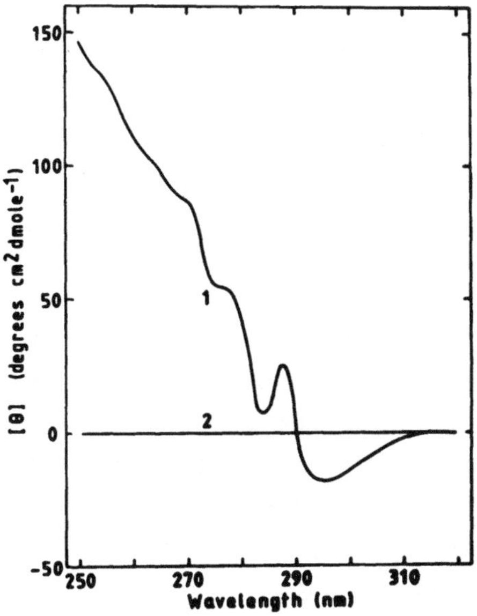

Figure 3. Circular dichroism spectra in the near UV-region of potato lectin in various conditions.

(1) Spectrum of native lectin in water.
The same spectrum was obtained when the lectin was renaturated either from 6 M calcium chloride or from 5 M guanidine-HCl.

(2) All conditions by which the S-S bridges of potato lectin are opened result in disorganization of the tertiary structure : the original spectrum completely disappears when the lectin is reduced and subsequently dialyzed against 10 mM potassium phosphate pH 7.3, or when it is carboxymethylated by iodoacetic acid.

urants (6 M $CaCl_2$ or 8 M LiCl) are removed by dialysis, the native conformation is completely restored, as well as the haemagglutinating activity (Fig.1 and Table 3). Similarly, the original conformation and activity are obtained when potato lectin is renaturated from 5 M guanidine-HCl by dialysis against 10 mM potassium phosphate pH 7.3 (Fig.1A).

Table 3.

Haemagglutinating activity of potato lectin (STA) in various conditions.

	Activity, %
Native STA in saline or in water	100
STA after renaturation from 5 M guanidine-HCl	100
Carboxymethylated STA	0
STA after renaturation from reduced (in guanidine-HCl) STA	0.5
STA after renaturation from reduced (no guanidine-HCl) STA	3
STA after renaturation from 8 M LiCl	100
STA after renaturation from 6 M $CaCl_2$	100

Structural features of reduced and carboxymethylated potato lectin. Reduction of the lectin is accompanied by an extensive conformational change. Both the positive and the negative circular dichroism bands decrease in value, while the position of the maximum and minimum are not affected : $[\theta]_{222.5nm} = 2050$ and $[\theta]_{200nm} = -11200$ degrees $cm^2 dmole^{-1}$ (Fig.2). Carboxymethylation of the lectin did not further change the circular dichroism spectrum.
Attempts to refold the reduced lectin to its original conformation by dialysis against 10 mM potassium phosphate pH 7.3 were unsuccessful : the circular dichroism spectrum was indistinguishable from that of the reduced protein, and nearly no haemagglutinating activity was recovered (Fig.2 and Table 3). Carboxymethylated lectin was also completely inactive and structurally indistinguishable from the reduced lectin (Fig.2). The same results were obtained whether the reduction and carboxymethylation was performed in buffer or in 5 M guanidine-HCl. When carboxymethylated lectin is incubated in 6 M $CaCl_2$, a spectrum typical for the unordered poly(L-proline)-II conformation is obtained (Fig.2). Upon removal of the $CaCl_2$ by dialysis, the original spectrum of carboxymethylated potato lectin is restored (Fig.2).
The effect of reduction and carboxymethylation on the circular dichroism spectrum in the near UV-region is shown in Fig.3.

Discussion

From the comparison of the circular dichroism spectrum of potato lectin with those of polymers which were shown to have a poly-(L-proline)-II conformation, it is obvious that an important part of the lectin molecule adopts a similar extended helix structure. This is not surprising, since 36-40% of the amino acids consist of proline, hydroxyproline and glycine (14, 17), residues which are known to favor this conformation. From the general aspects of the spectrum, it may be concluded that only limited amounts, if any, of α-helix structure is present, since the α-helix displays a strong negative band at 222 nm and a very strong positive one at 193 nm. In view of the conflicting literature data on the contribution of β-pleated sheet and random coil structures (38-40), and in view of the generally small $[\theta]$-values of these structures in proteins (40), no definite conclusions about the presence of β-pleated sheet segments can be obtained.

It was shown by Allen *et al.* (17) that all the hydroxyproline residues are clustered into one region, which also contains all the carbohydrate. The other, non-glycosylated part of the molecule is highly cross-linked by the presence of 13 S-S bridges. In conditions which may be supposed to destroy the conformational organization of the latter domain of the molecule (*i.e.* reduction, eventually followed by carboxymethylation), the observed circular dichroism spectrum may be completely attributed to the contribution of the extended helix, formed by the hydroxyproline-rich glycopeptide, which was isolated by Allen *et al.* (17) and which represents 30% of the total polypeptide moiety. Indeed, taking into account the apparent blue shift of the potato lectin spectrum when compared to the poly(4-hydroxy-L-proline) spectrum, and omitting the small contribution of β- or random coil structures, 26-32% of the lectin adopts the extended helix structure. In the presence of 6 M $CaCl_2$, the latter helical structure in the reduced or carboxymethylated lectin is further unfolded. This unfolding was shown to be completely reversible. If, on the other hand, the S-S bridges of the highly cross-linked domain remain intact, the extended helix segment seems to be much longer, as can be deduced from the circular dichroism spectrum of the native lectin. From the minimum at 200 nm, and when bearing in mind the same considerations as above, about 40% of the lectin should now have adopted the extended helix conformation. This prompts us to postulate that the numerous proline and glycine residues are also clustered in the protein molecule, and that they too are arranged in an extended helix conformation (the polyhydroxyproline-containing peptide as isolated by Allen *et al.* (17), together with the amount of proline and glycine, indeed represents about 50% of the polypeptide moiety). The unusually high [θ]-value of 8500 degrees cm^2dmole^{-1} at 222.5 nm, which exceeds the reported value (Table 1) for 100% extended helix structure (34), must be attributed to an influence of the structure of the cross-bridged domain on the circular dichroism spectrum of the helix.

In conclusion, we suggest that in native potato lectin two stretches of poly(L-proline)-II type of helix might exist : a first segment, which contains all the hydroxyproline residues and all the carbohydrate and which can persist in the form of an extended helix, even when the cross-bridged domain is unfolded, and a second segment, containing probably most of the remaining proline and glycine residues; a correct folding of the cross-bridged domain is however a prerequisite for the extended helical conformation of this stretch. From these considerations we may conclude that the non-glycosylated, highly cross-linked part of the lectin has at least two functions to play, *i.e.* *primo* to stabilize the extended helix to its whole extent in native potato lectin, and *secundo* it is implicated (as was shown by Allen *et al.*, 17) in the formation of the carbohydrate binding site (NAG_{4-5}). From the observation that, in 6 M $CaCl_2$, native potato lectin on the one hand, and reduced or carboxymethylated lectin on the other hand display a different circular dichroism spectrum, it may be concluded that the highly cross-bridged part of the lectin molecule stabilizes at least part of the extended helical structures. Furthermore, during the synthesis of the lectin, the cross-bridging of the non-glycosylated part of the molecule may be a prerequisite for the extended helix to be formed in its full length. Indeed, denaturation of the native lectin is fully reversible as long as the S-S bridges remain intact.

During the synthesis of collagen, the triple helix in procollagen is stabilized during the folding process by disulfide bridges between the chains (41, 42). Our results indicate that the folding of both proteins may be governed by the same mechanism although in potato lectin the S-S bridging is intracatenary.

Although much information is already available on the physico-chemical properties of potato lectin, its physiological function remains very poorly understood. Based on the preferential agglutination of avirulent strains of *Pseudomonas solanaceranum* (43), Sequeira and Graham suggested that potato lectin might play a role in the defence mechanism of the host. However, since the histological localization of potato lectin is not known yet, this conclusion based on *in vitro* activities of the lectin should be interpreted carefully. At present, we can only speculate about the physiological role of potato lectin. However, in view of its high hydroxyproline content it may be argued that the lectin represents an extractable, salt-soluble precursor of a cell wall glycoprotein. Indeed, hydroxyproline-containing glycoproteins are characteristic components of the extra-cellular matrix of both animals and plants (44). In higher plants, the primary cell walls contain a hydroxyproline-rich glycoprotein called extensin from which glycopeptides could be isolated and characterized (25, 44-55), which both chemically and structurally are identical with the glycopeptide of potato lectin (17, 18, 27). Unfortunately, extensin is covalently bound in the cell wall and can only be released from it in conditions which most probably break peptide bonds, making conformational investigations of the native molecule impossible (44). From lower plants such as *Chlamydomonas reinhardii*, a hydroxyproline-rich glycoprotein can be isolated in its native state (56) which structurally resembles potato lectin. Both proteins are built up of structurally distinct domains, to one of which all the oligosaccharides are attached (17, 56, 57); in addition both proteins have at least in part a rod-like poly(L-proline)-II helical backbone (this paper, 58).
The salt-extractable extensin precursor isolated from aerated carrot roots has also been shown to display a poly(L-proline)-II conformation (37). Most interestingly Brysk and Chrispeels (47) were able to isolate an intact precursor of extensin from

aged carrot disks. This glycoprotein could be released from both membranous organelles and from the cell wall by salt solutions and has many chemical characteristics in common with potato lectin; both have a high isoelectric point (around 9), both are highly glycosylated (40-60%) with arabinose being the most abundant sugar, both have a similar amino acid composition and subunit molecular weight. Most remarkably, Brysk and Chrispeels reported that the extensin precursor "tended to be bound to Sephadex and could not be eluted with low ionic strength buffers" (47). Although from the results described by these authors, non-specific adsorption on the Sephadex matrix can not be excluded, it is tempting to suggest that a specific carbohydrate binding site might be responsible for this interaction. Murray and Northcote suggested that potato lectin might be associated with intracellular membranes (18). More recently, Owens *et al.* (25) demonstrated that the arabinosyl : hydroxyproline transferase in potato tissue cultered cells is localized in the Golgi-apparatus and that deglycosylated potato lectin can serve as an *in vitro* acceptor for the glycosylating machinery. In addition, Kawasaki (59) could unequivocally show that the arabinose-containing cell wall precursors in suspension-cultured tobacco cells are glycosylated in the Golgi-apparatus and suggested that the precursors are transported from there to the cell wall in small vesicles derived from the Golgi-complex. It was further shown (59) that, in the early phase of the synthesis, the precursors are bound to the membranes and that they are released later on, as the glycosylation proceeds, so that the transport vesicles fraction contains molecules of the later phase of synthesis. From these results it may be suggested that the lectin too is transiently bound to the membranes of the Golgi-apparatus during glycosylation.

Based on the increased recovery of *Datura* lectin, in conditions which are supposed to disorganize membranous structures, Kilpatrick *et al.* (5) concluded that the *Datura* lectin, which shows extensive homology to potato lectin (60) and thus may reasonably be supposed to have identical or at least very similar physiological functions, is loosely bound to intracellular membranes. This conclusion was confirmed by Jeffree *et al.* (7) who, by means of immunofluorescent techniques, showed that the lectin was associated with the plasmalemma and membranes of cell organelles. It should be noted however that the cell walls were found to be completely free of lectin. This finding might be a rather strong argument for rejecting the hypothesis that the hydroxyproline-rich *Solanaceae* lectins such as potato and *Datura* lectin are soluble precursors of extensin. Other explanations can however be given for the lack of cell wall staining in the experiments of Jeffree *et al.*: it is possible that, as a result of steric hindrance, the primary antibodies fail to bind to the antigenic determinants which are situated on that part of the molecule which is inserted into the cell wall; alternatively the antigenic determinants might be destroyed upon covalent attachment of the lectin into the cell wall.

From our results presented in this paper and from the arguments presented above, it is tempting to suggest that potato lectin is not only built up of two structurally different domains, but that each of these two domains displays a specific physiological function. We suppose that the globular part of the molecule attaches the unglycosylated precursor onto the membranes of the Golgi-complex during glycosylation and transport, while the highly glycosylated, hydroxyproline-rich part of the molecule becomes covalently incorporated into the cell wall.
It may be reasonably supposed that in the soluble extensin-precursors from different plant families (or species), the hydroxyproline-rich and highly glycosylated part of the molecule is structurally rather conserved, while the globular part may be quite variable from one family to another. All these precursors must not necessarily behave as lectins : the bivalency, which is a prerequisite for a molecule to function as a lectin, is not required for attachment of the extensin-precursors to membranous structures of the Golgi-apparatus.

References

1. Horejsi, V., Kocourek, J.: Biochim. Biophys. Acta 532, 92-97 (1978)
2. Kilpatrick, D.C., Yeoman, M.M.: Biochem. J. 175, 1151-1153 (1978)
3. Kilpatrick, D.C., Yeoman, M.M., Kilpatrick, S.P.: Plant Sci. Letters 13, 35-40 (1978)
4. Kilpatrick, D.C.: Plant Sci. Letters 15, 279-284 (1979)
5. Kilpatrick, D.C., Yeoman, M.M., Gould, A.D.: Biochem. J. 184, 215-219 (1979)
6. Desai, N, Allen, A.K., Neuberger, A.: Biochem. J. 197, 345-353 (1981)
7. Jeffree, C.E., Yeoman, M.M.: New Phytol. 87, 463-471 (1981)
8. Crowley, J.F., Goldstein, I.J.: Methods Enzymol. 83, 368-373 (1982)
9. Kilpatrick, D.C.: Biochem. J. 185, 269-272 (1980)
10. Nachbar, M.S., Oppenheim, J.D., Thomas, J.D.: J. Biol. Chem. 255, 2056-2061 (1980)
11. Nachbar, M.S., Oppenheim, J.D.: Methods Enzymol. 83, 363-368 (1982)
12. Kilpatrick, D.C., Weston, J.W., Urbanik, S.J.: Anal. Bioch. 134, 205-209 (1983)
13. Lambrechts, E., Van Driessche, E., Kanarek, L.: Arch. Intern. Physiol. Bioch. 92, B39-B40 (1984)
14. Allen, A.K., Neuberger, A.: Biochem. J. 135, 307-314 (1973)
15. Delmotte, F., Kieda, C., Monsigny, M.: FEBS Letters 53, 324-330 (1975)
16. Van Driessche, E., Kanarek, L.: Arch. Intern. Physiol. Bioch. 83, 414-415 (1975)
17. Allen, A.K., Desai, N.N., Neuberger, A., Creeth, J.M.: Biochem. J. 171, 665-674 (1978)
18. Muray, R.H.A., Northcote, D.H.: Phytochem. 17, 623-629 (1978)
19. Desai, N.N., Allen, A.K.: Anal. Bioch. 93, 88-90 (1979)
20. Matsumoto, I., Mizuno, Y., Seno, N.: J. Biochem. 85, 1091-1098 (1979)
21. Jimbo, A., Seno, N., Matsumoto, I.: J. Biochem. 95, 267-275 (1980)
22. Kilpatrick, D.C.: Biochem. J. 191, 273-275 (1980)
23. Owens, J., Northcote, D.H.: Phytochem. 19, 1861-1862 (1980)

24. Ashford, D., Menon, R., Allen, A.K., Neuberger, A.: Biochem. J. 199, 399-408 (1981)

25. Owens, R.J., Northcote, D.H.: Biochem. J. 195, 661-667 (1981)

26. Ashford, D., Allen, A.K., Neuberger, A.: Biochem. J. 201, 641-645 (1982)

27. Ashford, D., Desai, N.N., Allen, A.K., Neuberger, A., O'Neill, M.A., Selvendran, R.R.: Biochem. J. 201, 199-208 (1982)

28. Desai, N.N., Allen, A.K., Neuberger, A.: Biochem. J. 211, 273-276 (1983)

29. Doi, A., Matsumoto, I., Seno, N.: J. Biochem. 93, 771-775 (1983)

30. Matsumoto, I., Jimbo, A., Mizuno, Y., Seno, N., Jeanloz, R.W.: J. Biol. Chem. 258, 2886-2891 (1983)

31. Van Driessche, E., Beeckmans, S., Dejaegere, R., Kanarek, L.: in: 'Lectins : Biology, Biochemistry, Clinical Biochemistry' (T.C. Bøg-Hansen, G.A. Spengler, eds.), Vol. 3, pp. 629-638 (1983) (W. deGruyter)

32. Jimbo, A., Seno, N., Matsumoto, I.: J. Biochem. 95, 267-275 (1984)

33. Mattice, W.L., Mandelkern, L.: Macromolecules 3, 199-201 (1970)

34. Mattice, W.L., Mandelkern, L.: Biochemistry 9, 1049-1058 (1970)

35. Mattice, W.L., Mandelkern, L.: J. Am. Chem. Soc. 92, 5285-5287 (1970)

36. Tiffany, M.L., Krim, S.: Biopolymers 8, 347-359 (1969)

37. Van Holst, G.J., Varner, J.E.: Plant Physiol. 74, 247-251 (1984)

38. Greenfield, N., Fasman, G.D.: Biochemistry 8, 4108-4116 (1969)

39. Saxena, V.P., Wetlaufer, D.B.: Proc. Nat. Acad. Sci. USA 68, 969-972 (1971)

40. Chen, H.H., Yang, J.T., Martinez, H.M.: Biochemistry 11, 4120-4131 (1972)

41. Bornstein, P.: Ann. Rev. Biochem. 43, 567-603 (1974)

42. Martin, G.R., Byers, P.H., Piez, K.A.: Adv. Enzymol. 42, 167-191 (1975)

43. Sequeira, L., Graham, T.L.: Physiol. Plant Pathol. 11, 43-54 (1977)

44. Lamport, D.T.A.: Symp. Soc. Developm. Biology 30, 113-130 (1974)

45. Lamport, D.T.A.: Nature (Lond.) 216, 1322-1324 (1967)

46. Lamport, D.T.A.: Biochemistry 8, 1155-1163 (1969)

47. Brysk, M.M., Chrispeels, M.J.: Biochim. Biophys. Acta 257, 421-432 (1972)

48. Heath, M.F., Northcote, D.H.: Biochem. J. 125, 953-961 (1971)

49. Lamport, D.T.A., Katona, L., Roerig, S.: Biochem. J. 133, 125-131 (1973)

50. Pope, D.G., Lamport, D.T.A.: Plant Physiol. 53, S-81 (1974)

51. Cho, Y.P., Chrispeels, M.J.: Phytochem. 15, 165-169 (1976)

52. Pope, D.G.: Plant Physiol. 59, 894-900 (1977)

53. O'Neill, M.A., Selvendran, R.R.: Biochem. J. 187, 53-63 (1980)

54. Akiyama, Y., Kato, K.: Agric. Biol. Chem. 41, 79-81 (1977)

55. Akiyama, Y., Mori, M., Kato, K.: Agric. Biol. Chem. 44, 2487-2489 (1980)

56. Catt, J.W., Hills, G.J., Roberts, K.: Planta 131, 165-171 (1976)

57. Roberts, K.: Planta 146, 275-279 (1979)

58. Holmer, R.B., Roberts, K.: Planta 146, 217-222 (1979)

59. Kawasaki, S.: Plant Cell. Physiol. 22, 431-442 (1981)

60. Kilpatrick, D.C., Jeffree, C.E., Lockhart, C.M., Yeoman, M.M.: FEBS Letters 113, 129-133 (1980)

THE OCCURRENCE OF LECTIN DURING THE LIFE-CYCLE OF PISUM SATIVUM L.

M. Hosselet[1], E. Van Driessche[2], M. Van Poucke[1], L. Kanarek[2]

1. Laboratorium voor Plantenfysiologie, Limburgs Universitair Centrum, Universitaire Campus, B-3610 Diepenbeek, Belgium

2. Laboratorium voor Chemie der Proteïnen, Vrije Universiteit Brussel, Paardenstraat 65, B-1640 Sint-Genesius-Rode, Belgium

In spite of a good knowledge of the physicochemical properties of legume lectins and their use as tools in glycoconjugate and mammalian cell research, their biological role in the plant remains unclear. In the last decade, several functions have been proposed, including roles as storage proteins (1-5), enzymes (6), enzyme regulators (7), sugar carriers (8), protective agents (9) or mitogenic stimulators of plant embryonic cells (10); other hypotheses are that lectins either play a role in cell wall structure (11-13) or are involved in the recognition of symbiotic *Rhizobia* (12-15). However, every proposed function has to be compatible with the distribution and localization of lectin in the plant. In Leguminosae, lectins have long been thought to occur in seeds and seedlings only (1-3,10,16,17) but the development of more sensitive techniques has also allowed to detect lectin or immunologically related proteins in other plant organs (18-20). This study shows the presence of low levels of lectin or an immunologically cross-reacting material in different tissues during the life-cycle of *Pisum sativum* L., using a sandwich type of enzyme-linked immunosorbent assay (ELISA).

Lectins, Vol. IV

Materials and Methods

Plants. Seeds of *Pisum sativum* L. (cvs. Wonder van Kelvedon) were planted in vermiculite supplied with a nutrient solution, containing 5 g NH_4NO_3, 25 g $Ca(HPO_4)_2$, 2.5 g K_2SO_4 and 2.5 g $MgSO_4$ per liter. They were allowed to germinate and to grow in a growth cabinet, using a 12 h photoperiod (160 μmol quanta/m^2.sec), a relative humidity of 60 % and a temperature of 25°C during the day and 20°C at night. Once a week the plants received Hoagland's nutrient solution.

Preparation of plant extracts. Weekly, two or three batches of 6-12 plants were dissected into roots, leaves, stems and, when feasible, flowers and/or fruits (for 1 week old plants, each batch consisted of 50 plants). The tissues were frozen in liquid nitrogen, ground to a powder in a mixer and 2 g powder was extracted for 30 min at 4°C with 6 ml 0.1 M potassium phosphate buffer (pH 7.0) containing 1 % insoluble polyvinylpyrrolidone (PVP) to precipitate phenols (21) and 5 mM thiourea to prevent their oxidation to quinones (22). After centrifugation (15000 g, 20 min) the supernatant was dialyzed overnight against PBS pH 7.0. Before submitting the roots to the extraction procedure, they were washed (23) to prepare the root surface lectin.

Preparation of antibodies. Monospecific rabbit anti-pea seed lectin antibodies were prepared as described earlier (4). Antibodies were precipitated from preimmune rabbit serum by addition of ammonium sulphate to a final concentration of 33 % saturation and further purified by affinity chromatography on Sepharose-protein A.

Lectin content was assayed by a sandwich type of ELISA. Microtiter plates (NUNC immunoplate I) were coated overnight at 4°C with 10 μg/ml monospecific anti-pea seed lectin antibodies. Plates were flicked off, washed twice with PBS pH 7.0 containing 0.05 % Tween 20 and twice with 0.15 M NaCl and then incubated with 1 % bovine serum albumin (1 h, 37°C) (24). After washing, samples and references of purified seed lectin (25) were applied and plates were incubated for 2 h at 37°C. Unbound lectin was removed by washing and anti-pea seed lectin antibodies coupled to alkaline phosphatase were added (Boehringer; dilution 1/50). After incubation for 2 h at 37°C, the plates were washed and 0.4 % (w/v) p-nitrophenylphosphate was added. The amount of p-nitrophenol formed after 30 min reaction at 25°C was measured by its absorbance at 405 nm in a multiscan (Titertek MC) and was shown to be linearly correlated to the amount of bound lectin in the range of 0-10 ng/ml. Each sample was tested 10-20 times in at least two independent assays.
Special attention was paid to the blanks. In each microtiter plate several cups without lectin were incorporated. The specificity of the assay was tested by replacing the specific coating antibodies by preimmune rabbit antibodies in the

presence of glucose (0.2 M). This prevented binding of lectin to the coating preimmune antibodies by its carbohydrate binding site (26), a type of interaction that was avoided in the tests by the purification method of the specific antibodies (4). However, to exclude a possible interference of glucose, blanks and tests were always compared in the presence of glucose.

Concentrations of pea seed lectin were determined by absorbance measurements at 280 nm, using $A_{1\ cm}^{1\ \%}=15.8$.

Results

During 7 weeks, plants were weekly dissected into roots, leaves, stems and flowers and/or fruits and the tissues were extracted and assayed for lectin content. At our growth conditions, flowers open in the fourth week after sowing and at 5 weeks 96 % of the plants bear pods. From this moment, increase in dry weight per plant is almost exclusively due to the development of the fruits (Fig. 1); at 7 weeks the plants show considerable senescence.

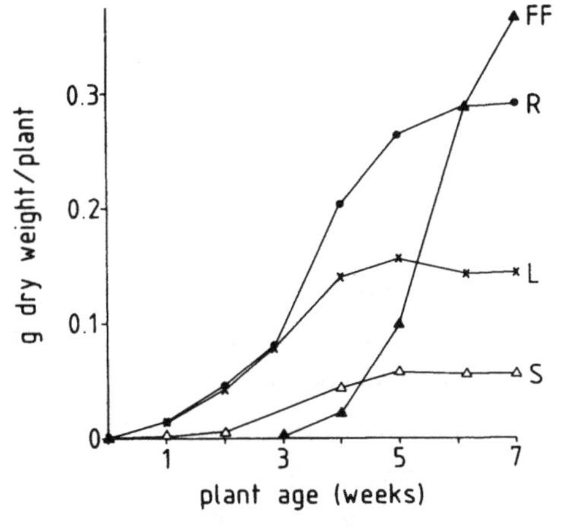

Fig. 1. The dry weight of organs of pea plants at culture conditions as described in materials and methods. Roots (R), stems (S), leaves (L), flowers and/or fruits (FF).

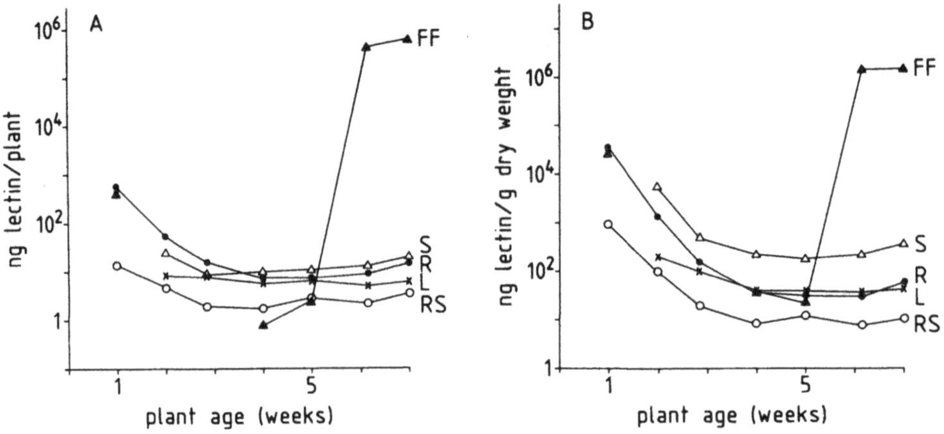

Fig. 2. Evolution of lectin content per plant (A) and on a dry weight basis (B). Symbols see Fig. 1. Roof surface lectin: RS. Stems on leaves of one week old plants were taken together:

Lectin was assayed by means of a sandwich type of ELISA that allowed us to detect quantities as low as 0.5 ng/ml. When the lectin content of roots, leaves and stems is expressed on a dry weight basis or per plant (Fig. 2), a dramatic decrease is observed after germination. However, small quantities of lectin are present throughout the life-cycle in all vegetative organs. On a dry weight basis, highest levels are detected in the stems, but when expressed as ng per plant the differences in lectin content of the various organs are small.

The amount of lectin isolated from the root surface decreases also during 3 weeks after sowing (Fig. 2); however, low levels of lectin above the detection limit of the assay are consistently detected throughout growth. The ratio of root surface lectin to total root lectin even increases from 2.4 % in 1 week old plants to as much as 20 % in 4 week old ones.

Lectin synthesis in developing seeds occurs about one week after the appearance of the pods. The amount of lectin then increases rapidly to the level found in mature seeds.

Discussion

Pea lectin has long been thought to occur in seeds and seedlings only. Using radial immunodiffusion, hemagglutination and polyacrylamide gel electrophoresis, a rapid degradation of lectin was observed during germination, and no lectin could be detected in 1 week old seedlings outside the cotyledons (16,17). With the more sensitive ELISA technique, it is now shown that the amount of lectin in roots, leaves and stems never drops to zero, but remains at a low level throughout the life-cycle. Compared with a functional test, an antibody-based technique such as ELISA has the additional advantage that it is not affected by lectin-carbohydrate interactions due to the presence of endogenous lectin binders (27) and/or lectin inhibitors released during the extraction procedure.

Lectin was also detected throughout growth of *Dolichos biflorus* and *Phaseolus vulgaris*, using radioimmunoassay and ELISA respectively (18,20). In *Phaseolus vulgaris* the lectin in the non-cotyledon tissue is identical to the seed lectin of the same species, but in *Dolichos biflorus* the lectin from the leaves and stems shows only partial identity to the seed lectin and exhibits a somewhat different carbohydrate specificity (19,28). Previously, it was shown that the endogenous root lectin of 4 days old etiolated pea seedlings is identical to the pea seed lectin (23). It is not yet clear whether the same applies to the lectin throughout the life-cycle. In fact, as long as no information on carbohydrate binding properties is available, the possibility remains that the detected material is not a lectin (29), but a cross-reacting molecule that is in some way related to the seed lectin.

The levels of lectin are very low. Comparing the total lectin content of a 4 weeks old plant (~25 ng) with that reported for the seed (~300 μg) (25), only 0.008 % is left. The question arises if such low levels have any biological significance. Several considerations can be made. First, the lectin can be present in very restricted areas of the plant, resulting in appreciable local concentrations. Second, the extraction method used is directed to soluble lectins, but the possibility remains that the lectin is predominantly located in the cell wall or in the membranes and is not released during buffer extraction (11,13,30-32).

Since pea seed lectin is localized in the protein bodies in mature as well as immature seeds, it has previously been supposed that the lectin is a reserve protein (4). The small quantities of lectin (or immunologically related protein) that are found throughout the life-cycle of *Pisum sativum* could point to a more active metabolic function at later stages of the development.

The presence of lectin on the root surface throughout the life-cycle is interesting in view of the possible involvement of lectin in the recognition and/or binding of symbiotic *Rhizobia*. The absence of lectin on the roots at the moment of *Rhizobium* infection was one of the objections against this hypothesis (17). In this context, it is interesting that the presence of NO_3^- in concentrations that inhibit *Rhizobium* nodulation has recently been shown to influence both root cell wall and root surface lectin quantities in pea seedlings (30). The fact that the lectin synthesis occurs relatively late in the development of the fruits is in accordance with previous results (2).

The detection of lectin or a cross-reacting molecule throughout the life-cycle of pea is not unique. It has also been described in *Dolichos biflorus* (18,19) and *Phaseolus vulgaris* (20). Further characterization of these molecules, as well as information on the histological and cellular distribution of lectins and immunologically related molecules may be helpful in understanding their biological role in the plant.

References

1. Rougé, P. (1974). C.R. Acad. Sc. Paris 278D, 3083-3086.

2. Rougé, P. (1976). C.R. Acad. Sc. Paris 282D, 621-623.

3. Gracis, J.-P., Rougé, P. (1977). Bull. Soc. bot. Fr. 124, 301-306.

4. Van Driessche, E., Smets, G., Dejaegere, R., Kanarek, L. (1981). Planta 153, 287-296.

5. Bollini, R., Chrispeels, M.J.(1978). Planta 142, 291-298.

6. Hankins, C.N., Shannon, L.M. (1978). J. Biol. Chem. 253, 7791-7797.

7. Bowles, D.J. (1982). La Recherche 13, 784-785.

8. Ensgraber, A. (1958). Ber. dt. Bot. Ges. 71, 349-361.

9. Etzler, M.E. (1981). Phytopathology 71, 744-746.

10. Howard, I.K., Sage, H.J., Horton, C.B. (1972). Arch. Biochem. Biophys. 149, 323-326.

11. Kauss, H., Glaser, C. (1974). FEBS Lett. 45, 304-307.

12. Kijne, J.W., van der Schaal, I.A.M., De Vries, G.E. (1980). Plant Sci. Lett. 18, 65-74.

13. Kijne, J.W., van der Schaal, I.A.M., Diaz, C.L., van Iren, F. (1983). Lectins. Biology, Biochemistry, Clinical Biochemistry vol 3, 521-529; T.C. Bøg-Hansen, G.A. Spengler eds. (W. de Gruyter).

14. Bohlool, B.B., Schmidt, E.L. (1974). Science 185, 269-271.

15. Kato, G., Maruyama, Y., Nakamura, M. (1979). Agric. Biol. Chem. 43, 1085-1095.

16. Rougé, P. (1975). C.R. Acad. Sc. Paris 280D, 2105-2108.

17. Rougé, P., Labroue, L. (1977). C.R. Acad. Sc. Paris 284D, 2423-2426.

18. Talbot, C.F., Etzler, M.E. (1978). Plant Physiol. 61, 847-850.

19. Talbot, C.F., Etzler, M.E. (1978). Biochemistry 17, 1474-1479.

20. Borrebaeck, C.A.K., Mattiasson, B. (1983). Physiol. Plant. 58, 29-32.

21. Loomis, W.D. (1974). Methods Enzymol. 31, 528-544.

22. Van Driessche, E., Beeckmans, S., Dejaegere, R., Kanarek, L. (1983). Lectins. Biology, Biochemistry, Clinical Biochemistry vol 3, 629-638; T.C. Bøg-Hansen, G.A. Spengler eds. (W. de Gruyter).

23. Hosselet, M., Van Driessche, E., Van Poucke, M., Kanarek L. (1983). Lectins. Biology, Biochemistry, Clinical Biochemistry vol 3, 549-558; T.C. Bøg-Hansen, G.A. Spengler eds. (W. de Gruyter).

24. Forghani, B., Schmidt, N.J. (1979). J. Clin. Microbiol. 9, 657-664.

25. Van Driessche, E., Smets, G., Dejaegere, R., Kanarek, L. (1981). Lectins. Biology, Biochemistry, Clinical Biochemistry vol 2, 729-742; T.C. Bøg-Hansen, G.A. Spengler eds. (W. de Gruyter).

26. Rougé, P. (1976). C.R. Acad. Sc. Paris 283, 1823-1825.

27. Rüdiger, H., Gebauer, G., Gansera, R., Schurz, H., Schimpl, A. (1982). S. Afr. J. Sci. 78, 346-347.

28. Etzler, M.E., Borrebaeck, C. (1980). Biochem. Biophys. Res. Commun. 96, 92-97.

29. Kocourek, J., Horejsi, V. (1981). Nature 290, 188.

30. Diaz, C.L., Lems-van Kan, P., van der Schaal, I.A.M., Kijne, J.W. (1984). Planta 161, 302-307.

31. Etzler, M.E., Gibson, D., Scates, S. (1979). Plant Physiol. Suppl. 63, 134.

32. Chrispeels, M.J., Bollini, R. (1982). Plant Physiol. 70, 1425-1428.

SOYBEAN SEED LECTIN (SBA) BINDS TO SOIL HUMIC ACIDS AND CLAYS

Henri Causse, Pierre Rougé, Louis Labroue and Jean-Pierre Spilthooren

Laboratoire de Biologie cellulaire, Faculté des Sciences Pharmaceutiques, Université Paul Sabatier, 35 chemin des Maraîchers, 31 062 Toulouse, France.

The possible role of lectins, either in the specific binding of nodulating *Rhizobia* strains to legume root hairs (1-4) or in the defense of plants against pathogenic soil fungi or bacteria (5), has been widely investigated during the last decade. Such a role presupposes that lectin is present inside the roots (4), or on the root surface (2), or in the vicinity of the roots (3), where it serves in the specific attachment of the symbiont or the phytopathogen. However, little attention has been paid to lectin release by germinating seeds in relation to the possible ecological role of such compounds during seed imbibition and early seedling growth.

Fountain et al. (6) have previously reported a lectin release from soaked soybean seeds and this finding was subsequently corroborated and extended by Hwang et al. (7) and by Causse and Rougé (8). Similarly, exudation of lectin by imbibed seeds (9) and intact roots or seedlings (10) of *Phaseolus vulgaris* L. has been evidenced. Dazzo and Hrabak (11) have also demonstrated the presence of Trifoliin A, a lectin involved in the attachment of *Rhizobium trifolii* to white clover root hairs, in clover root exudates. However, this exudation process is not restricted to lectins since release of other proteins (6,12), including various enzymes (13,14) and proteinase inhibitors (7, 15), has also been reported.

These results suggest that lectins are not merely intracellular or surface-attached components but that they are leached from germinating seeds and developping seedlings into the surrounding medium, that is in soil under natural conditions.

Lectins, Vol. IV

Although the mechanism by which lectins are leached out from seeds and seedlings is still poorly understood (16), it has been shown that SBA release from imbibed soybean seeds (8) is essentially passive in nature and that exuded SBA is mainly originated from the cotyledon protein bodies, where it is located within the mature seed (17). However, in order to play an ecological role in soil, exuded SBA must remain unaltered, i.e. in a functional state, for a sufficient length of time in the soil fraction surrounding soybean seeds or seedlings. As previously reported (8), SBA appears to be very resistant to hydrolysis either by trypsin or proteinase-containing soil extracts and thus, could remain functional in the vicinity of seeds and seedlings. It is necessary, however, that exuded SBA could be entrapped arround seeds and seedlings and one can suppose that such a result can occur by binding with humic acids and clays, the main soil colloids, whose binding and protective properties towards soil proteins and enzymes (18) are well known.

It was the purpose of the present work to study the possible *in vitro* interaction between SBA and soil humic acids and clays.

MATERIAL AND METHODS

Material: Soybean seeds (*Glycine max* (L.) Merr., var. Hodgson), were used. Bentonite (Prolabo, France), was throughout used as clay.

Protein and lectin isolation: Soybean seed albumin and globulin were isolated and separated by extensive dialysis against tap and distilled water of seed soluble protein previously precipitated by acid treatment, according to (6).
Soybean seed lectin (SBA), was isolated by affinity chromatography of a hexane-defatted seed meal extract on N-caproylgalactosamine-sepharose gel, as described in (19). The purity of the SBA preparation was checked by polyacrylamide gel electrophoresis and immunoelectrophoresis on agarose gel with antiSBA antobodies.

Humic acids isolation and fractionation: Humic acids were extracted from soil with 5% sodium hexametaphosphate (20) and 0.5 M NaOH (21). Humic acids of high and low molecular weight were subsequently separated by chromatography on Sephadex G-75: two elution peaks were obtained corresponding respectively to

the HMW- (1st peak) and LMW-humic acids (2nd peak). Humic acids content was estimated colorimetrically at 400 nm.
Humic acid insolubilized particles were obtained by precipitation of humic acids with 0.02 N $Al(NO_3)_3$ in 0.1M acetate pH 4.5

Immunochemical techniques: Specific antibodies against SBA were raised in rabbits using Freund's complete adjuvant, as previously described (22). Immunoelectrophoretic analysis was performed in agarose gel, on microscope slides, according to (23). Precipitin lines were fixed and stained with Amidoblack 10B according to (24). Radial immunodiffusion in agarose gel was carried out according to the procedure of Mancini et al. (25).

Biochemical techniques: Soluble protein and lectin content was determined colorimetrically using the micro-biuret technique of Goa (26) with bovine serum albumin (Sigma) as standard. Polyacrylamide gel electrophoresis was performed according to Davis (27) in 7% acrylamide gel rods, at a constant current of 2 mA/tube. Protein fractions were fixed and stained with Coomassie brilliant blue R-250 (BioRad) according to Chrambach et al. (28). SBA was covalently bound to CNBr-activated Sepharose 4B (Pharmacia) following the manufacturer's instructions. The amount of immobilized lectin was 5 mg/ml wet gel.

RESULTS

1. Interaction between SBA or seed proteins and humic acids:

Humic acid are retained by passing a humic acid solution through a column of SBA covalently bound to CNBr-Sepharose 4B and such a retention involves SBA-humic acids interaction since humic acids are not retained when filtered through a Sepharose 4B column under the same conditions. Table 1 indicates that similar results are obtained with both HMW- and LMW-humic acids.

In order to study the involvement of covalent or non-covalent bonds between SBA and humic acid fractions, elution of the retained humic acids from the SBA column has been performed with the following solutions: 0.1M galactose (a potent inhibitor of SBA); 1.0M NaCl; 50% ethylene glycol; and NH_4OH pH 11.0 (which cleave weak ionic bonds, hydrophobic bonds and noncovalent bonds respectively). Table 2 indicates that the *in vitro* interaction between SBA and humic acid fractions involves noncovalent bonds since NH_4OH pH 11.0 can elute all the fixed humic acids. Moreover, SBA binding sites seem not to be involved in

Table 1

Interaction between various humic acid fractions with a column of insolubilized SBA (1 ml of gel containing 5 mg SBA). Data are means of four separate determinations.

Humic acid fractions	Amount added on the SBA column	Amount retained by the SBA column	Retention %
Whole humic acids	202 µg	139 µg	69
High MW humic acids	202 µg	140 µg	69
Low MW humic acids	202 µg	127 µg	63

SBA-humic acids interaction since 0.1M galactose has no effect.

By mixing known amounts of seed albumin and globulin solutions with insolubilized humic acid particles, unretained proteins can be estimated colorimetrically in the supernatant obtained after shaking and subsequent centrifugation of the mixture. Data show that seed albumin and globulin interact with humic acids giving a saturation curve. In addition, all the seed albumin and globulin fractions react equally well with humic acids, as indicated by an electrophoretic comparison of unretained (supernatant) and untreated (starting protein solutions) albumin and globulin fractions.

2. Interaction between SBA or seed proteins and clay:

SBA interacts with bentonite in radial immunodiffusion of SBA and bentonite mixture as compared with SBA alone. By using this technique, results obtained with increasing concentrations of SBA show that binding of SBA on bentonite gives a saturation curve. By mixing known amounts of seed albumin and globulin solutions with bentonite, unretained proteins can be estimated colorimetrically in the supernatant resulting from shaking and centrifugation of the mixtures. Data show that seed albumin and globulin interact with bentonite giving saturation curves, as in the case of humic acids.

3. Activity of SBA adsorbed on humic acids or clay:

In order to check whether SBA linked to humic acids or bentoni-

Table 2

Elution of different humic acid fractions from a column of insolubilized SBS (1 ml of gel containing 5 mg SBA) by various eluting solutions. Data are means of four separate determinations.

Humic acid fractions	Amounts added on SBA column	Amounts eluted by			
		0.1M galactose	1.0M NaCl	50% Ethyleneglycol	NH_4OH pH 11.0
Whole humic acids	202 µg	0%	12%	18%	100%
High MW humic acids	202 µg	0%	12%	19%	100%
Low MW humic acids	202 µg	0%	26%	15%	100%

te remains functional or not, the following experiment is carried out. To humic acid insolubilized particles SBA is added and the mixture is thoroughly shaked. A 1% suspension of untreated rabbit erythrocytes is then added to the humic acid particles coated with SBA and the mixture is examined microscopically for agglutination. A similar procedure is performed with SBA-coated bentonite particles. In both cases, rabbit erythrocytes readily agglutinates the SBA-coated particles, which shows that SBA adsorbed on humic acid or bentonite particles remains, at least in part, functional.

DISCUSSION

The above reported results clearly indicate that SBA can interact unspecifically with humic acids and clays, the main soil colloids, by noncovalent bonds. Although obtained *in vitro*, one can suppose that such an interaction between SBA released from germinating seeds or seedling roots and humic acids or clays can occur in soils. The fact that SBA 1) is resistant to the action of soil proteinases (8), 2) can interact with humic acids and clays and 3) remains active after linkage with soil colloids, suggest that leached SBA can be involved in various glycoconjugate-mediated recognition processes occurring in soils. By entrapping microorganisms arround seedlings and roots the released lectin can be involved either in the defense of plants against phytopathogens present in soils or in the acceptance of specific symbionts, during the early stages of plant growth. However, the situation is much more complex since soybean seedlings exhibit simultaneously 1) SBA, which is still present in seedling tissues two weeks after the onset of germination (29), 2) another lectin, different from SBA, which is present in seeds (30) and 3) SBA exuded from germinating seeds (8) and seedling roots (31,32), which is present outside the seedling tissues. We don't know if these various lectins play a similar or different ecological role.

REFERENCES

1. Pueppke, S.G.: in Lectins, Biology, Biochemistry, Clinical Biochemistry, (Bøg-Hansen, T.C., and Spengler, G.A. ed.), De Gruyter, Berlin, 3, 513-520 (1983).

2. Kijne, J.W., van der Schaal, I.A.M., Diaz, C.L., and van Iren, F.: in Lectins, Biology, Biochemistry, Clinical Biochemistry, (Bøg-Hansen, T.C., and Spengler, G.A. ed.), De Gruyter, Berlin, 3, 521-530 (1983).

3. van der Schaal, I., Kijne, J., Diaz, C.L., and van Iren, F.: in Lectins, Biology, Biochemistry, Clinical Biochemistry, (Bøg-Hansen, T.C., and Spengler, G.A. ed.), De Gruyter, Berlin, 3, 531-537 (1983).

4. Hosselet, M., van Driessche, E., van Poucke, M., and Kanarek, L.: in Lectins, Biology, Biochemistry, Clinical Biochemistry, (Bøg-Hansen, T.C., and Spengler, G.A. ed.), De Gruyter, Berlin, 3, 549-558 (1983).

5. Mirelman, D., Galun, E., Sharon, N., and Lotan, R.: Nature 256, 414-416 (1975).

6. Fountain, D.W., Foard, D.E., Replogle, W.D., and Yang, W. K.: Science 197, 1185-1187 (1977).

7. Hwang, D.L., Yang, W.K., Foard, D.E., and Lin, K.T.D.: Plant Physiol. 61, 30-34 (1978).

8. Causse, H., and Rougé, P.: in Lectins, Biology, Biochemistry, Clinical Biochemistry, (Bøg-Hansen, T.C., and Spengler G.A. ed.), De Gruyter, Berlin, 3, 559-572 (1983).

9. Manen, J.F.: Thèse Doct. ès-Sc., Genève (1978).

10. Gietl, C., and Ziegler, H.: Naturwis. 66, 161-162 (1979).

11. Dazzo, F.B., and Hrabak, M.E.: J. Supramol. Struct. Cell. Biochem. 16, 133-139 (1981).

12. Amoros, M.: Thèse Doct. Spec., Toulouse (1967).

13. Soil Enzymes, Burns, R.G. ed., Academic Press, London (1978).

14. Dazzo, F.B., Truchet, G.L., Sherwood, J.E., Hrabak, M.E., and Gardiol, A.E.: Appl. Environ. Microbiol. 44, 478-485 (1982).

15. Wilson, K.A.: Phytochem. 19, 2517-2519 (1980).

16. Duke, S.H., Kakefuda, G., and Harvey, T.M.: Plant Physiol. 72, 919-924 (1983).

17. Horrisberger, M., and Vonlanthen, M.: Histochemistry 65, 181-186 (1980).

18. Van Sumere, C.F., Albrecht, J., Dedonder, A., De Pooter, H., and Pé, I.: in The Chemistry and Biochemistry of Plant Proteins, (Harborne, J.B., and Van Sumere, C.F. ed.), Academic Press, London, 11, 211-264 (1975).

19. Allen, A.K., and Neuberger, A.: FEBS Lett. 50, 362-364 (1975).

20. Schaefer, R.: Thèse Doct. ès-Sc., Paris (Orsay)(1966).

21. Levesque, M., and Schnitzer, M.: Can. J. Soil Sci. 46, 7-12 (1966).

22. Rougé, P.: Thèse Doct. ès-Sc., Toulouse (1977).

23. Scheidegger, J.J.: Int. Arch. Allergy 7, 103-110 (1955).

24. Uriel, J.: in Immuno-electrophoretic Analysis, (Grabar, P., and Burtin, P. ed.), Elsevier Publ. Co., Amsterdam, 30-57 (1964).

25. Mancini, G., Vareman, J.P., Carbonara, A.O., and Heremans, J.F.: in Protides Biological Fluids, (Peeters, H. ed.), Pergamon Press, Oxford, 370-373 (1964).

26. Goa, J.: Scand. J. Clin. Lab. Invest. 5, 218-222 (1953).

27. Davis, B.J.: Ann. N. Y. Acad. Sci. 121, 404-427 (1964).

28. Chrambach, A., Reisfeld, R.A., Wickoff, M., and Zaccari, J.: Anal. Biochem. 20, 1-15 (1967).

29. Causse, H., Lemoine, A., and Rougé, P., (in preparation).

30. Dombrink-Kurtzman, M.A., Dick, W.E., Burton, K.A., Cadmus, M.C., and Slodki, M.E.: Biochem. Biophys. Res. Commun. 111, 798-803 (1983).

31. Gade, W., Jack, M.A., Dahl, J.B., Schmidt, E.L., and Wold, F.: J. Biol. Chem. 256, 12905-12910 (1981).

32. Gade, W., Schmidt, E.L., and Wold, F.: Planta 158, 108-110 (1983).

FETUIN BINDING BY *Phaseolus coccineus*, VAR. "ALUBIA" SEED AGLUTININ.

Hector Nolasco and Jose-Luis Ochoa

Department of Biotechnology
CIB-The Center for Biological Research
Box 128, La Paz, Baja California Sur, 23000, Mexico.

Phaseolus coccineus agglutinin is a mitogenic glycoprotein (1,2) of about 112,000 Da and 20% in weight carbohydrate (1). Its hemagglutination capacity is the same towards the different human blood groups of the ABO system and is not inhibited by simple sugars or glycoproteins except by fetuin (1). Hence, it appears that only fetuin possesses some particular features which might play a role on its inhibitory activity towards this agglutinin. Fetuin is a glycoprotein of about 47,000-49,000 Da and 22% of its chemical composition is carbohydrate (3). The major asparagine-linked oligosaccharide portion which account for about 88% of the total carbohydrate contains sialic acid, galactose, mannose and N-acetyl-glucosamine as its sugar components and are distributed in three different large heterosaccharide chains. The minor serine or threonine-linked ones, contain sialic acid, galactose, and N-Acetylgalactosamine in three small units of 3 or 4 carbohydrate residues (3). Only the three small heterosaccharide units of fetuin are alkali-labile and can be split off by treatment with $NaBH_4$ under alkaline conditions (4). Sialic acid residues of retuin molecule can be selectively and almost quantitatively liberated by treatment with 0.025 N H_2SO_4 (3), neuraminidase (5), or by oxidation with sodium metaperiodate (3), indicating that they occupy a terminal position.

Lectins, Vol. IV

In this paper we report an attempt to elucidate the carbohydrate specificity of *Ph. Coccineus* agglutinin and of its binding to chemically or enzymatically modified fetuin by hemagglutination-inhibition test.

Materials and Methods

Trypsin, papain, chymopapain, neuraminidase and fetuin, were purchased from Sigma (St. Louis, MO, USA), protease and pepsin came from Calbiochem Behring Corp (La Jolla, CA, USA). Other chemicals of analytical purity from other commercial sources. Sephadex G-25 superfine was obtained from Pharmacia Fine Chemicals (Uppsala, Sweden). *Phaseolus coccineus* Var. "alubia" seeds were acquired from the local market.

Agglutinin Extraction: the procedure for the extraction of the agglutinin has been described before (6). Essentially, finely ground seeds were extracted with PBS for two hours under constant stirring at $4^{o}C$. The pH was then adjusted to 4 with 4M acetic acid and let stand over night at $4^{o}C$. The nearly clear supernatant was then centrifuged (19,000 rpm, 15 min, $4^{o}C$) and filtered through a glass fiber membrane filter GF/A, Whatman, England). The protein solution was finally fractionated by solid ammonium sulphate in the range 40-80% saturation. The material thus obtained was extensively dialized against PBS and then kept under refrigeration.

Agglutinin Purification: Phaseolus coccineus agglutinin was purified, as described before (1), by affinity chromatography on a stroma-column. Cell membrane residues were obtained by hemolysis of human fresh blood group A, (7), and packed in columns (6). The hemagglutinating activity eluted from the column by distilled water was pooled and concentrated by lyophilization.

Hemagglutination - Inhibition assays: a 2% suspension erythrocytes of the different human blood groups were used to determine the hemagglutination capacity of *Ph. coccineus* agglutination in the two-fold serial dilution procedure in microtiter plates (8). The red cells were enzymatically digested with protease, papain, chymopapain, trypsin and neuraminidase. 0.5 μg/ml of enzyme was added and the incubation carried out at $37^{o}C$ for one hour. After the enzyme treatment, the erythrocytes were washed with PBS to eliminate the digestion products and used directly in the hemagglutination assay.

Fetuin Treatment. Native fetuin was chemical or enzymatically modified as follows: A) treatment with $NaBH_4$ was done according to (9). 10 mg/ml of fetuin in 0.1 M $NaBH_4$-0.1 N NaOH were incubated at $23^{o}C$ for 24, 48 and 72 h. Afterwards, the mixture was neutralized with acetic acid to eliminate the $NaBH_4$ in excess before being filtered through a Sephadex G-25 superfine column (10 x 1 cm I.D.) equilibrated with PBS. B) Treatment with $NaIO_4$ was done according to (10). 10 mg/ml of fetuin were mixed with 0.075 M $NaIO_4$ in 0.05 M sodium acetate buffer pH 4.0. The mixture was incubated in an ice water bath at $4^{o}C$ during 24, 48 and 96 h at darkness. The reaction was terminated by the addition of 50 µl/ml of solubion of ethylene glycol to neutralize all the $NaIO_4$ in excess and then filtered through a Sephadex G-25 superfine column (10 x 1 cm I.D.) equilibrated with PBS. C) Treatment with H_2SO_4 was done as recommended by (11). 10 mg/ml of fetuin in 0.025 N H_2SO_4 were incubated in a water bath at $80^{o}C$ during 1 h. After being cooled, the mixture was neutralized with NaOH and filtered through a Sephadex G-25 superfine column (10 x 1 cm I.D.) equilibrated with PBS. D) Fetuin was also enzymatically treated with papain, chymopapain, protease, trypsin and pepsin according to the recommendation of the Worthington Enzyme Manual. 10 mg/ml of fetuin were mixed with 20 µg/ml of the corresponding enzyme in the buffers recommended in the manual. After incubation period at $37^{o}C$ during 24 or 48 h, the mixture was sieved through a Sephadex G-25 superfine column (10 x 1 cm I.D.) equilibrated with PBS.

Results

Table 1 summarizes the effects of the chemical and enzymatical modification of fetuin on its capacity to inhibit *Ph. coccineus* hemagglutinating activity. It is concluded that the inhibiting capacity of fetuin is fully maintained after it is subjected to H_2SO_4 treatment or slightly modified by $NaBH_4$-NaOH treatment. This indicates that at least in fetuin, the short carbohydrate chains containing sialic acid, galactose and N-acetylgalactosamine residues do not participate in the interaction with *Ph. coccineus* agglutinin. In the case of the oxidation with metaperiodate however, a complete loss on the inhibitory capacity of fetuin is observed (Table 1). This observation confirms the involvement of the carbohydrate portion of fetuin in the interaction with *Ph. coccineus* agglutinin and its susceptibility to extensive or servere oxidation treatment.

In Table 1 we also show that the treatment of fetuin by a number of proteolytic enzymes modifies its capacity to inhibit hemagglutination. Most severe is treatment with pepsin and protease where a complete loss of inhibitory capacity of the glycopeptides obtained was observed.

In view of the striking results obtained from the enzyme treatment of fetuin, it was considered of interest to determine whether a similar effect would be observed by digestion of erythrocytes with the same enzymes. From Table 2, a higher agglutination susceptibility of the erythrocytes as a result of their treatment with the enzymes is apparent. Also, $NaBH_4$-treated fetuin is an excellent inhibitor in this case and support the idea that enzyme treatment unmasks cell receptors resembling the structure and chemical composition of fetuin.

Discussion

As shown in this report, fetuin free of O-linked carbohydrates behaves in a way similar to native fetuin with respect to its inhibitory capacity against *Ph. coccineus* agglutinin. This fact may explain why mucin and other glycoproteins with mainly O-linked heteropolyssaccharides (1), lack inhibitory effect. It is therefore reasonable to assume that the large heteropolysaccharide chains of fetuin are the preferred binding sites for this lectin. We also observed that the terminal sialic acid residues of these chains are not required for such binding since their removal by either diluted H_2SO_4 or neuraminidase does not modify inhibiting capacity of fetuin (Table 1).

Fetuin was also digested with a number of enzymes with the main purpose of isolating the glycopeptides responsible of its inhibition power against *Ph. coccineus* agglutinin. The glycopeptides isolated by this means have shown to possess practically all the carbohydrate in three heteropolysaccharides each involving an attachment between the C-1 of an N-acetylglucosamine residue and the amide group of an asparagine residue (3). Therefore these oligosaccharide units are the alkaline-stable sugar chains of the fetuin molecule. However, as shown in Table 1 these glycopeptide fractions lack inhibitory capacity towards *Ph. coccineus* lectin. Thus, it would seem that for expression of full inhibition activity towards *Ph. coccineus* agglutinin, the molecular configuration and chemical composition of fetuin should be as close as possible to the native state.

In order to understand the effect of the enzymatic treatment of the red cells and their inhibition by $NaBH_4$-treated fetuin we ought to mention several explanations put forward to account for similar observations in the case of other agglutinins (13, 14). First, a reduction on the cell surface Z-potential that diminishes the electrostatic repulsion between adjacent cells as a result of the removal of charged group (i.e. sialic acid residues) (13); second, the "unmasking" of hindered receptors (14); and third, rearrangements of the receptors on the membrane plane originated by alterations on the dynamic equilibrium

Table 1

Capacity of chemically or enzymatically modified fetuin to inhibit Ph. coccineus agglutinin.*

Treatment	Hemagglutination Titer
Blank	16
Native	0
H_2SO_4	0
$NaIO_4$	16
$NaBH_4$	2
(24 h) Chymopapain	4
(24 h) Papain	4
(48 h) Chymopapain	16
(48 h) Papain	16
(24 h) Pepsin	16
(24 h) Protease	16
(24 h) Trypsin	8

**Protein concentration was adjusted to 10 mg/ml*

between cell surface interacting components (24). In any case, it is evident that $NaBH_4$-treated fetuin must resemble the structures located on the cell surface for which *Ph. coccineus* agglutinin shows affinity. The whole peptide backbone of native fetuin is important for expression of its inhibition activity, since the pool of the glycopeptides obtained by its digestion with a number of proteolytic enzymes possesses very poor activity. A similar observations has been done in the case of *Ph. vulgaris* receptor (16) where it was found that the trypsin digestion of the corresponding structure induced a 90% loss of its inhibition capacity. This has been interpreted as an evidence for the participation of the peptide backbone in holding the oligosaccharide chains in a favourable conformation for lectin recognition (16).

Acknowledgements: We thank Mrs. Maria Gomez and Ms. Reni Storgård for help in the preparation of the manuscript.

References

1. Ochoa, J.-L., Kristiansen, T., Pahlman, S.: Biochim. Biophys. Acta, 705, 396-404 (1982).

2. Calderon, R.A., Cordoba, F.: Eur. J. Immunol., 6, 522-525 (1976).

3. Nilsson, B.P., Norden, N.E., Svensson, S.: J. Biol. Chem., 254, No. 11, 4545-4553 (1979).

4. Spiro, R.G., Bhoyroo, V.D.: J. Biol. Chem., 249, 5704-5717 (1974).

5. Worthington: In: Enzymes, enzymes reagents and related biochemicals, Worthington Biochemical Corp.; Freehold, New Jersey, USA (1972).

6. Ochoa; J.-L., Kristiansen, T.: FEBS Letters, 90, 145-148 (1978).

7. Dodge, J.T., Mitchell, C., Hanahan, D.J.: Arch. Biochem. Biophys., 100, 119-130 (1962).

8. Lis, H., Sharon, N.: In: the Biochemistry of Plants, P.K. Stumpf and E.E., Conn., Eds., Academic Press, New York, pp 371-488 (1981).

9. Bayard, B., Kerckaert, J.-P.: Eur. J. Biochem. 113, 405-414 (1981).

10. Irimura, T., Tsuji, T., Tagami, S., Yamamoto, K., Osawa, T.: Biochemistry. 20, 560-566 (1981).

11. Graham, E.R.B.: In: Glycoproteins (A. Gottschalk, Ed.), Vol. 5, Part A, Elsevier Publ. Co., Amsterdam, 717-731 (1972).

12. Begbie, R.: B^ochim. et Biophy Acta 371, 549-567 (1974).

13. Pollack, W., Reckel, R.P.: Int'l. Allergy Appl. Immunol. 54, 29-42 (1977).

14. Sela, A., Lis, H., Sharon, N., Sachs, L.: J. Memb. Biol. 3, 267- (1970).

15. Ochoa, J.-L.: Pathol. Biol. 27, 103-113 (1979).

16. Kornfeld, S., Kornfeld, S.: J. Biol. Chem. 245 2536-2545 (1970).

LEAVES OF THE WILD GRASS *AGROPYRUM REPENS* CONTAIN AN N-ACETYLGALACTOSAMINE-SPECIFIC LECTIN

Bruno Cammue, Hetty M. Stinissen and Willy J. Peumans
Laboratorium voor Plantenbiochemie, KULeuven
Kardinaal Mercierlaan 92, B-3030 Leuven, Belgium

Gramineae lectins are a class of proteins differing from other phytohemagglutinins with respect to their biochemical, physico-chemical and biological properties (1,2). Within this group of lectins we may distinguish three subgroups, which according to their origin are designated as cereal, rice and *Brachypodium* lectins (2-5). All Gramineae lectins characterized hitherto have been isolated from dry embryos. They exhibit the same sugar-binding specificity to N-acetylglucosamine and have a common molecular structure (2-6). Basically, Gramineae lectins are dimers of two identical subunits (Mr 18,000) (2-6). In rice lectin, however, these subunits are further cleaved into two smaller polypeptides (4). In addition, they all have a very similar amino acid composition, characterized by a very high content of glycine and cysteine, and they are serologically related to each other (2,7). Within the subclass of cereal lectins, structural similarities are even more pronounced as lectins from different species readily exchange subunits *in vitro* giving rise to heterodimers (i.e. lectins composed of subunits from 2 different lectins) (8).

Lectins, Vol. IV

Like all Triticeae species, couch grass (*Agropyrum repens*) embryos contain a lectin which is structurally and serologically indistinguishable from wheat germ agglutinin and other cereal embryo lectins (7). In this report we present evidence that leaves of this wild grass species also contain lectin in contrast to wheat (9,10). Moreover, a comparison of the couch grass leaf lectin to the embryo lectin of the same species indicates that it represents a variant-type of cereal lectin.

Material and Methods

Couch grass (*Agropyrum repens*) is a perennial wild grass species, which is taxonomically closely related to modern cereals like wheat, barley and rye. For isolation of the leaf lectin, leaves were collected locally in the field during the period September-October (since at that time of the year the lectin content is usually the highest). Seeds were collected around the end of August.

Purification of *Agropyrum repens* leaf lectin (ARLL): Leaves (1 kg) were cut in small pieces and homogenized with a mixer in 5 l of 0.1 N HCl containing 5 mM thiourea. The homogenate was filtered through cheese-cloth, centrifuged (10,000 x *g* for 15 min) and the supernatant brought at pH 4.0 with 2N NaOH. After being left in ice for 1 h the extract was recentrifuged (10,000 x *g* for 15 min) and applied onto a column (5 ml bed volume) of immobilized N-acetylglucosamine (Selectine 1 from Pierce Chemical Co, Rockford, Il., USA) equilibrated with 50 mM Na-acetate buffer (pH 3.8) containing 0.1 M NaCl. Unbound protein was eluted with the same buffer and the lectin desorbed with 0.1 M N-acetylglucosamine.

Purification of *Agropyrum repens* embryo lectin (AREL): Couch grass embryo lectin was prepared from isolated embryos by affinity chromatography on Selectine 1 (cf. above) basically as described previously for the purification of cereal, *Brachypodium* and rice lectins (3-5).

Agglutination assays were done with trypsin-treated erythrocytes. Extracts were prepared by homogenizing plant tissue in 10 volumes of phosphate-buffered saline and centrifuging the homogenate at 10,000 for 4 minutes.

SDS-polyacrylamide gel electrophoresis was done on linear 12.5 - 25 % acrylamide gradient gels as described previously

(3). Mr reference proteins are indicated in the legend to Fig.2.

Double immunodiffusion was done as described previously (3). The partially purified anti-WGA was the same as that used in earlier experiments (7). After formation of the precipitin lines, gels were fixed and stained as described (3).

Subunit exchange experiments: Aliquots, containing equal amounts of agglutination activity of AREL, ARLL and CL_{AA} (is the WGA-isolectin coded for by the A-genome) were mixed two and two and dialyzed for 12 h against 0.1 N acetic acid and 30 mM Tris-acetate (pH 5.0) consecutively in order to allow subunit exchange (8). Then, the lectin mixtures were separated by ion-exchange chromatography on sulphopropyl-Sephadex as described previously (8).

Amino acid analysis: Purified leaf lectin samples were hydrolyzed (in sealed glass tubes under N_2 atmosphere) for 24 h at 110 °C in 6 M HCl containing 0.02 % β-mercaptoethanol. After hydrolysis, HCl was removed by evaporation, and the residue was dissolved in sodium citrate buffer, pH 1.9. Amino acid analysis was done with a Biotronik LC 2000 amino acid analyzer. This apparatus allows a separation of all 20 amino acids on a single column of Durrum resin, using an elution program of four different buffers. Amino acids were directly quantified using an Infotronics integrator. Tryptophan was determined after hydrolysis with 3 M mercaptoethanesulphonic acid. Cysteine was determined as cysteic acid after hydrolysis in performic acid.

Results

Occurrence of different lectins in couch grass embryos and leaves respectively. We observed that extracts from leaves of adult couch grass plants exhibit a relatively high agglutination activity. Indeed, as shown in Table 1, the agglutination titer of crude extracts from leaves was only 4-fold lower than that of embryo extracts. Taken into consideration that the water content of dry embryos (< 15 %) is much lower than that of leaves (90 %), the lectin content of the former is not higher – on a dry weight base than that of the leaves. Preliminary experiments were carried out with crude extracts to determine the carbohydrate-binding specificity and sero-

Table 1.

Blood-group specificity of crude extracts from embryos and leaves of couch grass and purified AREL and ARLL.

Blood-group	Agglutination titer of:			
	Leaf extract	ARLL	embryo extract	AREL
A	80	1280	360	320
B	20	320	360	320
O	20	320	360	320

The titer is expressed as the highest dilution at which the extract or purified lectin (20 µg/ml) still agglutinated a 1% suspension of trypsin-treated erythrocytes.

Table 2.

Sugar-specificity of ARLL and AREL in crude extracts and purified lectin solutions, with different blood-groups.

Extracts/lectin	Blood-group	Concentration required for 50 % inhibition of agglutination with		
		GlcNAc (mM)	GalNAc (mM)	$(GlcNAc)_4$ (µM)
Leaf extract	A	20	10	10
	B	20	> 100	10
	O	20	> 100	10
ARLL (20 µg/ml)	A	20	10	10
	B	20	> 100	10
	O	20	> 100	10
Embryo extract	A	20	> 100	10
	B	20	> 100	10
	O	20	> 100	10
AREL (20 µg/ml)	A	20	> 100	10
	B	20	> 100	10
	O	20	> 100	10

Other monosaccharides (including lactose, galactose, galactosamine, melibiose, raffinose) were not inhibitory. > 100 means that agglutination was not inhibited at a sugar concentration of 100 mM.

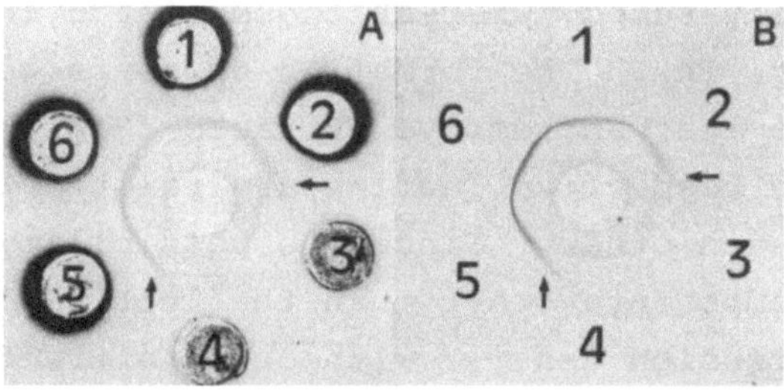

Fig. 1. Double immunodiffusion of AREL, ARLL and WGA against anti-WGA. In (A), crude extracts from couch grass embryos (1, 2) and leaves (3, 4), and wheat embryos (5, 6) were challenged against anti-WGA (central well). In (B), purified AREL (1, 2), ARLL (3, 4) and WGA (5, 6) were challenged against anti-WGA (central well).

logical properties of both lectins. As shown in Table 2, the embryo lectin exhibited specificity towards N-acetylglucosamine (GlcNAc) and GlcNAc-oligomers ($(GlcNAc)_n$) whereas the leaf lectin was better inhibited by N-acetylgalactosamine (GalNAc), when assayed with group A red blood cells. In assays with type B-or O-erythrocytes, however, the leaf lectin was not inhibited by GalNAc. In addition, when leaf extracts were assayed with erythrocytes of different blood-groups, titers were 4-times higher with group A- than group B- or O-red blood cells (Table 1). Embryo extracts, on the contrary, agglutinated all types of erythrocytes equally well (Table 1). To exclude any possible interference of components other than the lectins in assays with crude extracts, both sugar- and blood-group-specificity were also determined with affinity-purified AREL and ARLL. As shown in Table 1 and 2, these experiments confirmed that ARLL agglutinates preferentially group-A red blood cells and is better inhibited by GalNac than GlcNac in agglutination assays with erythrocytes of this particular blood-group.

Double immunodiffusion of embryo and leaf extracts against WGA-antiserum revealed that the precipitin lines of embryo lectin and WGA completely fused, whereas those of the leaf lectin and WGA did not (Fig. 1A). The formation of spurs between the precipitin lines of leaf and embryo extracts (Fig. 1A, arrow) indicated that the respective lectins are related to, but not identical with each other. Since the possibility could not be excluded that components other than the lectins interfered in double immunodiffusion assays with crude extracts, affinity-purified AREL and ARLL also were challenged with WGA-antiserum. As shown in Fig. 1B, the results obtained with the pure lectins confirmed what was observed in experiments with crude extracts.

Molecular structure of AREL and ARLL. Affinity-purified AREL and ARLL were analyzed by gel filtration on Sephadex G-100 and sucrose-density-gradient centrifugation. Both methods yielded the same results for AREL and ARLL, and indicated a Mr of the native lectin of 36,000 (results not shown). SDS-polyacrylamide gel electrophoresis showed that AREL and ARLL migrated with an apparent Mr of 23,000 and 25,000 respectively (Fig. 2). It should be emphasized, however, that Mr-determinations of Gramineae lectins by SDS-polyacrylamide gel electrophoresis

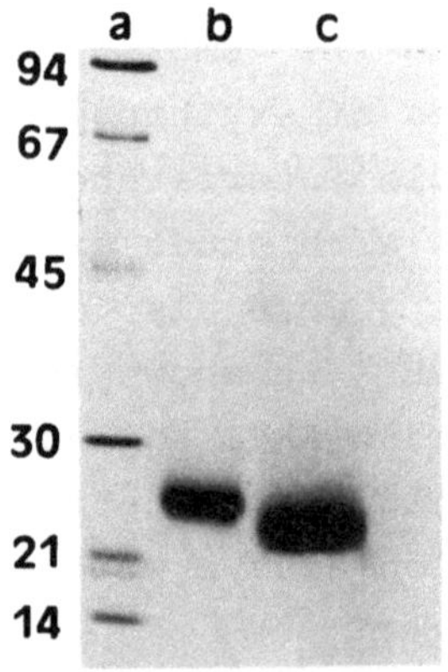

Fig. 2. SDS-polyacrylamide gel electrophoresis of ARLL (B) and AREL (C). Mr reference proteins are shown in lane A: lysozyme (14 kd), soyabean trypsin inhibitor (21 kd), carbonic anhydrase (30 kd), ovalbumin (45 kd), bovine serum albumin (67 kd) and phosphorylase b (94 kd).

Table 3.
Amino acid composition of ARLL.

Amino acid	Mol%	Residues/mol	Amino acid	Mol%	Residues/mol
AsX	10.8	21	Ile	0.5	1
Thr	2.6	5	Leu	3.9	8
Ser	7.2	14	Tyr	5.1	10
GlX	8.2	16	Phe	0.5	1
Pro	7.3	14	His	0.6	1
Gly	19.5	38	Lys	4.1	8
Ala	4.7	9	Arg	3.2	6
$\frac{1}{2}$Cys	19.6	38	Trp	0.4	1
Val	1.2	2			
Met	0.5	1	Total	100	194

yield overestimates because of their particular amino acid composition. For instance, WGA subunits, which have a real Mr of 18,000 migrate with an apparent Mr of 23,000 (5).

The amino acid composition of ARLL was determined. As shown in Table 3, it is especially rich in cysteine and glycine, each of which represents 38 residues of a total number of 194 amino acids. The amino acid composition of AREL could not be determined since this lectin was available only in very limited amounts. No covalently-bound carbohydrate was detected in ARLL preparations.

Further evidence for structural differences between ARLL and AREL. In another approach, we looked whether the couch grass lectins behaved differently in subunit exchange experiments with the CL_{AA} isolectin from WGA. Both AREL and ARLL were challenged with CL_{AA} under conditions allowing subunit exchange.

As shown in Fig. 3, AREL did exchange subunits giving rise to lectin molecules composed of one AREL subunit and one CL_{AA} subunit. ARLL, on the contrary, failed to exchange its subunits with CL_{AA}.

Discussion

Although there is evidence for the presence of low concentrations of WGA in roots and leaves of wheat plants (9, 10), Gramineae lectins are still considered as embryo-specific proteins (2). In this report it is demonstrated that couch grass contains a leaf-specific lectin which is related to but not identical with the embryo lectin (AREL) of the same species. Indeed, whereas AREL is by no means distinguishable from embryos of related cereal species, ARLL has several distinct properties. Structural differences are obvious because of the higher Mr of the ARLL polypeptides. Since the leaf lectin is not glycosylated its higher apparent Mr relative to that of AREL results most probably from a longer polypeptide chain. In addition, structural differences can be inferred indirectly from the results of subunit exchange and immunodiffusion experiments. The formation of spurs between AREL and ARLL upon double immunodiffusion against WGA-antiserum indicates that both lectins are structurally different, at least at their respective antigenic site. It is worthwhile to mention in this context, that a similar formation of spurs was observed between cereal and *Brachypodium* lectins (5), and cereal and rice lectins (4). The different behaviour of AREL and ARLL in subunit exchange experiments is another indication for structural differences between both lectins. Indeed, the failure of ARLL to form heterodimers with CL_{AA} subunits shows that it must be different structurally from AREL since the latter readily exchange subunits with CL_{AA}. Finally, a comparison of the agglutination properties of ARLL and AREL indicates that both couch grass lectins differ with respect to their sugar-binding- and blood-group specificity. Whereas AREL (and in fact all other Gramineae embryo lectins) are GlcNac-specific and exhibit no blood-group specificity, ARLL is not only inhibited by GalNac when assayed with blood-group-A erythrocytes but also has a four-fold higher agglutination activity with

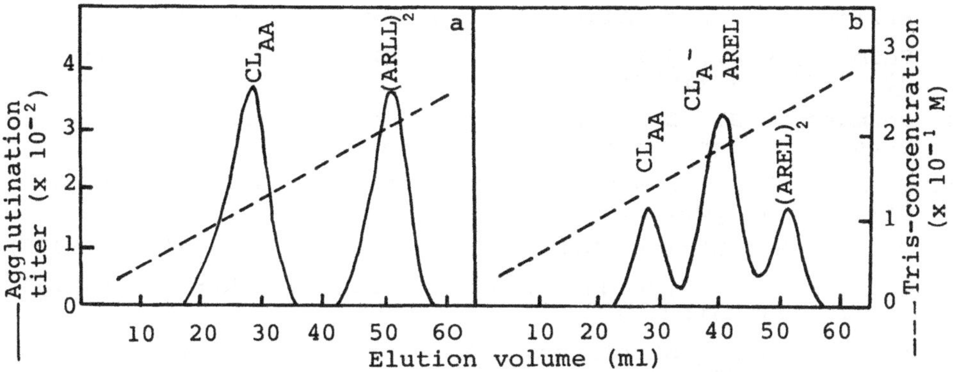

Fig. 3. In vitro subunit exchange between AREL and CL_{AA}, and ARLL and CL_{AA}. After being brought under conditions allowing subunit exchange, lectin mixtures were applied to a column (1 cm x 10 cm) of sulphopropyl-Sephadex and eluted with a linear gradient (75 ml) of increasing Tris-acetate concentration (0.03 - 0.3 M). Fractions of 1 ml were collected and quantitatively assayed for agglutination activity. The position of CL_{AA}, AREL and ARLL was determined in separate runs with single lectins (results not shown). Patterns shown in a and b were obtained with ARLL-CL_{AA} and AREL-CL_{AA} mixtures respectively. Lectin peaks are indicated as follows: CL_{AA}, $(AREL)_2$ and $(ARLL)_2$ refer to the homodimers of the respective lectins, whereas CL_A-ARL refers to the heterodimer composed of 1 CL_A and 1 AREL subunit.

type A than type B or O red blood cells.

Aknowledgements. This work is supported in part by grants from the National Fund for Scientific Research (Belgium), of which W. Peumans is Research Associate. H. Stinissen receives a fellowship from the Belgian Instituut tot Aanmoediging van het Wetenschappelijk Onderzoek in Nijverheid en Landbouw.

References

1. Goldstein, I.J., Hayes, C.E.: Adv. Carbohydr. Chem. Biochem. 35, 127-340 (1978).

2. Peumans, W.J., Stinissen, H.M.: Chemical taxonomy, molecular biology and function of plant lectins. Goldstein, I.J.,

Etzler, M.E. eds. Alan R. Liss, New York. pp. 99-116 (1983).

3. Peumans, W.J., Stinissen, H.M., Carlier, A.R.: Biochem. J. *203*, 239-244 (1982).

4. Peumans, W.J., Stinissen, H.M., Carlier, A.R.: Biochim. Physiol. Pflanz. *178*, 423-431 (1983).

5. Peumans, W.J., Spaepen, C., Stinissen, H.M., Carlier, A.R.: Biochem. J. *205*, 635-638 (1982).

6. Rice, R.H., Etzler, M.E.: Biochemistry *14*, 4093-4099 (1975).

7. Stinissen, H.M., Peumans, W.J., Carlier, A.R.: Planta *159*, 105-111 (1983).

8. Peumans, W.J., Stinissen, H.M., Carlier, A.R. : Planta *154*, 568-572 (1982).

9. Mishkind, M.L., Keegstra, K., Palevitz, B.A.: Plant Physiol. *66*, 950-955 (1980).

10. Mishkind, M.L., Raikhel, N.V., Palevitz, B.A., Keegstra, K.: Chemical taxonomy, molecular biology and function of plant lectins. Goldstein, I.J., Etzler, M.E. eds. Alan R. Liss, New York. pp. 163-176 (1983).

CHARACTERIZATION OF THE DIFFERENT LECTINS FROM VISCUM ALBUM (MISTLETOE) AND THEIR STRUCTURAL RELATIONSHIPS WITH THE AGGLUTININS FROM ABRUS PRECATORIUS AND RICINUS COMMUNIS

R. Samtleben, M. Kiefer

Central Institute of Molecular Biology, Acad. Sci. GDR,
1115 Berlin-Buch, GDR

P. Luther

Research Institute for Lung Diseases, 1115 Berlin-Buch, GDR

Lectins are carbohydrate-binding proteins which have been isolated from many plant and animal sources (1). Although many lectins have been characterized with respect to their physical and chemical properties, relatively little is known about their synthesis and function. A number of hypothetical roles for lectins in plants have been advanced, i. e. host-symbiont relations (2), involvement in the transport of carbohydrates (3) and seed formation and germination (4), plant defense system (5) and control of cell growth (6). To date, however, there is no convincing evidence to support any of these as a general role for plant lectins. In order to get more insight into the physiological aspects structural relationships among lectins have been investigated during the last years.

Legume lectins are found to be immunologically closely related and to possess a high degree of conservation in their primary structures (7). The possible relationships in structure between the toxic mistletoe leaf lectins and the toxic Abrus precatorius and Ricinus communis seed lectins which occur in unrelated plants is an interesting biological problem.

Lectins, Vol. IV

Materials and methods

Extraction and purification of the Viscum lectins. The two-year old outer shoots of the mistletoe were homogenized in a Waring blendor in a tenfold excess (v/w) of PBS (50 mM Na phosphate, 250 mM NaCl, 0.02 % NaN_3, pH 7.0). After extraction at 4 °C overnight the suspension was filtered through cheesecloth and centrifuged at 5 000 g for 20 min. The pellet was discarded. The supernatants were fractionated with ammonium sulfate and the proteins which precipitated between 30 and 70 % of ammonium sulfate were collected and dialyzed for 24 hr against PBS, pH 7.0. After dialysis the extracts were centrifuged at 100 000 g for 1 hr, and the supernatants were referred to in the following as crude extracts. Ricinus and Abrus lectins have been extracted by the same procedure from decorticated seeds.

Gel filtration. A column (1.6 x 70 cm) with Sephadex G 100 was equilibrated with PBS. The column was eluted with a speed of 6 ml/hr.

Electrophoresis. Polyacrylamide disc gel electrophoresis under nondissociating conditions was performed according to conventional methods (8) using 7.5 % polyacrylamide gel with 0.2 % methylenebisacrylamide in Tris buffer, pH 8.9; electrode buffer: Tris/glycine, pH 8.3. SDS-polyacrylamide disc gel electrophoresis was performed as described by Fairbanks et al. (9). The gels were stained either for protein with Coomassie brilliant blue or for carbohydrate with FITC-labeled lectins according to Furlan et al. (10).

Hemagglutination. Hemagglutination assays were performed in microtiter plates. Each well contained 25 µl of diluent (0.01 M phosphate buffer, pH 7.4, containing 0.14 M NaCl in the presence or absence of sugar). 25 µl of lectins in the same buffer were serially diluted. 25 µl of a 2 % suspension of human red blood cells in 0.15 M NaCl were added to each well and gently shaken. The plates were incubated at room temperature for 3 hr, and the last well showing hemagglutination was taken as the end point.

Preparation of antisera. A sample of lectin (20 µg) in 1 ml of 0.14 M NaCl, 10 mM Na phosphate (pH 7.4) was mixed with 1 ml Freund's complete adjuvant and injected subcutaneously into a rabbit at twelve dorsal sites. After two and four weeks the injections were repeated. Serum was collected one week after the last injection.

ELISA. Polyvinylchloride plates (VEB Berlin-Chemie, GDR) were coated with the antigen (lectin) in carbonate-bicarbonate buffer (pH 9.6) at 4 °C overnight (11). In the next step the antilectin (rabbit) was incubated (2 hr, 37 °C) in the presence of 0.2 M lactose in order to prevent unspecific binding. Preimmune serum was used as a control. The bound antibody was detected by horse-radish peroxidase labeled goat antirabbit IgG and ortho-phenylenediamine in phosphate-citrate buffer, pH 5.0, as a substrate. After 5 min the reaction was stopped with 2.5 M H_2SO_4 and the optical density was read at 492 nm.

Results

Isolation of two types of agglutinins. Crude extracts (see Materials and methods) were applied onto an affinity column of acid-treated Sepharose 4B (12) preequilibrated with PBS. The column was first washed with PBS to elute the proteins which were not absorbed on the Sepharose 4B and then with 0.2 M lactose in PBS to elute the toxic and strongly absorbed agglutinins. All of the agglutinins were eluted in the second peak with lactose. This peak which was demonstrated to be heterogeneous by PAGE was further purified by Sephadex G 100 column chromatography. Two peaks were obtained which gave the following results in the polyacrylamide gel electrophoresis. The protein from peak 1 was homogeneous when analyzed by PAGE under nondissociating conditions (fig. 1, lane 3). Under the same conditions peak 2 was shown to consist of four main and three minor bands (fig. 1, lane 5). A small third peak which is eluted with the void volume of the column consists of aggregated material which does not enter the electrophoresis gel (fig. 1, lane 4). The two peaks from the Sephadex G 100 chromatography were analyzed by PAGE in the presence of 0.1 % SDS according to Fairbanks. The molecular weight of the intact agglutinin from peak 1 was 65 000-M_r (fig. 1, lane 2). Peak 2 protein was split into two bands of M_r 60 000 and 58 000 (fig. 1, lane 6). When the agglutinins I and II were treated with 5 % 2-mercaptoethanol at 45 °C for 1 hr before electrophoresis, these proteins give rise to two subunits of M_r 34 000 and 29 000 or 32 000/31 000 and 27 000, resp. (fig. 1, lane 1 and 7).

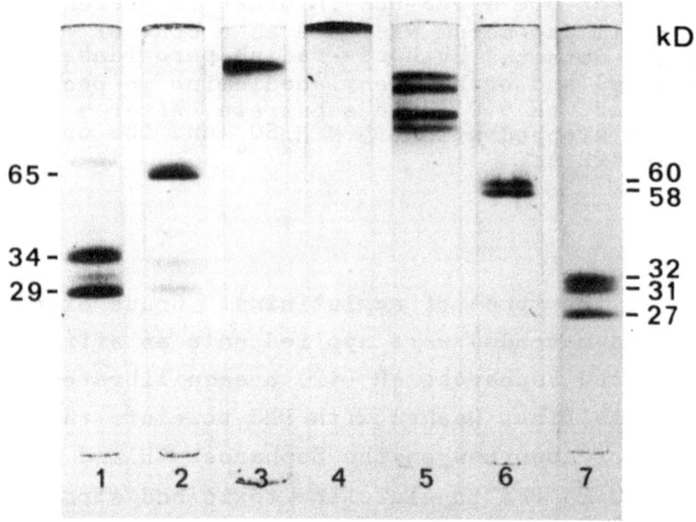

Fig. 1
Polyacrylamide gel electrophoresis. Lane 1: peak 1 in SDS-PAGE + mercaptoethanol, lane 2: peak 1 in SDS-PAGE, lane 3: peak 1 in the Maurer-system, lane 4: aggregated material from the prepeak (Maurer-system), lane 5: peak 2 (Maurer-system), lane 6: peak 2 in SDS-PAGE, lane 7: peak 2 in SDS-PAGE + mercaptoethanol.

Abrus and Ricinus agglutinins purified and analyzed under the same conditions gave the following molecular weight data (Table 1).

Table 1.
Molecular weights of Abrus, Ricinus and Viscum agglutinins.

Protein		Intact	Subunit A	Subunit B
Abrus agglutinin	APA I	65 000	30 000	34 000
Abrin	APA II	61 000	28 000	34 000
Ricinus agglutinin	RCA I	126 000	28 000	34 000
Ricin	RCA II	60 000	30 000	33 000
Viscum agglutinin 1	VAA I	65 000	29 000	34 000
Viscum agglutinin 2	VAA II	60 000	27 000	32 000
		58 000	27 000	31 000

The yield of lectins (type I + II) from 100 g of fresh mistletoe varied between 5 and 50 mg in dependence on the host tree, the type of the mistletoe and the season. The highest yields were found in winter with mistletoes from black locust, oak and birch trees.
Incubation of the SDS polyacrylamide gels with different FITC-labeled lectins according to Furlan revealed similar FITC-lectin binding to both subunits of all agglutinins from Abrus, Ricinus and Viscum. Good binding was found with Con A and with the agglutinins from Lens culinaris and wheat germ, no or very weak binding with the agglutinins from soybean, peanut and Helix pomatia. Affinity staining of the heavy subunits (B chains) was in all experiments greater than of the light A chains.
Hemagglutination. When compared with the corresponding type II lectins APA I, RCA I and VAA I are the stronger agglutinins for human erythrocytes (Table 2). The observed agglutination is without any blood group specificity. The best inhibition of hemagglutination was found with D-galactose for APA I, RCA I, VAA I and N-acetyl-D-galactosamine = D-galactose for APA II, RCA II and VAA II.

Table 2.
Hemagglutination of human type-0 red blood cells (MHC: minimal hemagglutination concentration)

Lectin	MHC	Lectin	MHC
APA I	0.5 µg/ml	APA II (Abrin)	16 µg/ml
RCA I	0.2 µg/ml	RCA II (Ricin)	8 µg/ml
VAA I	1.0 µg/ml	VAA II	32 µg/ml

Immunological reactivities. Affinity purified lectins were screened by Ouchterlony double diffusion with normal rabbit preimmune serum and antiserum prepared against pure VAA I, VAA II, RCA I and RCA II. We found the following lectins to crossreact: VAA I and VAA II or RCA I and RCA II.

However, strong interactions of these galactose-specific lectins with the agar gel interfered with the diffusion test. In our experiments with the more sensitive ELISA we detected immunological crossreactions not only between the type I and type II agglutinins of one species but also between APA I, RCA I, VAA I or APA II, RCA II, VAA II (Table 3).

Table 3.
Crossreaction between Abrus, Ricinus and Viscum agglutinins

	VAA I	VAA II	RCA I	RCA II	APA I	APA II
anti VAA I	++++	+++	+++	++	+	+/-
anti VAA II	+++	++++	++	+++	+/-	+
anti RCA I	+++	+++	++++	+++	++	+
anti RCA II	++	++	+++	++++	+	++

The immunological crossreactivity was measured by ELISA as described under Materials and methods. It was quantified into five steps: +/- 0-10 %, + 10-20 %, ++ 20-50 %, +++ 50-80 %, ++++ 80-100 % of the homologous immune reaction.

No crossreactivity was found under these conditions with Con A and with the agglutinins from peanut, soybean, wheat germ and Helix pomatia.

Discussion

By using affinity chromatography on acid-treated Sepharose and elution with lactose a mixture of two types of lectins was purified from the extracts of the mistletoe V.album. After separation of the two agglutinins on Sephadex G 100 they are strikingly similar in structure and mode of action to the two types of seed lectins from A.precatorius and R.communis. Because of this analogy to the Ricinus and Abrus agglutinins we prefer the designations Viscum album agglutinin I and II (VAA I and VAA II).

The molecular weight of the intact molecule is greater for the type I lectins, which are in all cases the stronger hemagglutinins. The type II lectins are smaller by about 4 - 5 kD. In high resolution SDS-PAGE the intact molecules of VAA I and VAA II are split into two bands which differ only slightly in electrophoretic mobility (in Fig. 1 the differences can only be seen for VAA II). Under reducing conditions type I and type II lectins are split into two constitutive chains: the A chain which inhibits protein biosynthesis in a cell free system (13, 14) is always smaller than the sugar binding B chain. Only the B chain could be resolved into two bands. There was no indication of an A chain heterogeneity. From the subunit analysis the following structural models are deduced:

VAA I: $A - S - S - B_1$ VAA II: $A' - S - S - B'_1$
$A - S - S - B_2$ $A' - S - S - B'_2$

Additionally, VAA II appears to be heterogeneous with respect to charge, as it does not elute from a DEAE-cellulose column as a single peak (data not shown). This charge heterogeneity does not reflect heterogeneity with respect to hemagglutination activity. Therefore, we define the different VAA II proteins as isolectins. They can be seen as 4 main and 3 minor bands in an electrophoresis which separates according to both size and charge (fig. 1, lane 5). The proposed structural model is in good agreement with Lin et al. (15) who recently purified two agglutinins and two ricins from the small beans of R. communis by an improved procedure based on ion exchange chromatography and gel filtration. The Abrus agglutinin preparations are found to be composed of isolectins, too (16). Despite extensive work in many laboratories, the fine structure of the lectin pattern of Ricinus and Abrus is still a matter of controversy.

Differences are found when different methods of preparation were employd or different kinds of seeds were used as starting material. RCA II prepared from the small beans of R.communis sanguinea was composed of four isotoxins only two ricins were found in the great beans of R.communis zansibarensis (R. Samtleben, unpublished data). The isolectin pattern of VAA II varies considerably, too. It is strongly influenced by the growth conditions of the mistletoe (supply of nutrients, growth in the shadow or in full light). Yet a reliable correlation is still missing.

VAA I is identical to lectin I previously described by Franz et al. (17) and to viscumin described by Olsness et al. (14). But, in contrast to these reports, we cannot from our experimental data correlate VAA I with APA II (abrin) or RCA II (ricin).

Further evidence for structural relations among the type I or type II lectins originates from the immunological experiments. As Table 3 indicates there is some species and type specificity. Antibodies to VAA I not only crossreacted with VAA II but to the same extent with RCA I. Anti-VAA II revealed the same behavior. With the two antisera to RCA I and RCA II a strong crossreaction can be detected against VAA I and VAA II, too. These results favor the hypothesis that VAA I is structurally related to RCA I and APA I, on the other hand, VAA II is related to the toxic lectins RCA II and APA II.

Antibodies raised against a given lectin will react with a different lectin only when the two proteins share common determinants. Different proteins can be functionally similar and yet antigenically totally unrelated. However, when proteins are found antigenically similar, there is a strong probability of both an evolutionary and a functional relationship. In our experiments it was shown that lectins from as different sources as Abrus and Ricinus seeds and Viscum leaves are structurally related proteins and are probably to some extent homologues. This is an indication for their possible role in very fundamental metabolic pathways common to all plants.

References

1. Goldstein, I.J. and C.E. Hayes: Adv. Carbohydr. Chem. Biochem. 35, 127-340 (1978)

2. Barondes, S.H.: Ann. Rev. Biochem. 50, 207-231 (1981)

3. Lis, H. and N. Sharon: Ann. Rev. Biochem. 42, 541-574 (1973)

4. Toms, G.C.: Chemotaxonomy of Legumes (J.B. Harbourne, D. Boulter and B.L. Turner, eds.) pp. 367-462, Academic Press, New York (1971)

5. Bird, G.W.G.: Brit. Med. J. 15, 165-168 (1959)

6. Nagl, W.: Exp. Cell Res. 74, 599-602 (1972)

7. Shannon, L.: In: T.C. Bøg-Hansen and G.A. Spengler, eds. Lectins-Biology, Biochemistry, Clinical Biochemistry

8. Maurer, H.R.: Disc Electrophoresis, de Gruyter, Berlin, New York (1971)

9. Fairbanks, G., T.L. Steck and D.F.H. Wallach: Biochemistry 10, 2606-2617 (1971)

10. Furlan, M., B.A. Perret and E.A. Beck: Anal. Biochem. 96, 208-214 (1979)

11. Voller, A., D.E. Bidwell and A. Bartlett: The enzyme linked immunosorbent assay (ELISA), Dynatech Europe, Borough House, Guernsey (1979)

12. Allen, H.J. and E.A.Z. Johnson: Carbohydr. Res. 50, 121-131 (1976)

13. Franz, H., A. Kindt, P. Ziska, H. Bielka, R. Benndorf and L. Venker: Acta biol. med. germ. 41, K9-K16 (1982)

14. Olsnes, S., F. Stirpe, K. Sandvig and A. Pihl: J. Biol. Chem. 257, 13263-270 (1982)

15. Lin, T.T.S. and S.S.L. Li: Eur. J. Biochem. 105, 453-459 (1980)

16. Lin, J.Y., T.C. Lee, S.T. Hu and T.C. Tung: Toxicon 19, 41-51 (1981)

17. Franz, H.: These Proceedings

APPENDIX

A TABLE OF LECTIN CARBOHYDRATE SPECIFICITIES

Albert M. Wu
Department of Veterinary Pathology, College of Veterinary Medicine, Texas A&M University, College Station, Texas 77843 U.S.A.

Anthony Herp
Department of Biochemistry, New York Medical College
Valhalla, New York 10595 U.S.A.

Many of the combining sites of the lectins have been well characterized during the past decade; however, the knowledge of receptor sites of lectins is still hampered by the paucity of suitable inhibitory reagent (oligosaccharides). The purpose of this report is to give an updated list of the specificities of the lectins, which have been characterized by hemagglutinaction inhibition, precipition inhibition and/or competitive binding assays.

1. Abbreviations: DGal, D-galactopyranose; DGlc, D-glucopyranose; DMan, D-mannopyranose; LFuc, L-fucopyranose; DGalNAc, 2-acetamido-2-deoxy-D-galactopyranose; DGlcNAc, 2-acetamido-2-deoxy-D-glucopyranose; DFruf; D Fructofuranose; Raffinose, DGalα1→6DGlcβ1→3DFruf; Stachyose, DGalα 1→6 DGalα1→6DGlcβ1→2DFruf; AcNeu, N-acetylneuraminic acid and Gly Neu, N-glycolyneuraminic acid. R, carbohydrate residues.

2. The binding properties of some lectins for glycopeptides of either the N-acetylactosamine or the oligomannosidic type derived from glycoprotein containing the N-glycosglamine linkage are illustrated in reference 3.

Lectins, Vol. IV

HPA, *Helix pomatia* agglutinin (Edible snail) (1-3)

DGalNAcα1→3DGalNAcβ1→3DGalα1→4DGalβ1→4DGlc
(Forssman specific pentasaccharide) >
DGalNAcα1→3DGalNAc >
DGalNAcα1→3DGalβ1→3DGlcNAc
(Blood group A specific trisaccharide) > DGalNAc.

DBA, *Dolichos biflorus* agglutinin (Horse Gram) (1,2)

DGalNAcα1→3DGalNAcβ1→3DGalα1→4DGalβ1→4DGlc
(Forssman specific pentasaccharide) >>
DGalNAcα1→3DGalNAc >>
DGalNAcα1→3[LFucα1→2]DGalβ1→4DGlcNAcβ1→6R
(A active pentasaccharide) >
DGalNAcα1→3DGalβ1→3DGlcNAc and
DGalNAcα1→3DGal > DGalNAc.

WFA, *Wistaria floribunda* agglutinin (1)

DGalNAcα1→6DGal >
DGalNAcα1→3DGal,
DGalNAcα1→3DGalNAcβ1→3DGalα1→4DGalβ1→4DGlc,
(Forssman specific pentasaccharide) and
DGalNAcα1→3DGalβ1→3DGlcNAc >
DGalNAcα1→3DGalβ1→4DGalβ1→4DGlc (Globoside) >
DGalNAcα1→3DGalNAc > DGalNAc, and
DGalNAcα1→3[LFucα1→2]DGalβ1→4DGlcNAcβ1→6R >
DGalβ1→4DGlcNAc >
DGalβ1→4DGlc (Lactose).

SBA, *Glycine max* agglutinin (Soybean) (2-4)

DGalNAcα or β1 linked,
DGalNAcα1→3DGalβ1→3DGlcNAc, and
DGalNAcα1→3DGal >>
DGalNAcα1→3[LFucα1→2]DGalβ1→4DGlcNAcβ1→6R,
DGalα1→6DGlc (Melibiose), DGalβ1→6DGlc,
Raffinose, stachyose, DGalβ1→4DGlc, and
LFucα1→2DGalβ1→4DGlcNACβ1→6R > DGal.

PLA, *Phaseolus limensis* agglutinin (Lima Bean) (2)

DGalNAc >> DGalα1 linked

VVA, *Vicia villosa* agglutinin (a mixture of two isolectins; B_4, major and A_4, minor) (5,6)

DGalNAcα1→3DGal > DGalNAcα1→6DGal >
DGalNAcα1→3DGalβ1→3DGlcNAc, and

DGalNAcα1→3DGalNAc >>
DGalNAcα1→3[LFucα1→2]DGalβ1→4DGlcNAcβ1→6R
DGalα1→3DGal.

VVA-B_4, *Vicia villosa* B_4 agglutinin (7)

DGalNAcα1→ser (or thr) >
DGalβ1→3DGalNAcα1→ser (or thr) >
NeuNAcα2→3DGalβ1→3DGalNAcα1→ser (or thr).

BSA-A, *Bandeiraea simplicifolia* agglutinin A_4 (8)

DGalNAcα1→3DGalβ1→3DGlcNAc
DGalNAcα1→3DGal
DGalNAcα1→6DGal, and DGalNAcα1→R >
DGalα1→3DGal, DGalα1→6DGlc(Melibiose), and
DGalα1→6DGalβ1→2DFruf (Raffinose).

BDA, *Bryonia dioica* agglintinin (9)

DGalNAc > DGalβ1→4DGlc > DGalα1→6DGlc >>
Raffinose > stachyose.

TMA, *Tridacna maxima* agglutinin (Roding Clam)(10)

DGalNAc>
DGalβ1→6Gal, DGalβ1→4DGlc, and DGal >
DGalα1→6DGlc, Raffinose and stachyose.

BPA, *Bauhinia purpurea alba* agglutinin (11)

DGalβ1→3DGalNAc >
DGalβ1→3DGlcNAcβ1→3DGalβ1→4DGlc
(lacto-N-tetraose)>
DGalβ1→3DGlcNAc, and
DGalNAcα1→3DGalβ1→3DGlcNAc
(Blood group A specific) >
DGalβ1→4DGlcNAc (N-acetyllactosamine)
DGalβ1→4DGlc > DGal, DGalα1→3DGal, and
DGalβ1→4DGlcNAcβ1→6DGalNAc.

PNA, *Arachis hypogaea* agglutinin (peanut) (12)

DGalβ1→3DGalNAc >>
DGalβ1→4DGlcNAc >
DGal and DGalβ1→3DGlcNAc.

MPA, *Maclura pomifera* agglutinin (13)

DGalβ1→3DGalNAc >>
DGalNAcα1→6DGal > DGalα1→6DGlc (Melibiose).

VGA, *Vicia graminea* agglutinin (Anti-Blood group N lectin) (14)

DGalβ1→3DGalNAcα1 linked and
Desialized blood group N glycopeptide >
Untreated N glycopeptide and
desialized M glycopeptide >>
Untreated blood group M glycopeptide.

SJA, *Sophora japonica* agglutinin (15)

DGalβ1→3DGaNAcβ1→N-tosyl-L-serine >
DGalβ1→3DGlcNAc >
DGalβ1→3DGlcNAcβ1→3DGalβ1→4DGlc
(Lacto-N-tetraose) >
DGalβ1→4DGlcNAcβ1→6DGal >
(Blood group I Ma specific trisaccharide)
DGalNAc >> DGalβ1→4DGlc >
DGal and DGalα1→6DGlc (Melibiose).

RCA, *Ricinus communis* hemagglutinin (Wu and Kabat unpublished data and Ref. 3, 16)

DGalβ1→4DGlcNAcβ1→6DGal >
DGalβ1→4DGlcNAc (N-acetyllactosamine) >
DGalβ1→6DGal > DGalβ1→4DGlc (Lactose).

Ricin (3, 16)

DGalNAc > DGalβ1→4DGlc (Lactose).

ECA, *Erythrina cristagalli* agglutinin (17, 18)

Tetraantennary and Triantennary oligosaccharides containing four and three DGalβ1→4DGlcNAcβ1 Linked branches >

DGalβ1→4DGlcNAcβ1→6DGal (Blood group 1 Ma specific)
DGalβ1→4DGlcNAc > DGalβ1→3DGlcNAc >>
DGalβ1→4DGlc (lactose) >>> DGalβ1→3DGal >
DGalα1→6DGlcβ1→2DFruf (Raffinose).

GCA, *Geodia cydonium* agglutinin (19)

DGalβ1→4DGlcNAc, DGalβ1→3DGlcNAc, and
DGalβ1→4DGlc > DGalNAc.

BSA B_4, *Bandeiraea simplicifolia* B_4 agglutinin (8)

DGalα1→3DGal, and DGalα1 linked
DGalα1→6DGlc > DGalα1→6DGlcβ1→2DFruf(Raffinose).

APA-I and II, *Axinella polypoides* I and II agglutinins (20)

DGalβ 1→6R > DGalα 1→6DGlc(Melibiose).

SNA, _Sambucus nigra agglutinin_ (21)

DGalβ 1→4DGlc, DGalNAc, and DGal >
DGalα1→6DGlcβ1→2DFruf > DGalα 1→6DGlc.

DGlcNAc binding lectin

WGA, _Triticum vulgaris_ agglutinin (Wheat germ) (3,16,22,23)

DGlcNAcβ1→4DGlcNAcβ1→4DGlcNAc >
DGlcNAcβ1→4DGlcNAc and sialic acid.

APA-I, _Aaptos papillata_ I agglutinin (Sponge) (24,25)

$(DGlcNAc\beta 1\rightarrow 4DGlcNAc)_2$ >
DGlcNAcβ1→4DGlcNAcβ 1→4DGlcNAc > DGlcNAcβ1→4DGlcNAc.

APA-II & III, _Aapots papillata_ II & III agglutinin (Sponge) (24,25)

DGlcNAcβ1→ 4DGlcNAcβ1→4DGlcNAc >
DGlcNAcβ1→ 4DGlcNAc.

LEA, _Lycopersicon Esculentum_ agglutinin (Tomato lectin) (26)

$(DGlcNAc\beta 1\rightarrow 4DGlcNAc)_2$ >
DGlcNAcβ1→4DGlcNAcβ1→ 4DGlcNAc >> DGlcNACβ1→ 4DGlcNAc.

DSA, _Datura stramonium_ agglutinin (Thorn-apple) (3,27)

$(DGlcNAc\text{-}MurNAc)_2$ >
$(DGlcNAc\beta 1\rightarrow 4DGlcNAc)_2$ >
DGlcNAcβ1→4DGlcNAcβ1→4DGlcNAc >
DGlcNAcβ1→4DGlcNAc, and DGlcNAc-MurNAc.

STA, _Solanum tuberosum_ agglutinin (Potato) (3,27)

$(DGlcNAc\beta 1\rightarrow 4DGlcNAc)_2$ >
DGlcNAcβ1→ 4DGlcNAcβ1→ 4DGlcNAc, and
DGlcNAcβ1→4DGlcNAc >
$(DGlcNAc\text{-}MurNAc)_2$ > DGlcNAc-MurNAc.

DMan and DGlcNAc binding lectin (3, 16, 28)

Con A, _Canavalia ensiformis_ agglutinin (jack bean) (16)

DManα1 linked > DGlcα1 linked > DGlcNAcα1 linked
The binding sugar requires the C-3, C-4, and C-6 hydroxyl groups of the DMan or DGlc ring [29,30].

Lentil, <u>Lens culinaris</u> or <u>Lens esculenta</u> agglutinin (16)

PSA, <u>Pisum sativum</u> agglutinin (Pea) (16)

DManα1 linked > DGlcα1 linked or DGlcNAc.

VFA, <u>Vicia faba</u> agglutinin (Faba bean) (16)

DManα1 linked > DGlcα1 linked >> DGlcNAc.

LFuc binding lectins

LTA, <u>Lotus tetragonolobus</u> agglutinin (3,31)

LFucα1→2DGalβ1→4[LFucα1→3]DGlcNAcβ1→6R
LFucα1→2DGalβ1→4[LFucα1→3]DGlcNAc
LFucα1→2DGalβ1→4DGlcNAc.

UEA-I, <u>Ulex europaeus</u> I agglutinin (3,32)

LFucα1→2DGalβ1→4DGlcNAcβ1→6R
LFucα1→2DGalβ1→4[LFucα1→3]DGlcNAcβ1→6R
LFucα1→2DGalβ1→4DGlcNAc (2'-fucosyllactose) >
LFucα1→2DGalβ1→4[LFucα1→3]DGlc (Lacto-difucotetraose).

UEA-II, <u>Ulex europaeus</u> II agglutinin (33)

LFucα1→2DGalβ1→4DGlcNAcβ1→6R
LFucα1→2DGalβ1→4DGlc (2'Fucosyllactose)
$(DGlcNAc\beta1\to4DGlcNAc)_2$> DGlcNAcβ1→4DGlcNAcβ1→4DGlcNAc
DGlcNAcβ1→4DGlcNAc >
DGalNAcα1→3DGalβ1→3DGlcNAc > DGlcNAcβ1→6DGal >
DGalβ1→3DGlcNAcβ1→3DGalβ1→4DGlc (Lacto-N-tetraose) >
LFucα1→2DGalβ1→3DGlcNAcβ1→3DGalβ1→4DGlc
(Lacto-N-fucopentaose I) >
DGalNAcα1→3DGal (Blood group A active dissacharide).

GSA-IV, <u>Griffonia simplicifolia</u> IV agglutinin (34)

LFucα1→2DGalβ1→3[LFucα1→4]DGlcNAcβ1→3DGalβ1→4DGlc
(Lacto-N-difucohexaose I) >
DGalβ1→3[LFucα1→4]DGlcNAcβ1→3DGalβ1→4[LFucα1→3]DGlc
(Lacto-N-difuiohexaose II) >
LFucα1→2DGalβ1→4[LFucα1→3]DGlc
(Lactodifucotetraose), and
LFucα1→2DGalβ1→4[LFucα1→3]DGlcNAc
(Lacto-N-difucotetraose) >>

DGalβ1→3 [LFucα1→4]DGlcNAcα1→3DGalβ1→4DGlc (Lacto-N-fucopentaose II.).

Sialic Acid Specific Lectins
LPA, *Limulus polyphemus* agglutinin (35,36)
sialic acid, DGlcNAc and glucuronic acid.
LAg-I, *Homarus amencanus* agglutinin (37)
AcNeu:GlNeu. and DManNAc.
LFA, *Limax flavus* agglutinin (Slug) (38)
AcNeu and GlyNeu.

Acknowledgements

This research is partially supported by grants from the Texas Agricultural Experiment Station (TAES H6194), US Department of Agriculture (USDA/SEA/AR58-519B-0880 and USDA 82-CSRS-2-1035), and USDA/SEA Formula Animal Health Funds (Project 6648). The authors would like to thank Lorie Honnas for her secretarial assistance in preparing this manuscript.

References

1. Baker, D.A., Sugii, S., Kabat, E.A., Ratcliffe, R.M., Hermenthin, P., Lemieux, R.U.: Biochemistry, 22, 2741-2750 (1983)
2. Hammarstrom, S., Murphy, L.A., Goldstein, J.J. and Etlzer, M.E.: Biochemistry 16:2750-2755 (1977)
3. Debray, H., Decout, D., Strecker, G., Spik, G. and Montreuil, J.: Eur. J. Biochem 117:41-55 (1981).
4. Pereira, M.E., Kabat, E.A.: Carbohydr. Res 37:89-102 (1974)
5. Tollefsen, S.E., and Kornfeld, R.: J. Biol. Chem. 258 5:5165-5171 (1983)
6. Kaladas, P.M., Kabat, E.A., Kimura, A. and Ersson, B.: Mol. Immun. 18:969-977 (1981)
7. Tollefsen, S.E. and Kornfeld, R.J.: J. Biol. Chem. 258:5172-5176 (1983)
8. Wood, C., Kabat, E.A., Murphy, L.A. and Goldstein, I.J.: Arch. Biochem. Biophys. 198:1-11 (1979)
9. Peumans, W.J., Nsimba-Labaki, M., Carlier, A.R. and Driessche, E.V.: Planta 160:222-228 (1984).
10. Baldo, B.A., Sawyer, W.H., Stick, R.V. and Unlenbruck, G.: Biochem. J. 467-477 (1978)
11. Wu, A.M., Kabat, E.A., Gruezo, F.G., Allen, H.J.: Arch. Biochem. Biophys. 204, 622-639, (1980).

12. Pereira, M.E., Kabat, E.A., Lotan, R., Sharon, N.: Carbohydr. Res. 51:107-118 (1976).
13. Sarker, M., Wu, A.M., Kabat, E.A.: Arch. Biochem. Biophys 209, 204-218 (1981).
14. Duk, M. and Lisowska, E., Kordowicz, M. and Wasniowska, K.: Eur. J. Biochem 123:105-112 (1982)
15. Wu, A.M., Kabat, E.A., Gruezo, F.G., Poretz, R.D.: Arch. Biochem. Biophys. 209, 191-203 (1981).
16. Goldstein, I.J., Hayes, C.E.: Adv. Carbohydr. Chem. Biochem. 35, 127-340 (1978).
17. Iglesias,J.L., Lis, H. and Sharon, N.: Eur. J. Biochem 123:247-252 (1982)
18. Kaladas, P.M., Kabat, E.A., Iglesias. J.L, Lis, H. and Sharon, N.: Arch Biochem Biophys 217:624-637 (1982)
19. Bretting, H., Phillips, S.G., Klumpart, H.J., Kabat, E.A.: J. Immunol. 127, 1652-1658 (1981)
20. Bretting, H., and Kabat, E.A. Biochemisty 15, 3228-3236, (1976)
21. Broekaert, W.F., Nsimba-Lubaki, M., Peeters, B, and Peumans, W.J. Biochem. J. 221, 163-169, (1984)
22. Peters, B.P., Ebisu, S., Goldstein, I.J., Flashner, M.: Biochemistry 18,5505-5511 (1979).
23. Bhavanandan, V. and Katlic, A.W.: J. Biol. Chem. 254:4000-4008 (1979).
24. Bretting, H.: Collog. Int. C.N.R.S. 291:247-254 (1978).
25. Bretting, H., Kabat, E.A., Liao, J. and Pereira, M.E.,: 15:5029-5038 (1976)
26. Nachbar, M.S. Oppehheim, J.D. and Thomas, J.D.: J. Biol. Chem. 255:2056-2061, 1980.
27. Desai, N.N., Allen, H.K. and Neuberger: Biochem J. 197:345-353, 1981
28. Allen, A.K., Desai, N.N. and Neuberger, A.: Biochem. J. 155:127-135 (1976)
29. Golstein, I.J., Hollerman, C.E. and Merrick, J.M.: Biochem. Biophys. Acta 97:68-76 (1965)
30. Golstein, I.J. Hollerman, C.E. and Smith, E.E.: 4:876-883 (1965)
31. Pereira, M.E., Kabat, E.A.: Biochemistry 13:3184-3192 (1974).
32. Pereira, M.E., Kisailus, E.C., Gruezo, F.G., Kabat, E.A.: Arch. Biochem. Biophys. 185, 108-115 (1978).
33. Pereira, M.E., Gruezo, F.G., Kabat, E.A.: Arch Biochem. Biophys. 194, 511-525 (1979).
34. Kaladas, P.M., Kabat, E.A., Shibata, S. and Goldstein, I.J.: 223:309-318 (1983).
35. Marchalonis, J.J. and Edelman, G.M.: J. Mol. Biol. 32:453-465 (1968).
36. Nowak, T.P. and Barondes, S.H.: Biochem. Biophys. Acta 395:115-123 (1975).
37. Hall, J.L., and Rowlands, D.T.: Jr. Biochemistry 13:828-832 (1974).
38. Miller, R.L., Collawn, J.F. and Fish, W.W.: J. Biol. Chem. 257:7574-7580. (1982).

THIRD REPORT OF THE INTERNATIONAL WORKING PARTY-IUIS-WHO SUBCOMMITTEE ON STANDARDIZATION OF LECTINS FOR DIAGNOSIS. PROGRESS OF THE STANDARDIZATION OF CONCANAVALIN A.

J.Breborowicz, President of the Subcommittee, Academy of Medicine, Poznan, Poland,
H.Franz, Vicepresident of the Subcommittee, Statlisches Institut fur Immunpraparate und Nahrmedien, Berlin, DDR,
M.Caron, U.E.R.Sante, Bobigny, France,
A.Faure, National Blood Transfusion Centre, Orsay, France,
K.Gryska, Academy of Medicine, Poznan, Poland,
J.A.Kint and L.M.J.Sabbe, Rijksuniversiteit, Gent, Belgium,
J.Kohn, Ludwig Institute, Sutton, Great Britain,
I.Lorenc-Kubis, University of Wroclaw, Poland,
H.Rudiger, University of Wurzburg, BRD, Germany,
K.Takeo, Yamaguchi University, Ube, Japan,
K.Toftager-Larsen, Sønderborg Sygehus, Denmark,
M.Wu, Texas A & M University, College station, U.S.A., and
T.C.Bøg-Hansen, Secretary of the Subcommittee, University of Copenhagen, Denmark

The following account is a short summary of work performed by members of the International Working Party on Standardization of Lectins for Diagnosis, now acting as a subcommittee of IUIS/WHO. As described in the first and the second reports (1,2), the International Working Party has given its priority to analysis and standardization of concanavalin A, con A. A complete documentation is planned for appearance in these proceedings volume 5.

Lectins, Vol. IV

Ten commercial companies (Boehringer, Chemapol, Flow Laboratories, Kem-En-Tec, Miles-Yeda, Pharmacia Fine Chemicals, Reactifs IBF, Sigma, Vector Laboratories and Worthington) and three research laboratories (The Department of Plant Pathology, University of Florida, USA, The Department of Protein Chemistry, Institute of Molecular Biology, The Free University, Bruxelles, Belgium, and The Protein Laboratory, University of Copenhagen, Denmark) supplied the subcommittee with 17 preparations of con A. Eight of these preparations coded A,C,E,K,L,N,O, and T were sent to the members of the subcommittee in 1982, and the next 10 samples coded 1-10 were distributed in 1983.

Since the con A samples were distributed, the members of the subcommittee have performed analyses. The con A samples were supplied either as freeze dried powders or in solution and contained various additives as specified in Table 1.

The samples were used for analytical studies either directly or they were dissolved and if necessary dialyzed with the buffer of choice, then centrifuged and used for experiments after determination of the protein concentration. Analyses were performed according to most of the techniques originally suggested (1) as listed below together with citation of the methods or an outline of the procedure together with a summary of the results, please also refer to Tables 2 and 3.

Methods and Results

1. Spectroscopic study of con A - sugar binding. Table 2 gives the content of functional con A as % of a standard.

2. Quantitative precipitin assay using mannan and Sophora japonica lectin as glycoconjugate according to (3). The result obtained with mannan is shown as % of a standard (Table 2).

3. Hemadsorption lectin test, HALT, according to (4). Results shown as the minimum concentration of protein in mg/ml required to observe hemadsorption (Table 2).

TABLE 1

Some characteristics of the con A samples

Sample	Form of sample	Additives	Appearance of solution	Estimated con A
A	lyophilized			100 %
C	lyophilized	85 % NaCl		15 %
E	solution	sat. NaCl		36 mg/ml
K	lyophilized		turbid	100 %
L	lyophilized			100 %
N	lyophilized			100 %
O	lyophilized			100 %
T	lyophilized		turbid	100 %
1	solution	4 M NaCl		25 mg/ml
2	lyophilized			100 %
3	solution	1 M NaCl 0.2 M glucose		18 mg/ml
4	lyophilized			100 %
5	solution	PBS pH 7.2 1 mM Mn^{++}, NaN_3	turbid	
6	lyophilized		turbid	100 %
7	lyophilized	54 % NaCl 32 % glucose		14 %
8	solution	80 % $(NH_4)_2SO_4$	turbid	13 mg/ml
9	solution	1 M NaCl 1 mM Mn^{++}, NaN_3		57 mg/ml
10	lyophilized		turbid	100 %

TABLE 2

Some studies of con A samples, please refer to the text for details.

Sample	1	2	3	4	5
A	100	89	6.5	normal	157
C	53	47	nd	nd	nd
E	105	60	2.5	normal	159
K	74	49	7.0	too high	137
N	90	64	6.2	normal	129
O	93	76	8.5	nd	133
T	54	62	8.0	too high	71
1	83	62	7.5	normal	125
2	102	60	9.0	nd	150
3	82	71	6.2	logn period	77
4	100	73	6.2	long period	149
5	94	62	5.5	long period	135
6	100	71	6.0	normal	140
7	102	56	nd	nd	nd
8	85	60	10.0	small	39
9	92	82	9.0	nd	nd
10	91	62	9.0	normal	136

nd = not determined

TABLE 3

Some studies of con A samples, please refer to the text for details.

Sample	6	7	8	9	10
A	cathodal[+]	-	nd	0.016	1.6
C	nd	nd	nd	0.040	3.0
E	cathodal	-	nd	0.12	1.54
K	cathodal[+]	- +	nd	0.029	7.27
N	cathodal	nd	nd	0.044	1.42
O	cathodal	-	nd	0.037	2.64
T	anodal	+	1.71	nd	nd
1	cathodal[+]	-	1.74	nd	nd
2	nd	nd	2.00	nd	nd
3	cathodal[+]	-	1.15	nd	nd
4	cathodal	-	1.67	nd	nd
5	cathodal[+]	-	1.40	nd	nd
6	cathodal[+]	-	1.54	nd	nd
7	nd	nd	1.53	nd	nd
8	cathodal	-	1.55	nd	nd
9	nd	nd	1.83	nd	nd
10	no migration	0	2.02	nd	nd

[+]Mainly cathodal migration plus anodal migrating aggregates

4. Nephelometric measurements of the reaction of con A with human serum. Con A was incubated with diluted human serum at pH 7.6. The results are recorded as normal, high, low, or abnormal signal (Table 2).

5. Mitogenic stimulation of human lymphocytes. Con A stimulated lymphocyte cultures were pulse labelled with tritiated thymidine. Results are shown as % of a standard (Table 2).

6. Electrophoresis in agarose gel pH 8.6. The migration and the appearance of aggregates is noted in Table 3.

7. Crossed affinity electrophoresis with human α-fetoprotein according to (5). Results express the activity as higher, as equal to, or as lower than that of a standard preparation (Table 3).

8. Calculation of the dissociation constant of con A and amniotic α-fetoprotein according to (6). Values of the dissociation constant, K_d, are given in mM (Table 3).

9. Calculation of the dossociation constant of con A and dextran according to (7). Values of the dissociation constant, K_d, are given in M (Table 3).

10. Estimation of Ca^{++} and Mn^{++} by atomic absorption spectroscopy (8). The results are expressed in meq of Ca^{++} per mg of protein and as Ca^{++}/Mn^{++} ratio (Table 3).

11. Radial affinity diffusion. Normal human serum was included in agarose, and con A samples applied to wells were allowed to diffuse towards the serum to form precipitates. Differences were noted.

12. Double affinity diffusion. Samples of con A and of glycoconjugates were applied in opposite wells in plain agarose gel. Precipitates were obtained with all samples, but the morphology of the precipitates differed.

13. Electrophoresis on cellulose acetate membranes. Differences were noted.

14. Affinity fixation after electrophoresis on cellulose acetate membranes. Differences were noted.

15. Electrophoresis in polyacrylamide gel, PAGE. Differences were noted.

16. Rocket affinity electrophoresis at pH 8.6 in agarose. All con A samples were active although the morphology of the rockets of samples T and 10 differed considerably from the others.

17. Isoelectrofocusing in agaorse gel was performed according to the LKB protocol with Pharmacia agarose and Pharmalytes. Three patterns were observed: I (samples X,O,3,5,8), II (samples T and 1), and III (samples A and K), Fig. 1. Sample X was a standard prepared in the laboratory.

Discussion

Although all the studied preparations contain active and functional con A, the results of the analytical work clearly indicate a great variation from sample to sample when tested in different ways. Therefore it is not possible to classify accurately the samples in order of their usability. Some of the preparations perform well in most of the tests. These include samples A,N,O,2,4, and 8. Other samples performed poorly in many of the tests. These include samples T,C, and 10. The rest of the samples could be classified preliminarily as intermediate.

Two reasons may be responsible for the unsatisfactory performance of a con A preparation; contaminating impurities and intrinsic properties of the con A molecules. That is it probably not contaminating impurities that couse the poor performance is shown by the electrophoretic analyses. All preparations appear to be rather pure. This leaves the intrinsic properties of the con A molecules to explain the unsatisfactory results: denaturation, loss of metal ions, aggregation, and fragmentation. It is important to realize that these properties are interconnected as well as they determine in part the monomer-dimer-tetramer equilibrium. Consequently they determine the activity of the lectin and hence the usability for diagnosis.

Denaturation of con A molecules takes place both during the purification procedures and during shipment and storage. It must also be realized that the con A can be denatured under the experimental conditions. This is particularly true in case of affinity electrophoresis in agarose at pH 8.6. Nevertheless, it is well established that many con A preparations remain active even under these conditions, while others readily precipitate in the agarose gel at this pH.

The conformation changes of con A molecules resulting in denaturation can be caused by various reasons. The most important of these is loss of metal ions. The content of Ca^{++} and Mn^{++} in the preparations varied very much. In this connection it is important to note that some preparations of demetallized con A can be reconstituted (7). For some analyses the preparations were treated in such a way that they might have been fully reconstituted with metal ions.

Little is known about the occurrence of con A fragments and aggregates in the preparations. Both aggregates and fragments might remain active, but with changes affinity. As the activity tests depend on the binding affinity to a different degree, some discrepancies might be expected.

Still another reason for a variable activity of the con A preparations is the source of the seeds used for extraction. While genetic variability of con A remains a possibility we have no data concerning this parameter.

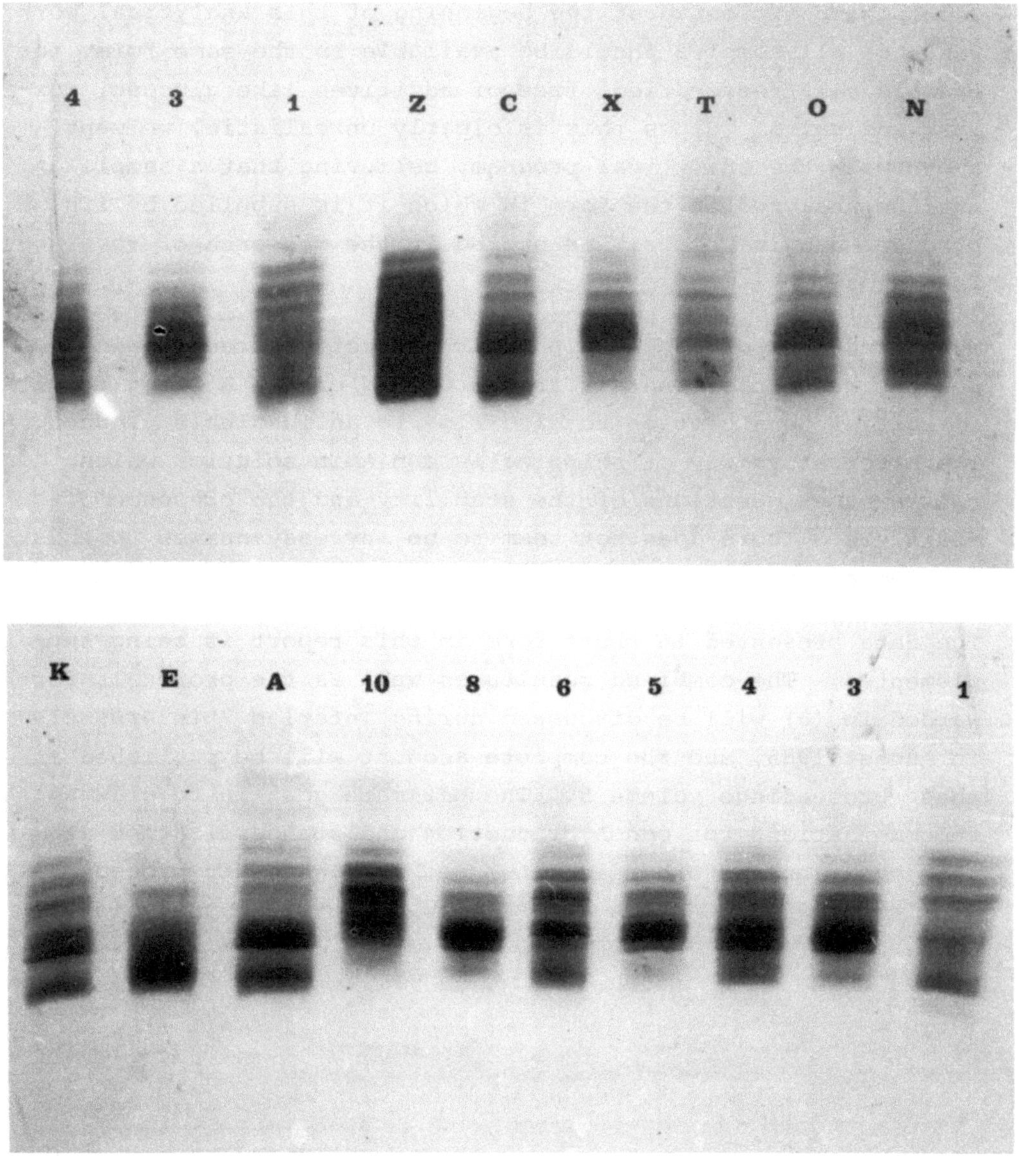

Fig. 1. Isoelectrofocusing in agarose gel. 5 ul samples of con A, 2 mg/ml were applied. Pharmalytes 3.5 - 10. Sample X was a standard prepared in the laboratory, see text for details, point 17.

One opinion expressed at the beginning of this analytical work was that all samples should be available in the same form, for example all freeze dried, free of additives like glucose, toxins, and salts. Since this is clearly unrealistic, we went ahead with the analytical program, believing that a sample must be analyzed in the form in which it is supplied by its producer and in which it is stored in the research or routine laboratory.

Most producers of con A ship their product as freeze dried powder. But unfortunately freeze drying of con A often results in partial denaturation to give rise to an insoluble product. The alternative way of shipping is con A in solution which raises other questions of the stability and the presence of additives. There does not seem to be any easy answer to this problem at the moment.

The data presented in short form in this report is being supplemented. The combined results as well as the proposals forwarded in (8) will be discussed during Interlec 7 in Brussels in August 1985, and the complete account will be published in these proceedings volume 5. The ultimate goat is to draw up recommendations for con A production and standardization at this occation, as other lectins used in diagnostic procedures (including PHA, WGA, LCA and PNA) await detailed analysis.

References

1. T.C.Bøg-Hansen, J.Breborowicz, H.Franz in (T.C.Bøg-Hansen, ed.) Lectins, Biology, Biochemistry, Clinical Biochemistry Vol. 2, W. de Gruyter, Berlin 1982, pp. 791-793.
2. T.C.Bøg-Hansen in (T.C.Bøg-Hansen and G.A.Spengler, eds.) Lectins, Biology, Biochemistry, Clinical Biochemistry Vol. 3, W. de Gruyter, Berlin, 1983, pp. 699-700.
3. A.M.Wu, C.E.Snyder, A.Herp, R.P.Hsieh. These proceedings pp. 277-288.
4. J.A.Kint in (T.C.Bøg-Hansen and G.A.Spengler, eds.) Lectins, Biology, Biochemistry, Clinical Biochemistry Vol. 3, W. de Gruyter, Berlin 1983, pp. 415-424.
5. J.Breborowicz, A.Mackiewicz, D.Breborowicz. Scand.J.Immunol. 14, 15 (1981).
6. K.Toftager-Larsen in (T.C.Bøg-Hansen, ed.) Lectins, Biology, Biochemistry, Clinical Biochemistry Vol. 2, W. de Gruyter, Berlin 1982, pp. 433-444.
7. K.Takeo, M.Fujimoto, R.Suzuno, T.Tanaka, K.Nakamura, A.Kuwahara. These proceedings pp. 213-220.
8. H.Franz, P.Ziska, C.Flemming, V.Horejsi, M.Ticha. These proceedings pp. 649-652.

QUALITY CONTROL OF CONCANAVALIN A: PROPOSAL FOR REQUIREMENTS

Hartmut Franz and Peter Ziska

Staatliches Institut für Immunpräparate und Nährmedien
Berlin, DDR

Christian Flemming

Institut für Technische Chemie
Leipzig, DDR

Vaclav Horejsi

Institute of Molecular Genetics
Praha, CSSR

Marie Tichá

Charles University, Department of Biochemistry
Praha, CSSR

Concanavalin A is a globular protein composed of identical 25 500 mol wt. subunits arranged in dimeric and tetrameric forms. Con A generally binds to molecules containing D-mannose or D-glucose residues. The tetrameric form includes 4 Ca^{++} and 4 Mn^{++} per molecule that are necessary for carbohydrate binding activity. The lectin is completely free of carbohydrates (1). Although Con A is used in a wide range as a tool until now commonly accepted requirements for quality control do not exist. Therefore we try to propose some tests for basic characterization of Con A. Of course we know that these requirements do not characterize Con A completely, but we assume, that Con A preparations which meet these requirements are suitable for the most range of application.

Lectins, Vol. IV

Proposals

Absence of carbohydrates: To 0.2 ml Con A solution (5 mg/ 1 M NaCl) are added 0.2 ml phenol solution (4 g/ml H_2O) and 1 ml conc H_2SO_4. There should be no visible colour reaction between Con A and the reagent (2) as a proof for the absence of carbohydrates or glycoproteins (contamination by the procedure of Con A isolation).

Specific absorption: The absorption of a Con A solution (1 g/100 ml 1 M NaCl) at 280 nm in a 1 cm path length should be 11,4 ± 0.1.

Hemagglutination activity: To 0.05 ml 2fold serial dilutions of Con A (1 mg/ml PBS) 0.05 ml suspension of 1 % washed rabbit erythrocytes is added (microplates). After incubation at 37°C for 1 h the degree of agglutination is assessed. The hemagglutination activity should be 1250 ± 250 units per mg Con A (3).

Hemagglutination inhibition by D-mannose: To 0.05 ml 2fold serial dilutions of 0.2 M D-mannose in PBS 0.025 ml Con A solution containing 4 hemagglutination units is added (microplates). After incubation at 37°C for 1 h 0.025 ml of a 2 % rabbit erythrocytes suspension is added. After incubation at 37°C for 1 h the degree of agglutination inhibition is assessed. 12 - 25 µmol D-mannose per ml should be needed for complete inhibition of 4 hemagglutination units of Con A.

Precipitation of polysaccharides: To 0.5 ml 2 fold serial dilutions of Con A (5 mg/ml PBS) 0.5 ml of liver glycogen (10 mg/ml PBS) is added (microtubes). After incubation at room temperature for 30 min the degree of precipitation is assessed. The precipitation titre should be at least 1:8.

Homogeneity in PAGE: Polyacrylamide gel electrophoresis is performed at pH 4.5 in 7.5 % gel (4). After electrophoresis, staining with Coomassie Blue R-250 and background destaining there should be seen only a major band as well as a minor band.

Contents of Ca^{++} and Mn^{++}: For determination of Ca the flame atomic absorption spectroscopy and for Mn the atomic absorption spectroscopy are recommended. But the spectroscopy of the colour reaction between Ca and murexide (5) and Mn with formaldehydoxime (6), respectively is possible. The metal content should be at least 4 mol Ca and 4 mol Mn per mol Con A at a pH>6.

Carbohydrate-binding activity: Polyacrylamide gel electrophoresis is performed at pH 4,5 in a 7,5 % gel (control gel) and in a affinity gel containing incorporated 0,5 % yeast mannan. After electrophoresis, staining with Coomassie blue R-250 and background destaining the single band of Con A observed on the control gel should be strongly retarded on the affinity gel (10-15 % mobility as compared to the control gel). Only trace amounts of weakly retarded or non-retarded bands should be present on the affinity gel.

Stimulation of lymphocytes: Spleen cells from mice are cultured at 1×10^7 viable lymphocytes per ml RPMI-1640 medium. Cells are assayed for DNA synthesis by pulsing with 2 µCi of ^3H thymidine per culture 48 h after addition of Con A. The concentration of Con A required for maximal response should be between 1 µg and 5 µg per ml (7).

Agglutination activity for yeast cells: A suspension of yeast cells (Saccharomyces cerevisiae) treated with glutaraldehyde in PBS with an extinction of $E_0^{578\ nm} = 0.75 \pm 0.02$ is put in a continously stirred cuvette. 25 µl Con A (3 mg/ml) is added to 17 ml of the cell suspension. The decrease of the extinction, caused by cell agglutination, is monitored. The plot of $(E_0/E_t)^3$ via time results in a straight line during a periode of mor than 30 min (E_t = extinction measured at the time t). The slope should be $(0.04 \pm 0.005) min^{-1}$. The maximum of the slope should be arrived at (0.085 ± 0.01) min^{-1}. Con A preparations with lower activity induce a significant slower velocity of agglutination.

References

1. Bittinger, H., Schnebli, H.P.: Concanavalin A as a Tool, Wiley, New York 1976
2. Dubois, M., Gilles, K.A., Hamilton, J.K., Rebers, P.A., Smith, F.: Analyt. Chem. 28, 350-356 (1956)
3. Allen, A.K., Desai, N.N., Neuberger, A.: Biochem. J. 155, 127-135 (1976)
4. Maurer, H.R.: Disc Electrophoresis and Related Techniques of Polyacrylamide Gel Electrophoresis, Walter de Gruyter, Berlin.New York 1981
5. Pollard, F.H., Martin, J.V.: Analyst 81, 348 (1956)
6. Bradfield, E.G.: Analyst 82, 254 (1957)
7. Wang, J.L., Edelmann, G.M.: J. biol. Chem. 253, 3000-3007 (1978)

COMPARATIVE STUDIES OF DIFFERENT TUMOR CELL LINES AND SOME D-TYPE RETROVIRUSES WITH LECTINS

M. Rudolph, R. Samtleben

Central Institute of Cancer Research and Central Institute of Molecular Biology of the Academy of Sciences of the GDR, 1115 Berlin-Buch, Robert-Rössle-Str. 10

The binding of lectins to cell surface glycoproteins can be used to determine the number and distribution of receptor molecules on the cell membrane (1-4). The usefulness of labeling different tumor cell lines with lectin/marker complexes for differentiation is evident (5-9). We used different lectins to find receptors on some tumor cell lines. In this respect the human rhabdomyosarcoma cell line A 204 was of special interest because this cell line is our fundamental one for virus production.

HIXSON et al. (10) have found strain-related variations in virus envelope oligosaccharide determinants. For that reason we used only this cell line in our comparative virus studies.

The group of type D-retroviruses was established when its prototype, the Mason-Pfizer monkey virus (MPMV), was first isolated in 1970 (11). Until now some other D-type viruses, morphologically resembling MPMV, have been isolated (12).

A special subclassification does not exist at present. Origin and relationship of some isolates are not clear and comparative studies are necessary. These viruses are of special interest because they seem to be restricted to primates. Furthermore, they have been isolated from different human cell lines and there is some evidence for their distribution in the human population.

Lectins, Vol. IV

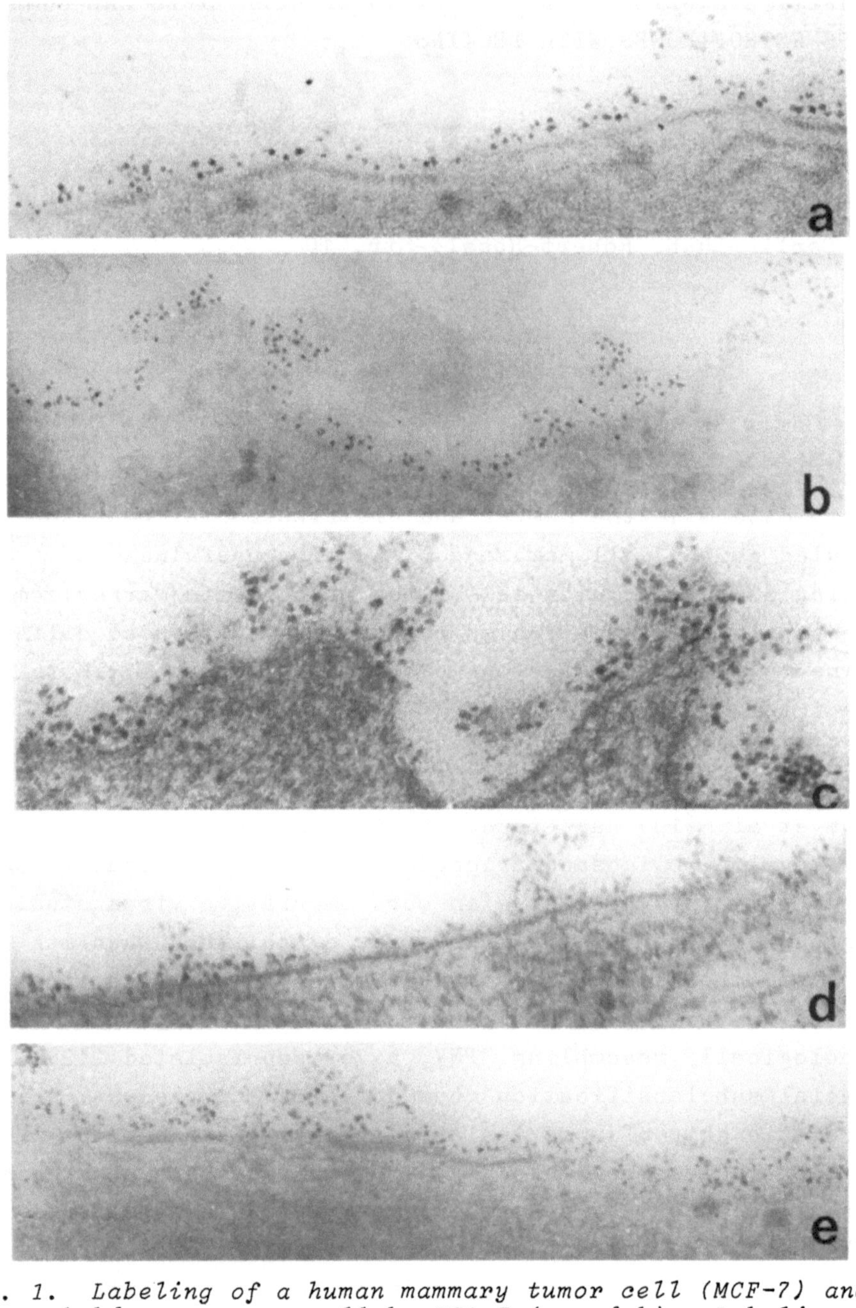

Fig. 1. Labeling of a human mammary tumor cell (MCF-7) and a human rhabdomyosarcoma cell by RCA-I (a and b). Labeling of MCF-7, A204, and PC-7/82 calls by con A.

We have compared in our studies the D-type particles MPMV and PMFV (permanent human fibroblast virus) (13).

Abbreviations used: MPMV Mason-Pfizer-monkey virus,
PMFV permanent human fibroblast virus

Materials and Methods

Cell cultures. The tumor cells were obtained from established cell lines of mammary tumor cells (MCF-7), human rhabdomyosarcoma cells (A204) and preneoplastic mammary cells (PC 7/82). The cells were maintained in RPMI 1640 or MEM (EAGLE) and 10 % NCS.
The A204 cell lines were infected with PMFV or MPMV. The infected and virus producing cells are used for the reactions with the lectin-marker complexes.

Lectins. All lectins were prepared by affinity chromatography: Con A, Lens culinaris and Helix pomatia lectins by binding to Sephadex G 100 and desorption by means of 0.2 M glucose, peanut agglutinin, soybean agglutinin, ricinus agglutinin and mistletoe agglutinin by binding to acid-treated Sepharose 4B (14) and elution with 0.2 M Lactose in PBS. The separation of the type I and II isolectins of the ricinus and mistletoe agglutinins was achieved by chromatography on Sephadex G 100. All lectins were pure in SDS gel electrophoresis according to Fairbanks (15).

Lectin-Ferritin Conjugates. All the lectins have been covalently coupled to ferritin by the two-step glutaraldehyde method of Kishida et al. (16).

Staining procedure. The unfixed cells were incubated with lectin-ferritin conjugate for 15 min at 23 °C. After in-

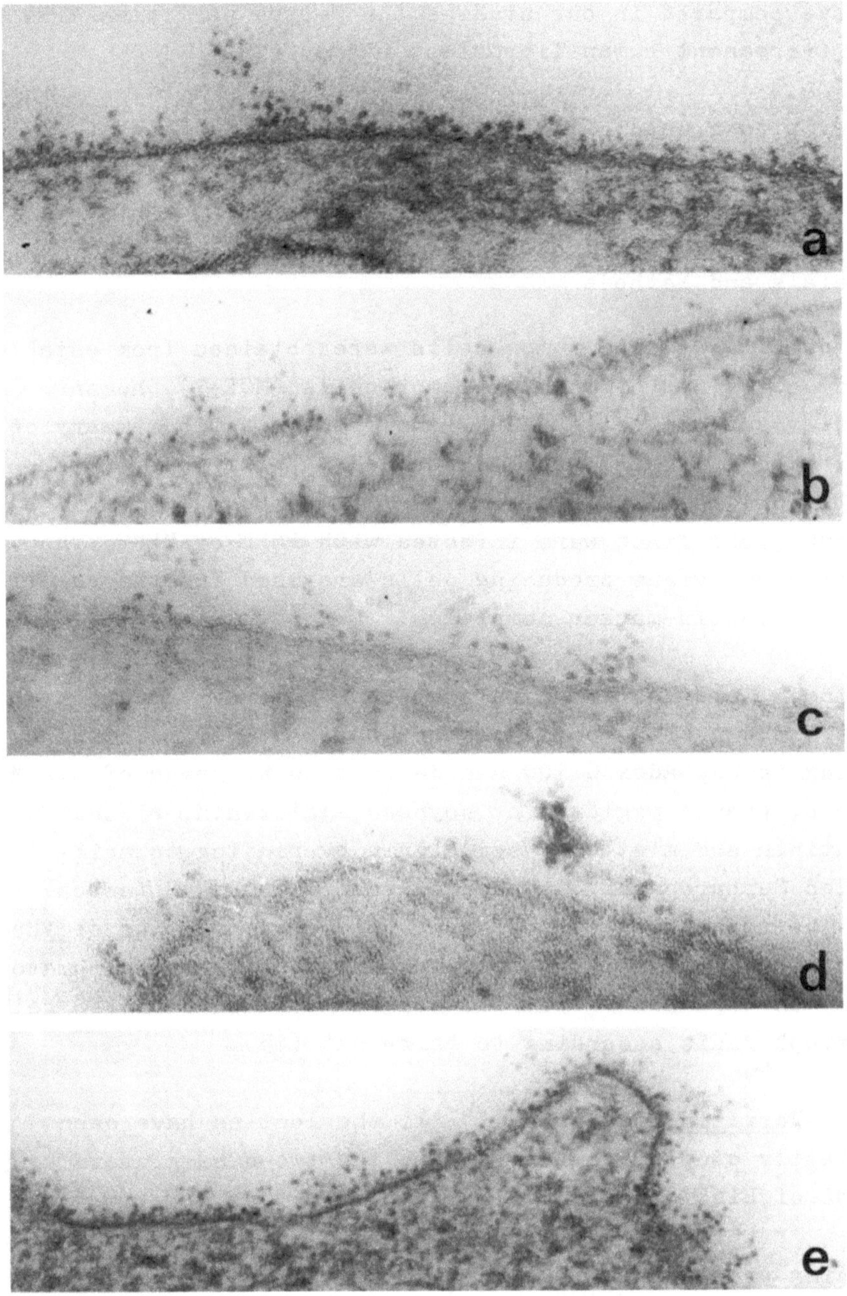

Fig. 2. Labeling of A204 cells with RCA (a), VAA-I, misteltoe I (b), con A (c), HPA, Helix (d), and LCA, Lens (e).

cubation the cells were washed with phosphate buffered saline or tris buffered saline, pH 7.2 and fixed in situ.

Controls. In all experiments the tissue cultures were incubated with peroxidase-labeled lectins mixed with 0,1 M solutions of the specific hapten sugars, α-methyl-mannoside for Con A and lens lectin, lactose for ricinus, mistletoe and peanut lectins, N-acetyl-D-galactosamin for helix and soybean lectin.

Electron microscopy. The cells were dehydrated, postfixed with OsO_4, embedded in Epon, sections were cut on an Ultratome III from LKB. Submicroscopic localization of lectin receptors was analyzed in noncontrasted ultrathin sections by JEM 100 B.

Results

Comparison of lectin receptors on the cell membrane from the different tumor cell lines

The cell membranes of tumor cell lines show distinct reactions with different lectins (Fig. 1, 2). Labeling of human mammary tumor cells (MCF-7) and human rhabdomyosarcoma cells (A 204) by ricinus lectin were homogeneous and strong (Fig. 1a, b). The cells of tumor cell lines MCF-7, A 204 and PC-7/82 labeled with Con A show a homogeneous strong reaction (Fig. 1 c - e). The examination of the A 204 cell line with five different lectins in one series is shown in Fig. 2 a - e. Very strongly labeled were the cells by Con A and Lens culinaris lectin.

TABLE 1

Lectin	Control		A204 cell line		PMFV		MPMV	
	Labeling of cell membrane	Labeling of virus	Labeling of cell membrane	Kind of labeling	Labeling of virus membrane	Kind of labeling	Labeling of virus membrane	Kind of labeling
Con A	–	–	++	column-like	++	column-like	++	column-like
lens	(+)	(+)	++	homogeneous	++	homogeneous	++	homogeneous
ricinus I	–	–	++	homogeneous	++	homogeneous	++	homogeneous
ricinus II	–	–	++	homogeneous column-like	++	homogeneous	++	homogeneous
mistletoe I	–	–	+–++	column-like	++	column-like	+	column-like
soybean	–	–	(+)	occasionally	+–++	homogeneous	(+)	occasionally
helix	–	–	(+)	occasionally column-like	+–++	column-like	(+)	occasionally
mistletoe II	(+)	–	(+)	homogeneous	(+)	homogeneous	(+)	homogeneous
peanut	–	–	(+)	occasionally column-like	(+)	occasionally	+	homogeneous

Table 1: Evidence of the different lectin receptors of two D type retroviruses (MPMV, PMFV) cultivated on A204 tumor cell line.

– negative reaction (+) weak reaction + good reaction ++ strong reaction

Comparison of lectin receptors on two different D-type retroviruses

The morphology of PMFV resembles that of the MPMV triangular core shell. The development of PMFV is similar to the MPMV: envelopment of A type particles and seldom C type budding: maturation of the triangular core shell of D type particles. We have compared the lectin-receptors on the virus membranes of MPMV and PMFV by means of 9 ferritin labeled lectins (see Tab. 1).
Principally we found that differences in the pattern of the lectin binding sites of MPMV and PMFV exist with three lectins: mistletoe, soybean and helix lectins. The majority of reactions was similar with both viruses. Virus particles of different stages of maturation reacted similarly (Fig. 3 a - e).

As the controls revealed, the observed differences in the binding of some lectins to MPMV and PMFV are specific. Additional unspecific binding (i.e. irreversibility with the appropriate hapten sugar) was only found with lens and mistletoe II lectins.

Discussion

The reaction of three tumor cell lines with Con A ferritin conjugate shows almost uniform labeling of the outer cell surface. These Con A-receptors are probably N-asparagine-linked oligosaccharide chains.
Remarkable differences exist between the reactions with different lectin-ferritin complexes at one and the same cell line (A 204). Lens lectin with a sugar specificity identical or similar to Con A gives a good labeling of the cell surface, too. The strong binding of ricinus lectin and (to a somewhat lesser extent) of mistletoe lectin I is an indication of non-

reducing D-galactose residues (perhaps in O-linked glycoconjugates). The very small binding of the helix, soybean, mistletoe II and peanut lectins can be explained by either nearly complete or complete absence of terminal N-acetyl-D-galactose residues in the oligosaccharide chains. Morphological comparison of MPMV and PMFV yielded a similarity of both viruses. Micheel et al. (17) and Uckert et al. (18) found differences in the antigenic reactions and in the protein pattern. It was of interest for us to find differences in the patterns of lectins receptors on the virus envelope, too.

Virus membranes contain specific proteins, mainly glycoproteins. These glycoproteins can appear as surface structures ("spikes") or integrated in the lipid bilayer of membranes.

According to Galehouse et al. (19) the lectin receptors of the virus envelope are synthesized by the enzymes of the host cell but the localization is determined by the viral genome.
In our studies we have cultivated both viruses on the same cell line (A 204) so we can exclude different cellular glycoproteins. Therefore, the observed differences in the lectin binding patterns of MPMV and PMFV with three agglutinins (mistletoe I, soybean and helix) are virus-specific. All three lectins bound stronger to PMFV.
The results suggest the possibility to classify these D type retroviruses by means of galactose or N-acetylgalactosamine specific lectins.
Furthermore the good helix and soybean lectin binding to PMFV gives experimental support that viruses can possess glycoconjugate receptors which are different in the quantity in relation to the host cells.

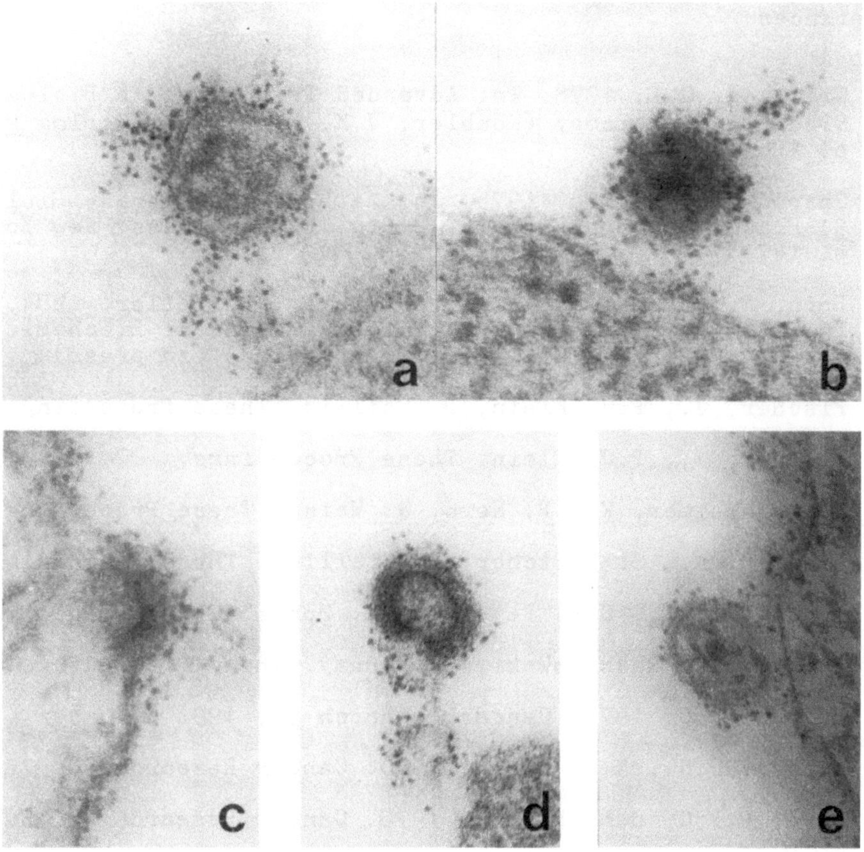

Fig. 3. *a. Labeling of permanent human fibroblast virus, PMFV, with con A (ferritin conjugated).*

b. Labeling of Mason-Pfizer-monkey virus, MPMV, with con A.

c-e. PMFV in different stages of maturation labeled with con A.

References

1. Nicolson, G.L. 1978. In: Advanced Techniques in Biological Electron Microscopy (Koehler, J.K.). Springer Verlag 2, p. 1.
2. Sharon, N., H. Lis. 1975. In: Methods in Membrane Biology 3, Plasma Membranes (Korn, E.D.). Plenum Press, New York, p. 147.
3. Born, A., K. Schwechheimer, H. Maier, P. Möller. 1984. In: Lectins-Biologie, Biochemistry, Clinical Biochemistry 4, (Bøg Hansen, T.C. and J. Breborowicz) (in press).
4. Fischer, J., P.J. Klein, B. Csillik. These Proceedings.
5. Fischer, J., P.J. Klein. These Proceedings.
6. Schwechheimer, K., P. Kern, G. Weiss. These Proceedings.
7. Raedler, A., S. Böttcher, F. Stalling. These Proceedings.
8. Örntoft, T.F., S.E. Petersen, O. Mohrs. These Proceedings.
9. Dus, D., C. Radzikowski, H. Debray. These Proceedings.
10. Hixson, D.C. 1979. Cancer Research 39, 199.
11. Chopra, H.C., M.M. Mason. 1970. Cancer Research 30, 2081.
12. Fine, D., G. Schochetman. 1978. Cancer Research 38, 3123.
13. Graffi, A., D. Bierwolf, R. Widmaier, E. Bender, V. Wunderlich, M. Rudolph, E. Mothes, A. Niezabitowski, G. Papsdorf. 1974. Dtsch. Ges.wesen 29, 1489.
14. Allan, H.J., E.A.Z. Johnson. 1976. Carbohydr. Res. 50, 121.
15. Fairbanks, G., T.L. Steck, D.F.H. Wallach. 1971. Biochemistry 10, 2606.
16. Kishida, Y., B.R. Olsen, R.A. Berg, D.J. Prockop. 1975. J. Cell Biol. 64, 331.
17. Micheel, B., L. Baumbach, V. Wunderlich, A. Niezabitowski, D. Bierwolf. 1979. Eur. J. Cancer 15, 101
18. Uckert, W., V. Wunderlich, E. Bender, G. Sydow, D. Bierwolf. 1980. Archives of Virology 64, 155.
19. Galehouse, D.M., P.H. Duesberg. 1978. J. Virol. 25, 86.

COMPARISON OF THE STRUCTURE OF A GALACTOSYL-BINDING LECTIN FROM THE TUNICATE DIDEMNUM CANDIDUM WITH OTHER PUTATIVE RECOGNITION MOLECULES FROM INVERTEBRATES AND VERTEBRATES.

Gerardo R. Vasta and John J. Marchalonis
Department of Biochemistry, Medical University of South Carolina, Charleston, South Carolina 29425, USA.

In addition to the classical immunoglobulin system, recognition functions have been ascribed to several major classes of molecules including complement components, MHC products, Thy 1 antigens, acute phase proteins, some humoral bactericidal substances from vertebrates and invertebrate and lower chordate lectins (1,2). These recognition functions include their involvement in interactions with cells or extracellular materials from the same organism (which could be considered self-recognition or endogenous functions) and interactions with foreign particles or cells (recognition of non-self). However, from the evolutionary point of view the question of homology among most of those molecules has not yet been elucidated. Modern tunicates are highly specialized and, hence, probably far removed from the main stem leading to the chordates. Moreover, since no immunoglobulins have yet been found in protochordate groups these taxa are important for the search of homologs or functional analogs of these molecules. In addition, studies on protochordate lectins might yield information on possible structural relationships between these molecules and the vertebrate putative recognition molecules described above.

Purification of Didemnum candidum lectins. Molecular and biological properties.

Our preliminary results (3) indicated that all ten species of North American tunicates studied exhibit agglutinins in their plasma that were detectable by their reactivity with vertebrate RBC. Agglutination, crossed-absorption and

Lectins, Vol. IV

hemagglutination-inhibition experiments using a panel of untreated and enzyme-treated vertebrate erythrocytes showed that all ten species possessed hemagglutinins susceptible to inhibition by defined sugars and glycoproteins and all species except one (Styela plicata) exhibited more than one plasma lectin. The specificities of lectins within individual species which included specificity for D-galactose, sialic acid and complex carbohydrate structures, did not correlate with taxonomic relationships of the species studied.

Studies on the binding of the tunicate lectins to glycoproteins rendered possible the isolation and purification of lectins from the tunicates Halocynthia pyriformis (4), Didemnum candidum (4-6) and Clavelina picta by affinity chromatography. In particular detail we studied the properties of the two galactosyl specific lectins from the tunicate D. candidum, DCL-I and DCL-II, and characterized their combining site and molecular structure (4-7).

Purification of D. candidum lectins (5) was carried out based on the fact that although both lectins DCL-I and DCL-II bind to asialo-fetuin only the major lectin DCL-I binds to acid-treated Sepharose 4B. D. candidum body fluids were passed through a column of acid-treated Sepharose 4B previously equilibrated with Tris HCl-buffered saline (TBS) pH 7.6 and recirculated until depleted of DCL-I lectin. The column was washed to the baseline with TBS and eluted with a concentration gradient of D-galactose in TBS (0 to 50 mM D-galactose). The unbound material was passed through a column of asialo-fetuin-Sepharose 4B. When the effluent was almost devoid of agglutinating activity (titre 1:2) for pronase-treated human RBC the column was washed to the baseline and eluted with a gradient from 0 to 50 mM of D-galactose in TBS. Both affinity chromatography procedures were monitored by the absorbance at 280 nm and fractions corresponding to the protein peaks were pooled and dialyzed against TBS.

The purity of both preparations was assessed by immunoelectrophoresis, size exclusion chromatography of the reduced and alkylated DCL-I and DCL-II in 6 M guanidinium chloride in a high performance liquid chromatography (HPLC) system and electrophoresis in polyacrylamide gels under reducing conditions in the prescence of sodium dodecyl sulphate (SDS-PAGE). Subunit molecular weights for DCL-I and DCL-II were approximately 14,000 and 14,500 respectively. The two molecules were serologically distinct using rabbit antisera and differed significantly in their agglutination profiles. They were inactivated by temperatures over 55°C and exhibited optimal agglutinating activity over a wide range of pH (from 5 to 11). Both lectins exhibited a single sedimentary boundary during sedimentation velocity and native molecular weights estimated by sedimentation equilibrium were 56,600 and 57,500 suggesting that each is a tetramer constituted by equal subunits (6). Circular dichroism spectra suggested a high proportion of β-structure in DCL-I. Carbohydrate analysis by gas chromatography of alditol derivatives of D. candidum lectins showed that only traces of D-mannose and D-galactose were present (<0.01%). Although size exclusion HPLC and SDS-PAGE experiments suggested subunit size homogeneity for DCL-I and DCL-II, isoelectric focusing in a pH 3-11 range showed that both lectins are constituted by a family of bands focussing at pH 3.6 to 5.2 and 9.2 to 10.2 respectively (6). The DCL-I molecule was characterized for its saccharide specificity (5) by binding and inhibition assays using characterized sugars and glycoproteins. Galactose and oligosaccharides bearing non-reducing terminal D-galactose were the best inhibitors. The inhibition analysis indicated that the DCL-I combining site is small and interacts only with hydroxyls on carbons 2, 3 and 4 of galactose and exhibits moderate steric hindrance for voluminous groups on C6 and the α anomeric linkage. The combining site would be smaller than the peanut lectin combining site for galactose since DCL-I does not interact with the subterminal monosaccharide hydroxyls on C4 and C6 as does peanut agglutinin (5). The molecular properties of D. candidum lectins are summarized in Table 1 and compared to galactosyl-binding lectins from the sponges Geodia cydonium and Axinella polypoides, the mollusk Tridacna maxima and the tunicates Halocynthia roretzi and Botrylloides leachii.

The mitogenic properties of D. candidum lectins were also investigated (7). DCL-I induces blastogenesis in neuraminidase-treated murine thymocytes and splenocytes. However DCL-I discriminates between untreated thymocytes and splenocytes in terms of lectin binding and mitogenic stimulation. Untreated splenocytes are stimulated to the same extent as neuraminidase-treated splenocytes but untreated thymocytes do not bind the lectin and are not stimulated to divide. Both the binding and stimulation of neuraminidase-treated splenocytes and thymocytes can be specifically and completely blocked by D-galactose, a potent inhibitor of DCL-I

TABLE 1

Molecular properties and carbohydrate specificities of selected galactosyl-binding lectins from invertebrates and protochordates.

Species	Lectin	Approximate Native MW	Subunit MW	Sedimentation coefficient	Isoelectric point	Carbohydrate content (%)	Carbohydrate specificity	Ref.
Geodia cydonium	I	36,500	13,800 13,000 12,200	—	5.6(4.4,7.4)	9.9	β 1-6 D-Gal	8
Axinella polypoides	I	—	21,000	2.6	3.8 , 3.9	0.5	β 1-6 D-Gal	9
	II	—	15,000	2.8	3.9	0.5	β 1-6 D-Gal	
Tridacna maxima	—	470,000	10,000 20,000 40,000	15.0	4.0 , 4.2	7.0	β D-GalNAc β D-Gal	10
Didemnum candidum	DCL-I	56,600	14,000	4.0	3.6-5.2	0.01	β D-Gal	4,5,6
	DCL-II	57,500	14,500	4.1	9.2-10.2	-	β D-Gal	
Halocynthia roretzi	—	600,000	41,000	24.0	—	-	α1-6 D-Gal	11
Botrylloides leachii	HA-1	152,000	28,200 29,600	5.7	4.0 - 5.0	+	β 1-4 D-Gal	12
	HA-2	65,000	33,400	4.3	6.0 - 8.0	-	D-Gal	

binding to erythrocytes, suggesting that the lectin binds to galactosyl residues from cell membrane glycoconjugates as a first stage of the blastogenic stimulation (7).

Structural relationships of DCL-I with other invertebrate lectins and putative recognition molecules from vertebrates.

Although the only rigorous quantitative criterion for assessing homology among molecules and the existence of families is the comparison of their amino acid or gene sequences, the comparison of partial N-terminal sequence can also supply important information provided that a high percentage of homology exists between the molecules considered. As an example the first 21 amino acids of the lectin DCL-I were compared (6) with the N-terminal sequence of immunoglobulin variable regions and immunoglobulin-related molecules such as β_2-microglobulin and Thy 1. In addition, other lectins such as limulin and concanavalin A, and C-reactive protein were compared. No gaps were introduced to maximize homology. The limulin and concanavalin A showed no identities in 21 residues with the DCL-I lectin. β_2 microglobulin, which is known to be homologous to γ chain constant regions, gave one identity out of 21 positions compared and the murine T-cell marker Thy 1 showed two identities in 21 positions compared. By contrast, identities in five positions were found between DCL-I lectin and various immunoglobulin variable regions. C-reactive protein and DCL-I lectin were identical in 6 positions. Based upon this straightforward comparison, it shows stronger homology to immunoglobulin variable regions than does the Thy 1 alloantigen which has been asserted to be related to V regions.

When data on primary structure are unavailable, a variety of structural and functional aspects can be used to infer temptative relationships. That is the case of statistical methods for the comparison of bulk amino acid composition such as the $S\Delta Q$ of Marchalonis and Weltman (13) which for j amino acids is obtained through the formula $S\Delta Q = \Sigma j(Xia-Xib)^2$ where X represents the mole percentage of the amino acid i for two proteins a and b. Comparisons on related and unrelated proteins suggested that a value of 30 $S\Delta Q$ units indicates a high likelihood of relatedness. The frequency of unrelated proteins having values smaller than 80 would be less than 5%. In addition, the $S\Delta Q$ index allows the comparison of molecules of different sizes. Our data indicated that the galactose-binding protein

DCL-I is very similar in amino acid composition to the sialic acid-specific lectin of the tunicate Halocynthia (3). It also showed similarities to two of the Botrylloides lactose-binding proteins (12) and to the lamprey Petromyzon marinus egg lectin, but the degree of relatedness is much less than that shown to the Halocynthia protein. The DCL-I lectin also showed close similarity in composition to the lamprey serum lectin (3) and also substantial compositional relatedness to the variable region of mammalian heavy chains and to the μ chain of lamprey immunoglobulin (6). Similarities in amino acid sequence and composition with immunoglobulin μ chains and C-reactive proteins from vertebrate species would suggest that some lectins from tunicates might be included in an extended family of recognition molecules which could have evolved from a common precursor ancestral to such diverse classes of molecules as immunoglobulins, C-reactive proteins, complement components, MHC products and humoral lectins from vertebrate species (1,3).

Acknowledgements: Part of the work reviewed was supported by Grant GM-30672 from the U.S. Public Health Service. G.R. Vasta is associated with the Consejo Nacional de Investigacions Científicas y Técnicas de la República Argentina. We thank Mrs. Joan Eynon for typing the manuscript.

References:

1. Marchalonis, J.J., Vasta, G.R., Warr, G.W. Barker, W.C.: Immunol. Today 5, 133-142 (1984).
2. Vasta, G.R., Marchalonis, J.J., Kohler, H.: J. Exp. Med. 159, 1270-1276 (1984).
3. Vasta, G.R., Warr, G.W., Marchalonis, J.J.: Comp. Biochem. Physiol. 73B, 887-900 (1982).
4. Vasta, G.R., Marchalonis, J.J.: In "Lectins", Vol. III p. 461-468, Ed. by T.C. Bøg-Hansen & G.A. Spengler, W. de Gruyter & Co., Berlin (1983).
5. Vasta, G.R., Marchalonis, J.J.: J. Biol. Chem. Accepted for publication.
6. Vasta, G.R., Hunt, J.C., Marchalonis, J.J., Fish, W.W. Submitted for publication.
7. Vasta, G.R., Marchalonis, J.J., Decker, J.M. Submitted for publication.
8. Müller, W.E.G., Conrad, J., Schröder, C., Zahn, R.K., Kurelec, B., Dreesbach, K., Uhlenbruck, G.: Eur. J. Biochem. 133, 263-267 (1983).

9. Bretting, H., Kabat, E.A.: Biochemistry 15, 1325-1332 (1976).

10. Baldo, B.A., Sawyer, W.H., Stick, R.V., Uhlenbruck, G.: Biochem. J. 175, 467-477 (1978).

11. Yokosawa, H., Sawada, H., Abe, Y., Numakunai, T., Ishii, S.: Biochem. Biophys. Res. Com. 107, 451-457 (1982).

12. Schluter, S.F.: PhD. Thesis, University of Adelaide (1982).

13. Marchalonis, J.J., Weltman, J.K.: Comp. Biochem. Physiol. 38B, 609-625 (1971).

Author Index

Allen, A.K. 481
Aminoff, D. 345
Andersen, M.M. 229
Arora, J.S. 547
Ascencio, F. 409

Barca, A.M.C. 531
Barth, J. 39, 179
Bayer, E. 109
Beeckmans, S. 453, 567
Begbie, R. 15
Berlan, J. 321
Bezouska, K. 353
Bjerrum, O.J. 241
Bladier, D. 345
Bleys, M. 193
Bøg-Hansen, T.C. 229, 241, 637
Bonnafous, J.C. 45
Borisch, B. 155
Born, I.A. 75, 127
Borojevic, D. 203
Böttcher, S. 137
Bräuer, R. 29
Breborowicz, J. 637
Broekaert, W.F. 481, 523
Brossmer, R. 425

Calderon de la Barca, A.M. 581
Cammue, B. 607
Carbajal, R. 531, 537
Caron, M. 637
Cartron, J.P. 447
Causse, H. 591
Chopra, S.K. 547
Christensen, B. 65
Cros, J. 289
Csillik, B. 117
Cuperlovic, M. 203

Danisova, A. 397
Debbage, P.L. 185
Debray, H. 65
Dejaegere, R. 567
Derwahl, M. 39, 179
Dobroszycka, W. 253
Dornand, J. 45
Doroszewski, J. 307
Driessche, E.v. 453, 567, 583
Dudova, V. 505
Duprat, A.M. 193
Dus, D. 65
Duszyk, M. 307

Entlicher, G. 491

Faure, A. 637
Favero, J. 45
Ferens, M. 515
Fischer, J. 95, 101, 109, 117
Flemming, C. 649
Flottes, S. 193
Franz, H. 53, 327, 463, 473, 637, 649
Fujimoto, M. 213

Gangal, S.G. 559
Gattegno, L. 345
Graham, C. 3
Groeger, B.K. 185
Gryska, K. 637
Guszezynski, T. 253

Harms, D. 155
Henzgen, S. 29
Herp, A. 277, 629
Holthöfer, H. 83
Horejsi, V. 297, 649
Hosselet, M. 583
Hsieh, R.P. 277

Janicot, M. 321
Jarnik, M. 491
Jarosova-Dejlova, L. 491
Jesenska, K. 491

Kamboj, S.S. 559
Kanarek, L. 567, 583
Karaskova, H. 353
Kasai, K.I. 369
Katnik, I. 253

Kaukel, E. 137
Kawalec, M. 307
Kern, P. 75
Kiefer, M. 617
Kieler, J. 65
Kilpatric, D.C. 3
Kindt, A. 53
King, T.P. 15
Kint, J.A. 221, 637
Klein, P.J. 95, 101, 117
Kocourek, J. 353, 377, 397, 491, 505
Kofronova, O. 353
Kohn, J. 637
Krajhanzl, A. 377, 397
Kubrycht, J. 353
Kuhlmann, W.D. 169
Kushner, I. 337
Kuwahara, A. 213

Labroue, L. 591
Lalaurie, M. 321
Leroy, J.G. 215
Lihme, A. 229, 241
Lindahl, M. 409, 425
Lorenc-Kubis, I. 259, 637
Luther, P. 53, 617

Mackiewicz, A. 313
Maier, H. 127
Maldonado, G. 537
Malmi, R. 145
Mani, J.C. 45
Marchalonis, J.J. 663
Marti, J. 45
Matousek, V. 297
Mellor, R.B. 267
Miquel, F. 321
Mohr, J. 327
Moller, P. 127
Montano, L.F. 537
Montreuil, J. 65
Morawiecka, B. 259, 515
Movsesijan, M. 203

Naaby-Hansen, S. 241
Nakamura, K. 213
Nemeth, A. 109
Neumann, D. 327
Niedbala, W. 313
Noack, S. 229
Nolasco, N. 599
Nosek, J. 377
Nsimba-Lubaki, M. 523

Ochoa, J.L. 409, 437, 531, 537, 599
Oda, Y. 369
Ørntoft, T.F. 89

Pancoska, P. 377, 397
Parra, C. 537
Peschke, P. 169
Petersen, S.E. 89
Peumans, W.J. 481, 523, 607
Ponder, B.A.J. 161
Poucke, M. 583
Prigent, M.J. 447

Radzikowski, C. 65
Raedler, A. 137
Rapola, J. 83
Ravens, K.G. 39, 179
Rayon, I. 537
Reen, R.S. 547, 559
Rouge, P. 193, 289, 591
Rudiger, H. 637
Rudolph, M. 653
Ruiz, B. 537

Sabbe, L.M.J. 637
Samtleben, R. 53, 617, 653
Sandhu, R.S. 547, 559
Schumacher, U. 39, 155, 179
Schwechheimer, K. 75, 127
Sehrt, I. 53
Simon, E.J. 289
Sinay, P. 447
Sjöberg, P.O. 409
Smets, G. 453
Snyder, C.E. 277
Söderström, K.O. 145
Somasundaram, S. 559
Sotonyi, P. 109
Spilthooren, J.P. 591
Stalling, F. 137
Stinissen, H.N. 607
Subrahmanyan, D. 559
Suzuno, R. 213

Taborsky, O. 353
Takeo, K. 213, 637
Tanaka, T. 213
Tatake, R.J. 559
Thoss, K. 29
Tichá, M. 297, 505, 649
Toftager-Larsen, K. 637
Trnka, T. 297

Uiterdijk, H.G. 161
Urbaniak, S.J. 3

Valencia, M. 531
Valette, A. 289
Vasta, G.R. 663
Vaysse, J. 345
Verez-Bencomo, V. 447
Volc, J. 505
Vorisek, J. 353

Wadström, T. 417, 425
Waldmann, G. 29
Weiss, G. 75
Welsch, U. 179
Werner, D. 267
Wiktorowicz, K.E. 313
Wu, A.M. 277, 629, 637

Zenteno, E. 437, 531, 537
Ziska, P. 473, 649

Subject Index

Aaptos lectins 633
Abramis lectin 377
Abrin 53
Abrus lectin 53, 559, 617
ABT reaction 109
Acid phosphatase 259, 515
Acteno carcinoma 155
Adenoma 89
Affinity chromatography
253, 259, 491, 505
Affinity electrophoresis
221, 229, 259, 377, 491, 505, 637
Affino-immunoelectrophoresis
221, 253
Agaricus bisporus lectin 3
Ageing 345
Agglutination 45, 397, 437, 447, 531, 547, 599
Aggregation 307
Agrocybe aegerita 505
Agropyrum repens lectin 607
Alpha-2-macroglobulin 221
Amaranth toxicity 531
Amaranthus leucocarpus lectin
531, 537
Amino acid composition 491
Amorphophallus lectin 547
Amphibian 193
Amyloid protein 337
Antibodies 15, 253
Antibodies, lectin 377
Antigen-mediated transformation 3
Antigens 29
Antiserum 15
APA (Aaptos papillata) 633
APA (Abrus precatorius)
53, 559, 617
APA (Axinella polypoides)
632, 663
Araceae lectins 547, 559
Arachis lectin, see PNA
Arisaema lectin 547
Arthritis 29
Ascaris suum 203
Asialofetuin 547, 663
Asialooligosaccharides 353
Atomic absorption 637
Autoimmunity 29
Autoradiography 3
Avidin-HRP 241
Axinella lectin 632, 663

Bacterial adherance 417
Bacterial fimbriae 417
Basophil granulocytes 53
Bauhinia lectin 559
Bauhinia purpurea lectin 277
BDA (Bryonia dioica) 631
Beta-galactoside binding 369
Binding constants 213, 353, 637
Binding protein 481
Binding sites 185
Biotinylated lectins 241
Blood brain barrier 185
Blood group A 505
Blood group antigens 169
Blood group N 633
Blood group substance A 327
Blood grouping 321, 447
Blotting 185, 241, 267
Botrylloides lectin 663
BPA (Bauhinia purpurea) 277, 631
Brain 75, 185
Breast carcinoma 155
Brem lectin 377
Bronchial glycoprotein 277
Bronchial lavage 137
Brush border antibody 83
Bryonia dioica agglutinin 629
BSA (Bandeiraea simplicifolia)
289, 631, 632
Butea lectins 547, 559

Cactus lectins 437
Caesalpinioideae lectins 547
Cancer marker 89
Cancer therapy 473

Capillaries 185
Carbohydrate immobilization 297
Carcinoma 89, 137, 155, 229
Cardiac glycosides 109
Cartilage 29
Cell lines 63
Cell surface 63
Cell-mediated immunity 29
Cellulose acetate membrane 221
Cerebal hemispheres 185
Cerebellum 185
Chains 53
Chemiluminescence 39, 179
Chick embryo skin 369
Cholla opuntia 437
Circa-annual rhythm 522
Circular dichroism 377, 397, 567
Clam lectin 409, 629
Clay 591
Clitocybe nebularis 505
Colloidal gold 327
Colon 89
Colorectal tissue 89
Con A, concanavalin A 29, 39, 45, 63, 75, 127 179, 213, 229, 241, 253, 259, 277, 289, 313, 559, 633, 637, 653
Con A uptake 179
Con A-FITC 109
Conformation 567
Couch grass, see Agropyrum
Cross-reaction 582
Crossed immunoelectrophoresis 253
Crotalaria lectins 547, 559
CRP, C-reactive protein 337, 663
Cytotoxicity 463

Datura lectins 559
Datura stramonium 3, 481
DBA (Dolichos biflorus) 161, 277, 630
DBA-binding sites 95
Delayed-type hypersensitivity 3
Delta opioid-binding sites 289
Dextran 637
Didemnum candidum lectin 663
D Dietary lectin consumption 3
Differentiation marker 83, 229
Diffusion constant 307
Dissociation constant 297, 637
Distribution 523
DNA content 89
Dog fish lectin 337
Dog lectin 337
Dolichos lectin 161, 277, 559, 630
Double affinity diffusion 637
DSA (Datura stramonium) 481, 633
Dumasia lectin 547

ECA (Erythrina cristagalli) 632
Echinoccocus granulosus 203
Echinocereus engelmanii 437
Eel lectin 321
Egg lectin 662
Elder 523
Electroblotting 241, 267
Electrofocusing 637
Electron microscopy 179, 185, 653
Electrophoresis 213, 221, 229, 245, 253, 259, 267, 377, 491, 505, 637
ELISA 473, 583
Embryo lectin 607
Embryonic cells 193
Endo-beta-galactosaminoglycan 369
Endogenous receptor 369
Enzyme-linked lectin assay 137
Erythrina lectin 547, 559, 632
Erythroagglutination 505
Erythrocytes 345
Extensin 567

Fabaceae lectins 547
Fasciola hepatica 203
Ferritin 229
Ferritin conjugates 653
Fetuin 599
Fibronectin 337
Fimbriae 425
Fimbrial lectin 417, 425
Fish oocyte lectin 377, 397
FITC conjugates 83, 109, 203
FITC blood group substance 327
Flammulina velutipes 505
Fomes fomentarius 505
Fruit bodies 505

Galactose oxidase 45
Galactosyl copolymer 297
Ganglion 117
Gasserian ganglion 117
Gastric concers 101
Gastric mucus 101
Gastrointestinal mucosa 169
Gaucher's disease 95
GCA (Godia cydonium) 277, 632
Gel electrophoresis 185
Gelatin 297
Genetic dimorphism 161
Geodia cydonium lectin 277, 632
Glycine lectins, see SBA
Glycoconjugates 145, 169
Glycolipids 95
Glycophorin 447
Glycoproteins, lymphocyte 3
Golgi 267
Grass 259
GSA(Griffonia simplicifolia) 634
Guinea pig cerebellum 289

Halocynthia lectin 663
Haptoglobin 221, 253
HEA 297
Helix pomatia, see HPA
Hemadsorption 637
Hemagglutination 437, 447, 531, 599
Hemoglobin 253
Hepatocytes 353
Histamine release 53
Histiocytes 101
Histochemistry 63, 83, 89, 95, 101, 109, 117, 127, 137, 145, 155, 161, 169, 179, 185, 193, 203, 653
Horse radish peroxidase, see HRP
HPA (Helix pomatia) 75, 83, 127, 161, 277, 630, 653
HRP conjugates 241, 653
Human opioid binding sites 289
Human testis 145
Humic acids 591

IgG 15
Immobilized carbohydrates 296
Immuno-gold labeling 453
Immunocytochemistry 75, 127, 453
Immunoelectrophoresis 229
Immunoglobulin 15
Immunohistology 169
Immunoperoxidase 89
Immunosuppression 537
Immunotoxin 463
In vivo binding 185
Inflammation 29
Inhibition 437, 447, 599
Inhibition of lectin binding 297
Interleukin 3, 45
Intestinal epithelium, mouse 161
Intracellular cAMP 39
Invasiveness 63
Invertebrate lectins 337, 409
Isoelectric focusing 637
Isoelectric point 491
Isoenzymes 259
Isolation and characterization 505, 523
Isolectins 3

Jimson weed, see DSA

K99 lectin 425
Kappa opioid binding sites 289
Krabbe's disease 95

Labelled lectins 193
Lablab lectin 559
Laccaria amethystina 505
Lactose-specific lectins 523
LAg-I (Homarus americanus) 635
LaTA 289
LCA (Lens culinaris) 29, 193, 241, 289, 634, 653
LEA (Lycopersicon esculentum) 633
Leaf lectin 607
Lectin carbohydrate interaction 297
Lectin enzyme interaction 515
Lectin ferritin 653
Lectin radioiodination 297
Lectin-mediated panning 137
Lentil lectin, see LCA
Lens culinaris, see LCA
Lens esculenta lectin 634
LFA (Limax flavus) 635
Light microscopy 185
Limulus lectin 337, 635

Liver, rat 185
Localization 453
Lophocereus shottii 437
LPA (Limulus polyphemus) 75, 635
LTA (Lotus tetragonolobus) 277, 634
Lycopersicon esculentum 3
Lymphocytes 3, 29, 313, 353, 559

Machaerocereus lectin 437, 537
Macrophages 45, 353
Mammary tumor cells 653
Mannosyl copolymer 297
Megapitaria squalida 409
Melanoma 101
Membrane flow 267
Membrane proteins 241
Membrane receptors 307
Merkel cell tumor 101
Mesothelioma 155
MHC 663
Microvessels, rat 185
Microvilli 63
Misteltoe lectin 53, 463, 473, 617, 653
Mitogen responsiveness 29
Mitogen stimulation 29, 313 637
Mitogens 3, 29, 537, 559, 637
Mitogenic index 559
Mixed lymphocyte reaction 3
MN blood types 447
Mobilization of reserves 515
Molecular structure 491
Monocyte lectin 345
MPA (Maclura pomifera) 277,631
Mu opoid binding sites 289
Mucous glycoproteins 101
Mushroom 505
Mycelial cultures 505

N-acetylgalactosamine-specific lectins 607, 630
N-acetylglucosamine-specific lectins 607, 633
Neoplastic transformation 89
Nephelometry 637
Neural induction 193
Neuraminidase 63, 89, 447, 547
NK cells 353
Nodules 267
Non-hemagglutinating carbohydrate 481
Nonspecific lectin 491, 505
Nutritional value 3, 15, 531

O-beta lactosyl gel 505
Oat cell carcinoma 137
Oligosaccharides 447
Oocyte lectin 377, 397
Opiate receptors 289
Oral immunisation 15
Ovalbumin 297

Pachycereus pringleii 437
Panning 137
PAP (peroxidase antiperoxidase) 75, 155
Paper chromatography 297
Parasite antigens 203
Parkia lectin 559
Parotid gland 127
PEA (Pea embryo agglutinin) 15, 297
Peptone hog gastric 297
Perca fluviatilis lectin 377
Perch lectin 377
Periodate oxidation 109
Peroxidase antiperoxidase 155
Peroxidase, see HRP
Petromyzon lectin 663
PHA (phytohaemagglutinin) 3, 15 29, 203, 241, 313, 530, 559
Phagocytosis 344, 537
Phallus impudicus 491, 505
Phaseolus coccineus lectin 599
Phaseolus lectins, see PHA
Phaseolus limensis 630
Photochemical cross-linking 369
PLA (Phaseolus limensis) 630
Plasma membrane 89, 185
Pleurotus ostreatus 505
Pleurotus spondoleucus 505
PNA (peanut) 45, 63, 83, 127, 241, 267, 277, 631, 653
PNA-binding sites 101
Pokeweed mitogen 3
Polycystic kidney 83
Polylactosaminoglycan 369
Polymorphonuclear leucocytes 353
Post synthetic modifications 259

Potato, see STA
Precipitation 637
Proliferation assay 313
Protease treatment 599
Proteoglycan 29
PSA (Pisum sativum) 29, 193, 289, 453, 583, 634
PWM (Poke weed) 3, 313

Quantitation 277

Rabbit cerebellum 289
Rabbit lectin 337
Radial affinity diffusion 637
Rat brain 185, 289
Rat lectin 337
RCA (Ricinus communis) 53, 75 185, 221, 296, 617, 632, 653
RCA-I 63, 241, 277, 307
Receptors lymphocyte lectin 3
Renal carcinoma 229
Rhamnopyranoside specific lectin 397
Rhizobium japonicum 267
Rhodamine-conjugated lectins 83
Ricin 53, 277, 632
Roach lectin 397
Russula nigricans 505
Rye germ agglutinin 515

S-S bonds 491
S-sulfonation 491
SAA (serum amyloid component) 337
Salivary gland 127
Sambucus edulus 321, 523
Sambucus nigra 523, 633
Sandhoff's diseases 95
Sarcoma 154
SBA (soybean) 39, 75, 127, 193, 267, 547, 559, 591, 630
Seed reserves 515
Seminoma 145
Senecent human red blood cells 345
Sensory terminals 117
Sepharose-fetuin 491, 599
Serum factors 29
Sialic acid specific 417, 424, 635
SJA (Sophora japonica) 277, 632
Small intestine 15
SNA (Sambucus nigra) 523, 633
Soft tissue tumor 155
Soil 591
Solanaceae 3, 481
Solanum tuberosum 567, 633
Sophora, see SJA
Soybean, see SBA
Spectroscopy 637
Spermatogenesis 145
Spermatozoa glycoproteins 241
Spinal cord 117
Squamous cell carcinoma 137
STA (Solanum tuberosum) 3, 241, 566, 633
Stability 566, 637
Stenocereus thurberi 437
Stimulation, lymphocytes 3, 29, 313, 353, 559, 637, 649
Storage disease 95, 523
Submandibular gland 127
Subunit arrangement 353
Subunits 377, 617, 491, 462
Sugar binding 297
Supressor cell 29
Surface antigens 3, 203, 229, 241, 63, 537, 547
Synovial sarcoma 155
Synthetic oligosaccarides 447

T-antigen 89
Tamm-Horsefall glycoprotein 83
Thomsen-Friedenreich antiben 89
Thymocytes 45
TMA (Tridacna maxima) 631, 663
Tomato lectin 3
Toxicity 15, 531
Transferase 267
Transformation, lymphocyte 3
Transitional uroepithelium 63
Trichinella spiralis 203
Tridacna lectin 629, 931, 663
Triticum lectin, see WGA
Trout lectin 337
Tuberculin 3
Tuberosum lectin 3
Tumor associated glycoconjugates 137
Tumor cells 652
Tumor markers 63,229
Tumorigenicity 63
Tumors 75

Tunicate lectins 663

UDP sugar transferase 267
UEA (Ulex europapaeus) 75, 117,
127, 277, 154, 634
UEA-I-binding glycoprotein 117
Uroepithelium cell lines 63

VAA (Viscum album) 53, 463,
472, 617, 652
Vascular endothelium 161
Vectorial cell surface label 3
VFA (Vicia faba) 241, 634
VGA (Vicia graminea) 447, 629
Virus 652
VVA (Vicia villosa) 630

WFA (Wistaria floribunda)
161, 277, 631
WGA (wheat germ) 3, 29, 39, 63,
75, 179, 203, 241, 327,
277, 289, 313, 633
WGA uptake 179
Wisteria floribunda, see WFA

Bei Fragen zur Produktsicherheit wenden Sie sich bitte an:
If you have any questions regarding product safety,
please contact:

Walter de Gruyter GmbH
Genthiner Straße 13
10785 Berlin
productsafety@degruyterbrill.com